CHARLES LYELL
ON NORTH AMERICAN GEOLOGY

This is a volume in the Arno Press collection

History of Geology

See last pages of this volume
for a complete list of titles

CHARLES LYELL ON NORTH AMERICAN GEOLOGY

Edited, with an Introduction
by Hubert C. Skinner

ARNO PRESS

A New York Times Company

New York / 1978

Editorial Supervision: ANDREA HICKS

Reprint Edition 1978 by Arno Press Inc.

HISTORY OF GEOLOGY
ISBN for complete set: 0-405-10429-4
See last pages of this volume for titles.

Manufactured in the United States of America

Publisher's Note: This book has been reproduced from the best available copies.

Library of Congress Cataloging in Publication Data

Lyell, Charles, Sir, bart., 1797-1875.
Charles Lyell on North American geology.

(History of geology)
Reprint of the 1843 ed. of Lectures on geology published by Greeley & McElrath, New York and of 45 articles published between 1841 and 1869 in various journals.
1. Geology--North America--Addresses, essays, lectures. I. Title. II. Series.
QE71.L93 1977 557 77-6524
ISBN 0-405-10446-4

INTRODUCTION

Charles Lyell, the eminent British geologist and author of the *Principles of Geology* (London, 1830-1833, *et seq.* eds.) visited the United States of America four times between July, 1841 and mid-1853. The first visit was in response to an invitation to deliver a course of lectures at the Lowell Institute in Boston. The lectures, given in early August, were so well received that each was delivered twice to accommodate the 4500 persons who received tickets to the series. Lyell gave similar well-attended lectures in New York and Philadelphia, and the New York lectures were published.

Charles Lyell and his wife travelled extensively in North America during the thirteen months of the first tour, as far northward as Quebec, Toronto, and Nova Scotia, and southward as far as the Carolinas. During these travels, they were kindly and graciously received, meeting with and engaging in field excursions with numerous American geologists including James Hall of Albany at Niagara Falls, John William Dawson at the Bay of Fundy, and Henry Darwin Rogers in the Appalachians. Later, at Boston in April, 1842, Lyell was present at the third meeting of the Association of American Geologists when the two brothers Rogers presented their classic description of the folds of the Appalachian Mountains and postulated their origin. The Lyells returned home in August, 1842.

In early 1845, the Lyells again came to America, a second tour during which they visited Georgia, Alabama, and Louisiana, travelled down river from New Orleans through the lower delta to the mouths of the Mississippi River and up river to Baton Rouge, Natchez, Vicksburg, and Memphis. Lyell visited the area affected by the New Madrid earthquake of 1811 and continued overland through the coal fields of Pennsylvania to Philadelphia.

Lyell's third visit to America, in late 1852, was to present a second series of lectures at the Lowell Institute, again attended by record numbers of people. In mid-1853, he returned a fourth time, as commissioner to the New York Industrial Exhibition and goodwill representative of the British government. He was accompanied on the two month trip by Lord Ellesmere, also an official commissioner.

Throughout his travels, Charles Lyell kept a daily journal and recorded his astute observations on the geology of areas he reached in a steady flow of papers to scientific journals. The best known works on America by Lyell are his *Travels in North America* (London, 1845, 2 vols.) and *A Second Visit to the United States of North America* (London, 1850, 2 vols.). Both of these travel narratives were published in several English and American editions, and each was translated into German. Less well known are the two editions of *Eight Lectures on Geology* (New York, 1842, 1843) and the more than two score papers on his scientific observations in America published in various American and British scientific journals. It is these lesser known lectures and papers which are collected here into a single volume.

It seems ironic that the quality of the scientific work by Charles Lyell, who published the first lucid and definitive textbook of geology and who incorporated into the later editions of his textbook and manual the latest discoveries and the best scientific opinions put forth during his lifetime, should be questioned today. However, shallow interpretations such as the statement that Sir Charles believed in strictly constant rates of geologic process or that he subdivided the Tertiary on statistical evidence alone are too frequently encountered. When he came to America, Lyell came seeking quantitative information on the dynamics of the geologic processes of erosion and deposition. He calculated or estimated the rates of erosion at the rim of the Niagara gorge, and of the great gully near Milledgeville, Georgia, as the combined result of his personal observations and data communicated to him by other scientists. He calculated the rate of advance of the delta of the Mississippi River and rates of relative change of the levels of the lands and the sea as exhibited by several levels of fossil forests in the bluffs of the Mississippi River above Baton Rouge. The Lyell papers on American geology collected here will serve to supplement the better known Lyellian works as sources for evidence of Lyell's intense insight and expertise in the understanding of geologic phenomena and geologic processes.

HUBERT C. SKINNER
Tulane University of Louisiana
May 7, 1977

CONTENTS

EIGHT LECTURES ON GEOLOGY: Delivered at the Broadway Tabernacle in the City of New York. New York, 1843

REMARKS ON SOME FOSSIL AND RECENT SHELLS, COLLECTED BY CAPT. BAYFIELD, R.N., IN CANADA (Reprinted from *Geological Society of London, Proceedings,* Vol. III), London, 1841

[CAUSE OF DIP OF NEW RED SANDSTONE IN EASTERN UNITED STATES] (Reprinted from *American Journal of Science,* Vol. 43), 1842

[ON SOME OF THE PHENOMENA CONNECTED WITH THE COAL MEASURES AND OLDER STRATA OF PENNSYLVANIA] (Reprinted from *Geological Society of London, Proceedings,* Vol. III), London, 1842

A MEMOIR ON THE RECESSION OF THE FALLS OF NIAGARA (Reprinted from *Geological Society of London, Proceedings,* Vol. III, Part II), London, 1842

ON THE TERTIARY FORMATIONS AND THEIR CONNEXION WITH THE CHALK IN VIRGINIA AND OTHER PARTS OF THE UNITED STATES (Reprinted from *Geological Society of London, Proceedings,* Vol. III), London, 1842

ON THE FOSSIL FOOT-PRINTS OF BIRDS AND IMPRESSIONS OF RAIN-DROPS IN THE VALLEY OF CONNECTICUT (Reprinted from *Geological Society of London, Proceedings,* Vol. III), London, 1842

[ON THE *STIGMARIA* CLAY, IN THE BLOSSBERG COAL-FIELD OF PENNSYLVANIA] (Reprinted from *The Annals and Magazine of Natural History,* Vol. X), London, 1842

ON THE RIDGES, ELEVATED BEACHES, INLAND CLIFFS AND BOULDER FORMATIONS OF THE CANADIAN LAKES AND VALLEY OF ST. LAWRENCE (Reprinted from *American Journal of Science,* Vol. 46), 1843

ON THE TERTIARY STRATA OF THE ISLAND OF MARTHA'S VINEYARD IN MASSACHUSETTS (Reprinted from *American Journal of Science,* Vol. 46), 1843

ON THE GEOLOGICAL POSITION OF THE *MASTODON GIGANTEUM* AND ASSOCIATED FOSSIL REMAINS AT BIGBONE LICK, KENTUCKY, AND OTHER LOCALITIES IN THE UNITED STATES AND CANADA (Reprinted from *American Journal of Science,* Vol. 46), 1843

ON THE UPRIGHT FOSSIL TREES FOUND AT DIFFERENT LEVELS IN THE COAL STRATA OF CUMBERLAND, NOVA SCOTIA (Reprinted from *American Journal of Science,* Vol. 45), 1843

ON THE COAL FORMATION OF NOVA SCOTIA, AND ON THE AGE AND RELATIVE POSITION OF THE GYPSUM AND ACCOMPANYING MARINE LIMESTONES (Reprinted from *American Journal of Science,* Vol. 45), 1843

NOTES ON THE CRETACEOUS STRATA OF NEW JERSEY, AND OTHER PARTS OF THE UNITED STATES BORDERING THE ATLANTIC (Reprinted from *Geological Society of London, Quarterly Journal,* Vol. I) London, 1844

ON THE PROBABLE AGE AND ORIGIN OF A BED OF PLUMBAGO AND ANTHRACITE OCCURRING IN MICA-SCHIST NEAR WORCESTER, MASSACHUSETTS (Reprinted from *Geological Society of London, Quarterly Journal,* Vol. I), London, 1844

ON THE MIOCENE TERTIARY STRATA OF MARYLAND, VIRGINIA, AND OF NORTH AND SOUTH CAROLINA (Reprinted from *Geological Society of London, Quarterly Journal,* Vol. I), London, 1845

OBSERVATIONS ON THE WHITE LIMESTONE AND OTHER EOCENE OR OLDER TERTIARY FORMATIONS OF VIRGINIA, SOUTH CAROLINA, AND GEORGIA (Reprinted from *Geological Society of London, Quarterly Journal,* Vol. I), London ,1845

[ON THE EOCENE OF ALABAMA AND GEORGIA] (Reprinted from *American Journal of Science* (2), Vol. I), 1846

COAL FIELD OF TUSCALOOSA, ALABAMA: Being an Extract of a Letter to Prof. Silliman (Reprinted from *American Journal of Science* (2), Vol. I), 1846

ON THE EVIDENCE OF FOSSIL FOOTPRINTS OF A QUADRUPED ALLIED TO THE *CHEIROTHERIUM,* IN THE

COAL STRATA OF PENNSYLVANIA (Reprinted from *American Journal of Science* (2), Vol. II), 1846

OBSERVATIONS ON THE FOSSIL PLANTS OF THE COAL FIELD OF TUSCALOOSA, ALABAMA (Reprinted from *American Journal of Science* (2), Vol. II), 1846

NOTICE ON THE COAL-FIELDS OF ALABAMA: Being an Extract from a Letter to the President (Reprinted from *Geological Society of London, Quarterly Journal,* Vol. II), London, 1846

ON THE NEWER DEPOSITS OF THE SOUTHERN STATES OF NORTH AMERICA (Reprinted from *Geological Society of London, Quarterly Journal,* Vol. II), London, 1846

ON FOOT-MARKS DISCOVERED IN THE COAL-MEASURES OF PENNSYLVANIA (Reprinted from *Geological Society of London, Quarterly Journal,* Vol. II), London, 1846

ON THE STRUCTURE AND PROBABLE AGE OF THE COAL-FIELD OF THE JAMES RIVER, NEAR RICHMOND, VIRGINIA (Reprinted from *Geological Society of London, Quarterly Journal,* Vol. III), London, 1847

ON THE DELTA AND ALLUVIAL DEPOSITS OF THE MISSISSIPPI, AND OTHER POINTS IN THE GEOLOGY OF NORTH AMERICA, OBSERVED IN THE YEARS 1845, 1846 (Reprinted from *American Journal of Science* (2), Vol. III), 1847

ON THE ALLEGED COEXISTENCE OF MAN AND THE *MEGATHERIUM* [Natchez, Miss.] (Reprinted from *American Journal of Science* (2), Vol. III), 1847

ON THE RELATIVE AGE AND POSITION OF THE SO-CALLED NUMMULITE LIMESTONE OF ALABAMA (Reprinted from *Geological Society of London, Quarterly Journal,* Vol. IV), London, 1847

ON THE FOSSIL FOOTMARKS OF A REPTILE IN THE COAL FORMATION OF THE ALLEGHANY MOUNTAINS (Reprinted from *The Athenaeum,* No. 1059), 1848

NOTES ON SOME RECENT FOOT-PRINTS ON RED MUD IN NOVA SCOTIA, COLLECTED BY W. B. WEBSTER OF KENTVILLE (Reprinted from *Geological Society of London, Quarterly Journal,* Vol. V), London, 1849

LOWER SILURIAN REPTILE IN CANADA (Reprinted from *American Journal of Science* (2), Vol. XII), 1851

ON IMPRESSIONS OF RAIN-DROPS IN ANCIENT AND MODERN STRATA (Reprinted from *Royal Institution of Great Britain, Proceedings,* Vol. I), 1851

ON FOSSIL RAIN-MARKS OF THE RECENT, TRIASSIC, AND CARBONIFEROUS PERIODS (Reprinted from *Geological Society of London, Quarterly Journal,* Vol. VII), London, 1851

ON THE DISCOVERY OF SOME FOSSIL REPTILIAN REMAINS, AND A LAND-SHELL IN THE INTERIOR OF AN ERECT FOSSIL-TREE IN THE COAL MEASURES OF NOVA SCOTIA: With Remarks on the Origin of Coal-Fields, and the Time Required for Their Formation (Reprinted from *Royal Institution of Great Britain, Proceedings,* Vol. I), 1853

ON THE REMAINS OF A REPTILE (*DENDRERPETON ACADIANUM,* WYMAN AND OWEN) AND OF A LAND SHELL DISCOVERED IN THE INTERIOR OF AN ERECT FOSSIL TREE IN THE COAL MEASURES OF NOVA SCOTIA. With J. W. Dawson (Reprinted from *Geological Society of London, Quarterly Journal,* Vol. IX), London, 1853

SPECIAL REPORT OF SIR CHARLES LYELL ON THE GEOLOGICAL, TOPOGRAPHICAL, AND HYDROGRAPHICAL DEPARTMENTS OF THE [NEW YORK] EXHIBITION: Presented to the House of Commons by Command of Her Majesty, in Pursuance of Their Address of February 6th, 1854, London, 1854

RAPPORT SUR LA PARTIE GÉOLOGIQUE DE L'EXPOSITION DE NEW YORK (Reprinted from *Annales des Mines, Paris,* Series 5, *Mèmoires, Partie Scientifique,* Vol. 6), Paris, 1854

EXTRAIT D'UN RAPPORT DE M. CHARLES LYELL SUR LA PARTIE GÉOLOGIQUE DE L'EXPOSITION DE NEW-YORK, EN 1853, PAR M. A. LAUGEL (Reprinted from *Société Géologique de France, Bulletin* (2), Vol. 12), 1855

ON CERTAIN TRAINS OF ERRATIC BLOCKS ON THE WESTERN BORDERS OF MASSACHUSETTS, UNITED STATES (Reprinted from *Royal Institution of Great Britain, Proceedings,* Vol. 2), 1855

[LETTER ON AGE OF RICHMOND COAL FIELD, VIRGINIA] (Reprinted from *Geology of North America*), Zurich, 1858

ON THE CONSOLIDATION OF LAVA ON STEEP SLOPES, AND ON THE ORIGIN OF THE CONICAL FORM OF VOLCANOES (Reprinted from *American Journal of Science* (2), Vol. XXVIII), 1859

[RATE OF FORMATION OF DELTA OF THE MISSISSIPPI]

(Reprinted from *Geological Society of London, Quarterly Journal,* Vol. XXV), 1869

LECTURES ON GEOLOGY
DELIVERED AT THE BROADWAY TABERNACLE
IN THE CITY OF NEW-YORK

IDEAL SECTION OF PART OF THE EARTH'S CRUST.

D. PLUTONIC......C. METAMORPHIC......B. VOLCANIC......A. AQUEOUS.

EXPLANATION OF THE FRONTISPIECE.

The Engraving in the Frontispiece is from LYELL'S Elements of Geology, and is designed to explain his theory of the cotemporaneous origin of the four great classes of rocks represented by the masses A, B, C and D—the relative position of which will be readily perceived. Thus, while the aqueous deposites, A, have been accumulating in successive strata at the bottom of the sea, the volcanic cone B has been piled up during a long series of eruptions; and the other igneous rocks represented by the black color have also ascended from below in a fluid state. Some of these last have been poured forth into the sea, and then mingled with aqueous sediment. On the right side of the plate a sheet of lava is represented reposing upon the strata A, and covered by other parts of the same sedimentary deposit. On pursuing downward either the small dikes or large masses of volcanic rock, we find them pass gradually into plutonic formations, or granites, D, which are colored black, and underlie all the rest. The different tints represent two granites, one newer than the other, but belonging, like the upper and lower strata of A, to the same geological period. These plutonic rocks are seen to be in contact with a zone of cotemporaneous metamorphic strata, C, which they penetrate in numerous veins. The formations C originate as fossiliferous strata of the periods *d*, *e*, *f*, *g*, *h* and *i*, but have acquired a new structure during the accumulation of the strata A, and belong therefore to A, if we regard them solely in the light of metamorphic rocks; for during the deposition of A, they assumed a crystalline form, and were thus regenerated. In that part of the plate which is left unshaded a more ancient series of mineral masses is seen, belonging also to the four great classes of rocks. The strata from *a* to *i* represent as many distinct aqueous formations, and are each distinguished by their peculiar fossils. The mass *v v* is of volcanic origin, and was formed when the strata *g* were in the course of deposition. The strata *m* are of metamorphic formation, which assumed a crystalline texture before the strata C were so altered as to belong to what Prof. Lyell calls the hypogene class. The rocks 1 2 are plutonic, of different dates, but more ancient than D D. For a full explanation of the theoretical views elucidated in this engraving, reference must be had to Lyell's Elements.

LECTURES

ON GEOLOGY,

Delivered

AT THE BROADWAY TABERNACLE,

IN THE CITY OF NEW-YORK.

BY CHARLES LYELL, F. R. S.

VICE-PRESIDENT OF THE GEOLOGICAL SOCIETY OF LONDON, ETC.

SECOND EDITION: REVISED AND CORRECTED:

WITH A

GENERAL INTRODUCTION:

TO WHICH IS ADDED A

SKETCH OF A LECTURE ON THE DIFFERENT RACES OF MEN,

BY J. AUGUSTINE SMITH, M. D.

REPORTED FOR "THE NEW-YORK TRIBUNE,"

BY H. J. RAYMOND.

NEW-YORK:

GREELEY & McELRATH, TRIBUNE BUILDINGS.

1843.

ADVERTISEMENT.

The Eight Lectures on Geology here submitted to the public are substantially those which have been delivered by Mr. LYELL before large and respectable audiences in the principal Cities of the Union. They present the conclusions to which Geologists have been led by the most recent Discoveries and Researches in that Science, by explorers in both hemispheres, including those who have labored in the employment of our several States. They may therefore be justly regarded as additions to the sum of Geological knowledge conveyed in preceding publications, and as such, valuable to the man of Science no less than to the inquirer and student. They were reported for The New-York Tribune, as delivered at the Tabernacle, by Mr. H. J. RAYMOND, Assistant Editor of that paper, whose Reports of Dr. LARDNER'S and other Scientific Lectures have already been received with decided and merited approbation. They have been carefully revised and prepared for this edition, which the publishers confidently trust will be found worthy of equal favor.

New-York, January, 1843.

INTRODUCTION

THE Eight Lectures which are contained in this collection, were originally delivered by Prof. LYELL at the Tabernacle in the winter of 1840; and are principally devoted to a presentation of some of the most recent observations that have been made in various parts of the world by eminent Geologists of Europe, among whom the lecturer, himself, is especially distinguished. Few who will read them will need to be told that as a complete science, the business of Geology is to explain the present structure, and past history of the planet upon which we live:—to develope with scientific precision the manner of its creation, the causes which have in any way modified its character and the methods of their operation. This, however, it has yet but imperfectly done. Though from the earliest ages of the world, theories concerning the origin of the Earth have been cherished and defended, none of them have attained either the precision or the rigorous demonstration demanded by science. They are based upon assumptions which were never proved to be anything more, or to have even probability enough to warrant the erection of superstructures so imposing as many of them were forced to support. All of them, however, even the Ancient Indian and Egyptian systems professed to be grounded upon observed *facts;* from which they were deduced by what was afterwards called the Process of Induction. Of course the theories differed among themselves according to the various phenomena from which they grew. They were divided however mainly into two—which were called respectively the *Neptunian* and *Vulcanian* theories:—the first maintaining that to the two fold action of water, the chemical and mechanical, all the changes of the Earth are to be ascribed; and the latter that the whole interior of the globe is composed of melted matter which, by its constant operation, gives to the earth its natural heat and by its occasionally violent action gives birth to mountains and causes other phenomena upon the surface.

Hutton at the close of the eighteenth century was the first to declare that Geology had nothing to do with the origin of things; he dispensed with all hypothetical causes and sought to explain the former changes of the earth's crust, by reference exclusively to natural agents. Since his time the aim of Geologists has been rather to collect facts than to construct theories: and in the observation of phenomena few branches of study have been crowned with such abundant success or have made so decided progress in so short a time. The same tendency to build up systems and propound theories concerning the origin and changes of the Earth, which so marked the early ages of geological inquiry, is of course to be observed now: is attended with the same rash conceit and often leads to the same sad results. But there is a very general feeling among men of true science that the facts yet collected are too few, that they do not harmonize sufficiently with each other, and that sufficient progress has by no means yet been made in the observation of phenomena to warrant the construction of a scheme which shall claim to present Geology as a clearly and well established Science.

The object of Geology, then, is now generally held to be merely the examination of the structure of the Earth, as it now exists; the character and arrangement of its materials and the action upon them of natural agents. The first thing which becomes evident is the comparatively slight depth to which it is possible for us to extend our observations; a single mile being the greatest depth that has ever yet been reached. All that we can do, then, is to examine the component parts of this crust. At first sight everything would seem to be entirely without regularity: but closer observation shows that the rocky as well as earthy materials of which it is composed are not all alike, and that they are disposed in some regular order, one above the other. For the sake of a convenient classification these rocks (which term includes all the mineral masses of the earth, whether hard or soft,) are divided into four great divisions, as follows:—

1. *Aqueous Rocks:* or those which are the products of watery action at or near the surface. They are found to be divided into layers, or stratified; and these, from their position, are supposed to have been deposited by the action of water. Embedded in these rocks are often found *fossils* or traces of the existence of some animal or vegetable body, of which shells and corals are the most frequent:

2. *Volcanic Rocks:* or those which have been produced by the action of heat They are not stratified like the others and contain no fossils: but seem to be made up of lava or melted matter which, after deposition, has cooled and become hard.

3. *Plutonic Rocks:* which differ from volcanic in having evidently been cooled, not in the open air, but under an enormous pressure. They are, like the others highly crystalline and destitute of organic remains.

4. *Metamorphic Rocks:* or the crystalline strata and slates, which contain no scoriæ and no traces of organic remains: yet are divided into beds corresponding in form and arrangement to those of aqueous formation.

Now each of these four great classes of rocks may evidently be considered in two points of view: 1. Simply as a mineral mass deriving its origin from a particular cause and having a certain composition, form and position in the earth's crust—or other characters both positive and negative such as the presence or absence of organic remains:—2. As a chronological monument, fixing by the remains it contains or the structure it exhibits a distinct event, (and the whole series thus attesting a succession of events,) in the former history of the globe and its living inhabitants.

It is the purpose of Geology to consider it from both these points of view.

Now the true method of Geological inquiry (which from necessity in inductive,) is to find as many distinct portions of the Earth as possible in which these rocks are most favorably exposed; to examine closely their relative position, their structure and the fossils which are found embedded in those known to be fossiliferous; and then, when the number of such facts, collected with care and scientific skill, and harmonized among themselves as nearly as possible, shall warrant the conclusion, to establish upon them some general theory which will, under proper philosophic restrictions, embrace and explain the whole.

The Lectures which follow embrace some of the most recent observations that have thus been made. The points of the earth selected are those known to be best adapted to the purpose; and the Lecturer, who made nearly all the explorations of which he speaks in his own person, is deemed one of the most capable and successful geologists in the world. The facts which he states may beyond all doubt be depended upon with entire reliance and perfect safety; with the conclusions which we draw from them, it is not expected that all will agree, though it is believed that few will fail to study them with interest and profit.

MR. LYELL'S LECTURES.

LECTURE I.

AUVERGNE.

Mr. Lyell opened his Lecture by saying that he had been invited to give a short course of Lectures upon one of the most extensive branches of Natural Science; and as he was to have but a few meetings, he should lose no time in prefatory remarks, but would proceed at once to the subject and endeavor as well as, he could to enable his class to comprehend the objects of Geology, the means of proof employed, and in what manner we attempt to interpret those monuments we term Geological. If any one should ask him in what way he could soonest arrive at an understanding of the subject so far as eight meeting would allow, he should take him at once into the field of observation. I shall go, said he, up this magnificent river—the Hudson—and point you first to the Palisades, which you can see from this City; I would show you the natural rock, called Basalt, with its columnar structure, and explain the reasons why we conclude that these rocks thus piled up have existed in a melted state in the interior of the Earth. I would show you the other rocks—as the Sandstone, which was once sand until it was consolidated—deposited, one above another, under the water until a flood of melted matter flowed over and made it solid. Going still farther up, I would show you the gneiss and granite of the Highlands; or I might come back to this very island on which New-York is situated.

But as it is impossible thus to carry you into the field, the next best thing I can do is to show you some representation of natural scenes, and explain their several parts. The scene to which I first call your attention is one of the most remarkable known in Europe—far more striking than any I have seen in the British Isles. The series of Geological phenomena, here exhibited, all belong to the same great class of rocks, but it is easy for the merest novice to see that they belong to different periods: their origin is to be referred to distinct epochs and to dissimilar causes. This is a scene of *Auvergne* Having visited all the portions of it I can vouch for the accuracy of the Geological representation which is enlarged from a drawing by my friend Mr. Scrope. To understand the geographical position of the region it will be necessary to refer you to a map of France. What is called the Paris Basin is a strata of comparatively modern origin: nearly two hundred miles South of Paris are situated the Extinct Volcanoes, of which the principal mass is Mount Dor, which is in the centre of the granite region—its rocks being of marine origin. And here let me say that by rocks we mean technically in Geology any set of mineral masses whether hard or soft—whether loose like sand or consolidated like granite—soft clay as well as hard shale. These rocks contain a variety of shells and corals, called marine as having been formed originally beneath the sea. This is the zone of the granite region in France. Mount Dor, which is near Clermont, is more than 150 miles South of Paris.

The next scene I shall show you is a vertical section of this region—just as if you should dig a railway—the rocks being represented to a certain depth, as far as half a mile, as though they were cut down vertically, just as you may see in a line of sea cliff. The upper portion is of basalt—similar to the Palisades—and the lower is of a different kind of volcanic rock—a Feldspar rock called Trachyte. As it would be impossible for you to follow me were I to attempt it, I shall not enter upon any detailed explanation of these terms. It may be well to say that more rocks are found in this mountain than could be represented in a drawing. There is a great prevalence of Feldspathic rocks and of heavier basaltic rocks containing a dark mineral called augite. You see the granite represented below the volcanic rocks as we have abundant evidence for thus representing it in that section. You see the granite peeping out from the Trachyte; we find dykes of other kinds of lava passing thro' the granite and then joining into sheets of lava, as we may call them, which we know once flowed in a melted stream.

We have the granite then as the fundamental rock: then come the different kinds of volcanic rocks: then at a greater distance the stratified rocks, which we know are of fresh water origin—formed at the bottom of lakes which have now disappeared just as the fires of the volcanoes are now extinct. The basalt appears, as I have said, in precipitous cliffs divided into vertical columns like the Palisades. In another portion is a mass of white calcareous marl abounding in shells and not far off is a smaller cone—a mass of volcanic matter of later date than the other. After the volcano had poured forth its lava until it had piled up a mountain 3,000 feet high, and a valley had been cut down to the subjacent granite through a great series of volcanic rocks, and after the lake had been drained by a river called the Coos, then there happened another volcanic eruption from the volcano called Puy de Tartaret similar to the one which was formed in the Bay of Baiae which it resembles in structure and size. Now when this was thrown up (which is but a type of a hundred similar cones in that part of France,) it stopped up the course of the Coos—blocking it up so as to produce a lake which once extended much farther, for the alluvial matter which was carried into it formed a large Delta which greatly curtailed its boundaries. At the same time that the volcano bursted forth in the midst of this ancient valley on flowed streams of lava from the other side, which you may see passing at the base of the hill on which the castle of Muro stands. Flowing down the valley for 13 miles in a narrow stream it dispossessed the river of its ancient bed, as has been the case in Iceland within the last century, when streams of lava have flowed from the sides of Hecla and Skapta Jokul and forced rivers to flow aside from their usual beds. So out of this flowed a stream which took its course along the base of the hills, cutting its way through the fresh water formations.

It is a matter of proof that these strata are really of fresh water formation—that they were formed in ancient lakes where were left valleys after the draining of the lakes, and that through the bottom of these flowed the lava. Thus we may trace the cappings up to Mount Dor, the ancient and grand centre of the first eruption. From this the higher cappings of basalt were produced; and to cause these phenomena and cut through the valleys so as to allow other valleys to be formed, how great a series of changes and how long a chronological order of events is implied! The hight of Mount Dor above the level of the sea is 7000 feet, and we can look back to the time when that country was only 3000 feet above the sea, and consisted exclusively of the fundamental granite. At that time we find there were depressions in which there were a number of lakes extending all the way from Paris to the source of the river Loire. That was the first state of things, and the mountains were featureless in appearance—with round swells but with no picturesque, uneven outline—and into the lakes sediment was carried by rivers, and the beds, of which I shall soon speak, were accumulated. Then came a period of the draining of the lakes, and then the excavation of the valleys; just as we have on the borders of Lake Superior and other Northern lakes, where the rocks are granite. If Earthquakes should occur and shake that region and alter the level of the country, the lakes would be drained and we should have the second state of things as in this part of France. If we follow the volcanic eruptions in the same district until the lava was piled 3000 feet high, we should have the third period.—If floods should produce deep valleys, and a new set of volcanoes should form small mountains at the bottom of the valleys, we should have the fourth period. The fifth period would arrive when rivers should have eaten out deep chasms even in the most modern lava, and formed precipices some two hundred feet high—which would involve a period of considerable duration.

While great changes are going on in the inanimate world others no less interesting are produced in the animate world; for we see entire fluctuations in quadrupeds and other beings which successively inhabit the region. It is impossible to represent in a picture, in their proper proportions, the hight of the various parts of the country. The fresh water formations do not appear sufficiently in depression to give the proper appearance of the ancient lakes.

There is no doubt that the superior upheaval of the Southern part of the country was what caused the drainage of the Lake and the excavation of the valleys which are cut through them, producing what is called the Lacustrine formation. If we simply take for granted the upheaval of the whole district, (which need surprise no one, for we shall have occasion to show that far more extensive upheavals above the level of the sea are *now* going on;) we need only suppose a series of movements such as are demonstrated to have taken place, to have raised the Southern portion at a greater rate than the Northern, to find an easy explanation for these geographical phenomena.

THE LACUSTRINE PERIOD.

I have now stated that cotemporaneously with these mighty changes, physical and geographical, there went on a series of revolutions in organic life. I am not going to enter in the present lecture upon the difficult problem of the origin of granite. We must take the granite as it exists in the country.—Hills are made of it, and lakes have been hollowed out of it. I will come at once, then, to the Lacustrine period. The lakes were inhabited by a different set of shell-fish, of moluscular animals, water-birds, crocodiles and turtles from any that now exist in that part of France. There are now no such animals as once abounded in the ancient lakes of Auvergne. We find their remains well preserved in the various strata. The tortoise which then existed is not now to be found. I may remark, too, that there have been discovered the skeletons of some thirty or forty quadrupeds to which Cuvier gave as many different names, as the Palæotherium, Anoplotherium, &c., which need not be mentioned. The Palæotherium is an intermediate species between the Tapir and Rhinoceros, and the Anoplotheria has some relation to the same animals, and also to the hog tribe. There are also several species of animals found in the neighborhood of Paris, which now exist—*though not in that region*—as, for instance, the species of Opposum, which now exists only in this country and Australia. On the whole, there are about fifty species of quadrupeds discovered, some of which are as large as the Rhinoceros. I need not say that a single jaw, or even sometimes a single tooth, is quite sufficient in the hands of a Naturalist to decide the species of this Mammalia. These animals were undoubtedly washed down by the swelling of the rivers, just as they are liable to be now. We have no hesitation in saying that these fty species of quadrupeds have now passed away from the globe. Some might doubt our ability to say this; but a little consideration of the fossils found will be sufficiently convincing. I have procured drawings of one or two shells which are there found—as, forinstence, the *Planorbis*, which abounds at present in the lakes of North America, and which once existed in the lakes of France, and are now found in the strata. The *Valvata*, a fresh water shell which is common here and in Europe, was also to be met in the old Lacustrine formation of France. Any naturalist will be easily convinced that these shells are now extinct in that region, and no one will be so skeptical as not readily to grant that we are safe in saying that the Palæotherium, the Magatherium, and many other large quadrupeds which CUVIER discovered and compared, are now lost, and have been permitted by the Author of Nature to die out from the globe.—This is the first state of the ancient world of this region, to which we are carried back by the geological monuments of this scene.

The next period of the ancient volcanic rocks of Mount Dor was when they produced a perfectly distinct race of quadrupeds, of which I delieve only one has been found to resemble those of the former period. All the genera are different, and almost all of this period are of existing genera, but not to be found in that part of Europe. We find the rhinocerous, the elephaet, the hippopotamus, the bear, and others, which never within historical times have peopled the cold region. The beaver which is found is not the same as that now common to Europe and North America—and so with a vast multitude of others.

But you may ask how we should have these fossils remaining in valcanic rocks—how they could be preserved if they were enfolded in a sheet of red-hot matter as from a furnace? It is easy to explain how this might happen. In all volcanoes, where the lavas are feldspathic, there are great floods; lakes which exist in the craters are suddenly voided, as happened in Ætna when floods issued from it which swept rocks and quadrupeds with ease. Showers of ashes also fall, and often surprise wild animals—destroy, bury and preserve them. In this way it is that we find the skeletons of so many quadrupeds entombed in the ancient ashes—conglomerates as they are called—the rocks being formed of the pebbles carried down by these floods.

But before I speak farther of the quadrupeds of that period, all of which are now extinct, let me describe more minutely the operations of the first period—the first gradual filling up of the lakes and the formation of the Lacustrine strata; then we will speak of the origin of the volcanic mass and of the animals buried in it. Let us first ask how these ancient lakes are filled up. We find around the borders, and under the other strata, beds of pebbles which have been derived from the rounding of fragments torn from rocks in the neighborhood; and there is this remarkable circumstance connected with it, that no geologist has ever detected a single pebble of volcanic origin in these conglomerates. Every one knows that if a river flows from a country abounding in particular rocks, it will bear off and carry along portions of them, discharging them near the shores of the sea. The fine mud is carried out to a greater distance. Now there are vast piles of volcanic rocks, some soft and others hard and capable of being rounded; though they are sometimes a hundred feet thick, not one is found in the old gravel beds of the Lacustrine formation. The inference is obvious—that *not one volcanic eruption had occurred*, any more than upon the borders of Lake Superior, which is a country without volcanic rocks. This negative fact alone proves that in the first period the region was tranquil and no volcanic eruptions had occurred.

The gravel beds at the bottom of the lake soon became fine sand and passed into fine marl: some sand was carried to a greater distance as you may now see falling clouds of it at the mouths of rivers settling to the bottom and being carried out. Marl which has a leaf-like appearance is often produced by the small shells of an insect called *Cydris*—a crustacean—or between an insect and a crustacean—which is very like a bivalve, such as fresh water muscles, with this difference, that it molts its integuments, while the others do not. Accordingly myriads of these shells fall to the bottom, and for hundreds of square miles you may find that the strata have plates and leaves as mica causes sandstone to divide; and this is produced entirely by the cases of that small species of animals. Many other shells show that this is a fresh water and not a marine formation, and this may shew how gradual was the process that went on, as does also another kind of rock called *Indusial Limestone*—which consists of thousands of cases of animals which have been embedded by the waters that flowed into the lakes. This is in the upper part of the marl, and we see it breaking out in the side of the hill at Clermont, when it occurs in round nodules. Every one of these consists of cases of insects encased by limestone. This insect, which is called *Phrygarea*, when it is in the *larva* state, and exists in the water in the form of a caterpillar, covers its case with shells to give it a proper weight; it often takes a hundred shells at once; as we see in the American Lakes they take small stones for the same purpose. They often seize upon these shells without the least ceremony, even when the fish are alive within them, and walk about in this way. It seems that they swam about in these lakes when they fell one after another encased in carbonate of lime, until you may now find the whole mass leagues in extent. This may give some notion of the time required, and the gradual process necessary to produce beds 1,200 feet thick. You may often find the seed vessels of plants petrified in these beds. In several lakes of New-York—in Seneca lake for instance, the whole bottom is covered by a green plant called *Chara* and *Gyrogonite* in such numbers that without other indications these alone would tell us of the origin of the formation.

It was formerly a matter of great difficulty with Mr. SCROPE and other naturalists to explain how Indusial Limestone if often found over wide areas; it was easy to see how it might exist near the shore, but not so easy to understand how it should be found at a great distance out; for the borders of these ancient lakes, with all their bays, can yet be traced as acurately as if they now existed.—How should these Indusial Limestones be found ten or twelve miles from shore? A few years since, while in Denmark, I believe I found an explanation. In company with Dr. BECK, an eminent naturalist of Copenhagen, I saw in the lake 20 or 30 miles in diameter a large band of rushes or reeds floating on the water; they had been torn up and floated out miles from the shore. And on every one of them was an immense number of the *Phrygarea,* wafted out with them; and I found that twice every year a crop of these rushes of different kinds is produced, covered by different kinds of Phrygarea, and carried out by storms to the middle of the lakes. Thus was removed the difficulty which before had existed with relation to this part of Auvergne.

THE VOLCANIC PERIOD.

The transition from the Lacustrine period to the Vocanic is not abrupt. Changes rarely occur so suddenly as are represented by geologists. None such have taken place in Auvergne. Though the two periods are very distinct, yet the volcano began its eruption before the lakes were drained.—Showers of ashes fell into the lakes so that volcanic *tuffs* are often found to alternate with the marl containing limestone which is the upper part of the formation. Here and there may also be found *dikes* or threads of basalt. The facts show that there were few outbreaks of this igneous matter before the lakes are drained, when great rivers were still carrying down mud to the lakes. But the lakes soon disappeared and then commenced the excavation of the valleys.

Of this excavation of the valley there is a beautiful example along the valley of the Coos. When about twenty miles beyond the termination of the stream of lava which comes down from Mount Dor and filled the valley—we find *breccias* in which are contained the bones of animals. A large part of this formation contains beds of pumice stone, which is sometimes found floating on the sea; it is produced by the trachyte that is thrown up, the gases expanding within it producing the pores.—There is no pumice in the bosaltic formation nor in the upper part of Mt. Dor; but in the older portions you may see thousands of them, showing how gradually the mountain was formed just as were the Lacustrine formetions.

In a general view of these subjects you may despair of understanding them; but when we examine the minute parts and sub-divisions—when you see in the volcanic rocks how one shower of ashes has fallen above another, you will begin to have hope of conprehending the successive steps by which these geological monuments have been built up. We see how the mountain was formed by the greater filling up of volcanic matter at that point than elsewhere; after that we see how the valley was again hollowed out, and then came the flood from Mount Dor bringing along the rhinoceros and other animals of the second period. Again was this cut out, leaving bones enough to determine its character; and lastly the river had its channel occupied by the most modern lava stream. About 40 different species of quadrupeds have existed there—of all sizes from that of the water rat to the great Mastodon.

ERUPTION OF COSEGUINA.

In the eruption of Coseguina, in Nicaragua, Central America, on the 18th of January, 1835, only seven years ago, there was a fall of ashes which reached several hundred miles, and some fell at Chiapa, 1,200 miles distant. But for 25 miles the beds were ten feet thick. Birds, cattle and wild animals were scorched to death in great numbers and buried in these ashes; fish were smothered in the rivers; birds fell from the air, and the destruction extended over a vast region of country. There is evident reason to believe that a similar eruption occurred during the formation of this ancient mountain, and that the creatures which lived there were buried in the fall of red-hot, melted matter from the air. Between the two eruptions there was probably a long interval, and the animals had time to recover their numbers.—There have been intervals of seventeen centuries in the history of active volcanoes. In the island of Ischia is an instance: and in the history of Vesuvius there was an interval of five centuries. We have a description of Vesuvius in 1681; it was then covered with wood; its crater was five miles in circumference, and over its edge, after descending a short distance, was a beautiful copse wood; grassy plains spread out below, cattle were grazing upon them, and a pleasant lake added charms to the scene. The peasants who lived about them had begun to look upon the stories of fiery floods having poured out of it as fables of the olden times. All was peace and security. But there came an earthquake, and in an instant all their forests and grassy plains and their animals were hurled into the air with the ashes and flame; then arose as from a caldron a fiery flood that filled the whole gulf, flowed down in seven rivers and took a direction directly over where Herculaneum had been destroyed sixteen centuries before by a similar flood of molten lava: and thus one town was buried beneath the other. After the heated gases have found relief, burst forth and been discharged, it appears that the duct or channel gets sealed up—the lava cools and consolidates, and it then requires no little force to burst it open again and cause a new discharge. Thus there is a long interval of rest until the gases again accumulate and have sufficient force to mako for themselves a passage.

MOUNT DOR.

Thus much for the Lacustrine period, and now a few words respecting the more modern volcano as a type of the class which are the result of only one, two or three eruptions, and are often formed on the peaks of Vesuvius and Ætna at the hight of from 500 to 700 feet in a few months. There is a considerable crater at the top of Mount Dor, covered with beautiful chestnut trees. Getting at the top of the crater you look down in a deep funnel. One circumstance of interest in regard to this is its loose, incoherent nature. There can have been no violent rush of water, of sufficient force to carry away animals, else much of this loose matter would have been carried away. Every thing about it looked so fresh that one would say it could have been formed but a short time ago, and yet we find no mention in history of any eruption in this part of France; and when we go down the valley and examine the lava stream whice we know to be its own, we open our eyes to the prodigious antiquity of this most modern volcano. By reason of its loose structure, as is also the case with Ætna, all the rain that falls upon it is imbibed at once, and no streams can be formed down its sides.

Following these valleys along you see at last the old river bed as it was in the olden time when there was no lake and the river flowed on uninterruptedly. It has a bed of pebbles—formed by rounding the fragments of basaltic trachytic rocks from the higher country and from the cappings of the hills. At a still higher point we find no volcanic pebbles, because there were no volcanoes: that then must have been once the lower portion of the country in order to be a water course, as we know it was.

Cæsar encamped his army but about ten miles from Clermont, and yet it does not appear that he was aware of any eruption having taken place in that vicinity, else he would have alluded to it in his Commentaries. And Apollonarius certainly would have recorded it had any such eruption taken place within his historical knowledge, as he resided in this vicinity and wrote a very minute account of its history; but it does not appear that he had the least idea of the origin of the lake. So that his silence is conclusive as to the high antiquity of the occurrence of even these modern events.

You have from these observations some idea of the great succession of events contemplated by Geology. If you ask me why all these hills and valleys may not have been formed at once in the origin of things?—why they may not have been called into existence at the same time as the nucleus of the planet itself?—I may answer, that even the brief skeleton of facts I have given will be sufficient to show that this could not have been the case; that these hills and valleys were not made at the same instant, but that they were the work of different causes, acting at different times—some being produced by the action of water, and others by fire; that they were produced at different periods of time by secondary causes, appointed by the Author of Nature to govern the successive changes in the inanimate and possibly in the animate creation, which he had in his wisdom employed in producing the successive changes of which the records are found in the very structure of the globe.

LECTURE II.

THE EARTH'S STRATA.

In the course of my last Lecture, while endeavoring to describe the geological phenomena of a region about the centre of France in Auvergne, I had occasion to introduce you to several classes of strata—the *granitic*, the *volcanic* and certain *aqueous* strata, constituting three of the grand divisions of the rocks of which the Earth's crust is made up. Let me remark that by the Earth's *crust* technically we mean that part of the exterior of our planet which is accessible to human observation; and the whole of this may be said to be made up of these three kinds of rock.

The *granitic* rocks are crystalline and contain no organic remains which have been mentioned as characterising the sedimentary or aquous strata. They are crystalline masses, and the true granite is unstratified. It has no pebbles nor any of the *tuffs* or porous rocks which are found in the volcanic class.

The *volcanic* rocks may also be said to be void of organic remains except those portions which I described as containing beds of sand and ashes showered down from the air or which fell into lakes; or except those *Breccias* and conglomerates which I described as having flowed down the banks of some volcano as at Mount Dor, in which quadrupeds and other animals have been buried and their skeletons found so as to determine what animals inhabited that country when the overwhelming catastrophe occurred.

MARINE STRATA.

The *aqueous* strata, or those which were formed in lakes and seas, are the next division. I shall only allude to these briefly by saying that they are known to be of fresh water formation from the fresh water shells, &c. which are found imbedded in them. These fresh water strata compose but a small portion of the strata of the Earth's crust; just as lakes now are of small extent in proportion to the seas on the Earth's surface, so were formerly the estuaries and lakes of small area compared to the ocean; and we find these sedimentary strata containing Lacustrine or fresh water remains much less extensive than the make strata. We find the strata of marine origin, containing sea shells and corals, at all distances from the present shore of the ocean, and at all hights above the level of the sea. This fact alone would render the results of geological investigation extremely interesting. We find the marine strata often far in the interior; and among the Alps there are met, 10,000 or 11.000 feet above the level of the sea, rock containing unequivocal remains of creatures that once lived in the ocean. I have myself gathered limestones containing marine shells far above the hight of perpetual snow in the Bernese Alps and at Mount Carmel, the highest point of the Pyrenees. And still more recently there have been found in the Himmelaya Mountains, 10,000 feet above the level of the sea, similar organic remains; and this not only in the mountains, but limestones containing these fossils are continually found through the whole table land of Thibet; so that this vast platform of Asia is made up of rocks formed—not at the most ancient geological period—gradually beneath the ocean, and which have since risen up to higher regions. I speak as if we had already proved that this process is carried on by the upheaval of the land and not by the subsidence of the sea. In my next Decture I will endeavor to convince you by distinct evidence of the mighty changes that have taken place, that this was the result of a rising up of the land, and not of a sinking down of the sea. Speaking in the language of that theory, these mountain masses, which now appear at hights of one, two and even three miles above the level of the sea, were formerly beneath the ocean, and were raised up during the first era of animals and plants, and not all at the same period.

Now the sedimentary or marine strata which enter thus largely into the masses of the Earth are divided into various groups, which we refer to distinct periods. They form a chronological series of volcanic action—a history of the Earth; and when we trace this series down from the most ancient to the most modern, it is not till we arrive at the latest of the great series that we come to those groups that environ the section of France which I have already described. Grand as was this series of events, it is not till we arrive at the latest of the larger groups that we come down to this Epoch. I shall not at the present meeting attempt to sketch the history of the different formations, but continue to describe most particularly this strata and those which are cotemporaneous with it—those containing the remains already alluded to, or those of a posterior date. Now the oldest of the series of which I spoke last, is the fundamental granite; and the fresh water formations are imposed upon it. In order to classify the different depositions which belong to this more-modern period of the Earth's history it is indispensable to attend to the organic remains, and particularly to the shells contained in it:—shells are more useful in this science than any other indications, and have been appropriately called the *medals*, by aid of which Nature has recorded the events of the most ancient history of the globe. They are found in such abundance in the fresh water and marine strata of all ages, that by comparing and contrasting them we may build up a chronological series, and find characters clearly to distinguish the different periods. In the fresh water formation in the neighborhood of Paris are found twelve hundred species of these shells—carefully distinguished from each other; as is well known to Conchologists, only some thirty of these species are known now to exist any where upon the globe. We are acquainted with about 10,000 species of shells; and of these only 30 resemble any of the 1200 found in this strata. You may ask how we know that these rocks were formed in the lakes in the province of Auvergne, at the same time that the others were formed in the sea on the site of Paris. The manner in which we make out this cotemporaneousness is briefly this:—Rivers carry down shells to the sea; the shells of the American Lakes, for example, are carried from Lake Erie and other lakes by the River St. Lawrence into the marine gulf of St. Lawrence. The Tiber carries down the fresh water shells of Italy to the Mediterranean, and deposits them in the Delta. So of the Adige and the Po; and if we examine the strata in the Delta we find that they are filled with fresh water shells and contain also marine shells. So in the neighborhood of Paris—shells of the Lacustrine deposit are mixed with marine shells. Quadrupeds, too, are carried down by rivers and deposited with the 10 or 1100 species of marine shells of which we were speaking; and by this means we first know that these quadrupeds lived, and these shells inhabited the lakes when the sea was inhabited by those testacea. In Vicenza, Italy, we find certain marine strata, and in Paris these alternate with volcanic formations, and thus we make out the cotemporaneousness—the Synchronism—of a certain volcanic formation, and a certain marine or Lacustrine strata.

PERIODS.

It is necessary to have some technical names for these periods, otherwise it will be impossible for you to follow me when I allude to them. By a few terms we may express in a word what it would be tedious otherwise to express. By the *Eocene* period, then, I mean the oldest, as when this fresh water strata of Auvergne and other districts contained the same shells of alternating strata. It takes its name from ηως—the *dawn*, and καινος, *recent;* because the fossiliferous strata of this period contain but few species that are now living, which may indicate the dawn of the existing species of testaceous fauna. The next period I call *Miocene*, from μειον, *less*, and καινος, *recent;* because it contains a *minority* of recent species—a much greater number than the Eocene, but still a minority. The next period is the *Pliocene*, from πλειον, *more*, and καινος, *recent;* because it contains a *plurality* of the existing species. That is, out of 100 species of shells found in the Eocene, one or two only will be of those now living. When the land had been raised by volcanoes—in the Miocene period, about one-fifth, or twenty out of the 100, may be identified with existing species. And if we come down to the more modern or Pliocene period, more than one-half the shells are identifiable—and sometimes 90 out of the hundred.

THE AUVERGNE DISTRICT.

When the volcano which formed Mount Dor in a period distinguished as recent, was filling the lakes in that region, by volcanic eruptions lakes and seas in other parts of Europe were filling up, and that at no great distance. If you travel from this mountain to the Loire near Tours you may find great numbers of marine depositions—strata rich in shells; and you may find, too, some of the very quadrupeds buried by the ashes and scoriae and preccias which floated down into the sea, and were buried near the shore in marine depositions; enabling us thus to prove that those strata were formed in the same period, and that the basin of the Loire was formed in the same Zoological period as this Mountain. So also in Bavaria are many Lacustrine deposites, and on the Rhine you may find the same quadrupeds associated with fresh water shells. Thus extending your observations from one part of Europe to another, you can prove that in the second period there were volcanoes making marine deposites. Thus is it on the borders of the Loire between Tours and Mount Dor; and four-fifths of the shells found there are of now extinct species—while the other fifth are still living. This period is called Miocene—as containing a minority of shells of existing species.

I might remark that these formations are found all through the United States. In my tour to the South I found in Virginia formations belonging to the first or Eocene period. [The whole group is called Tertiary, to distinguish it from the Secondary formation. A friend lately suggested to me that he could never remember the order of the different periods—Eocene, Miocene and Pliocene, until he observed that it was the natural order of their initials—E. M. and P.—in the alphabet. These little helps to memory are often of considerable value.] At Richmond, Va., at Wilmington, N. C. and for 40 miles on the Santee river and between Augusta and Savannah I found Eocene strata, containing shells of a species identical with those found in the Paris basin; the same shells are found all along your coast as on ours. You have on the James river beautifully developed the marine Miocene strata—and this fact has before been pointed out by RODGERS and other American Geologists, whose observations I am only confirming and perhaps adding something of my own. You see there the strata containing marine species in nearly the same proportion as those on the Loire.

But I must not spend time upon these matters, lest I pass over the main object of the present lecture. If you had examined a portion of this country when the Lacustrine and Eocene strata were largely developed, and should then go some 20 or 3 miles from Mount Dor and explore the valley of that part of France, when on all the hill sides you found nothing but fresh water formations, you might say that between the white marls that contain these shells and the other period there was nothing intervening except a thin stratum of vegetable soil; and that therefore (a mode of reason often followed by geologists,) there must have been a sudden transition from that state of the globe under which the *Palæotheria* lived, and when shells peopled the lakes—crocodiles, turtles, &c.—to the present; that there must have been a sudden annihilation—a sweeping away of existing races, and an introduction of those that now inhabit it. But when you pass to Mount Dor and observe the manner in which the volcanic strata overlie the fresh water, you will see how vast a period must have intervened between the present and the Eocene—giving time for another race of quadrupeds—the mastodon, hippopotamus, hyena, and other animals to supplant the more ancient. But you may ask might not *this* at least have been the last formation? Examine the great sheets of lava which are there. Then follow on and find where suddenly they are cut off and again resumed—a vast bed of pumice, and trachyte and basalt corresponding to a similar one on the other side of the valley. Consider that instead of this being the last in the series there has been an excavation of the great valley, and a throwing up of the Puy de Tartaret and the flowing out of a stream of lava for 30 miles, and then a subsequent erosion of the ravines by the lava. By this time, if you are a philosopher, you will have acquired more caution, and will not hastily assert that the Miocene period was the last; you will say rather that on extending the sphere of your researches you were obliged to abandon your first opinions, and so if you still farther enlarge the field of observation to other parts of Europe, you would see that there was another strata between the Miocene and our own—and such is the case.

THE SUB-APPENINES.

When you have entered Italy and followed the Appenines from the Po to Calabria you find a low range of hills called by Brocchi, an Italian Geologist, the Sub-Appenines, consisting of strata of a date posterior to the Miocene. These we call the Pliocene—which are divided into the Older and Newer Pliocene. I shall not dwell upon the first but allude to the Newer Pliocene, or more modern part of the group. In the South-Eastern part of Sicily we have an example of marine formations of the Newer Pliocene. There are rocks of marine origin in which you would at first say that all the shells were the same as those that now people the waters of the Mediterranean; and yet when you come to a more careful examination you will find that one-tenth are of a species never seen in that sea. These rocks are sometimes 2000 feet thick, and it would be easy, if I had time to describe them minutely, to show that they had been accumulated in the same slow and gradual manner as the fresh water strata of Auvergne were elaborated by the deposition of one layer after another, with this difference, that in the former the organic remains are marine, and in the latter they are those of lakes, and therefore a fresh water species. Another difference is that volcanic rocks are distributed nearly through the whole thickness of the formation in the South of Sicily. Near Syracuse also you may see a strata 2000 feet thick, alternately of volcanic and marine origin; at the bottom is a bed several hundred feet thick of volcanic ashes, then a bed of limestone, on which rests the great sheet of lava. In one of these strata at Vincenza an observer found a bed of oysters 20 feet thick. There had been time for this oyster bed to accumulate—for *Serpaluae* and other parasitic animals to attach to them until the bed grew 20 feet thick; and then a stream of lava rolled over it. These facts shew how gradually these strata were formed. Another remarkable circumstance is that these rocks rise 2000 and in Castrogiovanni, in Sicily, 3000 feet above the level of the sea. These have all been formed since the newer series of strata were deposited. Volcanic matter often rises in the sea, overflows the bottom and forms a temporary island, as Graham's island was formed five years ago. The waves soon destroy it, spreading the volcanic matter over the bottom of the ocean. All this takes place before there is any upheaval. Here, too, we have proofs of what an extended series of groups in the chronological table is to be formed subsequent to the Miocene period—the date of that volcano, before you think you have come to the end of the series of modern events—those which are considered as belonging to modern history.

MOUNT ÆTNA.

My principal object in the present lecture is to convince you how far from the end of the series these rocks were formed, although they have a respectable claim to antiquity. When we examine Ætna, and the strata on which it rests, we shall see another period, called the Post-Pliocene period, as being subsequent to the Pliocene. In this all the shells are of the existing species. Its formation has been sufficiently protracted to admit the elaboration of this vast series of volcanic movements. The locality which I shall select this evening is Mount Ætna—for that volcano rests on strata in which all the shells and corals appear to be identical with those now found in the neighboring seas.

The height of Ætna is nearly 11,000 feet. It was well described by the ancients. It is divided into three distinct zones, called the *fertile*, the *woody*, and the *desert* regions. The lowest, the fertile, is a most beautiful territory, embraced the delightful country around the skirts of the mountain. It is well cultivated, covered with groves, olives, wines, corn, lemon and other fruit trees, and is densely inhabited. When you pass above this, (and its extent may be judged when I say that the base of the mountain is ninety miles in circumference,) you come to the woody region, a thick forest of chestnut, oak, and pine trees growing upon lavas of different dates: and as you go higher you occasionally find groves of oak and beech. This region is rendered extremely beautiful by the great number of minor cones, or small volcunoes like Pay de Tartaret, which have broken out at different periods on one side of the mountain. Some of these are four or five hundred feet high, and one of them, Monte Minardo, is seven thousand feet high. There are few objects in Nature more beautiful than this multitude of minor cones that have broken out and covered the trees of this woody region. This zone is two, five and in some places six miles wider and when at its upper limits you are about five thousand feet above the level of the sea. From the upper confines of this forest region, in 1828, I took a sketch of the scene above us. We climbed up upon the eastern side. But from that point we have still six thousand feet elevation between the highest point and the foreground of this picture. The mountain is generally symmetrical, in the form of a flat cone, broken down upon the eastern side by a deep chasm called the *Val del Bove*, as may be seen in this figure:

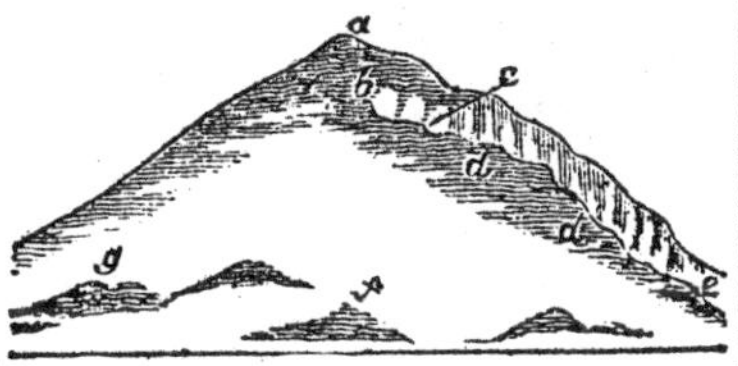

[Only that part of the drawing presented by Mr. Lyell, which is necessary to show the Val del Bove, is here copied. The following are all the explanations necessary:
a—highest cone of Ætna.
b—Montagnuola.
c—head of Val del Bove.
f and g—lateral cones.
At e, the extremity of the ravine, is a little village called Zaffarana, at the lower border of the woody region. *Rep*]

After entering upon the woody region, and rising about one thousand feet from the base, you reach a great precipice two or three thousand feet high; then comes a plain covered deeply with snow, and lastly the cone, eleven thousand feet above the level of the sea, from which steam or aqueous vapor is constantly emitted. When these views were taken, which was in October, there was less snow upon the mountain than at any other period of the year. One eruption from the summit occurs for every two of the cones on the flanks. When one of these happens the melted lava pours out over the snows, which are suddenly melted, and rush down the mountain, producing sometimes floods which sweep down the flanks and across the valleys.—Except when floods thus occur, there is a singular absence of running water on the flanks of the mountain. The silence which pervades on this account is quite remarkable; for no torrents dash from the rocks, nor is there any movement of running water as in most mountainous countries. Not a rill runs down the sides. All the rain that falls from the heavens, and all the water from the melting snow, is instantly absorbed by the porous lava.

There are numerous vertical dikes which traverse the mountain and sometimes extend for more than a thousand feet in a perpendicular. They consist of liquid matter emitted by fissures, which as it rises is consolidated, mixed with tuffs and scoriæ, and thus becoming harder than the cliffs around, resists decomposition and stands out in bold relief, while the rocks around are wasted away by the rapid freezing and thawing to which this region is exposed. There was an opportunity during the eruption of 1669 of observing the manner in which these dikes are produced. The mountain was rent asunder by fissures radiating from the centre—though some of them were parallel. These emitted a vivid light, showing the incandescent lava at a great depth below. The lava flows sometimes nine or ten miles down the split mountain, and when it is cooled and crystalized it forms one of these vertical dikes. Similar dikes are formed in Madeira, where basalt has been traversed by volcanic rocks.

It is supposed that as often as Mount Ætna is rent open, there may be a slight upheaval of the whole mass—an uplifting bodily of the whole mountain: and if we suppose that this has happened from the earliest periods, it may explain the great difficulty which has puzzled some geologists in endeavoring to explain why the more ancient part of the mountain has not been covered by posterior eruptions. It has been objected, that if the whole mountain was formed (as I believe it was) by successive eruptions, that the ancient parts of the volcano ought to have been buried long ago by the more modern eruptions. I ought pehaps to have remarked, that there is a general dip of the beds in all directions, which there are two ways of endeavoring to explain. One attributes it to sheets of lava flowing to a certain point; and the other to an upheaval in the centre, tilting the beds in all directions. Both of these may be in some degree true. I will now endeavor to show why the most ancient parts of the mountain would not be covered by the overflowing lava.

We know that there has been an upheaval of the whole mass of Ætna at no very remote period; because the mountain rests on a marine stratum which contains shells identical with those now living in the Mediterranean. A force which could carry up this marine strata which skirts the sides of the mountain must have been able to carry up the whole mountain; and if the lavas did not cover, at each successive eruption, more and more of this marine strata, we should be able to trace it to a greater hight up the mountain. As it is, we find it about 1000 feet above the level. If the ancient part of the mountain has been thrown up in this way, we readily find the explanation of the phenomena we are seeking. First, let me state that all geologists admit that the formations of alternate lava and scoriæ of which the mass of Ætna is constituted, is of supra-marine origin—that it is formed above the water; for it contains no marine shells—no beds of tuff—no stratified mass—no fossils; but has every indication of being a volcano found in the open air. At first, the volcano may have been of moderate hight—accumulating one bed upon another until it reached a hight of 4 or 5000 feet.—Then as it grew higher an eruption would take place, as it does now, chiefly on flanks.

I told you that an eruption generally occurs on the summit for every two on the sides: that is the proportion that has been observed during the historical period. The higher the mountain becomes, the greater, evidently, is the hydrostatic pressure of the column of lava that rises up: the higher the chimneys the greater the pressure of the lava rising within them upon the flanks. In this way the lava might continue to rise until at last, if it became more than two miles high, the pressure would have become so enormous as to burst through the flank of the mountain. Every geologist will admit that in a volcano such an outbreak may take place. At first, before this the accumulation went on by eruptions from the summit; but these eruptions being afterwards performed on the flanks about the woody region, the older part of the volcano was not buried.

Before I go farther, there is one point of great interest which I will notice: Ætna is, of all volcanoes that of which we have the most ancient records, and to a knowledge of which history and tradition carry us the farthest back. There is a tradition mentioned by DIODORUS SICULUS, that an eruption happened before the Trojan war, which compelled the Sicani to desert their district. THUCYDIDES tells us that in the sixth year before the Peloponnesian war, or in 435 B. C., an eruption occurred which was the third that had taken place there since the colonization of Sicily by the Greeks. (*a*) The second of these three eruptions was the one mentioned in that beautiful ode of PINDAR, which is worth referring to because it is the description by a poet which corresponds precisely with the appearance of the volcano twenty centuries afterward when Catania was destroyed in 1669. Part of the city had been overflowed during the eruption, alluded to by PINDAR. In that passage PINDAR speaks of "*the snowy Ætua, the pillar of Heaven—the nurse of everlasting frost, in whose deep caverns lie concealed the fountains of uuapproachable fire—a stream of eddying smoke by day—a bright and ruddy flame by night: and burning rocks rolle down with loud uproar into the sea.*" (*b*)

We have still the 'stream of eddying smoke' by day produced by an eruption. It had the same appearance when I saw it in 1828, and also when Mr. Scrope saw the stream nine months after it began to flow in 1819. He says that it was then moving on at the rate of one yard in an hour: during the day it constantly emitted a dense vapor: for aqueous vapor enters minutely into this lava, though no chemist has yet fully explained it. As often as the stream cracks, it emits several gases, but chiefly steam, which boiling up brings the lava to the surface, and causes it to overflow the crater just as steam in a kettle carries up the water, and causes it to spill over. So this steam and all the gases become entangled with the red hot lava, causing it to discharge large quantities of aqueous vapor during the whole process of conglomeration.—By night we saw the same cracks emitting a glowing heat, and the whole stream makes out the 'bright and ruddy flame.'

It is a singular circumstance, in regard to the theory of flowing lava, that, instead of being level, as you would naturally expect of a liquid, there are hillocks of considerable hight along its surface, in consequence of the uneven ground over which it flows. Another peculiarity of these streams is the fact of their becoming hard and solid while in motion. Instead of thinning out, as would be natural, at the edges, the upper surface becomes cool, and the sides likewise, and both grow hard; so that the edges are often rocky walls, inclined at angles of from 30 to 40 degrees. It is thus a flood of red hot lava flowing along a solid tube. Thus, when they come to a rise of ground, the streams rise as water does in a pipe—as has been the case in the Val del Bove. When the stream comes to a hill, it goes up the slant in a tube of its own making, and thus, too, it goes up the sides of solid walls. In 1669, when Catania was threatened with destruction by the approaching stream, a respectable citizen of the town, desiring to secure the city, took a troop of some fifty men, who, clothed with skins to protect them from the heat, and armed with iron crows and pick-axes, began to quarry the solid walls and let out the lava. They opened it at the village of Belpasso, and the stream immediately issued forth and took the direction of Paterno. But the inhabitants of that town, being alarmed for the safety of their own village, took up arms and prevented further operations. The torrent, therefore, moved on to Catania, and the inhabitants of that city, being terrified for their safety, turned out and threw up a rampart of stones, hewn from the the lava, sixty feet high. On went the burning torrent, and when it arrived at the wall, it continued to accumulate upon itself until it topped the wall, curled over, and fell in a fiery cascade upon the town; it passed on to the Mediterranean, which it entered with a loud explosion.—This corresponds precisely with the description of Pindar. The Prince of Biscari afterwards, at great expense, made a quarrying out of the lava, so as to show the manner of its falling over; and you may now see the upright wall, with the stream of lava curling over the top—appearing like a petrified cascade of lava, and remaining visible to this day. This drawing will give you some idea of the view of Ætna, and the situation of various localities near it:—

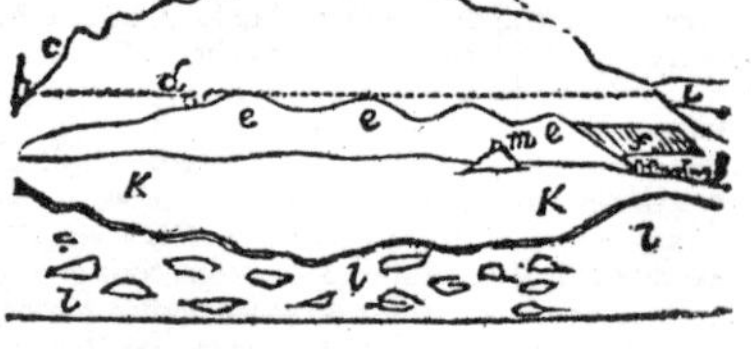

a—highest cone.
b—Montagnuola.
c—marine formation—rarely found above the dotted line *h i*.
f—escarpment of volcanic tuff, &c. N. W. of Catania.
g—town of Catania.
k—plain of Catania.
l—limestone platform of the newer Pliocene, from which the view is supposed to be taken.
m—La Motta di Catania.

(*a*) Thuycidides, Book III. Sec. 116.

(*b*) This passage may be found in the 1st Pythian Ode of Pindar, beginning

Χρυσεα φορμιγξ Απολλω-
νος και ιοπλακαμων
Συνδικον Μοισᾶν κτεανον, κ. τ. λ.

The description of Ætna, the whole of which is extremely beautiful, commences with the concluding lines of Dec. V.

κιων
Δ 'ουρανια συνεχει
Νιφοεσσ Αιτνα, πανετες
Χιονος οξειαφ τιθηνα,

and is continued through the seven succeeding verses. [*Reporter*.

From these facts you may get some idea of the slow manner in which the lava current goes on, at the rate of a yard an hour for days, months, and sometimes for years.

It is very generally conceded that the origin of the chasm on the side of the cone—the Val del Bove—was in the subsidence of the ground for some 3000 feet. We have an example at no very remote period when the Dutch possessed the Island of Java, of a similar sinking down in the volcano Papandayang—but a little before the commencement of the nineteenth century. A space of ground some fifteen miles long and six or seven wide gave way during an earthquake, and buried forty villages; and one part of the volcano fell in after another, until Papandayang lost 4000 of its hight.—The mass engulphed was larger than we need suppose was swallowed in order to produce the Val del Bove on the East side of Ætna. There are two mountains mentioned by Pliny, and since his time the chasm between them has been nearly filled by the deluge of lava.

The composition of all the lavas of Ætna, from the oldest to the most modern, is singularly homogeneous. That of the modern eruptions of 1811 and 1819, when analyzed, was found to consist of the dark mineral augite, and of a kind of feldspa, called Labradorite. It has large quantities of iron, some of which is titaniferous. The most ancient lavas of which thousands of sheets are found, consist of nearly the same materials. At those points where we can see farthest into the internal structure of the Mountain, we find that it is constituted of successive beds of lava and scoriæ—with large quantities of augite and feldspar, have a granitic structure. As we go farther down, we find masses of rock, which are still more granitiferous: they are not divided into strata, nor do they agree with the volcanic rocks; we thus reach the *Plutonic* rocks, and I must endeavor to make you familiar with the technical term. Volcanic rocks are frequently porous, because they are found near the surface, and the discharge of their cases causes pores in them, as in the *slag* of iron furnaces.—When they exist in a strata of considerable depth, the pores diminish and the rocks become stronger, and begin to crystalize, until we reach the bottom, when they have no pores. Crystalization has taken place under a pressure, and they have slowly cooled. The Plutonic rocks consist of certain porphyries, and have silex and alumine, but never scoriæ or volcanic sand or pebbles, nor are any signs of these found near the surface. As, therefore, Volcanoes had been thus named, because Vulcan had his forge under Ætna, so it was natural that these rocks should be called Plutonic, as being found in the realm of Pluto, where flowed the Phlegethon—river of fire—and the Lethe rolled its watery labyrinth, which may be mythical of the Lethean influence exerted by these rocks—for we find that when these granitic masses come in contact with a fossiliferous strata, they destroy every trace of organic remains. If we suppose all the different strata to come in contact with these, we shall find that we can trace the passage from those of the most fossiliferous state to that when fossils are rare, at last come to a pure crystal rock, such as is sought by the sculptor. The Carrara marble is an example of this kind. Formerly this was called a primitive rock—one of the oldest. Now it is known not to be of high antiquity, but to have been converted from fossiliferous rock by contact with granite, which has deprived it of all its corals. Its nodules of flint have been fused, and occasionally will be found crystals of augite, against which the Sculptor sometimes breaks the edge of his chisel. I give this as an illustration with which you all acquainted; it is only one of a thousand I might mention. Thus you see that rocks which are fossiliferous, when far from Plutonic rocks, may become non-fossiliferous when placed for ages in their vicinity.

It was formerly a great puzzle for chemists to account for the manner in which heat pervaded the rocks, because they are bad conducters. But they should recollect that in the interior of the Earth we have to deal not only with masses of melted matter, but with steam. Volcanoes, besides mineral matter, discharge various gases—but ninetenths of their matter is water. The steam, before it found vent, mingled intimately with the pores of the lava, and forced the sides to disengage. Any rock, no matter how solid and compact, when put in contact with an immense reservoir of this, would become porous and spongy, under so enormous a pressure. This may be seen at Corinth, in the Lipari Islands, in Greece and other places. This hypothesis must not be combated, then, by objections founded upon the results of subjecting rocks to the heat of our furnaces; for they exist under altogether different conditions in the interior of the Earth.

I may state, that, as these remarks relate to the most difficult theories of geology, it is not to be expected that you should at once follow them. As I omitted it at the first lecture, I will say a few words of the origin of granite rocks. The true unstratified granite is generally believed to be of igneous origin, formed by that part of the volcanic phenomenon which is far below where the eruption takes place. In Mount Dor, they are believed to be below the Miocene, and in Ætna, below the Post Pliocene; and so of all antecedent periods.

It is impossible, though we know more of the history of Ætna than of any other volcano, to form even an approximate estimate of the number of years required for the accumulation of such a mass of volcanic matter. There is no reason to suppose that the ancient eruptions were more violent than the modern. The sheets of lava separated by beds of scoriæ and breccia, in the ancient part of the volcano, do not appear to be of any greater thickness than those of the modern. There is one method of attempting the computation of the more recent part of the mountain—to consider the minor cones which adorn its flank. Of these there are eighty of the first magnitude. We cannot, from what we know of these, suppose that more than one-fourth of them were produced since the earliest periods mentioned by Diodorus Siculus and Thuycidides. There may have been more activity, or less violent convulsions; but if we suppose that, in 12,000 years all these eighty cones were produced; and if we strip them all off, we shall still have the great collossal mass of Ætna remoining behind. It would still be one of the loftiest mountains in Sicily. It is evident, at all events, that, between the formation of the latest period of the Puy de Tartaret and the oldest of Ætna, the whole of the older and newer Pliocene periods must have intervened.

In my next lecture I will endeavor to explain, from the remains of the temple of Serapis, some of the proofs of the upheaval of the earth.

LECTURE III.

UPHEAVAL AND SUBSIDENCE OF THE EARTH.

LADIES AND GENTLEMEN:

I have had occasion, in both my previous Lectures, to allude to the fact that various marine fossils—the fossil remains of animals that once lived in the ocean—have been found entombed in the rocks of various districts; not that we find the remains of marine creatures strewed over the surface of the Earth merely, as if the Sea had once overflowed the land; but we find them embedded in the midst of mountains, at all depths below the surface, and entering into the composition of the mountain masses themselves, making up their very materials, sometimes for two or three miles thick, all having been gradually elaborated under the water; thus furnishing the most indisputable proof that what is now dry land was once for ages under the water and formed the bed of the sea. Here the geologist finds himself reduced to the alternative of supposing either that the ocean was once higher than it now is, and that it has been lowered; or that the land, the solid land itself, has been raised up. The earlier geologists preferred the notion of a sinking, a general sinking, of the sea; for the ocean cannot be lowered in one place without a general subsidence throughout the whole extent. But when the structure of the Earth came to be more attentively studied and more thoroughly examined, geologists gradually came round to the opposite opinion, namely, that the land had been elevated; and this opinion was embraced for several reasons, but principally because it would account for all the observed phenomena, explain all the periods of stratification—those in which the strata are disturbed and fractured as well as those in which they are horizontal. For if they were all upon a level, then the going down of the sea might explain their appearance; but as we find them curved, bent and fractured, the other theory is the only one which will explain both appearances.

Another reason for embracing the hypothesis of an upheaval of the Earth is this—that we find in the crust of the Earth fresh water formations—you (are now familiar with this term)—beds formed in lakes and estuarias, lying beneath the deposites which are evidently of marine origin. For example, in some parts of Europe you find the white chalk—a marine, calcareous matter, produced beneath the sea mainly by the decomposition of shells; beneath that you find another stratum, evidently of fresh water origin. Now to explain how a marine stratum 1,500 feet thick could have been thus deposited upon a fresh water formation, by the theory of the rising and the lowering of the ocean, you would have to make the ocean first retreat, in order that there might be dry land; then rise again to deposit the mass of corals and shells which are found; then retire and again go down to explain the present dry land which now exists in that part of the Earth where this calcareous matter is found. You would have a vast number of successive retirals upon that hypothesis; and the ocean must rise several miles and stand there for ages to form the marine rocks of the Himmelaya and other mountain chains.

But besides that this theory of the movement of the land explains all these geological phenomena, we have the experience of history, which teaches that the solid land, through extended districts, does sink down in some places and rise slowly in others; whereas there is no instance on record of a general lowering of the whole ocean—there has been no general sinking of the sea throughout the world. I think I shall be able to show in this and the next lecture, that there has been and is now going on a magnificent example of this, sometimes of a sudden and at others of a gradual and insensible rising and sinking of the earth. I will here mention one example in Sweden, because originally I disputed the accuracy of the statement, until I visited the locality and satisfied myself of its truth. I found that the Northern part of Scandinavia, of Sweden and Norway, was rising five or six feet in a century, as is ascertained by facts that have been observed for several hundred years. As you go south to Gefle you find that the rising has diminished to three feet, and at Stockholm to only two or three inches in the hundred years, while still farther south there was no motion at all. It is thus as if the whole land were a level, which is stationary at one point, rising more and more as you go toward the end. In Greenland it is certain that since the early Danes colonized the island, built churches there and settled the country, there has been a gradual sinking down of the whole coast for several hundred miles; the churches in some places have been submerged, and the greater part of the lower stories and the pavements are now wholly under water. These movements of the earth are sometimes accompanied by volcanic eruptions, though not in Greenland and Sweden, but in South America they are. In Chili, for example, in 1835, that part of the coast near Conception was raised some nine feet, and remained permanently at that hight. The bed of the sea was raised, as were the island of Santa Maria and a great part of the neighboring coast; at the same moment the volcano of Osarno, in the Andes, burst forth and lava was seen to flow from its crater, A submarine volcano also burst forth 400 miles west of Osarno at Bacalao, and the Island of Juan Fernandez was also violently shaken. These facts show the vast extent of the subterranean volcanic activity accompanying the movement of the land. The land has relapsed in some degree, though not to its former position. In some places it maintains its upheaval.

NAPLES.

But you may see a more remarkable illustration of this upward and downward movement in the immediate neighborhood of Naples: and I mention it because evidence drawn partly from antiquarian researches and partly belonging to geographical proof, comes home more easily and more convincingly to the minds of those not accustomed to purely geological evidences. In my next lecture when I come to speak of the coral reefs, I think I say not too much when I say that I shall show that these vast areas of the Earth's surface—extending some 8000 miles from East to West, and 3000 or 4000 from North to South—are now undergoing in some sections a slow upheaval, and in others an equally gradual subsidence. To-day I shall confine my attention to a small space in the neighborhood of Naples, where the changes may be traced for the last 1800 years.

If you first look at a map of the country near Naples, you see extended into the Bay of Naples, south of Vesuvius, the Sorentine Promontory, off which is the celebrated Island di Capri. Pompeii you will see at the south side of Vesuvius, and Naples at the northwest. At the northwest extremity of the beautiful Bay of Naples you see the small bay of Baiæ, to which I would call your particular attention. Now movements have taken place, which have caused a sinking and then a re-elevation in the bay of Baiæ of twenty-five feet—and yet have not affected the neighborhood of Naples, which is only twelve miles distant. This view of the Bay [exhibited] was taken by Sir WILLIAM HAMILTON, from the south side of the Bay at Puzzuoli. On the plain between the sea and the dark cliff, represented below, is the town of Puzzuoli.

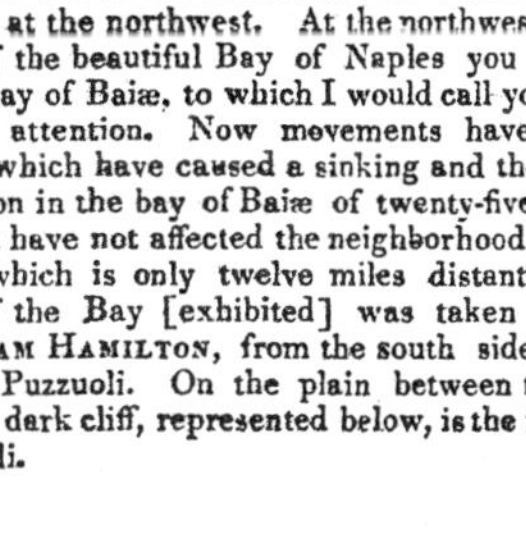

a Antiquities on the hill S. E. of Puzzuoli.
b. Ancient Cliff—now inland.
c. Terrace composed of ancient submarine deposites.

This inland cliff is eighty feet in hight. When I examined it in 1828, I found that it consisted of indurated volcanic tuff, containing some marine remains and the platform or terrace (*a.*) between the cliff and sea, contained evidences of having been formed under water. A wall had been built to protect the rich and valuable land of the terrace which slopes from the base of the cliff to the sea. But fortunately for me, a storm had swept away a part of this rampart and laid open a section so that I was enabled to examine the alternate beds of pumice and lapilli and volcanic matter, with strata abounding in various marine shells, such as cockles and the *Mytilus edulis* or eatable muscle. In one of these strata I found squares of a rich colored mosaic pavement and various sculptured ornaments. My friend Mr. BABBAGE found at a great hight barnacles and other shells attached to the face of the cliff. The appearance is as if the sea had beaten against the base, and thus undermined the cliff.

TEMPLE OF SERAPIS.

When we pass round the promontory to Puzzuoli and come in sight of the Bay of Baiæ we see what is called Caligula's Mole, a long line of arches and piers extending from the end of the town into the water. On the last of these arches but one, there was found, ten feet above the sea level, great numbers of flustra and other marine zoophytes attached to the building, which must once have been submerged ten feet lower than it now is and then raised to its present hight. On the sixth pier the same thing may be noticed. Breislac observes that as these arches now spring under water, and as it is certain that originally they were built so as to spring out of the water, though it is proved that the mole has been uplifted ten feet above its lowest level, still it has not yet been restored to its original position. If you stand on one of the arches and look toward the land you see another inland cliff similar to, but not so high, as the other. Between the cliff and the sea is a low terrace called La Starza, leading to the inference that the same movement which produced the other cliff, also extended to the Northwest and produced this with its plain and caused the cliff to become inland. Upon this plain is the temple of Jupiter Serapis. The cliff and plain are represented in this drawing:

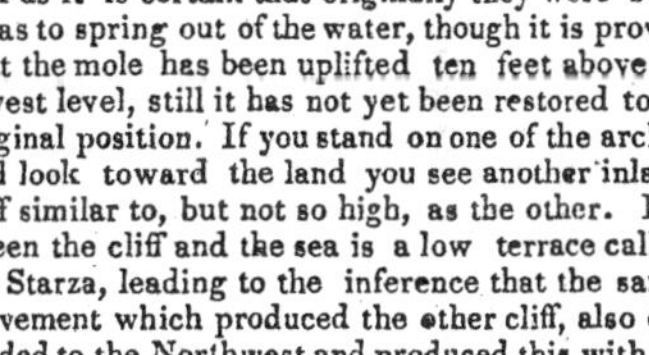

a. Cicero's Academia.
b. Ancient cliff—now island.
c. Terrace of Submarine deposites.
d. Temple of Jupiter Serapis.

I will now mention the manner in which this emple of Jupiter Serapis was discovered. In traveling along this terrace—which corresponds so exactly to that on the other side—some antiquaries in 1749, examining all the localities, saw concealed among some copse wood, the upper part of three columns—several feet above the ground. They were struck with the circumstance, and determined to remove the copse. They did so, and began to dig down to see how far the columns extended.—They dug down, accordingly, ten, twenty and thirty feet, and still found these great shafts: at forty feet they found no bottom; but at forty-six feet came to the pedestals, on which these columns stood. They were buried at this enormous depth, and when they had completed their digging, they found a large pavement, seventy feet in diameter, level at the base of the columns. Having found this, and still continuing to clear away, they discovered a large quadrangular building, the roof of which had been supported by forty-six of these magnificent columns, of which half were granite, and the rest marble—the marble in each case being a single block, whole from top to bottom. Now, in attempting to account for this, you might at first perhaps, suppose that the temple was buried like Pompeii, which was overwhelmed by a shower of mud and ashes—and was, after more than fifteen centuries, disentombed, and an immense amphitheatre dug out and exposed to view. You might think that this temple of Jupiter Serapis was thus buried, being in the region of the volcano of Solfatara, of which an eruption occurred in 1149. But we find that some of the beds, which have been dug through, contain marine shells, and fragments of works like those I mentioned as having been found in the terrace on the other side of Puzzuoli, and the situation of the temple is that represented in the drawing *(d.)* In endeavoring, too, to account for the burial of this temple by the rising and sinking of the sea, you have first to suppose, as the temple was built above ground, that the sea first went down—for the presence of marine remains in the lowest strata shews this: then that it was carried up again. But it is evident that in reality the temple was built above the water—that it went down and was buried in the sea, and that again it was raised. You may think all this very difficult to believe; but there are still greater wonders to account for, and which can be satisfactorily explained, when you come to examine the building.

The marble columns have been examined. For the first twelve feet from the bottom they are smooth and uninjured—just as they were originally in the interior of the temple. (I shall not stop to discuss the opposite opinions as to the use to which this temple was put. It is generally called the temple of Jupiter Serapis—from the ornaments found in it. Several hot baths are found also—and hot springs which still flow out from the ruin.) Above this point is a zone of about nine feet perforated by *Lythodomi*, a kind of marine bivalve which has the power of perforating rocks. The upper part of the column is smooth, having only marks of its exposure to the weather.

These animals which have thus perforated the columns are a species called *Mytilus Lytophagus*—or stone-eating moluscos. A diver at the Bay of Genoa once told me that some years before he had thrown to the bottom a piece of black marble to be bored by these animals: I prevailed upon him to get out in a boat, dive down to the bottom and bring up the marble. It was filled with these shells—about the length of your finger, buried in the solid rock. It has long been difficult to conceive how these tender shells and the still more tender animals that inhabit them, should be able to thus eat holes in the solid rocks. It was supposed by some that they had the power of turning around as on an axis, and that thus the shells scraped out the hole like a file. But so tender are the shells that this seems impossible. If we suppose the animals to secrete an acid capable of eating away carbonate of lime, of which the marble is composed, the question will occur, why should not the acid eat up the *shells*, which are made of the same substance? But we may well suppose that a feeble acid may act on the rock and not eat into the shell which contains the living animal—especially when we recollect that the shell is covered by an Epidermis or skin, upon which the acid will not act. It is probable that as they grow large—for the hole they make at first is small—they are enabled to eat away the calcareous matter by discharging the acid they secrete. They have drilled into the columns pear-shaped, cylindrical cavities, which it must have required a considerable time to excavate. Their numbers are so great, and they have so eaten into the columns as to diminish their diameter: and we find that some of the cavities which the *Lithodomi* have deserted have been occupied by a species of bivalve. The appearance of the columns is indicated by the following figure:

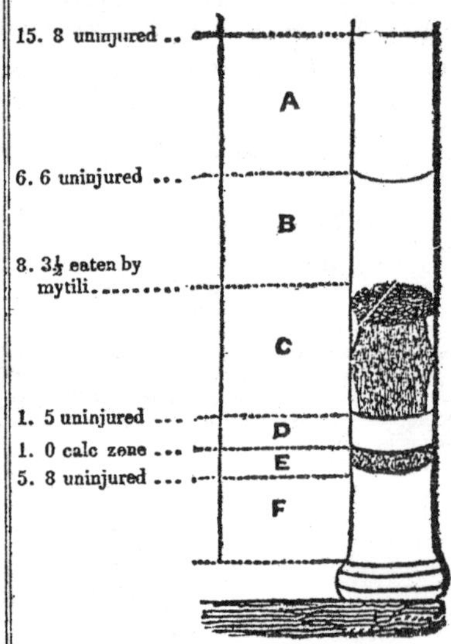

Total hight from bottom of Plinth 41. 1½.

When first these phenomena were observed in the last century, it caused a great variety of opinion: and this was not confined to scientific men, but even the poet GOETHE wrote an essay to explain how the salt water of the sea (in which alone lithodomi exist) might have peen barred out—enclosed in a barrier so as to account for this appearance. As there are no tides in the Mediterranean, how is the water to be carried up not only to the hight of twelve, but of nineteen feet? I do not say that if there were tides they would explain it; because the Lithodomi cannot live six hours out of water like bivalves, which take water into their shells: they must be always submerged.

But some geologists have asserted that there must have been a great change in the level of the sea; that it must have stood nineteen feet above its subsequent level. But it may be replied that there are a thousand proofs to be drawn from the history of all other parts of the Mediterranean, that no such change has ever taken place. It is certain that the temple existed in the third century: for in the atrium inscriptions have been found recording the reparations made by the Emperor Marcus Aurelius, and the additional ornaments given to it by another Emperor, Septimius Severus. We know then that it was used as a place of worship—or for whatever purpose it was built—down to the third century: and nothing is more certain than that between the third and the middle of the last century there has been no general rise of the Mediterranean.

Nevertheless, in spite of this difficulty, many maintained that the sea had gone down—so unwilling were they to entertain what they called the paradox of attaching constantly to the sea and mobility to the land. But after a time their opposition gave way. In 1828, the year when I had an oppertunity of examining the place, some excavations were made through the pavement, and at the depth of six feet below it the antiquarians came upon another pavement more rich and costly than this. This suggested the idea that there had been a sinking down previous to the changes we have noticed. The building being near the sea, as it gave way it was necessary to construct a new pavement six feet above the other; and then upon this the whole building was erected.

The next discovery was made by Niccolini, who was employed in 1807 to make drawings of the temples; he was in the habit of remaining there all day and yet never saw the pavement overflowed by the sea. Sixteen years after, he had occasion to go back to the temple, and to his surprise he found many parts of the marble pavement where he had stood in 1807 dry, covered at high tide by water. (I said there was no tide in the Mediterranean; and it is usually considered a tideless sea. Still there is a slight rise and fall of a few inches in the Bay of Baiæ.) This led Niccolini to make a series of experiments with the hydrometer; and he found that the ground in 1840 had been sinking at the rate of about three-fourths of an inch annually, so that in 1840, it would be two feet two inches lower than in 1800. When I was there it was always covered with water. When you know that the land is thus positively going down every year at a gradual rate you will be less skeptical in relation to the subsidence of the earth's surface. Thus we have not only to carry down the temple far enough to account for the nineteen feet we had before, but now we are obliged to add six feet farther for it to sink; we have now twenty-five feet of subsidence to account for.

Now, the question was, was this sinking gradual and successive, as it has been since 1800, or was it sudden? And to this we have a most satisfactory answer. In 1828 Mr. BABBAGE, the celebrated mathematician, and another of my friends, Mr. HEAD, carefully examined the temple and found the clearest evidence that the whole was gradually going down; and Mr. BABBAGE had the kindness to lend me the yet unpublished results of this examination. We find the columns eaten at the ends and upon the sides, and they wished to show how the columns must have stood to allow this. They found first, the whole column incrusted near the bottom by a black deposite two feet thick, containing *Serpulæ* and marine animals.—We may suppose that the temple had then begun to sink so that the sea should have covered the base of the columns where this incrustation was deposited. The top of the deposition represented the water level. Then came a shower of ashes, covering the column for six or seven feet, from an eruption of the neighboring volcano of Solfatara, probably, though there are many other volcanic, cones in that vicinity. Now this may have shut out the sea, and the water flowing into pools and ponds of the uneven surface may have caused the fresh water formations, as there are no serpulæ or marine animals to be found there. Then may have occurred another shower of ashes, perhaps from the eruption of 1149, when was a considerable earthquake, which may have thrown down the other columns which perhaps stood till that time. Suppose the marble ones, (for the granite are untouched,) then to have rested on the uneven surface of the layer produced by the second shower: the sea flowing in might have carried the lithodomi thither, and thus the columns might have been eaten on all sides and at both ends.

All these geological phenomena relate to the periods of the successive going down of the temple. No less than 27 pillars have been carefully examined and measured inch by inch and their original position carefully made out. In the year 1488 it appears that the rising began to take place, when the great eartequake occurred which destroyed Puzzuoli and formed those inland cliffs. We have a document, a charter of Ferdinand and Isabella of Spain, then sovereign of Naples, granting to the University of Puzzuoli the piece of land near the town of Naples 'where the sea is drying up,' *(che va seccando il mare;)* and this was followed a few years after by another charter dated in 1511, granting to the University, *solum desiccatum*, the *ground that is dried up* from the sea. So at this time there is evidence that the gradual retreating of the sea had begun. In 1530, nineteen years after, we have still the testimony of an Italian writer—Soffredo—that though the sea had dried up still it washed the base of the inland cliff and swept across La Starza so that he tells us people might have fished from the ruins. We are now led nearly to the epoch of the formation of Monte Nuovo—the new mountain produced by an eruption in 1538—only eight years after Soffredo wrote. And there is ample evidence in the documents collected by Sir William Hamilton that the whole upheaval of the land took place about that time. I shall refer to some account of that memorable event when a mountain 450 feet high was produced in a few nights. This mountain stands partly on the site of the Lucrine lake and partly on the site of a little village in the vicinity called

Tripergola. Sir William Hamilton has found two letters describing the eruption which formed the mountain, and I shall read from them some interesting passages. One is from Falconi, in which he says:

"It is now two years since there have been frequent earthquakes at Puzzuoli, Naples and the neighboring sections. On the day and in the night before the eruption of Monte Nuovo above twenty shocks, great and small, were felt. The eruption began on the 29th Sept. 1538. It was on Sunday about one o'clock in the night, when flames of fire were seen between the hot baths and Tripergola. In a short time the fire increased to such a degree that it burst open the earth in this place, and threw up so great a quantity of ashes and pumice stones mixed with water, as covered the whole country. The next morning the poor inhabitants of Puzzuoli quitted their habitations in terror, covered with the muddy and black shower which continued the whole day in that country—flying from death, but with death painted in their countenance. Some with their children in their arms; some with sacks full of their goods; others leading an ass loaded with their frightened families toward Naples; others carrying quantities of birds of various sorts, that had fallen dead at the beginning of the eruption; others, again, *with fish which they had found* (mark this,) and which were to be met in plenty on the shore, *the sea having left them dry for a considerable time.*"

This is the description of the eruption of the volcano that formed Monte Nuovo, by Falconi; there is another by Pietro Giacomo di Toledo, in which he says:—

"It is now two years since Campagna has been afflicted with earthquakes—the country about Puzzuoli more than other parts. On the 27th and 28th of September last, the earthquakes did not cease day nor night; the town of Puzzuoli, that plain between Lake Avernus, the Monte Barbaro and the sea was *raised a little* [a remarkable expression which he would not have used if it had been merely a shower of ashes which raised it, as sometimes happens] and many cracks were made in it, from some of which issued water: at the same time the sea adjoining the plain *dried up for about two hundred paces,* so that the fish were left on the land a prey to the inhabitants of Puzzuoli. At last on the 29th, about 2 o'clock in the night, the earth opened near the Lucrine lake and discovered a horrid mouth, from which ever vomited furiously smoke, fire, stones and mud composed of ashes, making at the time of its opening a noise like the loudest thunder. The stones which followed were by the flames converted to pumice, and some of those were *larger than an ox.* The stones were sent as high as a cross-bow can carry," and so he goes on to describe the shower of mud that built up a solid mountain which has a crater at the top as deep as the mountain is high. Toledo then proceeds to say that the eruption lasted two nights and two days—(the mountain was produced in two nights,) and that many persons were knocked down by the stones, and killed.

It appears to me evident, not that the sea had retired before the eruption, but that, when the tremendous explosion took place, the plain was suddenly raised, and there was then a drying up of the sea some time after; but a vast deal of rising must have taken place during those convulsions, as we know the hot lava to produce that liquid fiery matter that was thrown up must have melted near the surface, so that the stream, instead of flowing over the crator, should escape through the yielding beds below—between which it might flow latitudinally until it became high enough to pierce through, just as it is easier to thrust a knife between two leaves of a book than through them.

But I have too many facts to explain to you relative to these superficial changes having been caused by an upward and downward movement of the Earth, to enter this evening upon the various speculations respecting these matters. You perceive that we have carried the temple down twenty-five feet gradually, and that it has shot up nearly the whole distance, though not entirely, at one eruption.

Not far from the Temple of Serapis are the ruins of two temples, one of the Nymphs and the other of Neptune. The columns of the former stand erect in five feet water. The temple was doubtless submerged, and if there should ever be an upheaval of the bottom of the bay, this temple would probably be exhumed as was the other. As to the difficulty that arises from the columns not having fallen down like the others, it should be recollected that they went down slowly, only three-fourths of an inch in a year, and that before the great earthquake of 1149 occurred, they were buried twelve feet deep in incrustation or in ashes; being made of one block, they would be quite likely to remain erect.

One fact I forgot to mention: that these columns are a little out of a perpendicular—inclining slightly toward the sea. Originally they must have been perpendicular; thererefore the movement must have been greater toward the land; and this, too, may explain why the temples of Neptune and the Nymphs are lower down and are not yet raised. An antiquarian, named Capocci, has proved conclusively that while these movements occurred at Puzzuoli no changes have taken place at Naples, so that we have an oscillation of twenty-five feet up and down, while at twelve miles distance every thing is stationary: an important fact in the explanation of a great variety of geological phenomena, some of which I alluued to in my other lectures.

MONTE NUOVO.

When you examine the new mountain you find no indication that it is more modern in its origin than the others in its neighborhood. Indeed, as the same country is under cultivation, for the most part the vines are lifeless for much of the year, and the olives are of a pale green. But Mt. Nuovo is covered with evergreens, myrtles, olives and arbutus:—it is the most verdant spot in all that region: showing apparently that it is less modern than the barren hills in its neighborhood. Nothing can be more striking than the whole landscape—every part of the picture is in such perfect harmony with the rest, that you would not suppose different dates belonged to the different mountains. Yet I have found at a hight of 2,600 feet marine shells, identical in species with those which now live, buried in the strata of ancient, submarine tuff. The whole country, which is so modern, either consists of this volcanic matter which has been thrown up since it emerged from the sea, or of strata in which you find shells and zoöphytes identical with those that now live; and yet when the early Greek Colony first took possession of it 2,500 years ago, the appearance of the valleys and hills was much the same as you find it now. You can scarcely avoid being surprised at the prodigious antiquity, relatively to the historical period, even of the modern strata containing the same shells which I have spoken of in my last two Lectures, when alluding to the antiquity of the great mass of Ætna, while all its vast sheets of lava and scoriæ rest on marine formations as modern as the Bay of Baiæ. These species of plants and animals which inhabit the hills of that country are more ancient than the country itself.

Now you may say that this is an apparent paradox: yet you will easily comprehend it if you attend to such a mountain as Nuovo, which, since 1533, has been colonized by all the wild plants and animals of its neighborhood. So Ætna has been covered with vegetation older than the mountain itself; not only were these species of animals and plants alive before the country rose from the sea, but during that whole period when the vast thick strata of hills first began to be elaborated at the bottom of the ocean. So we may affirm that the *Fauna* and the *Flora* of this region are of a higher antiquity than the country itself. Were I to attempt to give you an idea of this period of time—the most modern subdivision of which, this tertiary and its succeeding periods, I have thus far considered—if I were to compare it to any thing of which we have a conception, I would say that this period, of which I have spoken thus far, may be compared to such distances as exist within our solar system—between the different planetary bodies.

Now when astronomers endeavor to measure the distances of the Solar System, they tell us that the Earth is one hundred millions of miles from the Sun; because they find that when six months have gone around, the Earth is in the opposite side of her orbit, which is two hundred millions of miles in diameter. They calculate, therefore, by the angles subtended, with the diameter of her orbit as the base, that a distance of one hundred millions of miles is between the Earth and the Sun. Then Jupiter is five times as far away; Saturn ten times, and so of the others.

Yet when they attempt to estimate the greater space which separates our Solar System from the nearest star, which again is probably the centre of a System as magnificent as ours, they are baffled in the endeavor, and can only find a distance which shall be the *minimum*—nor till lately have they arrived at any accurate calculation concerning it. But it seems, by observations on the parallax, a Prussian astronomer, BESSELL, has measured the distance of one star in the constellation *Cygni*. The angle subtended by the diameter of the Earth's orbit at that star is found to be *one-third* of a second, and what distance does this third of a second give you? Take the diameter of the Earth's orbit for a unit, and then 700,000 of these units will express the distance of one of those stars in Cygni from our Earth, and perhaps that star is the nearest, and separated from the others which we see by a distance equally vast.

Perhaps, should we compare time and space, this would be the kind of distance which should contain the myriads of organic remains of species distinct from the species immediately antecedent and following. The minor subdivisions of which I have spoken would be compared to the space o our own Solar System: the others, with the grander stellar distances.

Now it is a favorite speculation of astronomers, that all these worlds, separated so widely, may be inhabited; but this is mere conjecture—a probable conjecture, if you like, but still incapable of demonstration. But the geologist proceeds differently. He has indisputable proofs that there have existed on this planet a succession of inhabitants, and distinct races of animals and plants. And, though he does not measure the limits of time so accurately, and with such geometric precision, as the astronomer, still, by the vast series of events—by the methodical phenomena of the Earth—he finds established, by purely physical phenomena and proofs, the declaration of Revelation, that in the first time MAN had no existence; that Man had a beginning, and that other races existed anterior to him. Man had a beginning; and therefore the present state of the organic world has not gone on from eternity, as the ancient philosophers pretended; and we have been able to prove that beings lived, called by the Creator into existence, on this planet—to display the beautiful and perfect harmony of the Universe—to show that all is modeled on one plan; that different as are the various genera that have lived, they all belong to the same family. Geology shows that all things are the works of one Intelligence—one Mind—all links of one chain: that the Earth must have been admirably fitted for successive states which were to endure for ages. Thus do we learn to admire the variety and beauty of design displayed when we find traces and signs of the same design, the same unity of plan, the same harmony of wisdom through so vast a series as has been established by the Infinite and Eternal Creative power.

C

LECTURE IV.

CORAL REEFS.

I HAVE already said that in the structure of the crust of the Earth we have a great variety of groups in which the distinct strata may be expressed, if we group them largely—if we take wide and comprehensive divisions—by the different numbers and colors, as from one to eight in this figure [exhibited]; and that the different sets of strata—formations as they are called—are arranged one above the other in the order in which they are formed. Not that they are invariably all to be met in one place, for it is not often that we find more than one, or two, or three, of these strata in any particular district: but if these different sets are present, we shall find them arranged in the order of relative position in which they are represented—each corresponding to the different period of the Earth's history at which it was formed. So that they may be regarded as a great chronological table—as so many volumes of History in which the strata are the different pages; and upon them are written the names and characters of the plants and animals which lived and flourished at this period, with many other indications of climate, habits, &c. in the periods when the different tribes flourished. Now if we examine in succession these rocks we find that a great number of these of calcareous formation, or limestones, are composed in great part of *corals and shells;* for example, if we take the Chalk formation of Europe—represented in the figure as No. 2—we find a large portion of it to be a white calcareous rock made up of corals and shells almost entirely decomposed. I have myself seen in the island of Seeland, fifty miles from Copenhagen, at Faxoe, this very chalk formation pass into a mass of corals in which more than a hundred species of zoöphytes and shells are found. Between the zoöphytes is found chalky matter—like the ordinary chalk. This formation extends for more than a thousand miles in one direction, and eight hundred in the other—not continuously, but in large patches, which preserve much the same character. In this country too, as in New-Jersey, are found rocks of the same age, not with the white rocks of the same kind, but at different intervals down we find in the rock, corals and shells made up of an aggregate of extinct species, but in great part of genera, the same as now live in tropical seas.—And this is regarded as one proof that formerly the climate was much warmer than since; as it is only in warm climates that these stone building zoöphites increase and abound. There are zoöphytes of the same class in all regions of the globe: but these which form large masses of stone (which, when the animal dies becomes what we may term hard rock,) are found at present only in warm regions of the globe.

Now, I might trace the different formations in which they occur, and point out the limestones in which they abound. In what is called the Jurassique or Oölite formation, these corals are found through England, France and Germany, at different hights with great intervals of clay—in some cases six or seven hundred feet thick. The corals are found above the other except when they have been destroyed by the depositions of clay, for they can live only in pure water. It is clear that where these take place and in the intervals between the different layers of coralline formation there had been a sinking of the floor of the sea, since it is found that these zoöphytes cannot build in the deep and dark parts of the ocean—that they will not thrive in a depth of more than 120 feet below the surface: and therefore we find a mass of this coralline limestone with 1000 feet of clay above and then another layer of coral, and then 1500 feet thick of this cretaceous rock; it is evident that these were formed not at a depth of several miles but near the surface: and that they then sank down to allow the accumulations above.

If we come to the carboniferous or coal formation, (which I shall describe more particularly in my next lecture,) we find beds of coal separated by vast thicknesses of other strata, which can only be explained by supposing that there has been a sinking of the surface of the land. Below the coal are found the most ancient fossiliferous rocks, which have an antiquity far higher than the thick limestone, for the corals grow on spots which are now found covered in great part with extinct genera, or those which are only found in tropical seas, where coral reefs abound. In New-York, in particular, there are large developements of this formation, as I shall show when I come to speak of the district around Niagara, when I shall show the hights from which we are to suppose the bed of the sea has sunk, and the era when these marine plants and animals abounded. Now, if we find, on examination of existing coral reefs that have limestone in their structure, similar in their character over a vast extent of the globe, that there is now in progress a considerable going down of the bed of the sea; we shall then feel more confidence in the appearances and facts on which depend some of the most interesting problems respecting the origin of the materials of the Earth's crust.

I shall now proceed to give some account of the Coral Reefs, and of the manner in which they grow in the sea. I have been favored by Mr.

CHARLES DARWIN, who is about to publish a most excellent work on Coral Reefs, with this map—which will first be published in his book—drawn up after a personal examination of the region represented, and a most extensive reading of works upon the subject. On it are depicted all the spots known at present where Corals now grow. These portions [represented by the blue color] mark those places where there is reason to believe the bed of the sea is going down, as slowly, perhaps, as the floor of the Temple of Jupiter Serapis, described in my last lecture. By the red are marked the spaces where corals have been raised at various elevations above the level of the sea.

You will be able to see at a glance the vast extent of the region where corals abound. You will see, too, that it is chiefly within twenty degrees North and South of the Equator—in the warmest parts of the ocean. Sometimes it reaches beyond the twentieth degree but rarely so far as thirty.—Its greatest deviation is in the Bermuda Islands, which have a latitude of thirty-two degrees—which is the farthest point from the Equator where corals are yet known; and this is evidently connected with the course of the Gulf Stream which warms the ocean and raises the temperature beyond what naturally belongs to that zone. You see by this map that in the Indian Ocean in particular and in the great Pacific and South Seas is a prodigious growth of corals. And it is a very remarkable circumstance that almost all the islands on that part of the sea—all that are colored blue—consist of strips of coral of an annular form—more or less perfectly circular—and sometimes oval. And these strips have *lagoons* in the centre—small lakes of salt water. On one side of this narrow strip is an unfathomable ocean at but a short distance from the edge of the ridge of coral. It was formerly supposed that these coral reefs were built up from the bottom of the unfathomable ocean; but now since we know that these *lithophytes*, as they are called, cannot exist at a depth of more than 120 feet—the limit assigned by DARWIN (and some other considerable naturalists think that the limit is still more narrow,) we infer, as a fact of Natural History, that these were not built up from the bottom of the ocean.—And this is perfectly in accordance with the fact that in thousands of cases no soundings are to be found at enormous depths—only a fourth or half a mile from the outer part of these narrow rims of coral. One of these circular islands is represented in this picture, which was taken from an original drawing colored on the spot by Lieut. SMYTHE who accompanied Capt. BEECHEY in his voyage to the Pacific. It may be proper to say that the view is represented as having been taken from a higher point than the top of the mast from which it was actually taken, so that more of the inner part of the island can be seen than in the other case. The island is three or four miles in diameter, of crescent shape, and you see a narrow rim of coral covered with tufts of cocoa, and bread fruit and other trees. On the windward side the reef is higher than on the other; and on the leeward side is an opening of thirty or forty feet—so that ships can enter and lie in safety in the lagoon. This opening, most fortunately, is just on that side where it is most needed; for there, during terrible storms and tempests, vessels may enter into the tranquil lagoon, where the water is shallow, compared with the ocean without, and where, in consequence of its shallow depth, it is of a most beautiful green color. Many of you who have sailed across the Atlantic must have observed that, along the Banks of Newfoundland, where the water is comparatively shallow, it is of a deep green; and in those latitudes where there is white coral at the bottom and a burning tropical Sun over head, the vivid green of the sea water is described as most beautiful by those who have visited these islands. Down through its green depths, they tell us, you may see great herds of fish browsing upon the coral branches—for some species have strong, bony jaws by which they easily crush the coral, shell and all: you see them feeding upon the zoöphytes as the herds of buffaloes feed on the herbs and trees of your wide prairies. And even the prairie itself in spring time is not enameled with more beautiful colors or with a greater variety of flowers and plants than are those beautiful beds of coral, according to the descriptions of EHRENBERG and others who describe these lakes as like beds of tulips—so beautiful and variegated are their colors when seen through the still waters beneath that tropical sun.

I may here mention a fact of considerable geological interest relating to one of these fish, called *Sparus*. When their bodies are opened and their intestines examined, they are found to be filled with a very dry chalk—a soft, calcareous powder, which proves to be almost indistinguishable from soft, pulverulent chalk. I have myself seen some brought from the the Bermudas by Lieut. NELSON, which so nearly resembled European chalk that we were obliged to use great care lest the labels should get changed, and we should mistake it for the chalk with which we were comparing it.

The animals which form these reefs—for the whole is a rim not half a mile, and generally but three or four hundred yards wide, covered often by shells of *Echini*, or sea urchins and other shell fish—cannot build one inch above the level of the sea. They cannot allow themselves to be left bare at low tide; so that when the reef is so high that it remains almost dry, the corals leave off building. The heat of the sun then often causes the mass to braek, and the force of the waves tears off large branches of the coral, which are thrown up upon the reef, thus raising it above the reach of the usual tide. After this the white, calcareous sand thrown upon it by the wind lies undisturbed; and upon it are lodged the seeds of the cocoanut and other plants, which grow up, until at length the little island is overshadowed with luxuriant vegetation. Then come stray birds and build their nests there; insects float thither on wood which has drifted thousands and thousands of miles, and at length the island becomes inhabited. Here is a specimen of the *Meandrina* coral—[exhibited;] it is a small shell of a young animal; and you can conceive. when such is the size of this baby-coral, how vast must be the size of a number of zoöphytes of different genera—as the *Madyepora*, *Astrea*, *Porites*, (of which I here show you a specimen) the *Oculina*, and others, of which the remains are found in the older rocks, and which now abound in the tropical seas. These shells and zoöphytes constitute the mass of the materials of these reefs. As to the nature of these animal plants as they are called—and very properly, too, for they seem to form the intervening link between animal and vegetable organization—there is still great doubt. I shall only observe those called *Pollyps*, inhabiting this stony structure, have a number of *tentacula*, or feelers, and a great cup-shaped mouth, into which they force food seized by these tentacula, and which then closes, and they are able to digest their food. These assist in strengthening that part which is improperly called the root; for they are fixed at the lower point by a point which somewhat resembles a root, though, as it does not take in nourishment, as do the roots of vegetables, the analogy is not correct.

In the common red coral of the Mediterranean you see a solid internal skeleton, surrounded by a fleshy covering, which, in that case, is smooth; but when this, in which the animal resides, is taken off, you see a striated, fluted surface, to which attention must be paid in studying the fossils—since we have not the living animals. I shall not enter (as I have so much geological ground to go over,) upon a description of the different varieties of Pollyps. But you may ask whether they exist as separate, independent individuals—or whether the whole mass of coral is regarded as one and indivisible. The same kind of question may be raised with regard to trees:—whether each flower is a separate individual, or whether the whole is to be regarded as an aggregate. We may perhaps best consider them as an animal republic, in which all combine to build one habitation, while each preserves its individuality. The general opinion, both in regard to the plants, and zoophytes, is that each is a definite individual. Although so beautiful when in the water, take one of these stone building corals out, and you see nothing but a brown slime when the animal is collapsed.

It is a remarkable circumstance, that, although these little islands are scattered about so far from each other in the wide ocean, upon every one capable of supporting them were found a few families when first discovered. We should regard this as mysterious if we did not learn from the Voyages of Cook, Kotzebue, Flinders, and Lieut. Beechey, that canoes are frequently drifted 500 or 1,500 miles—drifted before the wind in one direction by the monsoon, until, hopeless of recovering their homes, they land upon some one of these islands—where they find cocoa-nuts and bread fruit, and fresh water too. This is singular, for you would think that there would be but little chance of finding fresh water upon these small islands: but if you dig into the sand, there it is—so that vessels at sea are often able to relieve their thirst.

CORAL ISLANDS.

The beauty of these islands is described in glowing terms by most of those who navigated those portions of the sea. Generally there is a constant white surf breaking over the outer margin, which is seen gradually to die away with the dark heaving waters of the ocean, which continue to boil and rage far beyond. Within, the water is green and tranquil; around it and outside the green line of blooming vegitation is a glittering beach of white sand. Beneath the water you may see fish and various kinds of zoöphytes. The very loveliness of these islands, moreover, gives them a romantic aspect. Sometimes they are seven hundred miles away from the nearest neighboring island, and more than a thousand from the continent, or even much farther. The structure of these islands, as well as their position in the unfathomed ocean, is remarkable. You find no part of them argillaceous or silicious; every portion is made up of an organic structure of which the tubes and stems are all you find. And what is still more wonderful, these islands, which appear so weak and fragile in their nature—these mere barriers of coral, are beaten upon incessantly by the swell of the great ocean, and yet are not annihilated. As DARWIN truly observes, if they were only rocks of common granite or quartz, they would inevitably yield to the prodigious force of these waves thus breaking over them. But here is a greater power than any mere inorganic structure: in its nature far superior to the mechanical force of the waves: it is the power of life, of vitality; these zoöphytes on the outer edge of the reef break the force of the waves by their yielding bodies; they bend like a willow before the storm, and thus conquer the power of the great restless ocean. Indeed that part of the reef against which the waves beat with most turbulence increases much the fastest; for these organic forces separate the atoms of carbonate of lime from the foaming breakers, and build them up into their symmetrical structure; myriads of these architects are thus at work, day and night—month after month.

But what causes the singular, ring-like shape of these islands, and why are so many just on a level with the sea, and neither higher nor lower? These questions have been asked ever since the peculiar formation and mineral character of these islands were ascertained. It was at first imagined, and I once maintained the opinion, that they were the craters of sub-marine volcanoes. When it was discovered that Flinders was wrong in supposing that they could be built at a great depth, in the cold and dark regions of the ocean, up from the bottom, it was natural to suppose that they might have grown upon the tops of some submarine mountains. But what mountains would give this *shape* to the corals? Or was it indispensable that they should be built in this peculiar form? Now when EHRENBERG examined the corals of the Red Sea he found that some banks were square while others were ribbon-like strips with flat tops and without lagoons; and yet they were of the same genera as those which *had* lagoons, the same as the *atolls*, as they were named from those observed in the Maldive and Laccadive Islands, the term expressing an island with a rim of coral outside and a lagoon within. It was thus found that this annular shape is not essential; but that the coral may form in a different shape. It was therefore natural to suppose that this form depended on the outline of the submarine bottom, and that they were based on volcanic craters, thousands of which were known to exist in that part of the sea. These craters would give the cup-shaped center and the rim might be covered with corals by these zoöphytes. In the South Shetland Islands and in Barren Island in the Bay of Bengal, are submarine volcanoes with craters in the center and a breach in the side so that you may sail within and see all around the walls of the crater. Now, when submerged, the corals might easily have been built upon them. In farther support of this theory it was observed that in Gambier's group of islands there were volcanic rocks with a lagoon in the center, just as modern volcanoes have sprung up in the Gulf of Santoria.

Notwithstanding all these arguments in favor of the theory of a volcanic origin, it was found necessary to abandon it entirely; because, though it would account for some of the facts, there were many others which it would not explain, and which could be explained by another theory. It was perfectly satisfactory so far as the rim and lagoons were concerned, and it also explained why the Ocean near by should be unfathomable, which at first seemed an argument in favor of the volcanic theory. Mr. DARWIN, after considering all the different facts I have related, observed that there were in some cases islands which had this same coralline ring, and yet were *not volcanic* but *granitic*: and yet the ring of coral surrounded a precisely similar lagoon.—How was this to be explained? Again, FLINDERS discovered a magnificent reef—on the North-East Coast of New Holland, more than a thousand miles in length; and he sailed for more than three hundred and fifty miles and yet found no passage through that narraw reef. Instead of being circular or oval or in any such shape, it was parallel to the coast of New Holland. In the island of New Caledonia, which is granitic, there was also a long ridge of coral 400 miles long and prolonged at each end under the water beyond these limits: between the coral reef and the shore was what might be called a lagoon. Now we must have a theory which will explain these facts. Of these encircled islands, as DARWIN called them—islands with encircling reefs around them—Vancouver's is one and Otaheite is another. Here we have a coral reef with a lagoon inside. In the figure we have given a section of one of these encircled islands.

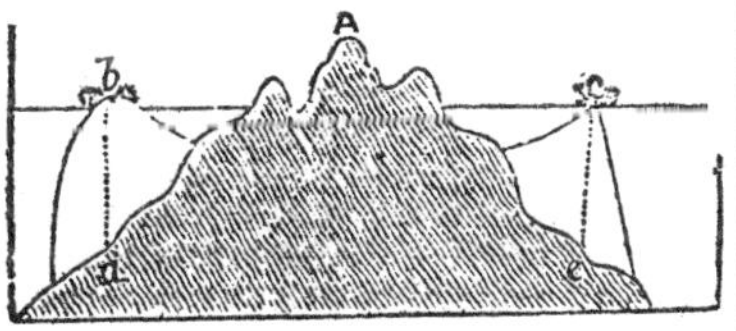

A represents the island in the centre; *c* and *b* are the points in the encircling reef upon which grew cocoa trees, &c., and inside, or between *c* and *b* and at the sides of the island, is a shallow sea communicating with the ocean through occasional passages, similar to those met with in the lagoon. Now, it was a rule laid down by DAMPIER, and other navigators, that near high and bold coasts we shall have a deep sea; while along gently sloping coasts the sea will continue shallow for some distance from the shore. It was therefore surprising that when we had an island like that represented in the figure, going down at so steep an angle, instead of deep water, we find it shallow, and we have a coral bottom. But go beyond the reef, and down goes the line some thousand feet. It is obvious that the coral alone is the cause of the shallow water; and we must account by means of the coral for the fact that at a distance from the island, instead of several thousand feet of water, as we are entitled to expect, we come suddenly to a bottom. Now, in what manner can these zoophytes, which cannot live at a greater deph than 120 feet below the water, be able to build up a reef two, three, four or more thousand feet high from the bottom? Because the belief of FLINDERS that these creatures could work in the deep, dark, cold ocean and build up their structures to the light and heat is proved not to be true; though he did not err as a geographer in his statements respecting these bold coasts.

DARWIN here suggested a simple hypothesis which clears up the whole difficulty. Suppose the coral to begin to grow at *e* or *d*, in the figure above; and suppose these points at *that time* instead of being 1500 or 2000 feet deep, to be within 120 feet of the surface, the sea then standing at a different relative level from that which it occupies now. In short the supposition of a *sinking down of the land* will explain it all. When the land began to sink the corals were building up, as is their nature, one generation forming on the skeleton of the other, just as peat grows, until it reaches a thickness of a hundred or more feet; each successive generation plants itself on the remains of the former. So with corals: each mass forms a foundation on which a new on is built. As I have shown, or as Niccolini has shown, the rate of subsidence in the Bay of Baiæ is three-fourths of an inch a year. Now common corals will grow up at that rate; and perhaps two inches a year would be a slow growth. Let me observe that we need only show that the greater part of them increase slowly: though passages of the Red Sea are known to have been filled up within a few years. This, however, may be easily explained, because the waves, &c. may have thrown up sand, and thus filled the passages. But the common growth of the coral is slow. Indeed, EHRENBERG thinks that some of the species, as the *Meandrina*, are as old as the most ancient trees of Europe, some of which are three or four thousand years old. DECANDOLLE has shown that some yews and other trees are two thousand years old, and some still more ancient. So Ehrenberg thinks that many of these corals in the Red Sea are several thousand years of age. But the vigor of their growth depends upon the warmth and carbonate of lime that reaches them.

Now we may suppose the land to be sinking down gradually as the corals are building up. What is the consequence? Instead of sinking down with the land, the corals will grow up by building upon the former masses—just as one pavement of the Temple of Jupiter Serapis was built upon the other, which had sunk six feet below the level of the Mediterranean. And when these corals come to the surface and can rise no higher, the waves then throw up calcareous sand upon them and form a ridge as in the lagoons. But while the corals are thus rising up, the mass always gaining in hight above its original base and yet remaining in other respects in the same position, it is different with the land: every inch that it sinks is irrecoverably lost, and the distance becomes greater from the island to the reef. If the island sinks down, it diminishes in size above the water; but the reef remains as it was, (though, from its slight inclination inwards, it may be a very little affected,) till at last the island peak disappears and is converted into a lagoon. All the land disappears, and the island becomes a perfect *Attol;* and if the subsidence still goes on, the lagoon will acquire considerable depth. This theory also is as satisfactory in explaining the *barrier reef* as in accounting for encircled islands, lagoons or *attols*. Let us take the following diagram as representing a barrier reef on the coast of Austravia:

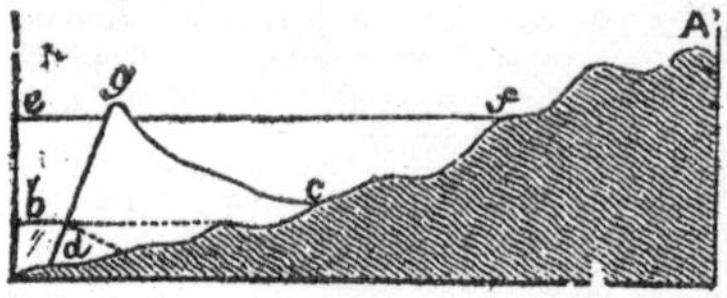

Here the land, instead of going down abruptly to the sea, has a gentle slope. When the reef began to grow at *d*, suppose the land stood 1000 feet higher than now; the level of the sea would be *b c*. But as the bed went down, the reef would be raised 1000 feet, or from *d* to *g*, and *e f* would represent the level of the sea. But as the distance *f g* is greater than *c d*, it is obvious that the higher up towards the surface the reef be built so much faster must the land be constantly retreating; and the rate at which this will go on is in proportion to the angle of descent. If at an angle of 30° one foot in vertical hight be lost, the change in distance will be about thirty feet. But in a gently sloping coast every foot in hight measures a vast number of yards; and the land will retreat at a greater rate. Thus the barrier reef is at a considerable distance from the coast—20, 30 and sometimes 70 miles off.

I ought here to say that corals will not grow too near the shore, in consequence of the sand and mud carried thither by streams and carried up by waves. It is only in clear, salt water that they lie; fresh water poisons them. On the other hand, they will not grow too far from land, because the sea is there too deep. Just as only a ring or zone around Ætna and Teneriffe is inhabited by human beings, because on one side is the sea and on the other volcanic ashes, &c.—so is it with these corals, which grow only in a narrow zone at a certain distance from the land. As the island goes down, the corals build up, and we have nothing left but the lagoon. We have islands also in an intermediate state, as Tahiti and Gambier's group, most of which, however, are perfect *attols*. Nine-tenths of them have arrived at this state.

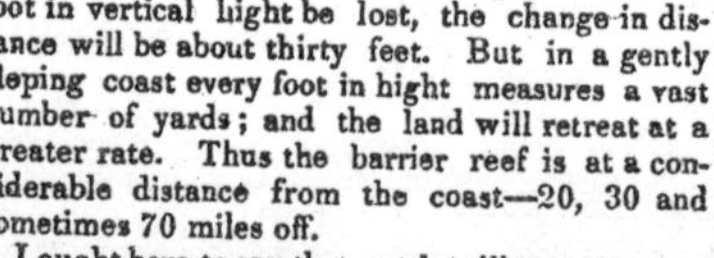

In this figure suppose A to represent an island 2000 feet high, and B a shoal in the sea, when *c c* is the sea level. Now, if the coral grow on both islands as they sink gradually untill it be 2000 feet thick, we shall have attols at the summits. If the top be washed, so that the corals wil not grow, a ring will be formed, and grow up, and we shall have a lagoon; A will become an encircled island for ages, until it is submerged, and then the rising coral will form a lagoon. So that this *lagoon island really represents a sunken mountain*. This may be seen at Tahiti, in Gambier's group, and in the Dangerous Archipelago—an area of 4000 miles in one direction, and 600 in the other. From this you may infer what must have been the vast differences in level of so vast an area. Take any continent containing a thousand mountains, and how very great a difference will there be between the elevation of the different summits! What a difference between the high and the low! And what an amount of subsidence must take place before they are all so sunk as that we shall see only the topmost, upon the same level! and yet to this are we led by the theory of subsidence; nor should we be terrified or frightened out of any sound theory by contemplating the vastness of the results to which it leads.

I may as well mention that this theory of subsidence was not invented for the purpose of explaining these phenomena. Long before DARWIN had made his examinations of these coral islands, in 1835, I published my opinion upon this point, that the sinking down of the Pacific *might be in excess*: that its depression might be greater than its upheaval. For in no other way could we explain how such a multitude of islands should exist first at a level, and now more than 60 feet above the surface. Henderson's Island is about 60 feet above, and the Tonga islands of Capt. Cook, are about 20 or 30 feet. The theory, then, was not made for the purpose of fitting the facts—though it is a perfectly legitimate reason for adopting a theory, that you find it will explain all the known phenomena which no other theory will explain.—Still, it is somewhat more satisfactory if the principle was not formed expressly to suit the facts of the case. I argued that if there was an equal amount of upheaval and depression—supposing a motion of oscillation—a movement up as well as down—to take place, large masses of coral will be raised above the level of the sea; and that, unless we assume that the downward movement is in excess we are impelled to conclude the coral would not have remained on the level. For if it sink 30 feet in a century, the coral would grow up in that time 30 feet:—if it rose ten feet we should then find the solid ground of the coral ten feet higher up. If you reflect you will see that it is quite impossible to have so vast a number of islands just on a level without supposing the sinking to be slightly in excess; and when you find that this will explain the formation of attols and encircled islands, you must be skeptical if you reject the evidence in favor of it; especially when it is ascertained that in Greenland the land is settling—that there is a small subsidence in the Bay of Baiæ—at the mouth of the Indus, and along parts of the coast of South America. But what shall we say when we are forced to conclude that, not only must there have been a sinking down of the floor of the ocean to bring all its mountains to the same hight, but that all must have gone down so gradually as *never to sink one hundred and twenty feet at once?* For if there had ever been a fall of 120 feet the corals never more would have grown; all would have perished, and no new ones would spring up. The whole process of subsidence then must have gone on at a very gradual rate—not more than a few yards at any one time.

But when islands are rising (and we find that a large proportion of the islands in these seas are volcanic islands, or that in some part of them are volcanoes in active operation,) in this case we have coral reefs, but they do not encircle the islands; but we shall have a formation called by Darwin, *fringing reefs;* reefs fringing the shore as near as the zoöphytes could live. There is a small island off Sumatra overspread, at all hights from the coast to 200 or 300 feet above, with coral shells of the *tridachna*. At the same time this upheaval has not carried the island as far above as it must before have been below the level.

The coast of South America is one where no coral grows. Why is it that zoöphytes will not live there is not perfectly explained, any more than why none grow in the Atlantic. The Atlantic is warm enough and has all the necessary conditions; but except on the borders of the gulf-stream no coral islands are found on the coast situated must favorably in the tropical region; at present we can no more explain this than why certain plants do not grow in the United States; why roses are not indigenous in the Southern hemisphere, &c. But in the geographical distribution of plants the Author of Nature has given certain laws; and so with the zoophytes. They do not flourish on the Atlantic coast. The coast of South America is a coast of upheaval so far as we can judge. There are volcanoes here, and animals and shells are found at various hights. Yet there are no coral islands. Then in Gambier's group, in Tahiti and the Friendly Islands and in the Dangerous Archipelago are attols which have flat tops; here we suppose there were stationary periods when the lagoons had time to fill up. Then in the Navigaters' and in Cook's island the oscillating movement causes attols and upraised reefs. Then in another region, in Sullivan's Island and the new Hebrides—for a long way from North-West to South-East, the upward movement is in excess. In New Caledonia we find an example of an excess in subsidence. Then in the volcanic islands, Java and Sumatra, there is a great line of upheaval under which the beds of coral and recent shells are raised higher.

One of the most remarkable ridges of *attols* is in the Laccadive and Maldive Islands. For several hundred miles in extent you find a series or circular assemblage of inlets all of coral, some of them 30, 40 or 50 miles in diameter. So you may trace them along the coast of Africa and the Red Sea, and go over a large area 2,000 or 3,000 miles from North to South, and from 7,000 to 8,000 from East to West, and find the same alterations.

It is often a subject of wonder to geologists to trace the same shells in different countries. Thus in Siberia and Russia MURCHISON has found shells identical with those in England; and the same species as the *Caryophyllia* and others may be found in the North of Scotland and in this State of New-York. Rocks are met containing this extinct species with fifty others; and they may be traced at least to Iowa and very likely to the Pacific. It is a matter for wonder and marvel that these same species should flourish over so large a space of the globe as from Siberia to Iowa: yet it is nothing to the extent over which masses of limestone are now forming. But if you penetrate to a little depth all the interstices will be found to be filled, the carboniferous lime will be found as solid as any limestone, and all its organic texure will have disappeared or can only be discovered by microscopic examination. So is it with the old rocks at Trenton and Niagara which are of most minute texture; so that it is not true that the extent of the old limestone was on a grander scale than what is now exemplified in the present state of the globe.

LECTURE V.]

ORIGIN OF COAL.

In speaking of the different strata of which the Earth's crust is composed—these at least which contain organic remains—I speke of them as so many volumes of History—as so many monuments of the ancient states of the globe; and of their different structures as being so many leaves of these volumes. All that I can do in this short Course of Lectures is to take down at random, first one and then another of these volumes, and endeavor not to give you any idea of the contents of the whole, but just to express something of the method employed in the attempt to decypher these ancient memorials of the Earth's History. Now, the volume which I intend to take down to-night is that which we term the *Coal formation;* and I shall speak of it only so far as to show the relative position, and the state of the different periods when they were deposited beneath the water.

Now when I term this formation *Coal,* I merely mean this assemblage of strata which rests on the Older Sandstone, and in which is found that valuable fuel we call coal; and although the quantity in which it is contained is very small in comparison with the bulk and volume of the other strata, there is still great interest and importance attached to it. We see that in going from the highest to the lowest beds yet discovered, the coal occupies quite an ancient position—one indicating a formation low down in the Sea—as we have above in the most modern formations. We have first the Post-Pliocene, then the Tertiary formation, then the Chalk, which is made up of calcareous matter, formed mostly, at least in Europe, from decomposed shells and coals, and of those green marls which are found in New-Jersey and are of such extensive use in Agriculture; then we have the Jura limestone or Oölite, in which also are masses of coral like the common coral reefs: below this are two other groups, of which I shall not speak at present, and lastly we come down to the Carboniferous or coal-bearing stratum which rests upon the thick sandstone beds, or the limestone containing corals, and which, like every other formation, contains species of animals, shells and plants of different species, from those immediately antecedent or following. Below this again we see limestone and shale, which enter most largely into the structure of the rocks of the State of New-York, and which abound in fossils.

Now a great change must have been experienced before the coal period, when the fossils were deposited. I am indebted to Mr. Sopwith, an eminent civil engineer, for copies of some models prepared by him of those sections, which are faithful and accurate representations of actual localities, as has been fully verified by Dr. Buckland and myself in examinations which we made last spring. The different strata of sandstone, shale and conglomerate of which the carboniferous formation is composed, are here represented. The sections represent facts ascertained in cutting perpendicularly through the Newcastle coal district. They are not hypothetical, but are founded upon exact measurement. In one of these sections you see the dip of the beds is at an angle of 20°, while the slope of the valley is 40°. In the other the dip is 50°, and the slope of the valley in the same direction is 20°. In these two cases, therefore, the relation of the slope of the valley and the dip of the beds is reversed. In both cases also, the slope of the valley and dip of the beds are to the South. To those who are not acquainted with these technical terms, I may say that the deviation from a horizontal plane of the beds is called the *dip;* while the *strike,* as it is called, is the extension of the strata in a direction at right angles to the dip. In this case, as the dip is to the South, the strike must be from East to West. The flexures of the valleys depend on their inclination relatively to the dip; and these two sections cut through beds of coal and shale and sandstone—the shale being indurated clay—are illustrations of cases in which the two strata come up to the surface according to the various relations of the slope of the valley and the dip of the bed. It is a rule among miners that when the dip of the beds is less steep than the slope of the valley in the same direction, then the Vs, as they are termed, will point upwards, those formed by the newer beds appearing in a superior position and extending higher up the valley. But when the case is reversed, and the dip of the beds is steeper than the slope of the valley, then the Vs point downwards, and those formed of the older beds appear uppermost.

These rules may often be of great practical service in many cases. For example, suppose a miner first to begin his operations in one valley with the structure of which he is familiar. If he should sink his shaft through the formations above he would come to the coal which is below. But suppose one unacquainted with these rules which I have been explaining, to go to another valley; and in England he might easily go to such a valley, for these cases, as I said, are not hypothetical. He might, continuing along the same side of the hills as he had seen in the other valleys, where he observed the same *out-cropping,* as it is termed, of the coal seams, suppose, reasoning from his former experience, that he might begin his workings in the bed at the higher part of the valley with the expectation of coming down to the other bed. But he would be disappointed, as you will readily see by observing that the uppermost bed is lower down in the valley, and the lower bed is the highest up. This you can easily trace with your eye upon these sections. An acquaintance with these ules and their application is of the greatest importance to those speculating in mining transactions. In the coal field of Pennsylvania, to which I shall presently allude, near Pottsville, I saw an exemplification this year of the two cases alluded to—when in the coal of the same valleys the V's in some cases pointed one way, and in the others in the opposite—the dip and slope being both towards the south. There is nothing more singular, or which has struck me so forcibly in respect to the coal fields of this country as their close resemblance to those of the north of Europe, and of England in particular. I have traveled on the north side of the Alps towards the south, and have been astonished to find minerals of fossil of entirely distinct genera from those met with in the Pyrenees. Nor have the chains of mountains any thing to do with this remarkable change—for the beds were formed at the bottom of the sea before the mountains existed. Observing this great change, then, in the short passage of a few hundred miles, it seems to be most surprising that in passing, at the distance of three or four thousand miles, from England to the Apalachian chain in Virginia, we should find the cool measures the same as those we left behind, represented in the red sandstone, and containing white grit and slaty shales, and clays not slaty, and beds of conglomerate containing quartz pebbles.

It is generally admitted by geologists that *all that fuel which we call coal is of vegetable origin.* If there has been any dispute with regard to this, it was settled when a portion of the Newcastle coal some years ago was submitted to a microscopic examination: After cutting off a slice so thin that it would transmit light, it was found that in many parts of the pure and solid coal in which geologists had no suspicion that they should be able to detect any vegetable structure, not only were the annular rings of the growth of several kinds of trees beautifully distinct, but even the medullary rays, and, what is still more remarkable, in some cases even the spiral vessels could be discerned. But besides these proofs from observing a vegetable structure in the coal itself, there has been found in the shales accompanying it, fern leaves and branches as well as other plants, and when we find the trunks of trees and the bark converted into this same kind of coal as we find in the great solid beds, no one will dispute the strong evidence in favor of the vegetable origin of this coal. If we find a circumference of bark surrounding a cylindrical mass of sand, we know that it has been a hollow tree filled up with sand, nor can there be any doubt that the coal is formed of vegetable matter. No less than three hundred species of plants have been well determined by botanists, some of whom have devoted a great part of their lives to this study. From this it is to be inferred that the carboniferous formation of Europe and America is made up of comparatively recent plants. I will allude to three or four of the most peculiar facts that lead to this conclusion.

In the first place the boughs and leaves of Ferns are the most frequently and strikingly met in America as well as in Europe. So perfectly have they been preserved that there can be no doubt that they are really ferns; and in some cases even their inflorescence has been preserved at the back of the leaves. Where we have not the flowers and prints remaining, we have found it possible to distinguish the different species of fossils and ancient ferns by attending to the veining of the leaves. At least one hundred species are determined in this way. The most numerous of these vegetable veinings are those which have been called *Sigillaria*—or Tree Ferns. Their stems are found to be fluted vertically and in the flutings are little stars—as it were—each of which indicates the place where the leaf was attached: and it is evident, as M. ADOLPHE BRONGNIART has shown, that they are recent Tree Ferns. One argument for believing this is that although the bark of these trees is so well marked that forty-two species have been described, yet there is never found any leaf attached; while we have in the same beds loose leaves in abundance which have no trunks. The natural inference is that they must have belonged to the arborescent ferns; as for instance the section named *Caulopteris* is admitted by all to have belonged to this species. This fact is also important because the tree ferns and especially the *Caulopteris* are now known to be exclusively the inhabitants of a warm and humid climate—much more hot and moist in those parts of the globe where coal now abounds. For we find coal not only in England and Nova Scotia, but as far north as Melville's Island and Baffin's Bay, in a climate where the growth of such fern plants is dwarfish and stinted. It is evident that when these vegetables existed there must have been a warmer, and probably a more equable climate than now even in warmer latitudes.

For even in the tropical zones, where we meet with large developements of the *Caulopteris*, their general growth is much smaller than these fossil remains. So is it with all the plants of the fir tribe; many of them of which we find fossil remains in the coal, now exist only in the Southern latitudes, where no coal is found. The *Araucaria* we now find in Chili, and other warm parts of the globe, but never at the North, where its fossils abound in coal. The gigantic plants of the *Equicetaceous* tube are also found to be much smaller now in hot latitudes than are their fossil remains. This would lead to the inference that the climate in Northern latitudes was then much warmer and more moist than it is now in any part of the globe. The same thing is made evident by a comparison of these fossil *Sigillaria* with those which now attain their greatest size in the islands of the Pacific. I have also found several plants, as the *Asterophyllites* in the Apalachian chain, this year, which I have also from Nova Scotia and Europe, and which cannot certainly be referred to any living families. These all, however, bespeak a terrestrial vegetation, though occasionally found mixed with marine shells and corals.

Another class of fossils common in coal shales is the *Lepidodendra*—somewhat allied in form to the modern *Lycopodiums*, or white mosses. Though the mosses of the present day are never more than mere shrubs even in the warmest regions, yet at the carboniferous period they attain an enormous developement, being 50, 60 or even 70 feet nigh.

There have been two theories to explain how these plants could have been carried into the sea, estuaries or lakes, and drawn beneath the water and accumulated in the strata so as to form coal. One of them asserts that the plants must have been drifted and buried in the water, since we find them intercollated between different slates or shales; just as plants lie between the leaves of a botanist's *herbarium*, and are pressed together, so have these ferns been found flattened between the seams of shale. They have been carried from the place where they grew, drifted out to a certain distance, water-logged and sunk in the mud and other strata deposited above them, so as to form this intercollation between the different leaves of clay.

But many believed, from seeing the roots, that the plants grew on the spot where we now find them. But when we come to observe that these roots terminate in different strata, it will seem evident that they were carried down, sunk and stuck in the mud as snags are now in the Mississippi.—In the quartzose sandstone at St. Etienne, near Lyons, are found a vast number of these *Lepidodendra* and *Sigillaria*. No one apparently can doubt that these drifted to their present position, and that they were afterward covered with sand brought down by rivers. Many appearances favor this hypothesis. Sometimes we find beds of marine shells, then vegetable matter, and then a mixture of fresh water and marine shells.

But though these facts may be thus explained, the discoveries that are being made lead geologists to come round more and more to the opposite view of the case—to the hypothesis which refers the growth of large beds of coal to the increase on the spot—after the manner of peat, as it is seen in cold and dark climates. This may appear contradictory to what I said with regard to a change of climate since the carboniferous era; but it is not necessarily so. The opinion of WERNER, confirmed by the speculations of BRONGNIART, led me to believe, contrary to my early impressions, that by far the greater part of the coal had grown in the spot where it is found. Accumulating like peat on the land, the land must have been submerged again and again, to allow the strata of sand and mud to be superimposed as we now find them.

In excavating for coal at Balgray near Glasgow, in 1835, many upright trees were found with their roots terminating in a bed of coal; and only three years ago, in cutting a section of the Bolton Railroad in Lancashire, eight or ten trees were found in a vertical position; they were referable to the *Lepidodendra* species and allied to the *Lycopodiums*, or club mosses. All were within 40 or 50 feet of each other, and some of them were 15 feet in circumference at the bottom. The roots spread in all directions and reached beds of clay, and also spread out into the seams of coal.—There is no doubt that these trees grew where they are found, and that the roots are in their original position. The seam of coal has possibly been formed of the leaves which fell from the trees. This is a singular fact: that just below the coal seam and above the covering of the roots was found more than a bushel of the *Lepidostrobus*—a fruit not unlike the elongated cone of the fir tree. It has always been imagined that the *Lepidostrobus* was the fruit of the *Lepidodendra;* but here they are found beneath other trees.

Under every seam of coal in Wales is found the fire clay—a sandy, blre mud, abounding in the plants called *Stigmaria.* First is the seam of coal, theh the fire clay, then another seam of coal, and then the sand stone. In one part of the Newcastle coal field about 30 species of *Sigillaria* were discovered: the trunks were two or three feet in diameter. They pierce through the sand in a vertioal direction, and after going for some 11 feet perpendicularly, the upper part bends round horizontally, and extends laterally into the sand—and then they are so flattened by the superincumbent strata, that the opposite barks are forced within half an inch of each other. The flutings are beautifully preserved in the flattened horizontal stems. Here we had an ancient forest growing in a bed of clay—buried in some way with sand to a certain depth, and then the upper part was bent and broken off by the water current, and buried in layers of shale and sand. There are many cases of this kind in Wales, where the roots of the trees evidently preserve their original position. Mr. LOGAN, an excellent geologist, has examined no less than 90 of these seams in Wales. They are so exceedingly thin that they are of but little value in an economical light—yet they are just as important for geological purposes, as if they were thick strata. Under every one of the 90 he has found the fire clay, a sandy mud containing the plants called *Stigmaria.* It was discovered years ago that this fire clay existed with the coal mine; but it was not known that it was the floor of every coal seam, and not the roof, which contained this plant in a perfect state. The *Stigmaria* appears in the under clay (to use the term employed by miners,) a cylindrical stem, from every side of which extend leaves—not only from the opposite sides, but from every side: they appear like tubercles, fitting on as by a joint.—They radiate in all directions in the mud, where they are not flattened like the ferns. Had they been, we might have had leaves in two directions, but not on every side. These plants resemble the *Euphorbiaceæ* in their structure, and in some respects are analogous to the Coniferous or fir tribes. In their whole structure they are distinct from all living genera or families of plants. In one instance a dome-shaped mass was found with stems and leaves—some of the branches being 20 or 30 feet in length and sometimes longer. It has been thought by Dr. BUCKLAND and other geologists, that these plants either trailed along in the mud at the bottom of swamps, or have floated in lakes like the modern *Stratiotes.*

After Mr. LOGAN had arrived at this remarkable fact, we become particularly desirous to know if the same fact was true in the United States.—When I arrived here in August, I had no idea how far it was true, yet it was known the *Stigmaria* did occur; and my first opportunity to inquire into the fact was at Blossburgh, in the Bituminous field in the Northern part of Pennsylvania. My first inquiry of the geologist was whether he found Stigmaria there. I was answered in the affirmative; and on asking if the plant occurred in the *under clay,* he said we could soon settle the point. He had one of the mines lighted up, and the *only plant we could find in the under clay was this Stigmaria:* it existed in abundance—its leaves radiating in all directions, just as in Wales, more than 4,000 miles distant. The same crutal appearance was preserved. In the roof of the coal seam were seen different species of ferns—*Sigillari* and *Calamites*, just as in North Carolina and in Wales. Afterward another opportunity occurred in the Pottsville region of Anthracite coal. Professor RODGERS, the State Geologist, who, though well acquainted with the strata of the district, was as anxious as I was to know if the rule would hold good, examined first at Pottsville and at Mauch Chunk, when the same phenomena were observed. In the first coal mine we came to, the coal had all been quarried away (for the work was carried on in open day) and nothing but the cheeks of the mine remained. The beds, as they have been horizontal, are now not vertical, but have gone through an angle of a little more than 90°, and turned a little over; so that what is now the under side was originally the upper; therefore the cheek on the left side was originally the floor of the mine. We now looked at the lower cheek; and the first thing wo saw was the Stigmaria, very distinct; on the other side, but a little way off, were ferns, Sigillarieæ, Calamites, Asterophyllites, *but no Stigmaria.* So it was at Mauch Chunk, where we found one 30 feet long, with leaves radiating in all directions. At this place there is a bed of Anthracite nearly 60 feet thick—a magnificent accumulation of vegetable matter, to which there is nothing comparable in Europe. Except in one place it is perfectly pure.

It has now been ascertained for many years that Prof. EATON was quite correct in affirming the Anthracite and Bituminous Coals to be of the same age. This is shown not only by their relative position with regard to the red sand-stone, but from the plants found in both being identical.

All the coal fields, therefore, may be regarded as one whole, and the question will occur, How did it happen that the great floor was let down so as to prevent the accumulation of coal and yet plants of so different textures should be found in it? It has been suggested that these plants grew in swamps; and it is possible to imagine that there may have been morasses fitted only for the growth of the species of plants called *Stigmaria;* and that this marsh filling up, this and the other plants became dry, and the leaves accumulated one layer above another, so as to form beds of coal of a different nature from those that preceded. You know it is a common thing for shallow ponds to fill up gradually with mud and aquatic plants and at last peat and trees are formed upon them. A corresponding change is constantly going on in different parts of Europe—the same transition from bogs and marshes to a soil capable of supporting various great trees is taking place, and then the ground is submerged; for always, again and again, we must refer to this subsidence of the soil.

Many of you, I suppose, have seen the morass called the Great Dismal in North-Carolina and Virginia; and you have probably had an opportunity, as I have, of crossing the northern extremity of it on a railway supported by piles, from Norfolk to Weldon. This is no less than forty miles from north to south, and twenty from east to west covered entirely with various forest trees, under which is a great quantity of moss; the vegetation is of every variety of size from common creeping moss to tall cypresses 130 feet high. The water surrounds the roots of these trees for many months in the year. And this is a most singular fact to one who has traveled only in Europe, that, as is the case in the United States, trees should grow in the water, or surrounded to a certain hight by water, and yet not be killed. This Great Dismal was explored some years since by Mr. EDMUND RUFFIN, author of a valuable Agricultural Journal. He first calls attention to the fact that a greater portion of the vast morass stands higher than the ground that surrounds it; it is a great spongy mass of peat, standing some seven or eight feet higher than its banks, as was ascertained by carful measurements when the railroad was cut through. It consists of vegetable matter with a slight admixture of earthy substance, as in coal. The source of peat in Scotland is that one layer of vegetation is not decomposed before another forms. So is it in Chili, Patagonia and Terra del Fuego. Thus also in different parts of Europe, in the Falkland Islands, as DARWIN has shown. Thus, too, is it in the Great Dismal, where the plants and trees are different from those of the peat in New-York. It is found on cutting down the trees and draining the swamp and letting in the sun, that the vegetation will not be supported as it was before beneath the dark shade of the trees. In the middle is a fine lake, and the whole is inhabited by wild animals, and it is somewhat dangerous to dwell near it by reason of the bad atmosphere it creates. It is covered by most luxuriant vegetation. We find in some places in England that there is a species of walking mosses, which are sometimes seized with a fancy to walk off from their places: the moss swells up, bursts and rolls off, sometimes burying cottages in its path. In some places this peat has been dug into and houses have been found radiating several feet below the surface—curious antiquarian remains. In the same manner the Great Dismal may spread itself over the surrounding country.

In speculating upon the probable climate of the Carboniferous period, it is believed that we have only to imagine a different distribution of the land over the surface of the planet than that which now prevails, to produce such a warm and humid climate as must have prevailed when these plants flourished which form coal. It is the existence of high lands near the pole which produces such great cold. If these mountains were to be transferred to the tropical regions, it would immediately lower the temperature of all climates of the earth. Now one who has attended to the study of rocks and fossils, sees at once that the present physical geography of the globe has no reference to its ancient condition. Seas once occupied a large portion of what are now continents, and we also find evidence of marked change in the Carboniferous and other strata. In the limestone accompanying the coal we find corals and shells, strongly indicating a higher temperature of the sea, as the plants shadow forth a higher temperature in the atmosphere.

I have been favored with a map illustrating these points by Prof. HALL, one of the State Geologists engaged in surveying the State, whose labors will soon be made public. And here I cannot avoid saying that I have been over much of the ground which they have surveyed, and it gives me great pleasure to bear testimony to the accuracy of their labors, to the great pains they have taken, and the science with which they have con ducted the survey. I look forward to the appearance of their work, embracing the results of their labors, as *an era in the advancement of science;* and the patronage which has been afforded by the different States of the Union to these surveys, is much greater, in proportion to the population, than any European power has ever extended to the advancement of geological science. When we remember, too, the complaints that may be heard in different parts of the State that the geologists have failed to discover any mineral wealth, even in an economical point of view, these scientific researches are of high value, though their greatest interest arises from the promotion of the knowledge of the structure of the globe.

But merely estimating the mischief they have prevented, we shall see an ample remuneration for all the expense attending the survey. I have been told that in this State alone more than a million of dollars have been expended since the Revolutionary War in boring for coal in formations where *it is impossible to find any*—below the carboniferous strata. I should not, to be sure, have ventured to generalize from Europe as a type and say that the rocks in the crust of the earth occupy the same relative position here, and that coal would be found always in this country under the same conditions as in Europe. But when for twenty years or more we find coal accompanied by the same plants, and that no valuable fuel has ever been found under any other circumstances, we should be safe in saying that none could be found in the older strata.—If we begin in the newer beds, we may come down to the coal, and find enough coal to pay the expense of boring for it. But if we begin in the srrata beneath the carboniferous, we should certainly never reach the coal until we had bored through the whole earth: we might find it at the antipodes, but not before.

Thus complaints are made against these geologists not only that they have found no coal, but that they have passed sentence of sterility upon the State, for they say that through all time no coal shall be found within its borders. And when we reflect on the enormous sums that have been wasted upon strata more ancient than the coal, in searching for coal, we shall see the great saving made in consequence of this survey; for when all its maps and sections are published, it will be seen how impossible it is to find coal in these more ancient beds. This is a kind of advantage which is never easily appreciated: because, to prevent mischief is never so clear and palpable a benefit to the multitude as to find mineral wealth. But one of the greatest advantages which have resulted from these surveys in England, and it will be among the greatest here, is the prevention of this rash and absurd speculation to find coal in strata below that in which those plants-known to be essential to the formation of coal are found to exist: and after examining the whole ancient strata, both in the United States and in Europe, there has never been found a single bed of coal where these plants do not exist.

LECTURE VI.

FOSSIL FOOT-PRINTS.

I HAVE been asked by several persons who heard my last lecture, upon the Origin of Coral, if I could explain the difference between Coral and Anthracite; and as I then had no time to touch upon this subject, although my last lecture was prolonged beyond the limit which I wished, I will now speak of it in a few words. It certainly is a good question: but several persons have asked me whether, conceding the vegetable origin of coal, it may not be the *difference in the wood* which causes the difference in the coal: whether one kind may not produce the Coal and the other the Anthracite.—Now there is no doubt that, in the strata of the Earth, there is some variety in the character of the various Coals discovered, which may be owing to a difference in the original texture and composition of the different plants from which they were formed. But this, nevertheless, is not the cause of the difference between Anthracite and Bituminous Coal, as they have been produced by the same plants.—There is no doubt that the Palm, and other Monocotyledonous plants, and bamboas, reeds, &c. would give a kind of coal to a certain degree different from that produced by the fir and other Dicotyledons. But as precisely the same assemblage of plants is found in the Anthracite and Coal, it cannot be in this way that the difference between the substances is caused; and the conclusion to which the geologist and chemist have come is this: that *Anthracite has once been Bituminous Coal*, but has lost its bituminous matter by its escaping; that the volatile part of it has escaped and the wood has become gradually converted into coal by the loss of its oxygen—vegetables being composed of carbon, oxygen, and hydrogen. The vegetable matter losing first its oxygen, loses also in the combination some portion of its carbon, forming a carbonic acid which escapes from the wood, often as a pure gas—as may be seen in bubbles which arise from the bottom of pools of water: it is thus converted into coal by the loss of its oxygen; and this conversion is so gradual that there are found in the Earth woods in every stage of the change—both in that state when they are called *lignites* and when they have become perfect coal. We find, too, all the intervening forms between wood and coal, and then all between coal and anthracite. This is caused again by the loss of the hydrogen which escapes in combination with carbon, in the form of carburetted hydrogen gas—the same substance that we are now burning here. You are aware that this gas cannot be produced from anthracite because in that the operation has already taken place: it is manufactured entirely from bituminous coal. This volatile, inflammable gas may be seen to escape from rents in the Earth—natural rents from which this carbo-hydrogen issues; and if the coal below be examined it will be found to be in process of gradual change. But there is one remarkable fact observed in this country: that in the coal region of Pennsylvania for example,—and this has been especially pointed out by Prof. RODGERS in his survey of Pennsylvania—the anthracite is found to be purest in the most disturbed part of the mountains; and it is half bituminous when you get into the regions that have been slightly disturbed—in the western part of the State; and when you reach the perfectly horizontal coal district—that where there has been no disturbance, no shattering and tearing up of the mountains—there the coal is purely bituminous, not having lost its hydrogen. Some have suggested that this is owing to the rending and fissuring of the rocks in the disturbed region, in which all the volatile matter would have escaped more easily. Of this there is no certainty. But the general fact is undoubted—that in proportion to the disturbance, overthrow and bending of the strata—some of which have been folded back upon themselves—do we find that the conversion into anthracite from pure carbom is most complete.

The next subject of which I shall treat is that of *Fossil Footsteps* found in the formation which we call the New Red Sandstone. Its position in the series is as follows:

1. Chalk.
2. Oölite.
3. New Red Sandstone.
4. Magnesia Limestone.
5. Coal.
6. Old Red Sandstone.
7. Silurian Group.
8. Granite.

In the upper part are found organic remains in great abundance and in the red and white sandstone there were found about seven years ago in Germany the remains of footsteps—prints apparently made by the feet of quadrupeds (rudely represented in this drawing) [exhibited.] Often five or six pairs of these are found in one track. They are found near the borders of Bohemia at Hessburg; and their discovery created at first a prodigious sensation in Germany because some imagined that the animals were a sort of *extinct form, perhaps, of man*, or some prototype of man. The hand-like form of these tracks, resembling in some degree the human hand, created great astonishment. The largest track was about eight inches long; before each of these large tracks is a small one which is referred to the fore foot of the creature. There were observed four toes and one other which resembled a thumb. In one track that was found on the right, in the next on the left, and so on; and to whatever distance they were traced they were found thus to alternate right and left. No doubt whatever is entertained that these represented the steps of some animal imprinted on layers of clay when in a soft state. They stood out in relief upon the under side of the slabs and sandstone when taken out; because this sandstone was once in the form of loose land, and was thus deposited in the hollows of the clay below; so that it would present the appearance of a cast moulded in the hollows of the adjacent clay, and we should have a perfect cast in relief of the footsteps of the animal as it walked on the clay.

Some conjectured that this animal belonged to the *Marsupial* or Kangaroo tribe, because the hind feet of this animal are remarkably large in comparison with its fore feet; and it has feet also very much in the form of the human hand. Others believed it to belong to the Crocodile tribe; and Prof. LINK early conceived that it belonged to the *Batrachian* or Frog family. While this discussion was going on in Germany—or but few months after this—similar footsteps were found in the Red Sandstone of the same age in England, at Storeton Hill in Cheshire, not far from Liverpool, on the banks of the Mersey. These salient footmarks were observed in no less than five distinct ranges or layers of sandstone; there were five thin layers of clay, and resting on each was a thin white or yellow quartzose sandstone, which presented these footsteps precisely like those found in Germany. It had been previously ascertained from the position of these beds, and the other fossil rocks in question, that they were of the same age with those in which the footprints were found in Germany. They were found of several sizes; one in which the larger footstep was eight inches in length; another in which it was twelve; this was called the *Ceiritherium Herculis*, on account of its great size. The tracks were about fourteen inches apart. In the Museum at Liverpool, is preserved a long slab of sandstone—as long as this staff (some ten feet) in which the prints are so beautifully distinct in relief that you would suppose they had been made by the chisel of the sculptor. In some slabs preserved in the British Museum you may see prints of this shape [exhibited] which are evidently casts of the small cracks below. The clay having shrunk, these cracks were opened by the heat of the sun: or so that when the loose sand was deposited in the footprints of the animals, it found them filled with these rents in the clay; and when they came to be taken up, these cracks are found to be in relief, as well as the footsteps. The same things are found in the same beds of England; and in this red sandstone, which is found all over Cheshire and other parts of the kingdom, are found ripple-marks—parallel ridges, marking the ripple on the beach of the sea at low-tide.

It was remarked when reasoning upon the subject in London, and in the discussions of the Geological Society, that these footsteps must have been made by an air-breathing animal: as no animal which existed under water could make such distinct foot-marks, not being heavy enough in that fluid; they were pressed down by the weight of an animal walking in the asmosphere, and therefore at low tide. And nothing is more common now than to observe the tracks of various animals upon the beach at low tide. I had an opportunity myself of observing this in the islands near Savannah in Georgia. At low tide I have seen the whole beach covered with the tracks of the Opossum, Racoon and other animals; and in many places after tracing these tracks some hundreds of feet, I came to places where the quartzose loose sand of the low cliffs of the islands had been blown by the wind only a few hours before while the tide was down; still there had been time for the tracks to be made and for the wind to blow the sand over, and in some places the tracks were thus obliterated, some entirely and others only in part. We have then only to suppose a submergence of a coast so situated carrying down this sand, afterwards that it was consolidated by ferruginous or other matter, and we should have a counterpart of this. Take up a slab of the common sand and it would present the same marks of the racoon and opossum.

At the same time that these fossil remains were discovered by Prof. KAUP, who devoted considerable attention to the subject, Mr. OWEN procured from Warwick several fossel teeth of a conical form; and being engaged at the time in investigating the internal texture of the teeth of various animals, he cut away a thin slice and subjected it to examination by a powerful microscope, and he was surprised to find that this tooth, found in the New Red Sandstone, No. 3 differed greatly from those of all the other reptiles found in No. 2, or the Oölite formation. This Oölite has been called the *age of Reptiles*, because we have in it so many of the *Terabratulæ*, and reptiles—flies, fish, lizards, *Ichthyosaurians* and *Saurians* of various kinds—not one of which has texture resembling in the least degree the texture of those found in the New Red Sandstone. He made a representation of a small part—one-fourth of a tooth, which he has allowed me to copy. [Exhibited.] You see that its texture recalls the appearance of the human brain; it has the same convolutions, formed by the dentine and pulp of the tooth; the animal that exhibited so remarkable a structure was believed to be distinct from any living species. Supposing this to be the case and becoming greatly interested in this discovery, Mr. OWEN applied to the German Professor KAUP, to have one of his fossil teeth sent from Germany that it might be dissected, and thus that it might be ascertained if it had the same structure as those found in England. This was done and the *same remarkable texture was observed.* Thus we have in both countries—five hundred miles apart—fossil teeth, found in the same rock and having the same structure, entirely distinct from that of reptiles in their formations. It was evident that they had belonged to reptiles—for they were different from any belonging to quadrupeds or fish. It was thus believed that they were a reptilian class of a new and before unknown family.

The next discovery made in the rocks of this very age, and in the same series of Sandstone in which the impressions of footmarks were found, was of parts of the scapula, femor and pelvis of more than one species of this animal. It had received the provisional name of *Cheirotherium* or *Beast of the Hand*, from the shape of the imprinted footstep.

The question now was whether the teeth and the bones could belong to the *Cheirotherium*.

Mr. OWEN remarked, after examining some of the bones of the hinder extremities and comparing them with the bones that evidently belonged to the anterior extremities—to the fore arm, found in the same place and probably belonging to the same animal—that *there was as great a disparity in the relative size of these bones as in the marks of the hind feet compared with those of the fore feet,* Soon after a portion of the jaw was found and Mr. Owen was able to construct a skeleton of this animal—to restore in imagination something of the form of the animal that once may have walked, so as to produce these lines of foot-marks, by bringing its hind foot forward so as to make an impression close to the last one of the fore foot.—This figure is a representation of the animal and the position of his tracks:

These parts were found when I left England nine months ago, and it is probable that more have been found since that time, for all these discoveries have be made within the last twelve or fourteen months.

It was therefore now clear that there had existed, at the period of the deposition of this stratum in which these impressions are found, several species of animals which correspond well, in relative dimensions, with the size of the several sets of footsteps found in Germany and England. At least three distinct sizes were found—teeth belonging to three distinct species. They were called *Labyrynthodon* from the labyrinthine convolutions in structure of the tooth.

The next step made in England were to ascertain whether this *Labyrinthodon* was an air-breathing animal or not. Now OWEN has shewn from the structure of some bones in the lower jaw, and also of one in the upper, that there was an orifice which implied a nasal cavity, placed as in air-breathing reptiles of the division *Batrachia*. For it is possible from the bones to discover whether the animal breathed by gills, or at least by bronchia, for in that those the opening would extend to the nostrils; whereas if the animal breathed air, the cavity would be placed much farther back—nearer the back of the head. We are able to show that this *Labyrinthodon* had been an air-breathing animal, and must have been able to walk along the side of the retiring water.

You are aware that reptiles are distinguished into four great divisions—

1. *Saurians,*
2. *Chelonians,*
3. *Ophidians,*
4. *Catrachians.*

The first embraces crocodiles. alligators, lizards, &c.; the second, tortoises, turtles, &c.; the third, snakes and boas, and the fourth frogs, newts and salamanders. Now, it is quite evident from its osteology that this creature belongs to the *Batrachian* order—that it is a gigantic frog, as large as a bull, and the Cheirotherum perhaps was as large as an elephant. The salamandrine reptiles are represented in the living order by creatures of a comparatively small size. In this country there is an aquatic salamander among the Alleghanies two feet in length—much larger than any that we have in the old world, except one which is found in the island of Japan, called the *Salamondræ Maxima,* exactly resembling a frog. It is very singular that this animal should have been found in a lake in a volcanic crater in that island. It is now in Leyden, where for eight or ten years it has been in the Museum. There were two procured, a male and female, about four feet in length. Only one was brought home: and to prove that they are carniverous I need only mention that on the passage the *male devoured the female,* so that we have only one in the museum. They are fierce, voracious creatures, and it is very singular that that they should be found in the crater of a volcano; because thre has always been a story—from the most remote period—as long ago, at any rate, as the time of Aristotle—that the Salamander was inconsumable; and there have been despositions in great numbers made before most respectable magistrates, some of whom are now living, that these creatures have been seen to come out of a fire in which they had been exposed to the white heat of coals for two or three minutes, perfectly uninjured. But after a full examination it turns out that this is entirely a fable.

I will now observe that although we have no complete specimen of either the *Cheirotherium* or the *Labyrinthodon,* we have still enough of their remains to leave no doubt of their character on the mind of any person. We have first in rocks of the same age in Germany and England the same footprints. We have in both countries certain teeth of a most remarkable structure, peculiar to that rock to which belong the animals which lived at the period when these footmarks were made. We find the hinder extremities in each species as much larger than the anterior portions as the posterior footsteps are larger than those of the fore feet.—And lastly we require air-breathing animals, and we find the structure of the nasal cavity such as to show that the *Labyrinthodon* was evidently an air-breathing animal.

You may be aware that in this country there have been found some remarkable tracks of birds in the Red Sandstone of the valley of the Connecticut; and Professor HITCHCOCK has embodied an excellent account of them in his final Report of the Geological Survey of Massachusetts, and has given some excellent plates not only of the extinct species, but also of the animals and birds that may now be found on the shores of the lakes of America. It appears to me that he has completely made out that several species of birds existed at that period which, so far as is known, is the same as Nos. 4 and 5 in which are found the footsteps in Europe; and he has found some as large there as an estrich, and others again of smaller size. The variety is considerable. He has found, too, the same ripple-marked figures—in which are seen casts of the skeletons of the cracks in the clay produced by shrinking. I have seen as near this as at Newark, examples of this sandstone ripple with casts of the tracks in the subjacent clay. In England there are places where occurs clay which has been pitted over by small depressions as though it were done by a shower of rain. You can see that if rain falls upon sand it will cause a pitted appearance; and of such we find slabs containing casts which have been called fossil showers of rain; and I am inclined to believe that this is the origin of the appearance. And I think that upon this point Mr. REDFIELD and myself will agree—that in Newark we have slabs in which can be seen the same marks, not so exactly representing a shower of rain as those in England, but still having a close resemblance to them. In this case the marks are in soft shale and we have pits; but when salient casts are found we have all the apparent marks of rain in relief like the footprints and casts of cracks in the clay. There are some beautiful remains of fish in the same strata in the valley of the Connecticut and in parts of New Jersey, some of which are upon the table. They are not in the same beds as the footprints, but in the same strata—formed deeper under water.

But as I must not dwell too long on any one subject—as the course I am delivering is so short, and as I am desirous of treating as many subjects as possible—and having already spoken of formation No. 3, I pass the coal—of the *Flora* of which I spoke in my last lecture—below the Red Sandstone to Stata No. 7, which we see abounding in fossils; and in no part of the world has it so magnificent a developement as in New-York, though it occurs in greater thickness in Pennsylvania. But there is at least no part of the world that has been investigated where we see the fossil character of the different periods so well made out as in New-York. When we come to the subdivisions of this one formation, you will see at once that it forms as many groups as are represented in the other table. The first of these divisions was formerly called Transition—but now the Silurian Strata; because it is found mostly in Siluria, a region on the borders of Wales, composed of the border counties of Wales and England. It has been well studied and elucidated by my friend Mr. MURCHISON. For this section [here exhibited] I am indebted to Professor EMMONS and Mr. HALL of the State of New-York. It exhibits portions of the results of their joint labors; and I have had the advantage of going over the region and seeing the beds in the order they are here represented.

First of all we have what is called the Mohawk group, which is 1200 feet thick in the state of New-York and thicker in Pennsylvania; for the thickness depends on local or accidental causes. If the material of which it is composed be coarse the strata will be thick. The fine sand will be floated farther out at sea. In this strata occurs the Trenton limestone—with trilobites and other species of fossils. We have in this strata some extinct species, as the *Lingula* and others, that must have lived in the seas. The *Graptolite* is found in the old rocks, and is a species of zoöphyte related to the *Pennatula*. It is an example of an extinct form only found in ancient rocks, never beyond the Transition or Silurian Strata. When we pass to the Red Sandstone we have a resemblance to the more modern rock, of which I have before spoken, in which are found the fossil foot-prints. It is called the Ontario sandstone because it is found along the borders of Lake Ontario. The lower part of the Cliff at Lewiston, in the Niagara District is the same bed seen horizontally; it is 60, 80, and in the lower part 120 feet thick. Of its exact position I will speak more at length in my next lecture; I only wish now to point out the order in which they follow. If you suppose yourself aloft in a balloon—looking down over that part of the country, it would present the appearance in this figure—in which a portion of the wood is supposed to be cut away so that the geological structure of the rock can be seen.

After the Ontario sandstone comes that which is called by Mr. VANUXEM the Protean group from the many different forms in which it appears. It contains the *Pentamerus*, an extinct species of bivalve shell. Sometimes it is also in the form of iron ore. This Protean mass is only about 50 feet thick at Queenston and Lewiston, though in some parts of the country it is probably thicker. It is found farther East and South. Passing, then, from the Mohawk limestone to the Ontario sandstone and thence to the Protean group, we next come to the shale and limestone of the Niagara district. These beds contain an assemblage of organic remains, as, for example, the Chain Coral—so called from the resemblance of its upper surface to a chain. It is a genus peculiar to the ancient rocks of Norway, England and the United States. When I say that each of these groups is distinguished by a different assemblage of animals, of which there are several hundreds in one, I would be understood to mean that some species have a wide, vertical range, as it is called. Some may live longer on the globe during its changes than others; and this would be expected when we consider that some species at present bear different climates with more facility than others; they are a sort of cosmopolitan species, which are found every where. Thus we find the same shells in the Hudson as in our seas and in the Baltic; the same may be found in the temperate and the tropical zones.

The same group which borders Lake Erie and is seen at Buffalo forms the Black Rock of Buffalo. It is called the Helderberg lime-stone, because in the Helderberg Mountains we have no other beds of limestone, and upon the fossils found in them I might employ the whole of eight lectures; and you might suppose that the whole of geology was embraced in their discussion. Still that is only one subdivision of this one stratum, and there are beds of Corals found in other lime-stone, and in the oldest Trenton lime-stone like the ancient reefs—the species being for the most part different in each. There are found in the same series of the Silurian strata in England corresponding lime-stones at different stages, and occurring in the position in which they originally grew. They have the point of attachment at the bottom of the stratum, and the upper part continues to be uppermost. Or if they are hemispherical, those which radiate, you must determine by the manner of radiation which is the base and which the upper surface of the coral, and thus determine whether it is in its original position or has been overturned. We often see them in their natural position, as they grew, one upon the other. In several cases we find them broken and thrown over, as if by the force of the waves. In both cases the same coral grew in one case at the surface in reach of the waves, and in the other near the surface; because it is known that corals do not grow in the deep, dark, cold, unfathomable abyss of the sea, but within 120 feet of the surface, enjoying the light and warmth of the sum. Some of these corals are as large as this table, as at Williamsville, near Buffalo. In following for a long distance, as may be done in England, Norway and Sweden, this strata of fine sand which has formed this group, you find no other fossils but the *Graptolite*, belonging to the zoophytes of which the living specimens are found in muddy bottoms. Beneath that fine mud is no example of fine sand being deposited, for it will not go on in deep water. If you consider the laws which regulate matter falling through fluids, you will easily see the truth of this: the doctrine of terminal velocities explains it. For example, if you let a bullet of lead one inch in diameter fall through through the air at the density at which it exists at the surface of the earth it will never fall beyond a given rate; it will fall faster and faster from the hand until it reaches a rate of 260 feet in a second, when it will never fall faster; for the fforce of gravity is a constant force, but the resistance of the air increases in a geometrical ratio. So in water; a bullet would never fall at a greater rate than 8½ feet in a second; while it would fall 260 feet in air it will only fall 8½ in water. If you reduce the ball of lead from an inch to a hundredth part of an inch in diameter it will fall only one inch in a second instead of 8½ feet; and so we shall find that if it be still farther reduced its maximum rate of descent will be much less.

But fine mud, such as large rivers deliver into the ocean, only falls at the rate of about two feet an hour. Now, suppose the mud to be directed into a current like that of the Gulf Stream for example, which runs at the rate of three miles in an hour—suppose it to be delivered into the ocean by the Amazon or some other great river, and thrown into this current, it would be carried 2000 miles by the Gulf Stream in a month, while it was sinking 600 feet, which is no depth at all in the ocean: and we have reason to believe that the influence of the Gulf Stream extends for at least 600 feet below the surface, as is shown by its action upon icebergs. I think Mr. REDFIELD will agree with me in saying that the action of the Gulf Stream on enormous floating masses of ice extends to that depth. When, therefore, the mud might be carried so far, I see no reason why it may not be carried from the Equator to the Pole by this influence.

I will now offer a few observations on the inferences which may be deduced from those fossils found in the ancient rocks, concerning the state of the globe at those very remote periods. Even the most trivial circumstances derive a great degree of interest from their mere antiquity—they become consecrated, as it were, by time. Those who walk in the streets of Pompeii will look with the deepest interest and the most lively curiosity upon the mere ruts left by carriages eighteen hundred years ago, which are now as distinct in their impressions as if they had rolled there yesterday: they will look with pleasure and interest on the scribblings in Greek of the soldiers, on the walls of the barracks, in this ancient city, which has remained buried under volcanic ashes since so remote a period. How much more interest will the naturalist feel when he examines the remains of creatures believed to have flourished on the earth—in water or on land—thousands of years ago—as he finds them in these rocks we have been considering.

D

Let us, then, endeavor to draw some inferences from the study of these animals, repecting the state of the planet where they lived—how far it was similar to, different from, that in which it now exists. Take first the *invertebrated*, and then the *vertebrated* animals. Here upon the table are the remains of *Trilobites*, of which there is a prodigious number in this country. It is an extinct form of *crustacea*—believed to be of the same form as that of the crab and living lobster, but of a genera now extinct. They are very characteristic of the ancient rocks in Europe and Asia: but they are never found in the newer rocks; Examine the eye—for it is a singular circumstance that the eye of some of these creatures is preserved in a most perfect manner in the fossil in the mud in which it was buried. You find that some species have a hemispheric eye—sometimes placed upon a pedal.—Here is represented one consisting of a great number of *facets;* the living dragon fly has no less than 14,000 *facets* upon each eye. Some of the living crustacea have a great multitude, and, to some extent they are analagous to the ancient.—These were intended—as we know by reasoning upon the habits and structure of living species nearly analogous—to enable them to see horizontally by these lenses. The *facets* are set so as to look in all directions except one which is supplied by the hemispheric form of the other eye. Hence it has been justly observed by Dr. BUCKLAND, from an examination of optical laws and the eye of the extinct animal, that we see there was the same retation of light to the eye, and of the eye to light at that remote period as now exists in every species. The ocean must then have been transparent as it is now; and must have given a passage to the rays of light, and so with the atmosphere; and this leads us on to conclude that the Sun existed then as now, and to a great variety of other inferences.

If we turn to vertebrated animals, therefore, we have several species of fish, of which the parts found were called *Ichthyodorulite* before they were aware of their origin. It was supposed that they were some part of a fish, but they are now known to be portions of the first spine or darsal fin, of the genus of fish, like the living genera called *Cestracion*. A tooth in the jaw of a different species has been found in England in the upper part of the Siberian series. In this country geologists have traced the existence of analagous species of fish a step lower. In several of these subdivisions there have been found remains, but never *vertebræ*, nor ribs, or other parts of the skeleton; but bony tubercles, &c. by which Mr. AGASSIZ has shown that they all belong to a cartillaginous fish, like the sturgeon; in which the external parts are bony, while the parts which in others are bony in these are soft; something like the Crustacea.—Still they are as perfectly vertebrated animals as any others.

What is implied here in the existence in the rocks from this remote period of these creatures is, that they may yet be traced back to an organic beginning. We find here a type of vertebrated animals —not of a high order, not yet fully developed, nor yet of the lowest oeder. They have the vertebral column, with the spinal marrow terminating in the brain; they have organs of motion in the orbit of the eye. There must, then, have been vision, locomotion, circulation of the blood, a nervous system—in short, we have sketched in this vertebrated animal of the earliest period a great outline of the very skeleton which now appears in reptiles and mammalia, and in man himself; and this acquires the highest interest when we find this first outline of that plan of organization already in existence, destined, after modification at different periods, to reäppear in the inferior animals, and in man himself.

It has been a subject of marvel and of incredulity to many to account for the existence of so many races of animals and plants which are now extinct, which all preceded the creation of man on the Earth. It is often asked for what purpose could they have flourished; and this difficulty arises from the habit in which we are too apt to indulge, of imagining that all the works of nature are produced either to satisfy the wants or to minister to the instruction and amusement of man. But those who take a wide and philosophical view of the present state of the creation must meet with the same difficulty. Look at the animalculæ in a drop of woter, and you might put the same question: Why do these exist in such myriads, of which nearly all are invisible? Look, too, at the thousands of shells and other fish with which the whole ocean teems, and which do not come under the knowledge even of the naturalist. When this question presented itself to the mind of our great Poet, in the exercise of his high imagination, he sees that beings might have looked over the world, though unbeheld by man. He even says that

Millions of spiritual creatures walk this earth,
Unseen, both when we wake and when we sleep,'

and had the existence of these countless creatures been known to him he might have imagined that they lived and moved under the eye of spiritual beholders, though man were not. And certainly the light now cast upon these periods by Science makes them as wonderful and sublime as could the imagination of him or of any other poet. For we find that these creatures, in such multitudes, and in such varieties, existed upon the Earth for ages, then disappeared—dying and giving place to other races which likewise for a time inhabited the globe. But they too perished, and then came the period of vegetation—of coal; and all the creatures that before lived and flourished in the water gave place to other and new varieties. So succeeded many other changes—each greater than the first in the catalogue. Yet through all these periods of stupenduous changes we find the same perfect unity of plan—the same infinite wisdom—the same great outline of organization. Throughout the whole universe and in all these ancient periods of the Earth's formation we see the same great laws prevailing—those laws of organic and physical life which govern all the revolutions of the animal tribes; at length we find superadded to these other osteological and physical organs; yet the type re-appears not only in the inferior animals but linked with the intellectual and reasoning powers of man; and we find him by the power of those thoughts that

——wander through eternity,'

taking a retrospect of all these great periods—examining their characters, one after another, until he arrives at the remotest period of animal existence and finds himself able to study their history and the character of their organization. Nor has he merely the power to decypher these records; but an exhaustible, infinite store of such memorials have been written upon the great rocks and placed, for his instruction, in the solid framework of the globe.

When I speak of the laws that govern these successive changes, I am aware that some persons look upon the idea, that such revolutions in the organic world are governed by laws, as favoring the doctrines of the Materialist. But we may regard these laws—when given either to the animate or inanimate world—as exponents fully to express the will of the Supreme Being. A law cannot be referred to any thing material; it belongs not to the material creation; it is higher than the world of matter—it is spiritual in its essence and leads us up to the contemplation of one Immaterial and Spiritual Lawgiver.

But I will not dwell longer upon this subject, but pass to one more strictly geological; to the conclusions which we draw from finding the older rocks in a horizontal position. It is striking to find in Europe, in great parts of Russia, in Sweden and in this country, these ancient strata in their original horizontal position. As to the manner in which whole portions of the country have been carried away and left great cliffs behind, I shall speak more fully in my next lecture. You may observe one strata resting on another on a level; and this may be owing to the depression of the sea, which has left them exposed, or, as I believe nearly all agree, that the beds themselves have been greatly lifted up out of the ocean and raised to their present hight, where they remain in their horizontal position. It is thought by some that there have been alternate periods of repose and violence all over the earth; but it may be laid down as a rule that the more ancient the rock the greater is the disturbance that has been suffered; because in the first periods the convulsions were more violent than those which occurred at a late epoch.

But whether we adhere or not to the opinion that there has been a great series of alternations—periods of violence and convulsion followed by long eras of tranquillity—one thing is certain, that, from the earliest period, we have no record of one such epoch—no knowledge of any such dissolution of strata—no such paroxysmal convulsion in the crust of the globe; and had there been any such occurrence, it would not be possible now to find the ancient fossiliferous rocks in their original position. As to the fact, it is true, that in general we find the older rocks have undergone the greatest convulsions; and this will be quite intelligible if we refer them to volcanic forces which always shift their points of principal developement.—There have been volcanoes, as I stated in one of my early lectures, in districts where they are now extinct—and where there is left no vestige of volcanic explosions. After a long period of disturbance and convulsion the fires were spent, and then succeeded a period of tranquillity. It may be shown, too, that in former periods there were more volcanoes than during the more recent; and, therefore, if we suppose a shifting of the principal points of developement, which were near the surface and attended by earthquakes, it would follow that in time they would extend over a great part of the Earth's surface, and we should find, as is the fact, in the more ancient rocks greater convulsions undergone. I should infer from the examination of all the districts in which these phenomena may be observed—from Iowa to this section and so through the Northern parts of Europe and Asia—that during all the successive periods there had been prodigious volcanic action; and that the areas of the volcanic districts during any one period had been limited to comparatively small portions of the surface of the earth.

With this observation I will conclude my present lecture; and in my next return to the consideration of the Niagara district.

LECTURE VII.

RECESSION OF THE FALLS OF NIAGARA.

You will have learned from what I said in my last lecture respecting the relative position and the age of the different rocks of which the Earth's crust is composed, that the rocks of which the Niagara district is composed, and of which I then said something, belong to a period of great antiquity in the Earth's history. Below them is only one stratum in which may be observed traces of fossiliferous remains, and we may regard this group of rocks as the most ancient respecting which we have any authentic information. Though we are in the habit, therefore, of calling this the New Word, from the great developement of ancient rocks and the remarkably perfect state and the richness of their fossils, it is to this New World that the geological antiquary is compelled to resort for a knowledge of the most ancient rocks. Not only the pyramids themselves, but even the lime-stone rocks out of which the pyramids are built, are things of yesterday compared with the ancient rocks found in this State of New-York. The rock of which the pyramids are made belongs to the upper part of what is called the *Secondary* series; the primary being the lower and formerly supposed to be older than any that contain fossil remains, though we now know that this is unfounded, and that they are igneous rocks that have come up at different periods. The fossiliferous rocks, however, were named Secondary as being supposed to be newer than the Primary. When some other fossiliferous strata were discovered in Germany—rocks with but faint traces of organic remains—WERNER, requiring for them some new name to distinguish them from both the others, called them *Transition* rocks, because they seemed to form a passage between the crystalline primary and the earthy, uncrystalline, secondary rocks of Germany. In the first were found no organic remains, while the latter teemed with them. But Werner went farther, and, finding that the primary and transition strata in the district he examined were highly inclined, while the newer fossiliferous rocks were horizontal, he named these *flotz* or *flat* rocks. His nomenclature would then run thus

Horizontal,
Transitive,
Secondary.

"A strange transition this," says an eminet English geologist, "from primogeniture to horizontality;" and in fact its strangeness is not its only objection; for, as I told you in my last lecture, the fossiliferous rocks are, in many districts, no more horizontal than the transition rocks; and this may be seen through this country to Iowa—through northern Russia and Sweden. The others are not vertical there, as Werner found them in parts of Germany.

The appellation of Silurian has been given to these rocks by my friend Mr. MURCHISON, because they characterize the ancient district of Siluria in Wales and on the borders of England. We have therefore in this Niagara district a country composed of the most ancient rocks yet discovered in the Earth's crust, and yet rich in organic remains. I have here a map of the district between Lake Ontario and Lake Erie. You see represented the Protean group, as it has been named by Mr. VANUXEM, and above that is a bed of soft shales some eighty feet in thickness, and upon the top of that is the Niagara lime-stone. That limestone extends for seven miles from its termination in the escarpment to a point near the rapids. The River Niagara flows out of Lake Erie in a tranquil, lake-like state, and runs gently along all the way by Grand Island, being on a level with Lower Canada on one side and New-York on the other. So nearly level is it that were the river to rise thirty feet, a considerable district of country on its border would be under water. From Lake Erie to the rapids—a distance of sixteen miles—the fall of the river, I believe, is not more than twenty feet; so that it quite resembles the expansion of an arm of Lake Erie. The hight of Lake Erie above the level of the sea is 565 feet; its hight above Lake Ontario is 334 feet. This fall is divided first into the fall of twenty feet in the sixteen miles from Lake Erie to the rapids; then comes a fall of forty feet in half a mile at the rapids; at the Falls it plunges at once 164 feet; then, between the base of the Falls and Lewiston, a distance of seven miles, it rushes rapidly along and falls 101 feet. It is evident that all attempts at representing the beauty and grandeur of this scene by any drawing must be perfectly unavailing. I have endeavored in this sketch to give you a notion merely of the geological situation of the country. It will of course seem to those acquainted with it most inadequate and in false proportion; but it must be remembered that it is a bird's eye view—such as you would actually get if you were raised above it in a balloon. The first idea of this drawing was suggested by Mr. BAKEWELL, son of the eminent English geologist of that name, who visited the falls ten or twelve years ago. He published his drawing in the Gentleman's Magazine. On my arrival in this country, I considered in what manner I could introduce into his view the geological structure of the country. and being able to make myself master of the rocks in that region, by having Mr. HALL, one of the State Geologists, for a traveling companion, sooner than I could otherwise have done, I prepared such a one, which has been copied, as you now see it, by Mr. Russel Smith of Philadelphia—who, though unacquainted with the country, entered so completely into the conception that he has given an excellent representation of its prominent geological features.

Three miles below the Fall is the Whirlpool; and just below that is a deep ravine, called the Bloody Run, from an Indian fight which occurred there: there is also another ravine some ten miles below, of which I shall have occasion to speak.—The first feature which strikes you in this region is the Escarpment, or line of inland-cliff, one of which runs to a great distance East from Queenston. On the Canadian side it has a hight of more than 300 feet. The other is found at the junction of the Black Rock of Buffalo with the shales—and gypseous marls, as they are called. I shall not stop to describe mineralogically these various groups. They are sets of soft marls which contain gypsum and so much salt that they are often called saliferous gypsums.

The first question that occurs when we consider the nature of the country is, how the cliffs were produced: why do we so suddenly stop from this range to the gypseous marks and then as suddenly to the subjacent shale and sandstone? We have similar lines of escarpment in all countries, especially where the rock is limestone; and they are considered to be ancient sea cliffs which have become more gentle in their slope as the country has emerged from the ocean. If we examine what is now going on on the sea-side, and reflect what would happen if the sea cliffs should be raised up as the country around was gradually raised (for I trust you are now familiar with this notion of upheaval) you will readily see how these escarpments are formed. There are similar appearances at Boulogne, in France, where we have alternate beds of limestone and clay, and between the base of the cliff and the present sea shore, (for the sea has evidently retired half a mile,) we see the edges of the strata *cropping* out between high and low water mark. Fifty years ago this cliff was 3,000 feet high, as we see from its present position. I have myself seen frigates floating in places where we know historically that land existed but a few years since. The waves have beaten against the base of the rock, and broken off fragments which have fallen down and been swept away by the tide in the shape of mud and pebbles, until a whole yard of the coast has given way in a single year. In the course of time a considerable portion of the rock is swept away. The shale and sandstone would thus be swept away, and the formation of the inland cliffs would be easily accounted for.

You may perhaps ask if the Ontario may not once have stood at a higher level and the cliffs have been produced by its action instead of that of the ocean. Some of you may have rode along the Ridge-road—as it is called—that remarkable bank of sand which exists parallel, or nearly so, to the present borders of Lake Ontario, at a considerable hight above it. I perfectly agree with the general opinion respecting this: that it was the ancient boundary of Lake Ontario. In some parts of it fresh water shells have been found.—You cannot explain the escarpements by the aid of the action of the lake, for it extends farther and not in the same direction. It may be traced to the Hudson River, and is not peculiar to any locality; but may be traced in all parts of the globe. When the land emerged gradually from the sea, as it is now doing, the sea would naturally create these sea cliffs, and during the upheaval they would of course become inland. In Europe proofs that limestone rocks have been washed away are abundant. In Greece in the Morea this is especially conspicuous. We have there three limestones one above the other at various distances above the sea. Along the line you may see literal caves—worn out by the action of the waves. In many of them are fragments of the limestone which have fallen down and are perforated by the *lithodomi*, and you can find among them shells of the *Strombus*, and various kinds of shells. The action of the salt spray which has also effected a sort of chemical decomposition, is also easily to be observed. So completely is this the case with each of these lines that you cannot doubt for an instant that here is a series of inland sea-cliffs; and this phenomena being so certain in the Morea leads us by analogy to infer that these escarpements of the Niagara district were produced by a similar cause.

Other proofs of this denudation exist in the surface of every country, and especially in the coal district, of England. There in the coal field of Ashby de la Zouch in Leicestershire you see coal-beds pushed up 500 feet higher on one side than they are on the other—or there has been a letting down of the other side. This is called by the miners a *fault*. You would expect to find a corresponding inequality in the surface; you would think you ought to find a line of hills 500 feet high; but there is none. The whole mass has been carried away, and the surface is as smooth and unbroken as in any other part of the country. In this way we can see that thousands of feet of earth must have been removed for an extent of twenty or thirty miles. When we examine the subterranean structure of the country we find proofs of such a denudation; and this it may be said has occurred in the Niagara district, either forcing up the cliffs or letting down the lower country between the base of the escarpement and Lake Ontario. But an examination of the country will satisfy you that the cliffs could not have been thus produced. The rocks are continued on the other side. There are first the 150 feet of shale, then 25 feet of white limestone and then the gray and mottled limestone, giving three great divisions and a great number of subdivisions, in which the beds can be traced and described by their organic remains.—If these cliffs were produced by a *fault*, as in England, the beds would not be found upon the opposite side.

But, without dwelling longer on this denudation, I must pass to the more immediate subject of the present Lecture. I have endeavored to show that these lines of escarpment were originally sea-cliffs, formed when the district was gradually emerging from the ocean. It is not disputed that there is some change going on at the Falls even now. There occurs, as we know, occasionally a falling down of fragments of rock, as may be seen in Goat Island. The shale at the bottom is destroyed in consequence of the action of the spray and frost—the limestone being thus undermined, falls down—and it has been believed that in this way there has been a recession of fifty yards in about forty years; but this is now generally admitted to have been overstated. There is at least a probable recession of about one foot in a year; though part of the Fall may go back faster than this, yet if you regard the whole river even this will probably be something of an exaggeration. Our observations upon this point are necessarily imperfect; and when we reflect that fifty years ago the country was perfectly wild and inhabited by bears, wolves, and here and there a hunter, we shall think it surprising that we have any observations at all, even for such a period back. We have an account of the Falls, given in 1675 by Father HENNEPIN, a French Missionary, who gives an exaggerated description of them, and yet one which is tolerably correct. He published with his Travels a plate representing the Fall; but it greatly exaggerated its hight compared with its width. He describes Goat Island just as it is found now. He estimates the hight of the Falls at double what it actually is, which, after all, remembering that he did not measure them, is not so gross as might appear; and any one who has witnessed them will readily excuse him for having given way to a little exaggeration in attempting to describe the grandeur and magnificence of the scene, without the slightest intention to deceive.

As you will see by this copy of his picture, he represents a cascade as falling from the Canada side across the other two. He says that between Lake Erie and Lake Ontario there is a vast and wonderful waterfall: after speaking of this he says that there is a *third cascade at the left* of the other two, falling from West to East—the others falling from South to North. He says in another place: 'I wished a hundred times some one had been with me who could describe the wonders of this frightful fall.' He several times alludes to the third cascade, which, he says, was smaller than the other two. Now those who consider that because Father HENNEPIN gave the hight of the Falls at 600 feet small value is to be attached to his testimony respecting any part of the country, do him injustice. I think it perfectly evident that there must have been such a third cascade, falling from West to East, as that to which he alludes.

A Danish naturalist, in 1750, came to this country, and visited the Falls, of which he has also given us a description, which was published in the Gentleman's Magazine in 1751. He also gives a view of the Falls. In its general features his description agrees well with that of Father Hennepin. He went seventy-three years after him, and there was *then no third cascade.* But the point where Father H. had put this cascade he has marked, and says that 'that is the place where the water was *forced out of its direct course by a prodigious rock which turned the water and obliged it to fall across the Falls.*' He goes on to say that only a few years before there had been a downfall of that rock, which was, undoubtedly, part of Table Rock—and after that the cascade ceased to flow. Now, it does not appear whether he had ever seen Hennepin's account or notes. He only mentions the fact that there had been a third cascade; and it is a striking confirmation of the general accuracy of Father Hennepin's description. We find these two observers, at an interval of seventy years apart, remarking on the very kind of change which we now remark as having taken place within the last fifty years; an undermining of the rock, and a falling down of the limestone, and a consequent obliteration of the fall. Every one who has visited the Falls, on inquiring of the guides about the changes that have taken place, may have been told that the American Fall has become more crescent-shaped than it was thirty years ago, when it was nearly straight. The centre has given way, and now there is an indentation of nearly thirty feet. The Horseshoe Fall, also, has been considerably altered. It is not of so regular a crescent shape as formerly, but has a more jagged outline—especially near Goat Island; it has less of the horseshoe-shape, from which it derives its name, than when it was given.

It is quite evident that things there are not stationary, and the great question is, whether by this action the whole fall has been produced in this manner. I have visited this year the Falls of the Genesee, both at Portage and Rochester, and obtained many facts, especially at the upper fall, of this recession on a small scale. I made like observations at Le Roy, at Jacock's Run near Genesee, and in other places where it is impossible to go far back; for there time immemorial is only about ten years. But the people in that vicinity will tell you there has been a change of a few feet or yards within this time. Mr. Hall observed a recession of several feet since he surveyed the same district a few years before. It is highly probable, therefore, that there has been an action of this sort constantly going on.

From representations made by other travelers, I was desirous of ascertaining whether fresh water remains were found on Goat Island as had been said: for it would be striking if on this Island there should be, in a stratum of twenty-five feet of sand and loam, pebbles and fresh water shells. They were found there, and I made a collection of several species of shells found on the Island; among them were the *Planorbis*, a small *Valvata* and several other kinds. They were of genera found living either in the rapids, in the river above, or in the Lake.

In digging a mill-race there only a few years since, there were found a great number of shells, ad also the tooth of a Mastodon, some twelve or thirteen feet below the surface. It was the common Ohio Mastodon, and must have been buried beneath these twelve or thirteen feet of fresh water deposits—one layer at a time, each containing different shells. In answer to my question, whether similar shells were ever found lower down? the guide said he would take me to a place half a mile below, where the strata had been laid open. We found there deposited in the rock a small quantity of fresh water shells, showing that this old deposition extended down to that distance. Here we have proof that the river once stood at a higher level and in a tranquil state; and there is every appearance of the rock having been like a solid barrier to hold the waters back in a lake-like state, so that they might throw down these fresh water deposites at that hight. You will understand this better if you consider that if the Falls go on receding, no matter at what rate—an inch, a foot, a yard in a year—in the course of time the whole must recede considerably from its present condition. What proof should we have of this afterward? You will easily see that if the river should cut its way back to a certain point, the effect would be to remove the rocky barrier, the limestone of the rapids, which had been sufficient to pond the river back. But if the river cuts its way back, this barrier could no longer exist; the channel would be deepened, and the deposites existing high and dry upon the land would become proof of the recession. This kind of proof we have that the Falls have receded three miles from the whirlpool, the limestone having been higher at the whirlpool than the river at the Falls. It may be well to say that the beds all dip to the South, at a rate of about twenty-five feet in a mile. In seven miles the dip causes a general rise of the platform to the North, so that, when at the top of the cliff, you are at a greater hight than the level of Lake Erie; and if the Falls were formerly at Lewiston, their hight was probably nearly double what it now is. Mr. Hall suggested that at that time the whole fall was not at one place, and I think it quite likely that that was the case. There is reason to believe that one fall was upon the quartzose sand below, and the other on the Protean bed. The upper part would of course recede faster than the lower, because it is softer, as is seen to be the case at Rochester; but the limestone becoming thicker and harder, would recede more slowly. There may have been several falls, as at Rochester—each one of them being less high than at present, and yet the whole being nearly double its present hight.

I told you that the river fell about 100 feet between the base of the Falls and Lewiston—so that the beds slope at that rate. This slope of the river, and then the upward slope of the platform, are the reasons why the Falls now are of less hight than formerly. So when we carry ourselves back in imagination to the time when the river had not receded so far, we have a barrier of limestone much higher. The valley in which the river then flowed must have been much narrower than its present ravine. The distance now from the Canada to the American side is about three-quarters of a mile, whereas half a mile below it is only half that width. Farther investigations, by tracing the fresh water deposites lower, will give more precise information.

You might suppose that, if we find the remains of a Mastodon in a fresh water deposite so lately laid dry as that near the village of Niagara and only twelve feet below the surface, the Mastodon had lived in the country at a modern period; you might think that perhaps a few centuries would have been sufficient for the accumulation of twelve feet of shelly sandstone and limestone, and that it may have been recently that this Mastodon was buried when the barrier was at the whirlpool before this twelve feet of fluviatile strata were deposited. Yet these strata are elder than the whirlpool.

Among the objections to the supposition that the ravine was cut out by the Niagara, one is that at the place called the Devil's Hole or the Bloody Run, the ravine must have been cut by some more powerful cause than by a slight stream. But this I regard as no objection at all; for on examining the nature of the soil, &c. I am convinced that even the small stream which now flows would have been perfectly competent to have cut out the ravine, and that we need look for no more powerful cause. Suppose the Falls once to have been near Lewiston: it would recede differently at different times—faster when the soft shales were at the base; at other times slowly, when the hard sandstone was to be cut through. First of all came the quartzose sandstone for a certain distance; then the Falls receded slowly, but more rapidly when it came to the soft shales. Then comes the sandstone again at the base which now extends to the whirlpool, and here the movement was slow. It probably stood for ages at the whirlpool. Then for another period it receded more rapidly; and it is probable that for the last mile its recession has been comparatively slow, because the Protean group and about twenty feet of sandstone, making about fifty feet of hard rock at the base, were to be cut through. It is certain that the movement now is at a faster rate, as the shale is exposed. If it recedes one foot in a year, then in about five thousand years it would recede a mile; and as the upward slope of the bed of the river is about fifteen feet in a mile, and as the bed's dip to the South is about twenty-five feet in a mile, we must have about forty feet for the loss in hight of the Falls by the receding of one mile. Another 5,000 years would cause the loss of another forty feet, and then eighty feet would have disappeared, and the cataract would fall over a solid mass of limestone only eighty feet high. Thus, at the end of 10,000 years, when the Falls shall have receded two miles, they would be eighty feet high. The recession then would be extremely slow, as the base would be of solid limestone.

But all these calculations would be easily vitiated by disturbing causes. Thus, by interfering with the body of water above the fall—by carrying them away, as is now done by the Erie Canal and the Welland Canal in Canada, we should have a different state of affairs. All the water taken from the upper lakes, as by the Illinois Canal, &c.

cheats the Niagara of its waters and acts as a disturbing force. Every mill-race built above the Falls has the same effect: and though this may seem to be a trifling matter, still, in the progress of population and civilization, such things may be frequently perpetrated, and thus, in the end, have a serious influence

It has been estimated that about 15,000,000 of cubic feet fall over the cataract in every minute: this was ascertained by an engineer under the direction of Mr. RUGGLES. By all the causes I have mentioned, perhaps one four-hundredth part of this may be diverted into other directions; and this is certainly an appreciable quantity, and might have no inconsiderable effect in the progress of the recession. I only mention this as one of the disturbing causes which may vitiate all the calculations of which I spoke. The movement of the whole country which I have before alluded to may be another cause of disturbance. It is extremely probable that, during the period when the Falls were receding from the whirlpool, there may have been an upward, or downward, or perhaps an oscillating movement of the whole country. This would leave whole cliffs exposed, as has been done in other localities in the St. Lawrence, of which you have here a representation—where columns of limestone are standing, perforated at different hights by the lithodomi. The same motion may have extended to the great lakes, and have affected this whole section of country. You will see, therefore, what a variety of different causes we have to regard in making any estimate of the former state of this District.

LECTURE VIII.

BOULDERS AND ICEBERGS.

THE subject of which I shall treat to-night is what we term the *Boulder Formation*, or sometimes the Norfolk Drift. The term boulder is applied to any large mass found resting upon the superficial gravel brought from a distance. By some, chiefly the writers of the last forty years, this formation is called diluvial; because they believe that this superficial gravel, and sand, and mud, in which are found these rounded fragments of rock, have been brought thither by a rush of some mighty deluge, either at one time or at different periods. But those who have thought that they saw reason to refer a large part of this to other causes prefer the term *Drift*—as not choosing to commit themselves to any particular theory except that which is certainly known to be true, that these boulders have been *drifted*, by some means or other, to a considerable distance from the parent rocks, from which, as fragments, they have been torn. You have so many examples in this country, of these foreign rocks scattered through beds of sand and mud, that it is not necessary for me to enter upon the description of any particular localities in Europe. If you pass by the great excavations that have been made for streets in Brooklyn, you may see some forty or fifty feet in thickness of what we call rubbish, an unstratified, confused mass of clay and sand, containing fragments of rock of various kinds. This may be seen near the Navy Yard, and in all parts of the suburbs of the city. The same kind of formation may be seen in various parts of the North of Europe, as well as in different districts of this Continent. In Europe, it is particularly noticeable in the country bordering the Baltic, beginning with Finland, and through part of Russia and Poland, to Pomerania, Prussia and Denmark, through the lower part of Sweden. The whole country consists of land at a moderate elevation, covered, to depths that have never been pierced, with this boulder formation, sometimes a thousand feet thick, and often, indeed, still more. It contains no strata; and you would become sensible, after having made a geological survey, how very rare are unstratified rocks which are not crystalline, like granite. Sometimes we find the boulder formation, entirely unstratified, passing into another strata which is arranged in layers. The absence of fossils—of organic remains—is another characteristic which makes it difficult to decide the nature or origin of this formation;—whether it be fresh water, formed in lakes, or a marine formation, formed in the sea—is a matter of great doubt from the absence of all organic remains.—Sometimes, indeed, we have found shells and bones washed out of the older rocks; for, relatively speaking, this is a moderate deposite—being strewed over the strata containing fragments of them all, and occasionally of their imbedded fossils. Sometimes, too, we find them alternately of stratified and unstratified rocks.

In tracing along this remarkable deposite through the borders of the Baltic, we sometimes find fragments of rock which must have traveled hundreds of miles from their point of departure, and, as a general thing, we shall find that they grow larger in size as we approach the region from which they were derived. This I found to be the fact in going north from the margin of the Rhine to Holstein, and Denmark, where I found fragments of Scandinavian rocks, in Sweden nine and sometimes forty feet in diameter; and at last the whole country was made up of these rocks.—Thus by tracing the stream along, we shall find that as it diminishes in size the stones continually diminish in their individual dimensions. This may be seen any where between the Thames and the Tine, and by following it out in any region you will become convinced that there has been a general drift from the north. You may travel for eight or nine hundred miles over the plains of Russia, and you will find these *Erratics*, as they are called, associated to such an extent with the rocks in the neighborhood, or immediately subjacent, that they have acquired the color and mineralogical character of the rocks of each country. If, for example, you trace the boulders to the red sandstone of New Jersey, you will find them red. So at Brooklyn, you will find that in great part they are red. Yet here in the red base you find scattered fragments of the trap of the Palisades huge masses of granite from the Highlands, and some of the green serpentine of Hoboken, all mixed together, and yet the whole reddened by the colors and marked by the character, of the adjoining sandstone. So in Europe, the boulders are white in the chalk of Scandinavia; black in the carboniferous formation near Edinburgh, where the bituminous shale of the coal formation enters largely into their composition. Sometimes you find them entirely angular, as if they had not suffered any of that mechanical rubbing against adjoining rocks, or against each other, which has perfectly rounded other masses equally large.—Some of these are so large that it is difficult to imagine that any force of water, which is the agent usually assigned, could be sufficient to roll them over and over as is necessary. Take some of the gneiss for instance: the edges of some huge masses have been cut off, so that the whole is as perfectly rounded as any of the smaller pebbles. Oftentimes, too, these huge boulders have been carried across seas—as from the Scandinavian rocks to the south side of the Baltic. We have them too, perched upon the Jura peaks, having evidently been carried from the higher Alps over a valley fifty miles wide, and deposted upon fossiliferous limestone rocks, which have nothing in common with those of the Alps. This is the kind of appearance which has so perplexed Geologists, and to explain which, shall be one purpose of this evening's lecture.

I will first, however, allude to one other appearance which distinguishes the boulder formation.—When it rests upon hard rocks—rocks which are capable of taking and retaining a polish—rocks which have not wasted away by disintegration—we find, upon removing the sand, &c. the solid rock below sometimes polished so as almost to answer the purpose of a looking-glass; at other times we find it scattered and ridged with long parallel stripes, perfectly straight for hundreds of yards, and sometimes for a quarter of a mile, occasionally deviating from being parallel to each other, but still retaining the same general direction from north to south, or sometimes 20 or 25 degrees toward the east or west of that direction. This is evidently a common characteristic of these erratics; and any theory advanced to explain them must comprehend that appearance.

Another grand fact which is now established respecting their geographical distribution is, that they are found in the Northern Hemisphere, both in Europe and this country, extending from the poles, and they diminish in quantity as they approach the warmer and equaterial regions, and at length disappear when we come near the tropics. We find them in Europe from the north of Sweden and Norway to the coast of England, in latitude 50°. We find them here still farther south, but the Long Island deposite is one of the last grand deposites, and that on the Susquehanna is, I believe, the most southerly; and in traveling toward Georgia, or even on the James River and the Potomac, you will be struck with the absence of these large erratic fragments. It is the same in Russia, in traveling from north to south; and it is only when there is a chain of mountains like the Alps, in latitude 46°, that any exception is observed. From these mountains, as from the Jura chain, these boulders seem to radiate as from a centre. You may see them traveled to Lombardy and toward the Italian side. Even in the Grampian mountains of Scotland, you may see them scattered over the hills on every side; so also with the small Cumberland chain. Thus, mountain chains seem to have exerted the same kind of influence as the poles; for this general theory is found to be true, not only of the north pole, but also of the south. When you pass from the southern part of the United States through Mexica to Peru, at Quito you find no boulders, except at the foot of some mountain chain where we may easily suppose the melted snow and other causes sufficiently obvious, account for their presence. Passing to Chili, it is not till you reach latitude 41°, that you begin again to meet these boulders, and then they continue to increase to Terra del Fuego, where they are as magnificent in their developement as in New England or in Sweden.

Another very remarkable appearance in regard to the stratification of this formation, is the contortion and disturbance of some of the beds. In parts of the strata in Scotland, for example, you find masses of the unstratified boulder with pebbles below of various kinds, as fragments of granite, gneiss, &c., in which parts shall be twisted so that a vertical section would pass through the same bed three times. You find alternate layers, first pebbles of a particular kind and color; then sand, then loam, and then gravel—all loose, but so that you may trace the same bed for several yards, one layer being deposited above another in a nearly horizontal position; and we find them sometimes folded together—bent back upon themselves. This appearance was of a most perplexing kind, and evidently implied a lateral thrust by which the pliant beds were brought into the folded position, though those below had suffered no disturbance.

In some cases we have a mass of chalk resting on another bed, in which one has been pushed out of its original position, and the gravel and sand folded around it. In other cases, as in part of the northern coast of England, for twenty miles this unstratified *till*, as it is locally called in Scotland, is covered with a layer of horizontal loam in which are curved strata. This folding and bending of the beds in a circle sometimes, has been effected, and the horizontal layers below are not all disturbed; so that it cannot be a motion from below which has caused it—a subterranean upheaval or subsidence, which I have before explained, caused by volcanic action, by which we explain mountain chains, &c. This cannot be introduced here, because in that case the lower beds would have been disturbed as much as the higher. This may be seen in the section laid open in Brooklyn; none of the contortions there have been so violent as some on the Norfolk coast of England and Scotland; but the same general disturbances may be observed.

As to the age of these boulders, you find them both here and in Europe, standing over rocks all of the most modern tertiary strata. I had an opportunity in Sweden of showing how modern some of the earatics are, by finding fragments of gneiss sixteen feet in diameter, resting upon a layer of sand; then came a bed of blue marl, containing an immense number of shells of the eatable muscle, from which the blue color of the marl is derived. That muscle is now a living species in the Baltic, and is found at Upsala, near the ancient University. Several other shells are also found peculiar to the Baltic. The water of the Baltic contains only one-fourth part as much salt as the water of the ocean; and the shells found in its brackish waters, though not of different species from those of the ocean, are yet of a dwarfish form, and of a different shape from those that live in the sea. There are also found fresh-water shells which have been brought down by rivers. We may observe, then, how very modern is the transportation of some of these blocks; for not only do we trace them to the times when those species existed the same as those which now live, but when they lived placed under those geographical relations which have modified the character of the waters.

I do not say that they are all as modern as that, for some of them contain shells that are partly extinct, or at any rate which do not live in the same region. In the St. Lawrence Capt. BAYFIELD came across boulders containing a small assemblage of shells which he sent to me many years ago; and Dr. BECK, a Danish Geologist, observed that there were fauna more ancient than those from the Baltic, which I received at the same time. There was a small number of species peculiar to the Southern regions, and the same as are chiefly found in Greenland and other districts near the Pole, as if the country had formerly been colder and the boulders had been dropped down by icebergs; for Capt. BAYFIELD frequently saw immense rocks carried by icebergs and let down, deposited with strata containing shells the same as those I spoke of in treating of the Niagara district. Thus we arrive at the conclusion that this boulder formation is one of the most modern deposites, geologically considered—sometimes extremely modern, and in other points ascending a little higher, just to the period when the same shells existed, nearly all of them belonging to living species: the newer Pliocene period as I have before designated the era—when perhaps 90 out of the hundred shells found are of living species. Down to the latter part of this newer Pliocene period it appears that these erratics extend. We have to account then for this formation being of such vast thickness and unstratified; for rocks having been transferred hundreds of miles over lakes and valleys; for their being nearly non-fossiliferous; for their being found chiefly in the Arctic and Antarctic regions or near mountain chains; and to explain how they are so often found contorted and disturbed while the strata they overlie are still horizontal.

In the hope to explain the greater part of these phenomena, I propose first to treat of *Glaciers* and then of *Icebergs*. You are aware that in lofty mountains, especially in the high latitudes, the snow never disappears during the whole year.—There is constantly falling snow which the summer heat is never sufficient to melt; and in Switzerland where the Alps are three miles in hight above the level of the sea, although in latitude 46°, their peaks are covered with perpetual snow; which comes down and fills the valley for ten or fifteen miles: then this ice becomes consolidated, being melted during the day and frozen in the night—so that it is pushed down toward the valley to a point 3,000 feet above the level, where the heat becomes so great as to arrest its progress, and it melts, and gives rise oftentimes to a considerable stream of water. The cause of the motion of these glaciers has been a matter of considerable controvesy. Gravity is admitted on all hands to be one considerable cause. It is suggested by SAUSSURE, and by the earlier writers is believed to be the principal cause; but this is denied by M. AGASSIZ in the history of his late exploration of the Alps. That this snow, if it goes on accumulating upon the mountains which are so steep, will by its own gravity fall down, is unquestionable. You see this to be the case in *avalanches*, as they are called—sliding down of large masses of snow in warm weather, which continues until the valley is choked by this descending mass, which crushes the trees and vegetation that lie its path. These become of such enormous size that they are sometimes 100 to 200 feet thick, and in particular places 500 or 600, though it is supposed now that 120 or 180 feet expresses their average depth. In Switzerland, where the glaciers of the Arve, the Lauter Arve, and the Shrecken meet, the former being merged in the other, along the middle is a remarkable ridge of rocks—many of them angular and some rounded. Now the first question is, how came these blocks in the middle of the valley where it is two or three miles wide. The glaciers descend from the region of perpetual snow to a hight of, say 8,000 feet above the level of the sea. How came all these rocks in the middle? You might imagine that from the steep sides there would be fragments detached. Avalanches might cause this, or frost penetrating the rents and freezing the water, would occasionally force them out and cause the rocks to descend. Sometimes lightning strikes the Alpine peaks, and shivers off large masses of rocks which descend. So that we should not be puzzled to find along the base of lofty cliffs, two, three or four thousand feet high, these fragments of rock. This would be perfectly intelligible. But how should they get into the middle of the valley, and why are there five distinct parallel ridges of these stones? SAUSSURE was at first completely baffled in accounting for this. But having once found the explanation, it was so easy that it became surprising how it could have been missed.

Prof. AGASSIZ found in exploring the higher regions that this was a *necessary consequence* of the junction of the two glaciers. It is easy to see why these lateral *moraines*, as they are called, should exist—the rocky fragments being deposited along these glaciers by their rubbing against the sides of the mountains. But suppose one of these immense masses of ice to be descending the valley of the Arve; and here comes a tributary to join it from the Lauter Arve—the rocks instead of being deposited in lateral moraines by rubbing against the mountain sides will be brought into the middle of these two frozen rivers, thus forming a central or medial moraine. Now as the glacier moves along, (and in a hot day you may see the motion daily—although an inch, or perhaps half an inch an hour would be a rapid movement,) —you may see sometimes fragments falling down, rubbing one against the other, and great rents often traverse the ice with a noise like thunder.—By this rubbing against the sides of the mountain, the rocks become rounded. Many of the fragments fall through the fissures to the bottom, and some are caught in the middle—the fissure penetrating only twenty or thirty feet. Sometimes, however, they fall to the bottom, and then the ice resting upon them grinds them along the rock, which becomes polished—those at least capable of receiving a polish—and scratched and furrowed as we afterward find it. All this may be seen by the occasional melting back of the glacier. So at the termination of the glacier it presents a beautiful green arched cavern, out of which a torrent of water rushes down the valley. Frequently the glacier melts back from the extremity, and thus gives an opportunity to see what has taken place under it, and you will find the bottom oftentimes most beautifully polished. In some of the boulders you will find quartz pebbles, and these have scratched and made furrows upon the limestone and other rock along which they have moved—just as a diamond scratches glass. In other places you will find still deeper furrows nearly parallel to each other. You will also see—as I have had occasion to refer to the prodigious power of these ice masses—rocks that have been ground down to the finest impalpable powder; and nothing can exceed the fineness of this mud which is formed from the powder thus produced by these masses of ice one or two hundred feet thick, equal in weight to five and twenty or even fifty feet of solid rock.

The downward motion of the glaciers is partly due to gravity. But Prof. Agassiz says that still more is due to the alternate melting and freezing of the water. The ice is in fact a great sponge, and not only may you see water in the day time held up in the clefts—as many of you who have traveled in Switzerland can testify—but the whole surface is a spongy mass which imbibes the water during the day, which every night is frozen by the same frosts, and thus occurs a universal dilatation of the whole mass; the water in all the rents freezing causes an expansion, and as this cannot push aside the mountains on the flanks, the only vent for the force is downward—in which direction it has the effect to force the huge mass down at the rate of one, two, three or four inches an hour, according to the heat of the summer and the amount of alternate melting and freezing that goes on, and also according to the farther distance which the glaciers have reached. I may mention that every one of the moraines between the central one and the sides is produced just like the large one—by the junction of the tributaries which come one after another down for many miles.—Thus the different moraines may be traced—one to the Schrecken and others to various tributaries which join higher up. Those at the sides move faster than the central ones, because the reflection of the heat from the boundary rocks is in addition to the direct heat of the Sun, causing the ice to melt away faster: and thus the ice is drawn from the middle to the sides, and the moraines become more and more scattered. At last we have only one great lateral moraine with smaller medial ones, and when we behold the beautiful arch at the termination we shall find in the middle of it no fragment at all. There was some difficulty in accounting for this, because it was supposed that the fragments had been caught in the fissures.—But why should the extremity be so beautifully free from them? The answer is this: that when a block falls into a fissure it works its way up—not by rising against gravity—but in this way: as the glacier goes down, it continually diminishes at the surface—the upper surface melting away; and the block which had dropped down to a certain distance must continually get nearer the surface.—There too this block protects the ice from the rays of the Sun, and you see the mass below unmelted. If the pebble be small it soon becomes heated through, and thus forms a pool or hollow. Thus if the rock be small, we shall have a hollow; if large, the opposite—or the rock will be mounted up on a pedestal. The wind also is one cause of evaporation. The ice wastes away like camphor, without passing through the liquid form. The general waste of superficial ice tends to bring up the fallen mass toward the surface.

I have said that there was great difficulty in seeing how such large fragments could be so perfectly rounded. Sometimes we have masses perfectly angular—twenty feet in one direction, and twelve or fifteen in the other. There is one, well known to travelers, in the central moraine of the glacier of the Arve. It was here that a hut was built in which a family lived in the summer. The rocks are rounded here, as is said by some, not by the action of the ice, but of the river that flows beneath the ice. But Professor Agassiz says that, as a proof that it is the ice, and not the river, to which the rounding of the rocks is attributable, you may go up to points above where any stream commences, and still you will find that the motion of the ice above has rounded masses of all sizes. This is found so far above the line of perpetual snow that it cannot be attributed to the action of running water.

I mentioned a hut, built by Professor Huge, some time ago. There are some remarkable observations concerning it, which show the rate of motion in these glaciers. It was built in 1827, on the glacier of the Arve. When Professor H. went back nine years after in 1836, he found that the hut had gone down without being otherwise disturbed or injured, 2,200 feet. He went again four years after, and found that it had gone down at twice the same rate. Taking an average of the whole for the thirteen years, it was found that it had gone down at the rate of eight inches in twenty-four hours. In the first part of the distance it went eight, and in the second sixteen; and the motion was entirely in the summer. In the winter, this glacier was stationary, showing again, as Professor AGASSIZ remarked, that the motion was chiefly owing to dilatation—to this alternate melting and freezing. This is certainly a strong argument in favor of that theory, that it is chiefly during this congelation and melting that the chief motion is observed.

There have been periods when the glaciers made less advance than at others, as between the eleventh and fifteen centuries; and again in the seventeenth and eighteenth a general motion forward occurred. This period of retrocession and advance is a striking meteorological phenomena showing the cycles of climate. When there is a great fall of snow during the winter, which melts in summer, there is an advance.

In Chili, which has the same latitude as the Alps in Switzerland, we have glaciers descending to the sea; but at the Alps they only descend within 3,000 feet of the sea level, and this too, although the Andes are only 7,000 feet high—half the hight of the Alps in the same latitude. The reason of this singular phenomenon is that to which I have alluded—that the summer heat is less intense in the Alps. In Europe we have to go to latitude 67° before we find a single glacier reaching the sea. But in the Southern Hemisphere, in latitude 46°—in Chili—we find this occurring twenty-one degrees nearer the equator; so that there is here an actual generation of icebergs in a region which is almost the limit to which the floating icebergs reach. That alternate period of advance and retrocession among the Swiss glaciers is one of the most remarkable facts, and explains many geological phenomena. In observing the terminal moraines of the Alpine glaciers, we often find a huge mass left at the end of one summer. Then when the glacier advances again, it pushes forward the moraine of the last year into that of the year before, and that into the third, and so on until we have four or five together, forming a huge mountain. You may see many of these ancient moraines covered with houses and lofty trees, and various kinds of herbage; and as I witnessed in 1838, when the glacier was advancing, if it approach this ancient moraine it destroys the forests upon it, forcing in the walls of the houses and crushing them by a slow and almost imperceptible, but at the same time irresistible power; and after treading down these lofty trees for a series of years, it will again retreat; and then the wood will grow again, the inhabitants again build their houses, and forget the disaster which once rendered them desolate. So that the trees show by their age how long it has been since the ice visited that part of the country.

These phenomena have been well described by M. CHARPENTIER, who remarked that we always have an unstratified mass of large boulders in the same district, the angular and rounded being mingled together. This shows that they cannot be attributed to any action of water, for water exerts an assorting power, carrying the finer materials farther than the coarser, and would carry the small stones to farther points than the large ones. Each different size would, supposing the whole to be attributed to the action of water, be arranged in different layers. But ice would carry them all indifferently to the same place, and we should find them unstratified—a promiscuous, confused mass, and that is the character of all the moraines. Now we are not to jump to the conclusion that all the boulders of Long Island are attributable to glaciers. I believe that they are not, but still to the action of ice. The pebbes found at North-Haven and along the Connecticut Valley, in the boulder formation are rounded on three sides but flat on the other—resting on polished rock; and all the furrows are parallel over a large extent of country. This parallelism does not bespeak the action of water; for in that case there would be none of the scooping out that we see in the action of glaciers; the motion would not be always in the same direction; but if the fragment of a rock becomes frozen in, it is kept in one position, and we should have, therefore, straight parallel grooves.

But not to dwell longer on the action of mere glaciers, let us pass to the consideration of icebergs. We know that icebergs carry fragments of rock in the same way as glaciers; that is, fragments of rock rest on glaciers when they come to the sea, and are then conveyed away by the floating iceberg, as well as by the moving glacier on land. This has been observed even in latitude 46° in Chili. SCORESBY tells us that he met in lat. 69° an iceberg in the Atlantic with 100,000 tons of rock upon it. But in 1839 there was met in the South Atlantic an iceberg 1,300 miles from any known land, from which projected a block twelve feet thick: how much rock was buried beneath the surface was not known. I do not say that this was 1,300 miles from any land—but only from any *known* land. Now as this was floating at a considerable rate from South to North—as it melted, the rock would fall to the bottom of the sea, and if the bed should be raised some day, we should have boulders at an immense distance from their starting point. The shores of the Antarctic regions are thus covered with coast ice a mile or two in thickness—stranded ice containing great quantities of rock. Thus as the glaciers descend to the sea, they float off, the rocks fall to the bottom, and the floor of the ocean is thus strewed over with them, and if the ice melt in still water the formation would be unstratified, because all the rocks fall perpendicularly through the water. But if there were a current, then an assorting power would be exercised, and we should have regular strata.

I now come to a remarkable feature to which I must allude, as I spoke of it at the beginning of my Lecture, but of which I can only speak briefly. I mean the appeaarnce of Alpine rocks on the peaks of the Jura. The Western valley is composed of talcose granite and gneiss. Then going farther East we come to the Bernese Alps, where the mountains are composed of crystaline limestone, gneiss and other rocks, frequently of highly crystaline marble, fragments of gneiss, &c. Next are the Alaps of the smaller cantons—Glaris, Schwytz, Uri, Zug, &c. Now the phenomenon alluded to is this:—We have the great valley of Switzerland between the Jura and the Alps. The Alps are from 10,000 to 15,000 feet high, and the Jura only one-third as high. Now we have the same blocks in the plain below, in Lake Geneva and others, perched upon the Jura Mountains, at all hights and of all sizes—one in particular, celebrated under the name of Pierre à Bot, near Neufchatel, no less than forty feet in diameter, composed of gneiss from the Alps. The whole chain of the Jura is composed of fossiliferous limestone, entirely different from the Alpine rocks; so that we have no other resort than to suppose that these rocks must have originated in the Alps. All agree that those on the central part of the Jura came from the Bernese Overland; then, again, the block of slate of the Glaris region sent over its erratics to the Eastern Jura, near Basle, and those on the Western Jura have come from the region of Mont Blanc and the Valais.

SAUSSURE, VAN BUCH and other writers supposed that these phenomena were produced by some grand rush of waters in consequence of the rise of the Alps from the Ocean; a most violent hypothesis, contrary to facts now known.' If we suppose that this was the cause of these rocks being carried across this great valley, how is it that they did not all fall down into the valley? How is it possible that they should be whirled sometimes for several thousand feet? This has never been explained to the satisfaction of any one; though to be sure this is rather more satisfactory than no hypothesis. But the remarkable fact is that we have on parts of the Jura fragments of rock such as we must suppose glaciers must have carried straight across the valley fifty miles wide; so that the Western stream seems to have followed the course of the Rhone, the Central that of the Arve, and the Eastern that of the Reuss and the Limmat. But to suppose all this is in defiance of all analogy of the motion of glaciers; for we must remember that they could have an inclination from one bank to the other of only two degrees; so that we should have them moving along a dead flat, which is contrary to all the laws of glaciers, that they should walk across this level and lodge these blocks upon the Jura peaks. I should rather believe this, however, than that they are owing to a diluvial rush, which is utterly inconsistent with all the laws of the transportation of matter.

There is another hypothesis suggested by the occurrence of glaciers in Chili in the same latitude as the Swiss Alps. We may suppose, from geological changes known to be going on in the upheaval of the land, &c., that the Alps were less lofty than they now are, during the boulder formation, and the Jura also. The sea may then have covered the bottom of that deep valley, and the Alps—at one half their present hight—would have stood as high as the Chilian Andes. Suppose they sent down glaciers to the sea, as we know they do in South America, where Sir GEORGE AYRES has seen them with blocks of syenite and granite.—These might be stranded in the great valley, which, in that case, would represent the channel that now separates the Island of Chiloe from the main land. The Island of Chiloe is about 100 miles long, and may well represent the Jura Chain. The Andes represent the Alps, and the channel the intervening valley. The Alps, at half their present hight, would be sufficient to give rise to glaciers, which, descending to the sea, would become icebergs and float blocks across the channel. In point of fact there are found on the Island of Chiloe blocks of syenite in one part and of granite in another, which might have been carried from different points of the sides opposite. Now if an upheaval were to take place on the Chili Coast, which should lift up the Andes—dry up the intervening channel—and lift up Chiloe so that it would appear to be a chain of mountains, we should have the same puzzling appearance to future geologists. It would be wondered how these blocks of granite and syenite could have crossed the valley from the Andes and perched themselves on the Chilian Mountains. This hypothesis is infinitely more satisfactory than the one which attributes the phenomena to the extension of glaciers from the Alps to the Jura, or to the diluvial rush of SAUSSURE. There have been found also moraines on the Jura which have polished and scratched the surface—implying that formerly there were glaciers here also. This may have happened when the climate more resembled that of Chili.

I will now state one more fact of some interest here. I observe that Mr. MATHER, the State Geologist, to whom was committed the investigation of this part of the country, says that on the Eastern extremity of Long Island are blocks that have come from the neighborhood of the Palisades or from different parts of the Highlands. Traveling farther on, he saw another group which came from the Connecticut region opposite; then to trap, porphry and other rocks from New-Haven. Stil farther were Rhode Island rocks, and so in different parts were groups corresponding with strata opposite. In these cases we have not to call in the action of glaciers nor of icebergs to explain this. It may have been caused by the action of coast ice. Ice might float thither, carrying these blocks, and, being stranded, lodge its load; then the next year might come a little more. Captain BAYFIELD, after fixing the position of one block, found it carried away several yards. The experience of DEASE and SIMPSON shows the immense power of these icebergs. The Utica, too, which has just arrived, talks of icebergs in this latitude 400 feet high above the water, and as there are eight cubic feet below for one above the surface, we may judge of the enormous size of these icebergs, which were several miles in circumference. They grate along the bottom of the ocean, ploughing up mud and sand with a force sufficiently easy to move a building like this, or even the whole city New-York before them! This is the kind of action that produced those contortions of which I have spoken. The top of a sand-bank in the bottom of the sea has been exposed to a violent thrust of this kind, and while the strata below remained horizontal, those above would be forced back and folded over upon themselves.

I will now leave this subject half-told, as I have been obliged to leave so many others of equal importance; and still more at which I have not been able even to glance. In concluding, I may take this opportunity—as this is the last Lecture I shall have the pleasure of delivering in the United States before my return to Europe, being about to resume my geological tour and to visit various parts of the West and of Canada—not only of thanking you for the great attention you have been so kind as to show me, but also to acknowledge the welcome which has attended me throughout my tour from Lake Erie to the Savannah River; and which has made me in this country feel as much at home as though I were in England; and I assure you that I shall always look back to the time spent here with a *home-feeling*, which will always make it difficult for me to regard America as a foreign land.

SKETCH OF A LECTURE

DELIVERED BY J. AUGUSTINE SMITH, M. D. BEFORE THE LYCEUM OF NATURAL HISTORY, IN THE CITY OF NEW-YORK, DECEMBER 9, 1842.

DIFFERENT RACES OF MEN.

Dr. Smith December 9 closed the series of Lectures on Geology which he has been delivering at the Lyceum of Natural History. His main subject was the differences perceived, in many particulars, between Men inhabiting different portions of the Earth's surface. The first general difference remarked by scientific men between different species of animals was the fact that in the various grades the head was placed upon the neck in a different position. In man, for instance, the head is directly above the neck, and so nicely balanced that it vibrates with equal ease in any direction.—As we descend to the lower animals we find the neck more and more behind and upon a level with the head, until we find in birds that it is directly behind it.

Soon after this was made known, it was remarked by a painter that, in painting portraits of negroes, the only difference between their heads and those of white men, as represented at that day, was in the color. But he perceived that this did not in fact represent negroes; and his observation led him to the construction of the *facial angle*, which, although it measures directly only the projection of the forehead, is always found to indicate the capacity of the skull. This angle is defined by the intersection of two lines: one drawn from the lower part of the ear to the nostril, supposed to be horizontal; and the other from the nostril to the forehead. Now in the different animals, the angle included between the two lines thus drawn is found to vary, in adults of our race, averaging 85°, though in children of three or four years old it often reaches 90°, and becoming more and more acute as we descend to the lower animals. So generally has it been understood that this angle measures the capacity of the skull, and therefore the powers of the intellect, that the ancient Greeks, in their representations of gods and heroes, made it 100°; and thus was the portrait of Alexander painted when he was deified.

It was for a long time supposed that man had a larger brain than any other animal: but it was soon found that he was excelled in this by the elephant and the whale. Then it was believed that its size in proportion to that of his body was the greatest, but it was soon ascertained that, while the brain of man is in bulk to his body as 1 to 30, that of the Canary bird is to its body as 15 to 30; and yet the bird is not remarkable for its sagacity. The proportion of an alligator's brain to his body is that of 1 to 1000, and yet Audubon tells us that this creature has an ordinary amount of intellect. The true anatomical difference between man and other animals is, that *his nervous system is more entirely collected in a mass than theirs.*

The face in man is smaller in proportion to the size of his head, than that of any other animal. It bears the proportion of about one to four, while if we descend the proportion becomes much greater, and in the horse it is enormous.

Anatomists divide mankind into different classes. Some French authors distribute them into 12 or 15 different races, while others make only *six*. Dr. Smith said he had contented himself with *four*, while Cuvier had but *two*, namely, those who are white and those who are not white. Dr. S. said he followed essentially the differences of the globe; and maintained that there were *four* races of men namely, the Caucasian, the American, the Asiatic and the African.

The Caucasian is the race to which we belong; and it might justly be said to stand at the head of all the races of the earth, possessing the highest intellectual power, the elements of all improvement, and by far the most mental and moral strength of them all. The hair of the race is long and of different colors in different individuals; the eyes are also of different hues; the skin is thin and the complexion more beautiful than that of any other; and we have physically fewer points or projections about us—the cheek bones not being prominent, and the whole frame being rounded off in the finest style. The Caucasian race has included all the most distinguished nations of the earth, the Phœnicians, the Jews, the Arabs, Greeks and Romans; and even on opening the mummies of Egypt, the Caucasian, with the African and Asiatic, was found among them.

Of the other races fewer words are necessary. The Asiatics are by no means so well made: the face being broader, the cheek bones higher, their bodies more feeble, and their intellect but of moderate power.

The Indians of our Continent are very strongly

marked: their hair is long and harsh, their physical powers are upon a small scale, and they are, in intellect, by no means the equals of the Caucasian race.

Last of the four came the negroes—the Ethiopian race. Their color and the character of their hair are well known, as well as the thickness of their lips. They inhabit the torrid parts of Africa, and their mental powers are upon an inferior scale.

Coming to a more particular description of these races, the *facial angle* in the

Caucasian race is..................85 degrees.
Asiatic ....78 "
American Indian ....73 "
Ethiopian ....70 "
Ourang Outang ....67 "

As we descend to the inferior animals, as already said, this angle becomes more and more acute; and in the duck, &c., is of only a few degrees. This rule, though it will not hold good with regard to individuals, is always true in respect to different races. The intellect of a race is always in proportion to the obtuseness of that angle. The bulk of the brain varies as this angle; and if we take a thousand individuals who have large heads, and another thousand whose heads are small—all belonging to the same race and receiving the same education, the large heads will have more intelligence, as a general thing, than the small ones, though this will not hold true of the heads of the different individuals of either class. The dimensions of the Caucasian head are 87 cubic inches;

Of the Asiatic ....82 " "
The Mongolian and Malay....81 " "
The American Indian........80 " "
And the Ethiopian..........78 " "

The senses vary in the opposite direction. They are more developed in the African than in the Caucasian race. The negroes' taste is better and far more uniform—all negroes disliking *mutton* for instance; their smell is better, and so of their other senses.

As a general thing the amount of intellect is indicated by the bulk of brain in the front part of the head: but this must not be regarded as affording ground for the theory of Phrenology,—since the posterior part of the African's head is smaller than ours, and yet his passions and senses are stronger. As to the moral qualities being always located between the animal and intellectual, we have no evidence that this is the fact. There are some fifteen or sixteen circumstances in which the anatomical structure of the Ethiopian differs from that of the Caucasian race; and in every instance they approximate to the inferior animals. There is not the slightest doubt that the anatomical structure of the negro is less favorable than ours. Dr. Smith said that it was his decided impression that in intellect they are much inferior to us, though in many moral qualities they are our superiors. They are grateful and affectionate, and have, what the Caucasian race, especially of this country, have not, a strong, unconquerable fondness for the place of their nativity. Hume, Dr. S. thought, was undoubtedly right in saying that the African race was far inferior, by nature to the European, for the only discovery for which we are indebted to them is that of the *drum*. Moreover wherever we find the two races living together the *Africans always occupy the inferior station*. Thus it is in St. Domingo where the mulattoes, though fewer in number, keep in subjection those who are merely negroes.

Still this can never justify any people in keeping them in Slavery: for Slavery, Dr. SMITH said, so far as he could see, had never any right but the right of force. There is no more reason, in the fact that they are black, why we should keep them enslaved than there would be for them to enslave us because we are white. Of the mischiefs of Slavery Dr. S. said he was also thoroughly convinced; though he thought the Abolitionists were not quite aware of what they were doing: for he held it to be undeniable that if the negroes should be freed they were as sure to be exterminated as the sun to rise; and this upon the principle that as Society advances there must inevitably, in the course of time, be a struggle for the means of subsistence: and whenever and wherever that struggle comes, the weaker party—as the negroes certainly are—must be driven to the wall. Among the marshes of the South, the climate of which they endure better than we do, they might prevail; but here at the healthy North they must inevitably perish.

We seem to owe our origin to the mixture of three branches of the Caucasian race—namely, the Celts, who inhabited the British Isles and Armorica and Bretaigne; the Northmen, who invaded Europe in the tenth century; and the Germans, who are subdivided again into the Saxons and the Goths. Besides these, two other classes—namely, the Teutonic, who inhabited what now is Prussia, and the Sclavonic, who dwelt in Poland, Bohemia and Russia, contributed to our existence. The Caucasians generally, and the Celts in particular, are indolent, captious and pugnacious to a high degree. If they can act the part of robbers, they will be sure to do it; and this wherever they obtain a foothold. The Germans were more pertinacious and more industrious, but still possessed of the same tendency to plunder. This is the reason why the Indians of our Continent have been continually pushed backward, and why they must inevitably, in the end, be exterminated. From the moment our people got a footing on this Continent, the destiny of the Indians was sealed. They were the most worthless of all the inhabitants of the globe; and it is utterly impossible to civilize them. Dr. S. said he spoke not as a moralist, nor said aught of the right or wrong done. He merely mentioned a necessary fact.

The truth is, said he, mankind are every where the same. Give them power and they will always seek conquest. Thus has it been even with the Colony of negroes planted at Liberia. We receive by despatches from the Governor news that they are gradually pushing their frontier farther towards the interior; and this is precisely what might have been expected. Dr. Smith said he had always believed that the English would plant themselves in China; though he confessed he had been disappointed in having lived to see it. And when all Australia and New Zealand are peopled with such a race, they will attack the Japanese, force them to trade with them and eventually get possession of their country. In truth nothing but force can keep our race in order: and other nations must compete with us in the arts of war or they must yield to us. This continual extension of the power of the Circassian race, if it has its evils, has also its advantages; and the world is improving under it. There never was a time when there was so much intelligence, so much virtue, so much happiness as now. A gradual course of improvement is in progress. War, for instance, is less frequent than of old—not because there are not men enough ready and willing and able to fight; but because it is becoming too expensive; nations cannot afford themselves the amusement. And so will it continue to be. The English language, English liberty and the English religion are destined to overspread this whole Continent, from its Northern to its Southern extremity.

These remarks, Dr. SMITH said, would doubtless suggest the question, whether these differenent races were different in their origin, or whether they all descended from one pair. This point, Dr. SMITH said, he argued about a year ago with a profound anatomist, who contended that all were not descended from a single pair. Dr. SMITH said he asked him what he would do with the plain declaration of the Bible, that all men were descended from Adam and Eve. He said he believed every word of the Bible, but not my deduction from it. But this, Dr. S. said, was not satisfactory; because a conclusion, legitimately deduced from certain premises, has all the force of those premises. He had always believed that the fact was as stated in the Bible, and it was sustained, he said, by analogy, and contradicted by no known fact; although there are difficulties which cannot as yet be explained. He had argued this point some time since with a Professor at Princeton, who had charged him with skepticism, because he had said that he could not believe that solar heat could have changed the size of the cheek bones of different men, as they are found to differ. Dr. S. said he did not say that men were not all descended from a common stock; but that there were difficulties in the way of this belief, which, in the present state of science, could not be satisfactorily explained.

The analogical argument would also show that Nature often produces the most different effects from one cause. And there is no fact to contravene this, in this case. CUVIER's notion, Dr. S. said, goes for nothing, when he defines species as animals that have descended from one common stock or that are as much alike as if they had descended from a common stock. This is a mere begging of the question. There is no instance on record in which a black man has become white: and whites, when removed to torrid climates, may change the color of their skins, but their skeletons remain the same. This phenomenon then, though not accounted for, should not lead us to doubt that all men are descended from Adam and Eve; but we should wait for the true solution of the difficulty, which future science will doubtless give.

Taking a retrospect of the ground gone over in the Lectures now brought to a close, Dr. S. said that we had found a few elements, sixteen in number, to have been created at a period immensely remote—"in the beginning;" that upon these elements certain laws were imposed which, in the lapse of time, produced effects such as we have seen. Though occasional interpositions may have occurred, the general current of events has been the same—ever onward to the same result. First we have the nebula; then the state of fusion; next the atmosphere, till we reach the rock, and then the first moss that springs up upon it, until we reach the inferior animal, and vegetables, and the grand whole is thus worked out. I said some time since, said Dr. S., that I would give you a demonstration of the being of God infinitely above that of Paley; and here it is. In the lapse of ages the same power produced all these grand results; one power designed it—one arranged, one directed the whole.

Still it seemed to him that the argument of adaptation must puzzle the disciples of LAPLACE, who supposed all things to have sprung from what he calls the fertility of matter. It is evident that if this were the case, the fertility of matter in Jupiter or Vesta must have acted differently from what it has here; since we know the same animals which it has produced upon the Earth could not live there. A man weighing 150 pounds on the Earth would weigh 1500 in Jupiter, and if he should once lie down he could never get up again. At Vesta, on the other hand, he would weigh only 10 pounds, and would be blown about by every wind. The only animal that could flourish there, said Dr. S., would be the *flea*, whose strength is so enormously great in proportion to his size. All creatures, then, are precisely adapted to the globe on which they are placed.

REMARKS ON SOME FOSSIL AND RECENT SHELLS COLLECTED BY CAPT. BAYFIELD, R.N. IN CANADA

A paper was then read, entitled, " Remarks on some fossil and recent shells, collected by Capt. Bayfield, R.N., in Canada," by Charles Lyell, Esq,, V.P.G.S.

Several eminent conchologists having observed that the English crag contains shells, which seem to indicate a somewhat colder climate than that which now prevails in our latitude; and it having been supposed that a similar inference may be deduced, with still greater certainty, from the abundant occurrence of many arctic species in the marine newer pliocene strata of Scotland and Ireland, Mr. Lyell was induced to examine carefully a collection of shells procured by Capt. Bayfield, and consisting partly of fossils from the most modern tertiary deposits bordering the Gulf of Saint Lawrence, and partly of recent testacea from the gulf itself.

The shells were obtained principally at Beauport (lat. 47°) 2 miles below Quebec and 100 feet above the St. Lawrence, but similar species are met with on the north side of the St. Charles, 3 miles from Beauport, and at Port Neuf, 40 miles above Quebec, in the latter instance at heights varying from 50 to 200 feet above the level of the river.

The deposits near Quebec fill a valley formed in a horizontal limestone, containing Trilobites and Orthocera, and they resemble those forming in the bed of the St. Lawrence. They consist of strata of sand, gravel, and stiff blue clay, the last composing the bottom of the series, and the first the uppermost part. Numerous boulders occur at different levels, not resting upon each other, but dropped apparently at widely distant intervals of time, from masses of ice on which it is supposed they had been floated. Some of the shells are broken, but many are perfect, and have both their valves together; and it is impossible to imagine that the clay, sand, gravel, and boulders could have been drifted together, into their present position, by a violent rush of water, as the fragile *Terebratula psittacea* is found perfect, and with its interior appendages complete.

On first examining the shells, which are found principally in the upper sandy bed, Mr. Lyell was struck with their great resemblance to those which he had collected at Uddevalla in Sweden. The *Saxicava rugosa*, so predominant there, is particularly mentioned by Capt. Bayfield as the most abundant shell in the tertiary strata of the St. Lawrence; and the *Natica clausa* and *Pecten Islandicus* are very common at each locality. The fossils of Beauport, however, considered as a whole, by no means agree with the marine shells inhabiting the Gulf of St. Lawrence, but, as far as they have been examined, possess a decidedly arctic character, the species ranging from the Gulf to the border of the north polar circle, or being found in the newer pliocene of Scotland and Sweden; and on the contrary many of the most conspicuous of the living testacea of the St. Lawrence are wanting in the tertiary deposits.

The following list of some of the fossil species is given by Mr.

Lyell on the authority of Dr. Beck: *Mya truncata* (var.), found fossil in Bute, and living in the St. Lawrence; *Mya arenaria* and *Saxicava rugosa*, recent in the Gulf of St. Lawrence; *Tellina calcarea*, fossil at Bute; *Tellina Grœnlandica*, which exists in the Gulf of St. Lawrence and at Icy Cape; *Mytilus edulis*; *Pecten Islandicus*, found living in the North Sea, and fossil in Scotland; *Terebratula psittacea*, which occurs on the coasts of Greenland and the Feroe Islands; also at places intermediate between them and the entrance of the Baltic; *Natica clausa*, recent in Greenland and fossil at Uddevalla; *Scalaria Grœnlandica*, *S. borealis*, *Tritonium fornicatum*, *T. Anglicanum*, all now existing in the Greenland seas, the last being considered by some authors as a variety of *Buccinum undatum*, and the *T. fornicatum* being also found living on the Irish coast, and fossil at Dalmuir and in Scotland. On the other hand, many of the shells living in the Gulf of St. Lawrence and most conspicuous for their size, are wanting in the collections of fossils hitherto obtained, as the *Mactra solidissima*, *Erycina Labradorica*, *Purpura*, allied to *P. Lapillus*, *Natica Heros*, and *Rostellaria occidentalis*.

The torrents and rivers which flow into the St. Lawrence wash down annually into that estuary great numbers of tertiary fossil shells, so that they become mingled with the living testacea. The latter, however, may be generally distinguished by retaining their colour, animal matter, or ligaments; but it is more difficult to distinguish those shells which have been derived exclusively from the tertiary beds. Nevertheless, Mr. Lyell has little doubt in assigning to them the specimens of *Balanus Uddevallensis* and the Fusus allied to *F. lamellosus*, which have been dredged up off Cape Bic, as they are all in the same condition as the Beauport fossils.

The climate of Canada being now excessive, it is natural to find in the Gulf of St. Lawrence many northern and arctic species, without any mixture of tropical forms, for the latter cannot resist severe cold, though they range far towards the southern polar latitudes, where a low mean annual temperature prevails. Mr. Lyell, therefore, conceives that during the period immediately antecedent to the present, the climate of Canada was even more excessive than it is now; and that the shells resembled still more closely the small assemblage now living in high northern latitudes. He is also of opinion, that this extreme cold may have coincided with the era of the principal transportation of erratic blocks, an inference supported by the masses of rock irregularly dispersed among the clay. He further believes, that a more equable though cold climate may have preceded immediately that condition; and that there may have been more than one oscillation of climate at the modern period, the last having been connected with the geographical changes which upheaved the shelly deposits of Canada 200 feet above the level of the St. Lawrence, and converted them from submarine deposits to dry land.

[CAUSE OF DIP OF NEW RED SANDSTONE IN EASTERN UNITED STATES]

After Prof. Hitchcock's observations respecting the bird tracks of the Connecticut valley, *Mr. Lyell* alluded to the subject of the cause of the present dip of the formation, expressing the opinion that it is due, in part at least, to an uplifting of the strata.

Prof. H. D. Rogers mentioned the reasons which had induced him to attribute the dip of the beds in the other great tract of this new red sandstone, or that which ranges southwestward from the Hudson, to oblique deposition. A uniform northerly dip of about fifteen degrees prevails entirely across the basin, even where it is twelve or twenty miles in breadth, and coëxists with a manifest shallowness of the deposit. This want of vertical depth is seen in several places in Pennsylvania, where denudation has exposed, in the interior of the tract, large patches of the older Appalachian strata, upon which this new red formation rests unconformably. No traces of dislocation occur to lead to the inference that the shallowness of the basin is deceptive, and that the present want of depth in the deposit has been caused by a succession of upthrows with denudation. The steady northerly dip is very rarely influenced, either in amount or in direction, by any of the numerous dykes of trap which penetrate the formation.

Prof. Rogers next mentioned facts which go to shew that the formation of the Connecticut valley and the equivalent one of the Middle States, are in all probability, accumulations in two originally distinct estuaries. He mentioned as one evidence, the independent direction of the dip in the two basins, and stated that the absence of a parallel order of succession in the members of the two formations, tends likewise to strengthen this opinion.

Mr. Lyell conceived the steepness of the dip, which sometimes amounts to twenty degrees, but more especially its direction,—transverse to the course of the ancient estuary, to present a difficulty. Prof. R. endeavored however to shew that the present dip might have been the original one, by suggesting, first, that there is evidence in the nature of the materials of the great southern basin for believing that they entered the estuary *laterally* on the outcrop side, by streams flowing from a country of decomposing, talcose, chloritic and hornblende rocks; secondly, that if the channel was near the same shore, the velocity of the tide might have prevented any horizontal deposition far out from the margin; and thirdly, that a gentle and steady *rising* of the region would, in conjunction with the proximity of the channel, tend to maintain both the slope of the sediment and the lateral advance of the shore which the hypothesis requires.

Mr. Benjamin Silliman, Jr., referring to the formation of the Connecticut valley only, considered a part of the present inclination of the beds to be the result of upheaval, connected with the outbursting of the trap.

Prof. W. B. Rogers made some remarks corroborating the views of Prof. H. D. Rogers in their application to the middle secondary or new red sandstone strata of Virginia and North Carolina. He described this group of rocks, consisting of shales, slates, sandstones and conglomerates of various tints of red and gray, as extending with some considerable interruptions in a nearly S. S. W. direction, entirely across the State of Virginia, and for some distance into North Carolina. With but a few local exceptions he had found the dip throughout this belt to be N. W. or N. N. W. Though destitute of the wide and prolonged ridges of trap met with in the corresponding districts of Pennsylvania and New Jersey, this region includes a great number of small dykes and knobs of trappean rock, penetrating the sedimentary strata, but in no instance causing any well marked change of dip in the adjacent beds. The materials of these strata Prof. R. stated to be very clearly traceable to the region of gneissoid and schistose rocks lying to the southeast of the tract, and in some cases, as in the limestone pebbles included in the conglomerate, could even be referred to the individual beds from which they had been torn.

He supported the opinion maintained for some years past by Prof. H. D. Rogers, that the inclination of the strata is not due to a tilting action subsequent to their deposition, but is the simple consequence of the influx of detrital matter from the southeast, and its deposition in a series of northwest-dipping planes. As greatly favoring this view he mentioned the fact generally observed in this belt throughout Virginia, that the strata become steeper in their inclination as we proceed towards the northwest; whereas the reverse should have been expected from a force tilting them from a horizontal or gently inclined position into the present northwestern dips. This opinion he conceived, was still further confirmed by the appearance of the strata in some parts of the district, where in consequence of the removal of the sedimentary rocks from a narrow space entirely across the tract, he was able to trace the beds from their outcrop nearly to the bottom of the trough in which they were deposited. In this case, he found the inclination of the beds to continue unchanged downwards, instead of becoming more gently inclined towards the bottom, as might be expected on the hypothesis of an originally horizontal position with a subsequent uptilting movement.

Mr. Redfield spoke of fossil rain-marks of a very perfect character which he had observed with Mr. Lyell, at the quarries of new red sandstone near Newark, N. J. He also gave notice of the discovery of a new species of fossil footmark in the new red sandstone of Connecticut. The specimen which he had seen was found in the well known quarries at Portland, (formerly Chatham,) by Mr. Russell, one of the proprietors, and is now in the possession of Dr. Barratt of Middletown. These footmarks are wholly unlike the Ornithichnites described by Prof. Hitchcock, some of which have been found in the same quarries; but they have some little resemblance to the *Cheirotherium minus*, which is figured in the Bulletin of the Geological Society of France.

[ON SOME OF THE PHENOMENA CONNECTED WITH THE COAL MEASURES AND OLDER STRATA OF PENNSYLVANIA]

A letter addressed to Dr. Fitton, by Mr. Lyell, and dated Boston the 15th of October, 1841, was then read.

Mr. Lyell's attention, between the period of his arrival in the United States and the date of his letter, had been principally devoted to the grand succession of Silurian, Devonian, and Carboniferous strata in the state of New York and on the borders of Pennsylvania, having been accompanied during a portion of his tour by the States' Geologist, Mr. J. Hall; but he had also visited, in company with that gentleman, the Falls of Niagara and the adjacent district, and he states, that he purposes to communicate a paper on the phænomena of the recession, drawn from new arguments, founded on the position of a fluviatile deposit below the Cataract. He expresses his intention of also communicating a notice of five localities of Mastodon bones which he had visited, digging up some remains himself, and collecting the accompanying shells, which he says, seem to have been neglected. He had likewise examined, accompanied by Prof. Silliman and his son, the new red, with intrusive trap, in Connecticut; and, assisted by Mr. Conrad, he had collected fossils in every member of the cretaceous system in New Jersey*. The principal object, however, of the present communication is, to point out the extension to the United States of Mr. Logan's generalizations on the beds of fire-clay containing Stigmaria, formerly laid before the Society in a paper on the coal-field of South Wales. Mr. Lyell had met Mr. Logan at New York, previously to that gentleman's visit to the anthracite coal-field of Pennsylvania, and he adverts to the delight which Mr. Logan must have felt in witnessing the occurrence of beds of Stigmaria fire-clay to an extent far exceeding what could have been expected. On the confines of the states of New York and Pennsylvania, Mr. Lyell found remains of Holoptychius and other fishes in the old red sandstone, and at the bottom of the overlying coal series a thick quartzose conglomerate; and he says that the coal-measures, with their imbedded plants, bear an exact analogy to British coal-measures, both in detail and as a whole. In investigating the coal district of Blossberg, Mr. Lyell had for a guide Dr. Saynisch, president of the mines. The first point which they examined presented three seams of bituminous coal resting on fire-clay containing Stigmariæ, with the leaves attached to the stems, and extending in all directions through the clay; and they observed, in a gallery lighted on purpose, that the stems seen *in situ* were very nearly all parallel to the planes of stratification, only one being in an oblique position. Every stratum underlying a coal-seam examined by Mr. Lyell, presented the same phænomena, except one, and in that case the bed was so sandy that it could not be considered as a fire-clay. The thickness of these Stigmaria deposits varied from one foot to six feet. The roof of the Blossberg coal-seams consists usually of bituminous slates, but occasionally of very micaceous grit, and it contains great varieties of

* Mr. Lyell mentions incidentally having observed between Easton and Trenton, on the Delaware, and in 40° of north latitude, that all the trees were barked on one side, at the height of twenty-two feet above the present level of the river, owing to a freshet and stoppage by ice in the spring of 1841. The stuccoed parts of the houses were also strangely scraped; and in one place the canal, the towing-path of which is twenty-two feet above the river, was so filled with gravel that carriages did not cross by the bridges.

ferns, as well as other plants, agreeing, generically at least, with those common in the British coal-measures.

Mr. Lyell next examined the anthracitic coal-district at Pottsville, on the Schuylkill, in the southern part of the Alleghanies. This district had been examined and described, as well as modelled, by Mr. R. C. Taylor, and the model had been inspected by Mr. Lyell previously to his visit. The whole of Pennsylvania has been mapped by Prof. H. D. Rogers, by direction of the State Legislature. Mr. Lyell refers to this survey, and he states that, by consulting Prof. Rogers's map, it will be found that the Alleghanies, or more properly the Appalachians, which, viewed geologically, are 120 miles broad, consist of twelve or more great parallel ridges, or anticlinal and synclinal flexures, having a general north-north-east and south-south-west strike, but in Pennsylvania a nearly east and west strike prevails. The strata are most tilted on the southern border of the chain, where their position is often inverted, and the folds become less and less towards the central ridges and troughs, which again increase in breadth the more northward their position, till at last the beds are almost horizontal. The oldest formations also are chiefly exposed in the most southern or disturbed regions, where syenite and other plutonic rocks are intruded into the lower part of the Silurian series. It has long been observed, that the anthracitic coal is confined to the southern or Atlantic side of this assemblage of small parallel chains, and that the bituminous occurs in the more inland or less disturbed region; the conclusion, therefore, Mr. Lyell states, seems inevitable, that the change in the condition of the coal was a concomitant of the folding and upheaval of the rocks. The conversion, moreover, is most complete where the beds have been most disturbed; and there are tracts in Pennsylvania and Virginia, near the centre of the chain, where the coal is in a semi-bituminous state. Chemical analysis, likewise, has shown that a gradation from the most bituminous to the most anthracitic coal may be found in crossing the chain from north to south*. The associated shales, &c., of the disturbed regions exhibit no alterations.

It has also been supposed that the anthracite belonged to the transition, and the bituminous coal to the secondary period; but this belief, Mr. Lyell says, has been gradually abandoned, as the knowledge of the geological position and the fossil plants of the coal-districts have become better known. Both the anthracitic and the bituminous coal overlie the old red sandstone, and contain the same ferns, Sigillariæ, Stigmariæ, Asterophyllites, &c.; and they are as abundant and perfect in the anthracite as in the bituminous coal.

At the first point where Mr. Lyell, accompanied by Prof. Rogers, examined the Pottsville coal-measures, the strata are nearly vertical, being cut off by a great fault from the less inclined beds which form the northern prolongation of the measures. They present thirteen beds of anthracite, the lowest of which alternate with

* See papers by Prof. H. D. Rogers, Dr. Silliman, &c.

the uppermost strata of the coarse underlying conglomerate. The southern wall of an excavation from which the coal had been removed, and which wall occupied the place of the underclay, presented impressions of the stems and leaves of Stigmaria; and on the more solid and slaty beds of the opposite wall, or original roof, there were leaves of Pecopteris, reed-like impressions, and Calamites. In the slightly inclined northern continuation of the coal-measures, Mr. Lyell observed in the Peachmount vein, three miles north-east of Pottsville, a bed of anthracite eight feet thick, overlaid by the usual roof of grey grit, and underlaid by blue clay or shale with Stigmariæ. Impressions of ferns were likewise noticed in the coal itself. Only one instance was met with in the Pottsville coal-district, by Mr. Lyell and Prof. Rogers, of a Stigmaria, placed at right angles to the plane of stratification.

The Pottsville, or southern anthracitic coal-field of Pennsylvania was illustrated by a section resulting from the former labours of Prof. Rogers, under whose guidance Mr. Lyell examined the country. The following remarks may explain the general structure of the country; the names applied to the formations are not, however, those previously employed by the American geologists, but those suggested by Mr. Lyell, in conformity with the conclusions at which he arrived after his tour in New York, and a comparison of the strata of that state with their British equivalents. The contrast between the relative importance of most of the Silurian and Devonian groups in Pennsylvania and in New York, Mr. Lyell states, is very great, arising from a larger portion of sandstones and grits in the Pennsylvanian rocks. The section extends from north of Pottsville to the country ranging immediately south of Orwigsburg. To the south of the vertical coal-measures and the subjacent conglomerate there are displayed successively—1st, a vast series, composed of red shales 3000 feet thick, of grey sandstone 2400 feet thick, and of red sandstone 6000 feet thick, the whole being considered by Mr. Lyell as portions of the old red sandstone; and 2nd, of olive-coloured shale containing Devonian fossils. The dip of the strata is either nearly vertical or inverted. Still further south, and a short distance north of Orwigsburg, the olive-coloured shales are succeeded by very highly inclined or inverted beds of upper Silurian rocks flanking a protruded band of lower Silurian strata; and lastly, on the southern confines of the section is a trough of the Devonian olive-coloured shales resting on the upper Silurian strata.

Beautiful exhibitions of the underclay with its associated plant, and of the overlying roof with its distinct remains, were observed by Mr. Lyell and Prof. Rogers at Tamaqua, in the southern coal-field. The thinning out of the grits and conglomerates of the west causes the beds of anthracite to be brought more nearly together in this district; and Mr. Lyell says, the decrease in the thickness of the intervening strata prepares the observer for the union of several of the seams still farther east, and for the enormous thickness of the anthracite at various places near the village of Mauch Chunk, or Bear Mount, particularly at the well-known Lehigh-Summit Mines. At this point a mass of anthracite forty feet thick, deducting three intercalated fire-clays and a fine thin vein of impure coal, is quarried in open day, a covering of forty feet of sandstone being entirely removed. In the south mine, where there is a sharp anticlinal fold in the coal, the Stigmaria-clay, four feet thick, was well seen, with nearly forty feet of coal above it and four below. In the Great mine Mr. Lyell observed the following section :—

Top, yellow quartzose grit.	
Coal, two or three inches of the uppermost part of the bed being in the state of dust, as if they had been crushed or rubbed by the yellow quartzose grit......	5 feet.
Blue fire-clay with Stigmariæ......................	15 inches.
Coal, including two or three seams of an impure slaty nature	25 feet.
Blue fire-clay with Stigmariæ......................	2 feet.
Coal, with an intervening layer of hard, bituminous slate	8 feet.

The anthracite, as in other parts of these coal-measures, often exhibits a texture exactly like that of charcoal; and frequently impressions of striated leaves, exactly resembling, as pointed out by Prof. Rogers, those of liliaceous plants, particularly the iris.

Mr. Lyell, accompanied by Prof. Rogers, afterwards examined the Room Run mines, on the Nesquahoning, where he saw a splendid exhibition of Stigmariæ in a bottom clay, one stem, about three inches in diameter, being no less than thirty-five feet in length. In the roof of slaty sandstone were impressions of Pecopteris, Glossopteris, and other ferns.

At Beaver Meadow, or the middle coal-field, a bed of anthracite is overlaid as well as underlaid by Stigmaria blue clay; the upper fire-clay, however, soon thins out, and is replaced by sandstone. No coal rested upon it, but Mr. Lyell observes that the carpeting of coal may not be always large enough to cover the flooring of fire-clay, or some change of circumstances or denudation may have interfered with the usual mode of deposition. Upon the whole, Mr. Lyell says, the accumulation of mud and Stigmariæ was, in Pennsylvania as in South Wales, the invariable forerunner of the circumstances attending the production of the coal-seams. The two extreme points at which he observed the Stigmaria-clay, Blossberg and Pottsville, are about 120 miles apart in a straight line, and the analogy of all the phænomena at those places, and still more on both sides of the Atlantic, is, he says, truly astonishing. In conclusion, Mr. Lyell states, that he had just received a letter from Mr. Logan, announcing the existence of the bottom clay, with Stigmariæ, in Nova Scotia; and that Mr. Logan had visited Mauch Chunk.

A MEMOIR ON THE RECESSION OF THE FALLS OF NIAGARA

PROCEEDINGS

OF

THE GEOLOGICAL SOCIETY OF LONDON.

VOL. III. PART II. 1842. No. 85.

Jan. 19th.—"A Memoir on the Recession of the Falls of Niagara," by Charles Lyell, Esq., V.P.G.S., was read.

The general features of the physical geography of the district traversed by the Niagara between Lakes Erie and Ontario, Mr. Lyell says, have been described with a considerable approach to accuracy by several writers. Prof. Eaton, in a small work published in 1824*, gives a correct section of the formations between Lewistown and the Falls of Niagara, and also refutes the hypothesis of the Lewistown escarpment being due to a fault by an exposition of the true structure of the country. Mr. R. Bakewell in 1830†, published an account of the country adjacent to the Falls, and Mr. De la Beche in 1831‡, endeavoured to point out the gradual manner in which the receding Falls, if they should ever reach Lake Erie, would discharge the waters of the lake; Prof. D. Rogers also in 1835 § showed distinctly, that, as the Falls retrograde, they would cut through rocks entirely distinct from those over which the waters are now precipitated, and correctly represents the superior limestone at Buffalo as newer than the limestone of the Falls, though he omits the intervening saliferous formation. Mr. Conrad likewise, in his Report for 1837 ‖, first assigned all the formations of the country to the Silurian system; but to Mr. James Hall (1838) ¶ is due the merit of having shown the true geological succession of rocks of the district.

The contents of the memoir may be divided into two parts: I. an account of the successive strata of the Niagara district; and II. a description of the phænomena exhibited by the Falls.

I. His sketch of the geology of the district, the author states, is derived either from the published surveys of Mr. Hall, or from the information he obtained while travelling with that gentleman in the State of New York during the autumn of 1841; and he acknowledges the great advantage he derived from the facilities thus afforded him. The strata between Lakes Erie and Ontario appear to belong to the middle and lower portions of the English Silurian system, and

* Mr. Lyell's attention was called to this work by Mr. Conrad.

† Loudon's Magazine of Natural History, 1830.

‡ Manual of Geology, three editions, 1831, p. 55; 1832, p. 55; 1833, p. 60.

§ Silliman's Journal, vol. xxvii. p. 326.

‖ States' Report of the Geology of New York.

¶ Geological Report of the State of New York for 1838.

they are divisible into the following five principal formations: 1st. the Helderberg limestone; 2nd, the Onondago salt group; 3rd, the Niagara group; 4th, the Protean group; and 5th, the Ontario group,

1. *The Helderberg limestone,* which has derived its designation from the range of mountains of the same name, and is the newest formation of the country, is exposed where the Niagara flows out of Lake Erie, and on account of the organic remains with which it abounds, it is considered to be the equivalent of the Wenlock rocks of Mr. Murchison's Silurian system. The correctness of this stratigraphical position Mr. Lyell has verified by an examination of the succession of formations from the coal-field on the borders of Pennsylvania to the group in question, the intervening deposits consisting, first, of old red sandstone, having at its bottom a large development of shales and sandstones called the Chemung and Ithaca formations, but containing organic remains which resemble those of the Devonian system; and then 1000 feet of Ludlowville shales with fossils analogous to those of the Ludlow rocks of Mr. Murchison. The superposition of this vast horizontal series is beautifully exposed in the banks of the Genessee and other rivers; and near Le Roy as well as elsewhere, the Helderberg limestones crop out from beneath them. On account of the middle portion containing nodules and layers of chert, the whole deposit was first called the corniferous formation by Prof. Eaton. In this part of the State of New York, and still further to the west, in Upper Canada, the limestone is only 50 feet thick, whereas at Schoharie in the Helderberg mountains, 300 miles to the eastward, its thickness is 300 feet.

2. *The Onondago salt group.*—This series of beds, Mr. Lyell says, is extremely unlike any described member of the European Silurian group. With the exception of a stratum of limestone at the top containing Cytherina, it consists of red and green marls with beds of gypsum, the former being undistinguishable from the marls of the new red system of England; and they are also destitute of fossils. Salt springs are of frequent occurrence, but no rock salt has been discovered in the group. The breadth of the zone of country occupied by the deposit is not less than 16 miles, and Mr. Hall infers from it and the slight southerly dip of the strata, that the entire thickness in the neighbourhood of the Niagara is at least 800 feet, an estimate confirmed by the nearest sections eastward of the river. In some parts of the State of New York the thickness is not less than 1000 feet. Along the Niagara the formation has been greatly denuded, and is covered by superficial drift, except at a few places.

3. *The Niagara group.*—This series of beds commences near the rapids, above the great cataract. It comprises, 1st, the Niagara, or Lockport limestone, and 2ndly, the Niagara, or Rochester shale; and it contains in both divisions fossils identical with those of the Wenlock limestone of England, with others peculiar to North America. The limestone at the rapids and the Falls is 120 feet thick; the upper 40 feet, being thin-bedded, have given way to the frost and the action of the stream, but the lower 80 feet, being massive, forms at

the cataract a precipice, beneath which occurs the shale, also 80 feet thick.

4. *The Protean group.*—Under the water at the base of the Falls crop out the higher beds of this formation, the name of which has been derived from the variable nature of its component strata. In the district more particularly described in this paper the group is only 30 feet thick, but farther to the eastward it attains thrice those dimensions. On the Niagara it consists of 25 feet of hard limestone, resting on 4 feet of shale; while at Rochester, eighty miles to the eastward, it comprises, among other beds, a dark shale with graptolites, or fossiliferous iron ore, and beneath them a limestone full of *Pentamerus oblongus* and *P. lævis*, considered by Mr. Conrad to be one species. On account of the occurrence of this shell, the whole of these strata have been separated from the Niagara series.

5. *Ontario group.*—About half a mile below the Falls the uppermost beds of the Ontario group crop out. At the whirlpool they have a thickness of 70 feet, and at Queenstown of 200, but to the latter dimension must be added 150 feet of inferior beds, exposed between Queenstown and Lake Ontario. The entire group consists of

1. Red marl with beds of hard sandstone in its upper division	70 feet.
2. White quartzose strata, so hard as to form at Queenstown a ledge projecting beyond the face of the escarpment	25 —
3. Red marl and sandstone	250 —

Other divisions of the group, concealed beneath the waters of the lake, may be studied in the cliffs of its eastern and north-eastern shores.

Mr. Lyell next proceeds to give a brief account of the geographical distribution of the formations or groups. The strike of the beds being east and west, and the dip very slight towards the south, the sections exposed along the Niagara afford a key to the structure of a large portion of the State of New York, the same deposits having been traced eastward through a region 40 miles in breadth by 150 in length, and westward to a much greater distance. The Helderberg and the Niagara limestones constitute platforms which terminate in parallel escarpments, from twenty to twenty-five miles apart, about sixteen miles of the intervening space being occupied by the saliferous group. The Helderberg escarpment, to the east of Buffalo, is 50 feet high; but in the neighbourhood of the Niagara it has been denuded and is half buried beneath drift; it is however resumed in Upper Canada, and eastward it may be followed to the river Hudson. The Niagara limestone escarpment presents at Lewistown and Queenstown a cliff 300 feet high, which may be traced eastward nearly 100 miles and westward for a much greater distance. The limestone series, however, constitutes only the uppermost third of the escarpment, the remainder being composed of the Protean and the Ontario groups; the whole section being as follows:—

1. Niagara limestone, lower beds	30 feet.
2. Niagara, or Rochester shale	80 —
3. Protean beds	30 —
4. Ontario group: red marl, with hard beds in the upper part	70 —
5. ————— : quartzose grey sandstone, with Lingulæ, &c.	25 —
6. ————— : red marl.	100 —
	335 feet.

Though only the lower beds of the Niagara limestone occur in the escarpment at Lewistown, yet, in consequence of the gentle rise of the strata to the north, the summit of these lower beds is at a higher level than that of Lake Erie. The whole of the Niagara platform is covered irregularly with hillocks of drift, beneath which the limestone is polished and furrowed.

From the foot of the Queenstown escarpment to Lake Ontario, a distance of six or seven miles, is a low tract, consisting of sandstones belonging to the Ontario group, and dipping like the preceding beds slightly to the south.

A section which accompanied the memoir to illustrate the preceding details corresponds, the author says, in all essential particulars with one previously published by Mr. Hall; but the whole succession of beds has been verified by Mr. Lyell in more than one line of section, from north to south. He is induced to believe, from a comparison of English Caradoc and Llandeilo fossils with suites of organic remains examined in America, that a series of beds which underlie the Ontario group, and termed by American geologists the Mohawk group, may be older than the lower Silurian rocks, and wanting in England.

II. *On the Recession of the Falls.*—The following measurements, Mr. Lyell says, are of great importance in speculating on the past or future recession of the Falls. The distance from the point where the Niagara flows out of Lake Erie to the Falls is sixteen miles, thence to the limestone escarpment seven miles, and from this point to Lake Ontario about seven more. From Lake Erie to the commencement of the rapids, fifteen miles and a half, the river falls only 15 feet; but from the top of the rapids to the great cataract the descent is 45 feet; and the height of the Falls is 164 feet, perpendicular. From the base of the Falls to Queenstown, seven miles, the difference of level in the river is about 100 feet; but from that place to Lake Ontario, seven miles further, it is only 3 or 4 feet. If the Falls were ever at Queenstown, they must, the author observes, have been about twice their present height, having lost a small portion of the difference by the southern inclination of the strata, and rather more than 100 feet by the rise of the bed of the river.

With respect to the opinion of the Queenstown escarpment being due to a fault, Mr. Lyell states, that the strata on the banks of the Niagara, both above and below Queenstown, presenting the same relative position as at Lockport or Rochester, the escarpment must be entirely due to denudation; and he has no hesitation in attributing this escarpment, as well as the Helderberg, to the action of the sea; these great inland cliffs having far too great a range to have resulted from a former extension and higher altitude of Lake Ontario.

The next question, whether the ravine through which the Niagara flows is to be regarded as a prolongation of the Queenstown escarpment and referable to the same period, or has been cut through by the river, is, the author states, of greater difficulty. From his own observations, he concludes that the ravine has been formed by the river; but he assumes, that a shallow valley pre-existed along the line of the present defile, resembling the present one between Lake Erie and the Falls. His reasons for conceiving that the river has been the excavating agent, are, 1st, the ravine being only from 400 to 600 yards wide at the top, and from 200 to 400 at the bottom, between Queenstown and the Whirlpool; 2ndly, the inclination of the bed of the river, $14\frac{1}{2}$ feet per mile, being everywhere cut down to the regular strata; 3rdly, the fact that the Falls are now slowly receding; 4thly, that a freshwater formation, which the author ascribes to the body of water which flowed along the original shallow valley, exists on Goat Island and half a mile lower down the river, and could not have been deposited after the Falls had receded farther back than the Whirlpool. Mr. Lyell considers that the indentation of about two acres on the American side of the Niagara, and not referable to the action of that river, is no objection to the theory of the recession of the Falls, because he conceives that the stream flowing down it could have effected the denudation, aided by atmospheric agents; and because a similar objection might be founded on a ravine on the Canada side opposite the Whirlpool, where several parallel gullies have been deeply eaten into by streams. The characters of this ravine were carefully examined by Mr. Lyell and Mr. Hall, and appear to have escaped previous observers. What was anciently a ravine joins the defile of the Niagara at this point, but it is entirely filled with horizontal beds of drifted pebbles, sand and loam; the first, near the bottom of the deposit, having been cemented into a conglomerate by carbonate of lime. This is the only interruption of the regular strata along the course of the Niagara; and Mr. Lyell observes, it is desirable to ascertain if it be a prolongation of the ravine which intersects the great escarpment at St. David's, west of Lewistown.

The author states, that he is by no means desirous of attaching importance to the precise numerical calculations which have been made respecting the number of yards that the Falls have receded during the last half century, as there are no data on which accurate measurements could be made; and because fifty years ago the district was a wilderness. Mr. Ingrahaw of Boston has, however, called his attention to a work published by the French Missionary, Father Hennipen, in which a view is given of the Falls as they appeared in 1678. Goat Island is represented dividing the waters as at present; but besides the two existing cascades, a third is depicted on the Canada side, crossing the Horse-shoe Fall at right angles, and appears to have been produced by a projection of the Table Rock. In the description Father Hennipen states, that this smaller cascade fell from west to east, and not like the other two, from south to north.

Seventy-three years afterwards, in 1751, a letter on the Falls, by Kalm, the Swedish botanist, was publisned in the 'Gentleman's Magazine.' It is illustrated by a plate, in which the third Fall is omitted; but the writer states in a note, that at that point the water was formerly forced out of its direct course by a projecting rock, and turned obliquely across the other Fall *.

Mr. Lyell then proceeds to show what are the geological evidences of the former prolongation of the river's bed, on a level with the top of the ravine through which the Niagara now flows. The existence on Goat Island of strata of marl, gravel and sand, containing fossil freshwater shells, was known before Mr. Bakewell's paper on the Falls was published, and they have been more recently described by Mr. Hall †; and Mr. Lyell states, that he was very desirous of ascertaining how far they extend on the banks of the river, or whether they could be detected below the present Falls. On the south-west side, in a cliff 12 feet in perpendicular height, a bed of gravel, 7 feet from the surface, contains eight species of fluviatile and one of terrestrial shells, determined for the author by Dr. Gould of Boston, the whole of the former now living in the waters of the Niagara, and some of them even in the rapids. At the south-west extremity of Goat Island this deposit must be 24 feet thick, and it rests on the Niagara limestone. On the right bank of the river, opposite the island, are two river-terraces, one 12 feet above the stream, and the other 12 feet higher; and both have been cut out of this freshwater formation. In making a mill-dam some years ago, the same species of shells as those on Goat Island were thrown out, and Mr. Lyell had still an opportunity of collecting them. He was also shown a tooth of the "*Mastodon Americanus*," which, with another tooth and a bone of the same animal, were discovered in the deposit 13 feet from the surface. From information given to the author by Mr. Hooker, the guide, the formation was found half a mile farther down the river, at the summit of the lofty precipice, 6 feet deep and composed chiefly of gravel. It contained in abundance *Cyclas rhomboidea*, *Valvata tricarinata* and *Planorbis parvus*. This patch of gravel demonstrates, therefore, the former position of the river at a level corresponding to that of the present summit of the cataract, and half a mile below the existing Falls. It proves however, Mr. Lyell says, much more; for in order that such a fluviatile deposit should have been accumulated in water tranquil enough to allow those shells to exist, there must have been a barrier farther down; and he is of opinion it may be safely placed as low as the Whirlpool, or three miles from the present Falls. If

* The author has observed distinct signs of recession in strata of the Silurian and Devonian epochs at the Falls of the Genessee in Rochester and at Portage, at the Fall of Allen's Creek below Le Roy, near the town of Batavia, and at the Falls of Jacock's river, three miles north of Genessee.

† Report for 1838.

this be admitted, then, the author says, "we may be prepared to concede that the still narrower ravine beyond the Whirlpool was excavated by the river cutting back its course."

A similar terrace, consisting of the Goat Island deposit, is distinctly seen also on the Canada side, and at about the same level between the Falls and the Whirlpool; but its extent, height and fossil contents have not been investigated.

If, Mr. Lyell observes, the river continue to intersect its way back, the sediment now depositing in its bed, above the Falls, will be laid dry in places, and cut into in the same manner as the Goat Island deposit.

Assuming that the cataract was once at the Queenstown escarpment, allowance must be made, in speculating on the probable time which has elapsed in cutting the ravine, for a very different rate of retrocession at different periods, dependent on the changes in the formation intersected, especially of those which successively constituted the base of the precipice. At Queenstown and Lewistown the fundamental rock, at the period when the Falls were there, was a soft red marl, and the river acted upon the same deposit for about three miles, where the rise in the channel, combined with the dip of the strata, caused the superincumbent hard quartzose beds, 23 feet thick, to form the base of the precipice. From this point the retrocession must have proceeded much more slowly for about a mile, or to the Whirlpool, where a small fall of 6 or 8 feet still marks the place of the highest beds of the sandstone. After, Mr. Lyell says, the cataract had remained nearly stationary for ages at this point, it next receded more rapidly for two miles, having soft red marl 70 feet thick to erode its way through; but beds of greater solidity, consisting of grey and mottled sandstone and Protean limestone, amounting in all to 30 or 40 feet, then offered a greater resistance, and continued to retard the backward movements of the Falls, the Protean limestone occurring at the base of the present precipice.

Lastly, the author offers some observations respecting the future retrocession of the Falls, quoting the opinions entertained by Mr. J. Hall (Report for 1838) on the effects which the strata above the existing cataract will have on the progress of the river, and pointing out results similar to those given by Mr. De la Beche in his 'Manual of Geology.' But all predictions, Mr. Lyell says, regarding the future history of the Falls may be falsified by the disturbing agency of man. Already a small portion of the waters of Lake Erie is carried off to supply the Welland canal, and another canal on the American side of Niagara; and numerous mill-races have been projected and others will be required along both sides of the river, as the population and wealth of the country increase. Many cities also, situated to the eastward of the great escarpment and at a lower level, may in aftertimes borrow water from Lake Erie, especially as the continued felling of the forests causes streams which were formerly constant to become dry in summer; and it must not be forgotten that Lake Michigan has lately been made by a cutting to feed the Illinois river, and that whatever quantity of water is abstracted from the upper lakes is taken away from the Niagara.

ON THE TERTIARY FORMATIONS AND THEIR CONNEXION WITH THE CHALK IN VIRGINIA AND OTHER PARTS OF THE UNITED STATES

A paper was then read, "On the Tertiary Formations and their connexion with the Chalk in Virginia and other parts of the United States," by Charles Lyell, Esq., V.P.G.S.

Having examined the most important cretaceous deposits in New Jersey, Mr. Lyell proceeded, in the autumn of 1841, to investigate the tertiary strata of Virginia, the Carolinas and Georgia, with a view to satisfy himself, first, how far the leading divisions of the tertiary strata along the Atlantic border of the United States agree in aspect and organic contents with those of Europe; and, secondly, to ascertain whether any rocks containing fossils of a character intermediate between those of the cretaceous or the eocene beds really exist. The conclusions at which he arrived, from his extensive survey, are given briefly as follows:—1. The only tertiary formations, which the author saw, agree well in their zoological types with the eocene and miocene beds of England and France; 2. he found no secondary fossils in those rocks which have been called upper secondary, and supposed to constitute a link between the cretaceous and tertiary formations.

1. *Virginia.*—The tertiary strata bordering the James River, Mr. Lyell says, have been well described by Prof. H. D. Rogers and Dr. Rogers*; and, he adds, they are also noticed in Mr. Conrad's excellent work on the tertiary strata of the United States. At Richmond, Mr. Lyell examined the remarkable bed of infusorial clay described by Prof. Rogers†, consisting of an impalpable siliceous powder derived from cases of microscopic animalcules. It varies in thickness from twelve to twenty-five feet, and is interposed between eocene greensands and miocene clays; but Mr. Lyell agrees with Prof. Rogers in considering it as probably belonging to the former epoch.

Similar eocene greensands, very much resembling the cretaceous greensand of New Jersey, occur at Petersburg, thirty miles south of Richmond, and are overlaid by a large deposit of miocene marls abounding in testacea different from those of the subjacent sands. Among the fossils of the latter deposit are a Venericardia scarcely distinguishable from *V. planicosta* of the London clay, also an Ostrea almost equally near to the *Ostrea bellovacina* and the *O. sellæformis*, so widely disseminated through the eocene formations of South Carolina and Georgia, is found in the uppermost beds of the formation at Coggin's Point, on the James River. The part of Virginia to which these remarks refer, is a flat region, forty or fifty feet above the level of the sea. The miocene strata which compose the upper beds consist sometimes almost exclusively of shells, and in the neighbourhood of Williamsburg Mr. Lyell collected eighty species, which bear a great resemblance generically, and in their relative numerical force, to collections from the Suffolk crag and the Faluns of Touraine.

* American Phil. Trans., New Series, vol. v. p. 319 *et seq.* 1835, and vol. vi. p. 347 *et seq.* 1839.

† States' Report, 1840.

Among these Testacea are several species of Astarte, some very analogous to those of Suffolk, the *Voluta mutabilis*, which resembles the *V. Lamberti*, also *Conus diluvianus*, *Lucina squamosa*, and *L. divaricata*. Mr. Lyell says there are many other analogies among the Mollusca, besides the occurrence of several corals, Echinodermata, fishes' teeth and bones of Cetacea; but he shows that the most important point of comparison is in the proportion of recent to extinct Testacea. Out of eighty-two species which he collected at Williamsburg, sixteen are considered by Mr. Conrad to be recent, and found for the greater part living on the coasts of the United States. The existing species, therefore, are in the proportion of one-fifth of the whole, which agrees well, says the author, with the average per-centage in the shells obtained by him in 1840 from the Faluns of Touraine. The entire number of American miocene shells known to Mr. Conrad is 238, of which thirty-eight have been identified with recent species.

North Carolina.—In the neighbourhood of South Washington, on the north-east branch of Cape Fear river, Mr. Lyell found the dark, bluish marls of the cretaceous series, to which his attention had been directed by Mr. Hodge's paper in Silliman's Journal*. They closely resemble, in composition and organic contents, those in New Jersey, and abound with *Belemnites mucronatus*, *Exogyra costata*, and a species of Gryphæa resembling *G. columba*; Mr. Lyell also found in them *Ostrea vesicularis* and *O. pusilla* of Nillson, likewise *Anomia tellinoides*, a species of Plagiostoma, and several new shells. These marls extend to the south of Lewis Creek, for several miles along the banks of the north-east branch of Cape Fear river, nearly to Rocky Point, where they are covered by the Wilmington limestone and conglomerate. This formation, which is overlaid by miocene strata, and ranges to Wilmington, as well as along the coast to Cape Fear river, has been considered by Mr. Hodge, and other geologists, to be an upper secondary deposit, or interposed between the eocene and cretaceous series; but Mr. Lyell could find in it no organic forms which supported this opinion, nor could he learn that any had been discovered. On the contrary, the only determinable species apparently agree with the *Lucina pendata*, an Alabama shell, and *Pecten membranosus*, both eocene fossils. The organic remains at Wilmington are only casts, but are referable to the genera Cardium, Nucula, Corbula, Cardita, Venus, Arca, Natica, Oliva, Cypræa, Conus, Calyptræa, and Siliquaria. Associated with these remains of Testacea are a species of Lunulites and several other corals, the claws of Crustacea, and teeth of the Lamna family. Many of these fossils occur at Rocky Point, including *Pecten membranosus*, with a Lunulite and a Vermetus subsequently found by the author in the limestone of the Santee canal in South Carolina.

South Carolina and Georgia.—Charleston stands on a yellow sand, beneath which is a blue clay containing the remains of Testacea that inhabit the adjacent seas; and Dr. Ravenel informed Mr. Lyell that he had found in it the *Gnathodon cyrenoides*, not now known to occur in a living state nearer than the Gulf of Mexico. The author

* Vol. xli. p. 332, 1841.

could not ascertain whether this post-pliocene formation rises above high-water mark; but he states that, on the Cooper river thirty miles north of Charleston, there occurs beneath the superficial sand and mottled clay a freshwater formation, in which Dr. Ravenel has found the remains of the Cypress, Hiccory and Cedar, which must have grown in a freshwater swamp, although the formation is now six feet below the level of high water. No shells have been noticed in the deposit, but they are also commonly wanting in the marsh accumulations of that region. As the salt water of Cooper river must now cover much of this deposit, a very modern subsidence, Mr. Lyell says, must have taken place along the coast. At Dr. Ravenel's plantation in the low country near the mouth of Cooper river is a pulverulent limestone, artificially exposed, which Mr. Lyell thinks may be an eocene formation, though its fossils differ from those of other deposit of that epoch.

Between this point and Vance's Ferry, on the Santee river, is a continuous formation of white limestone, which Mr. Lyell examined with Dr. Ravenel at Strawberry Ferry, Mulberry Landing, the banks of the Santee canal, Wantout and Eutaw. It varies in hardness, and consists of comminuted shells; but it very rarely exhibits any laminæ of deposition, and even where it attains a thickness of twenty or thirty feet there would be a difficulty in determining whether it were horizontal, if a bed of oysters, like that at Vance's Ferry, did not occasionally occur. At the Rock bridge near Eutaw springs, the limestone composed of comminuted shells, corals, the spines of Echini, &c., resemble so precisely the upper cretaceous formation at Timber Creek in New Jersey, that Mr. Lyell at first felt no doubt of the identity of the two formations, although the organic contents of the limestone prove that it belongs to the tertiary series. This resemblance has led to the admission into Dr. Morton's excellent work on the fossils of the cretaceous group, of the *Balanus peregrinus, Pecten calvatus, P. membranosus, Terebratula lachryma, Conus gyratus, Scutella Lyelli,* and *Echinus infulatus**, though they do not really belong to the chalk series; and to several other similar mistakes whereby, Mr. Lyell observes, beds of passage have been erroneously supposed to exist. Among the most widely distributed of the limestone fossils is the *Ostrea sellæformis*; and he searched in vain at various points throughout a distance of forty miles for an admixture of characteristic cretaceous and tertiary organic remains, though the chalk formation, containing Belemnites and Exogyræ, occurs between Vance's Ferry and Camden. The Santee limestone, he is of opinion, cannot be less than 120 feet thick at Strawberry Ferry, being vertically exposed to the extent of seventy feet in the banks and bottom of Cooper river, and to the height of fifty feet in the neighbouring hills. Its upper surface is very irregular, and is usually covered with sand in which no shells have been found. Mr. Lyell followed the limestone north-westwardly for twelve miles by Cave Hall and Struble's Mill to near Half-way Swamp. At Stoudenmire or Stout Creek, a tributary of the Santee, it has disappeared beneath a newer

* See pl. 10. of Morton's Synopsis.

tertiary deposit of considerable thickness, consisting of slaty clays, quartzose sand, loam of a brick-red colour, and beds of siliceous burr-stone. Mr. Lyell is not aware of any published description of this formation, though he afterwards met with it on the Savannah river. In both localities some of the clays break with a conchoidal flinty fracture when dry, and even occasionally pass into a stone closely resembling menilite. The fossils which he found were in the state of casts. He does not determine whether this formation should be regarded as an upper division of the eocene group or not; but he has little doubt that it is of the same age as the burr-stone series of Georgia. In the notice of the cretaceous and tertiary strata of the Southern states, drawn up by Dr. Morton from the notes of Mr. Vanuxem, the tertiary limestone and the burr-stone sand and clay are included in the same group, and Mr. Vanuxem informed Mr. Lyell that he had not been able to determine their relative position; but from what Mr. Lyell saw on the Savannah river, he infers that the burr-stone formation is above the limestone. One of the strata at Stoudenmire is extremely light and of white colour and resembles calcareous tufa, but according to the analysis of Prof. Shepard it contains no carbonate of lime; Mr. Lyell, therefore, states it may probably be of infusorial origin.

At Aikin, sixty miles west of Orangeburg, and near the left bank of the Savannah, an inclined plane in a railway has been cut through strata 160 feet in thickness, consisting of earth and sand of a vermilion colour and containing much oxide of iron; also of mottled clays and white quartzose sand with masses of pure white kaolin. These strata are within ten miles of the junction of the tertiary formation and the great hypogene region of the Appalachian or Alleghany chain, and their materials, Mr. Lyell states, have evidently been derived from the decomposition of clay-slate and granitic rocks. No fossils were observed by him in the deposit at Aikin. A similar formation is extensively developed at Augusta, where the Savannah divides the states of South Carolina and Georgia, and it must, in some places, be more than 200 feet thick. Three miles above the town are the rapids, which descend over highly inclined clay-slate and chlorite schist, overlaid unconformably by tertiary beds. This point is the western boundary of the supracretaceous series; and Mr. Lyell observes, that on all the great rivers of the Atlantic border from Maryland to Georgia, and still further south, the first falls or rapids are along a line at which the granitic and hypogene rocks meet the tertiary, and which is nearly parallel to the Atlantic coast, but at the distance of 100 or 150 geographical miles. This great feature, Mr. Lyell states, was first pointed out by Maclure, but he adds that portions of the tertiary formations usually cover the hypogene rocks for a certain distance above the Falls, and that their outline is very irregular and sinuous. On Race's Creek near Augusta, the highly inclined clay slate, containing chloritic quartzose beds with subordinate strata much charged with iron, are decomposed to the depth of many yards into clays and sands which resemble so precisely a large portion of the horizontal tertiary strata of the neighbouring country, that the disintegrated materials might be mistaken for them, if the veins of quartz which often traverse the argillaceous beds at a considerable angle, did not continue unaltered. The only point at which Mr. Lyell saw any organic remains in beds associated with these upper tertiary red strata was at Richmond in Virginia, where he obtained casts of decidedly miocene fossils; but as he observed on the Savannah river thick beds of sandy-red earth beneath the burr-stone of Stony Buff, he concludes that the same mineral character may sometimes belong to the upper division of the eocene group. At the rocks six miles west of Augusta, the tertiary beds derived from the hypogene rocks have the appearance of granite, and have been called gneiss by some geologists. They exhibit occasionally a distinct cross-stratification, and include angular masses of pure kaolin.

Though the Savannah, in its course from Augusta to the sea, flows for the greater part in a wide alluvial plain, and has a fall of less than one foot in a mile, yet Mr. Lyell descended it to obtain information, by means of the Bluffs, respecting the superposition of the several masses, natural sections being otherwise difficult to obtain. After passing cliffs of horizontal strata in which the brick-red sand and loam prevail, the first exposure of a new deposit was observed at Shell Bluff, forty miles below Augusta. The height of the section was 120 feet, and its extent more than half a mile. The lowest exposed strata consisted of white, highly calcareous sand, derived chiefly from comminuted shells, but the beds passed upwards into a solid limestone, sometimes concretionary, and containing numerous casts of shells. In one place a layer of pale green clay showed the horizontal character of the formation. The upper part of this deposit is more sandy and clayey, and incloses a bed of huge oysters, *Ostrea Georgiana*, occupying evidently the position in which they lived. The total thickness of these lower strata is eighty feet. The upper portion of the cliff is composed of forty feet of the red loam which prevails at Aikin and Augusta, and yellow sand. Mr. Lyell did not find any fossils in this deposit, but he believes that it belongs to the burr-stone formation, and therefore to be an upper eocene accumulation. At his first inspection of the casts contained in the limestone, he inferred that they belonged to eocene species, without any intermixture of cretaceous or miocene forms; but it was not till he had the advantage of Mr. Conrad's assistance that he was able to determine the following twelve species which are well known to be characteristic fossils of the eocene beds of Claiborne and Alabama:—

Oliva Alabamiensis.	Corbula nasuta.
Calyptræa trochiformis.	——— oniscus.
Dentalium alternans.	Nucula magnifica.
Venericardia planicosta.	Crassatella prætexta.
Cytherea Poulsoni.	Ostrea sellæformis.
——— perovata.	——— Alabamiensis.

The same shelly, white, calcareous beds, overlaid by red clay and loam, are exhibited at London Bluff, nine miles below Shell Bluff, and a horizontal bed of the large oysters is exposed in a cliff two miles farther down the river. At Stony Bluff, on the borders of Scriven county, the calcareous deposit is no longer visible, the cliff being composed of siliceous beds of the burr-stone and millstone series, resting upon brick-red and vermilion-coloured loam. This section, Mr. Lyell states, is of great importance, as it concurs in proving that the millstone of this region, with its eocene fossils, is an integral part of the great red loam and sand formation usually devoid of organic remains. The burr-rock of Stony Bluff abounds with cavities and geodes partially filled with crystals of quartz and agates. In the fragments scattered over the adjacent fields Mr. Lyell observed casts of univalves. At Millhaven, eight miles from Stony Bluff and five from the Savannah river, these siliceous beds again crop out and afford casts of the genera Pecten, Eulima or Bonellia, and a Cidaris. It had been pierced through to the depth of twenty-six feet, and was associated with red loam, white sand and kaolin, affording further evidence of these deposits belonging to one formation.

One mile west of Jacksonborough, in the ford of Briar and Beaver Dam Creeks, is a limestone passing upwards into white marl which appears to have been deeply denudated, and is overlaid by sand that belongs to a formation of sand, loam, and ferruginous sand-rock, referred by Mr. Lyell to the red loam and burr-stone series. The limestone and marl, although rarely exposed in sections, are considered to constitute very generally the fundamental strata of the region on account of the not unfrequent occurrence of lime-sinks or circular depressions, formed in the beds of loam and sand by subterranean drainage. The fossils procured from the limestone of Jacksonborough by Mr. Lyell, as well as those presented to him by Col. Jones of Millhaven, were for the greater part well-defined casts, and were specifically new to American palæontologists; nevertheless he has no hesitation, from their general aspect, to regard them as belonging to the eocene period. The genera enumerated in the paper are, Conus, Oliva, Bulla, Voluta, Buccinum, Fusus, Cerithium ?, Trochus, Calyptræa, Dentalium, Crassatella, Chama, Cardium, Cytherea, Lithodomus, Lucina, Pecten, and Ostrea. The Trochus is considered identical with the *T. agglutinans* which occurs in the Paris basin; and the Lithodomus to be undistinguishable from the *L. dactylus* of the West Indies, one of the few eocene Parisian fossils identified by Deshayes.

All the Bluffs examined by Mr. Lyell on the Savannah river below Briar Creek belong to the beds above the limestone, and are referable chiefly, if not entirely, to the burr-stone formation. In white clays exposed a few hundred yards below Tiger Leap in Hudson's Reach, the author found impressions of Mactra, Pecten and Cardita, also fragments of fishes' teeth, particularly of the genus Myliobates, likewise several teeth of the genus Lamna, and one belonging to a Notidamus or a nearly allied genus. At Sisters Ferry he observed not only the brick-red loam, with the red and grey clay and sand, but a highly siliceous clay, which though soft when moist, exhibits a conchoidal fracture when dry, and resembles flint; in some spots the clay also passes into a kind of menilite.

In conclusion, Mr. Lyell offers the following general observations. The part of South Carolina and Georgia which lies between the mountains and the Atlantic, and of which he examined a portion near the Santee and Savannah rivers, has a foundation of cretaceous rocks containing Belemnites, Exogyræ, &c., overlaid first by the eocene limestone and marls, and secondly by the burr-stone formation with the associated red loam, mottled clay, and yellow sand. According to Mr. Vanuxem's observations, a tertiary lignite deposit sometimes intervenes between the cretaceous and eocene series. The remarkable difference in the fossils of the eocene strata at different points, as the Grove on Cooper river, the Santee canal, Vance's Ferry, Shell Bluff, Jacksonborough, and Wilmington, might lead, Mr. Lyell states, to the suspicion of a considerable succession of minor divisions of the eocene period. That the whole are not precisely of the same age he is willing to believe, but he is inclined to ascribe the difference principally to two causes: 1st, that the number procured at each place is small and therefore represents only a fractional portion of the entire fauna of the period, so that variations in each locality may have arisen from original geographical circumstances; and 2ndly, no great eocene collection has been made from any part of the United States.

Some of the burr-stone fossils occur in the limestone, and Mr. Lyell thinks the former may bear to the latter a relation analogous to that which the upper marine sands of the Paris basin bear to the calcaire grossier.

With respect to the conclusion stated in the beginning of the paper, that he had been unable to find any beds containing an intermixture of cretaceous and tertiary fossils, Mr. Lyell says, it would require far more extended investigations to enable a geologist to declare whether there exist in the Southern states any beds of passage, but he affirms that the facts at present ascertained will not bear out such a conclusion.

The generic affinity of the cretaceous fossils of the United States to those of Europe is stated to be most striking, and Mr. Lyell observed in Mr. Conrad's collection from Alabama a large Hippurite, a point of analogy not previously recorded.

The proportion of recent shells in the eocene strata of the United States appears to be as minute as in Europe, and the distinctness of the eocene and miocene testacea hitherto observed to be as great. Mr. Lyell says, it is also worthy of remark, that the recent shells found in the American miocene beds are not only in the same proportion to the extinct as those of the Suffolk crag, or the Faluns of Touraine, but that they also agree specifically in most cases with mollusca inhabiting the neighbouring sea; in the same manner as the recent miocene species of Touraine agree for the greater part with species now living on the western coast of France or in the Mediterranean, and as the recent testacea of the crag are identifiable with species belonging to the British seas. This result appears to Mr. Lyell to confirm the accuracy of conchological determinations; for if, on the contrary, it should be maintained, that the number of recent species is so enormous, and different species resemble each other so closely as to have produced identifications from the mere difficulty of effecting discriminations, he would suggest that in that case, according to a fair calculation of chances, nine-tenths of the American miocene species hitherto identified ought to have been assimilated to exotic shells, instead of having been found to agree with some portions of the limited fauna at present known on the American shores. The same argument, he adds, is clearly applicable to the identifications which have been made of fossil and recent shells in the European tertiary formations.

ON THE FOSSIL FOOT-PRINTS OF BIRDS AND IMPRESSIONS OF RAIN-DROPS IN THE VALLEY OF CONNECTICUT

6. A paper "On the Fossil Foot-prints of Birds and Impressions of Rain-drops in the Valley of the Connecticut." By Charles Lyell, Esq., V.P.G.S.

The deposit in which these impressions, long known on account of the researches of Prof. Hitchcock, occur, is situated in a trough of hypogene rocks, about five miles broad, the strata, which consist of sandstone, shale and conglomerate, dipping uniformly to the east at angles that vary from 5° to 30°. Mr. Lyell first examined the red sandstone at Rocky Hill, three miles south of Hartford, in Connecticut, where it is associated with red shale and capped by twenty feet of greenstone. Many of the beds are rippled, and cracks in the shale are filled by the materials of the superincumbent sandy layer, showing, the author observes, a drying and shrinking of the mud while the accumulation of the strata was in progress. The next quarries he examined were at Newark in New Jersey, about ten miles west from New York city. The excavations are extensive, and the strata dip, as is usual in New Jersey, to the north-west, or in an opposite direction to the inclination in the valley of Connecticut, a ridge of hypogene rocks intervening. The angle is about 35° near Newark. The beds exhibited ripple-marks and casts of cracks, also impressions of rain-drops on the upper surface of the fine red shales. Mr. Lyell states, that he felt some hesitation respecting the impressions first assigned to the action of rain by Mr. Cunningham of Liverpool *, but he is now convinced of the justness of the inference, having observed similar markings produced on very soft

* See *antè*, p. 99.

mud by rain at Brooklyn in Long Island (New York). On the same mud were the foot-prints of fowls, some of which had been made before the rain and some after it.

Mr. Lyell next visited the red and green shales of Cabotville, north of Springfield in Massachusetts, where some of the best Ornithichnites have been procured, chiefly in the green shale. The dip of the beds is 20° to the east, a higher inclination, the author says, than could have belonged to a sea-beach. He observed in the same quarries ripple-marks as well as casts of cracks, and he was informed that the impressions of rain-drops have likewise been found.

In company with Prof. Hitchcock, Mr. Lyell afterwards examined a natural section near Smith's Ferry, on the right bank of the Connecticut, about eleven miles north of Springfield. The rock consists of thin-bedded sandstone with red-coloured shale. Some of the flags are distinctly ripple-marked, and the dip of the layers on which the Ornithichnites are imprinted, in great abundance, varies from eleven to fifteen degrees. Many superimposed beds must have been successively trodden upon, as different sets of tracks are traced through a thickness of sandstone exceeding ten feet; and Prof. Hitchcock pointed out to the author that some of the beds exposed several yards farther down the river, and containing Ornithichnites, would, if prolonged, pass under those of the principal locality, and make the entire thickness throughout which the impressions prevail, at intervals, perhaps twenty or thirty feet. Mr. Lyell, therefore, conceives that a continued subsidence of the ground took place during the deposition of the layers on which the birds walked.

It has been suggested, but the opinion has not been adopted by Prof. Hitchcock, that the eastward slope of the beds represents that of the original beach. With a view to this question, Mr. Lyell examined the direction of the ripple-marks, and found that it agreed with the dip, or was at right angles to the supposed line of beach; but he adds, though this agreement presents a formidable objection to the suggestion above alluded to, if the ripples were produced by waves, yet it does not disprove the opinion, as the ripples do not exceed in dimensions those which are produced by sand blown over a muddy beach, and often distributed at right angles to the coast-line. Instances of this effect of the wind Mr. Lyell has remarked along the shores of Massachusetts. Nevertheless he is of opinion that the rippled layer of sandstone in question contains too much clay to have resulted from blown sand, and he is disposed to think that in most of these localities the strata have been tilted, instances of such disturbance having been pointed out to him by Prof. Hitchcock in the state of Massachusetts, and by Mr. Percival near Newhaven in Connecticut. In reference to this subject, he says, that a few miles from Smith's Ferry a conglomerate, several hundred feet thick, containing angular and rounded fragments of trap and red sandstone, the base being sometimes a vesicular trap and trap tuff, passes upwards into the very flags on which Ornithichnites occur; and from this he infers, that there were eruptions of trap, accompanied by upheaval and partial denudation, during the deposition of the red sandstone.

With respect to the impressions having been made by birds, Mr. Lyell states, that until he examined the whole of the evidence he entertained some scepticism, notwithstanding the luminous account given by Prof. Hitchcock. In proof of their being the foot-prints of some creature walking on mud or sand, he mentions, 1st, the fact of Prof. Hitchcock's having seen 2000 impressions, all, like those he had himself examined, indented in the upper surface of the layer, the casts in relief being always on the lower surface; and 2ndly, that where there is a single line of impressions the marks are uniform in size, and nearly uniform in distance from each other, the toes in the successive steps turning alternately right and left. Such single lines, Mr. Lyell says, indicate that the animal was a biped, and the trifid marks resemble those which a bird leaves, there being generally a deviation from a straight line in any three successive prints; and his attention having been called to indications of joints in the different toes, he afterwards clearly recognised similar markings in the recent steps of coots and other birds on the sands of the shores of Massachusetts. Prof. Hitchcock has shown, that the same impression extends through several laminæ, decreasing in distinctness in proportion as the layer recedes from that in which it is most strongly marked, or in proportion as the sediment filled up the hollows and restored the surface to a level; and Mr. Lyell states, that he has observed a great number of instances of this fact.

He also says, that he can scarcely doubt that some of the impressions on the red sandstone of Connecticut are not referable to birds, but he believes that the gigantic ones described by Prof. Hitchcock are Ornithichnites. At Smith's Ferry they are so numerous that a bed of shale many yards square is trodden into a most irregular and jagged surface, so that there is not a trace of a distinct footstep; but on withdrawing from this area to spots where the same tracts are fewer, the observer, Mr. Lyell says, is forced to admit that the effect in each case has been produced by this cause.

On examining the shores on some small islands about fifteen miles south-east from Savannah, the author was struck with the number as well as the clearness of the tracks of raccoons and opossums imprinted in the mud during the four preceding hours, or after the tide had begun to ebb. At one spot, where the raccoons had been attracted by the oysters, the impressions were as confused as when a flock of sheep has passed over a muddy road; and in consequence of a gentle breeze blowing parallel to the line of cliffs composed of quartzose sand, the tracks had in many places already become half filled with blown sand, and in others were entirely obliterated; so that if the coast should subside, the consolidation of this sand would afford casts analogous to those of Storeton Hill in Cheshire, yet the impressions had been made and filled in a few hours.

When considering the broad question whether the fossil foot-prints were made by creatures walking on mud or sand after the ebbing of the tide, Mr. Lyell reminds his readers of the fact that in the United States, as in Saxony and Cheshire, the tracks in sandstone and shale are accompanied by littoral appearances, as ripple-marks, the casts of cracks in the clay, and often by the marks of rain.

In regard to the age of the red sandstone of the valley of the Connecticut and New Jersey, the author states he has nothing to add to what had been previously advanced, by which its position had been shown to be between the carboniferous and cretaceous series. In the neighbourhood of Durham, Connecticut, he had collected in the sandstone, fishes of the genera Palæoniscus and Catopterus, but no other organic remains, except fossil wood.

In conclusion, Mr. Lyell remarks, 1st, that the Ornithichnites of Connecticut should teach extreme caution in inferring the non-existence of land animals from the absence of their remains in contemporaneous marine strata; 2ndly, that when this red sandstone of Connecticut was deposited, there was land in the immediate vicinity of the places where the Ornithichnites occur; and that but for them it might naturally be inferred that the nearest land was several miles distant, namely, that of the hypogene rocks which bound the basin of the Connecticut. Now, the land that caused the sea-beach, Mr. Lyell says, must have been formed of the same sandstone which was then in the act of accumulating, in the same manner as where deltas are advancing upon the sea.

In a postscript, Mr. Lyell states, that subsequently to writing the paper he had read the luminous report of Mr. Vanuxem on the Ornithichnites described by Prof. Hitchcock, and though it agrees in substance with his own account in some particulars, yet that he has left his notice as it stood.

[ON THE STIGMARIA CLAY IN THE BLOSSBERG COAL-FIELD OF PENNSYLVANIA]

GEOLOGICAL SOCIETY.

Nov. 17, 1841.—A letter addressed to Dr. Fitton, by Mr. Lyell, and dated Boston the 15th of October, 1841, was read.

Mr. Lyell's attention, between the period of his arrival in the United States and the date of his letter, had been principally devoted to the grand succession of Silurian, Devonian, and Carboniferous strata in the state of New York and on the borders of Pennsylvania, having been accompanied during a portion of his tour by the States' Geologist, Mr. J. Hall; but he had also visited, in company with that gentleman, the Falls of Niagara and the adjacent district, and he states, that he purposes to communicate a paper on the phænomena of the recession, drawn from new arguments, founded on the position of a fluviatile deposit below the Cataract. He expresses his intention of also communicating a notice of five localities of Mastodon bones which he had visited, digging up some remains himself, and collecting the accompanying shells, which he says, seem to have been neglected. He had likewise examined, accompanied by Prof. Silliman and his son, the new red, with intrusive trap, in Connecticut; and, assisted by Mr. Conrad, he had collected fossils in every member of the cretaceous system in New Jersey*. The principal object, however, of the present communication is, to point out the extension to the United States of Mr. Logan's generalizations on the beds of fire-clay containing Stigmaria, formerly laid before the Society in a paper on the coal-field of South Wales. Mr. Lyell had met Mr. Logan at New York, previously to that gentleman's visit to the anthracite coal-field of Pennsylvania, and he adverts to the delight which Mr. Logan must have felt in witnessing the occurrence of beds of Stigmaria fire-clay to an extent far exceeding what could have been expected. On the confines of the states of New York and Pennsylvania, Mr. Lyell found remains of Holoptychius and other fishes in the old red sandstone, and at the bottom of the overlying coal series a thick quartzose conglomerate; and he says that the coal-measures, with their imbedded plants, bear an exact analogy to British coal-measures, both in detail and as a whole. In investigating the coal district of Blossberg, Mr. Lyell had for a guide Dr. Saynisch, president of the mines. The first point which they examined presented three seams of bituminous coal resting on fire-clay containing Stigmariæ, with the leaves

* Mr. Lyell mentions incidentally having observed between Easton and Trenton, on the Delaware, and in 40° of north latitude, that all the trees were barked on one side, at the height of twenty-two feet above the present level of the river, owing to a freshet and stoppage by ice in the spring of 1841. The stuccoed parts of the houses were also strangely scraped; and in one place the canal, the towing-path of which is twenty-two feet above the river, was so filled with gravel that carriages did not cross by the bridges.

attached to the stems, and extending in all directions through the clay; and they observed, in a gallery lighted on purpose, that the stems seen *in situ* were very nearly all parallel to the planes of stratification, only one being in an oblique position. Every stratum underlying a coal-seam examined by Mr. Lyell, presented the same phænomena, except one, and in that case the bed was so sandy that it could not be considered as a fire-clay. The thickness of these Stigmaria deposits varied from one foot to six feet. The roof of the Blossberg coal-seams consists usually of bituminous slates, but occasionally of very micaceous grit, and it contains great varieties of ferns, as well as other plants, agreeing, generically at least, with those common in the British coal-measures.

Mr. Lyell next examined the anthracitic coal-district at Pottsville, on the Schuylkill, in the southern part of the Alleghanies. This district had been examined and described, as well as modelled, by Mr. R. C. Taylor, and the model had been inspected by Mr. Lyell previously to his visit. The whole of Pennsylvania has been mapped by Prof. H. D. Rogers, by direction of the State Legislature. Mr. Lyell refers to this survey, and he states that, by consulting Prof. Rogers's map, it will be found that the Alleghanies, or more properly the Appalachians, which, viewed geologically, are 120 miles broad, consist of twelve or more great parallel ridges, or anticlinal and synclinal flexures, having a general north-north-east and south-south-west strike, but in Pennsylvania a nearly east and west strike prevails. The strata are most tilted on the southern border of the chain, where their position is often inverted, and the folds become less and less towards the central ridges and troughs, which again increase in breadth the more northward their position, till at last the beds are almost horizontal. The oldest formations also are chiefly exposed in the most southern or disturbed regions, where syenite and other plutonic rocks are intruded into the lower part of the Silurian series. It has long been observed, that the anthracitic coal is confined to the southern or Atlantic side of this assemblage of small parallel chains, and that the bituminous occurs in the more inland or less disturbed region; the conclusion, therefore, Mr. Lyell states, seems inevitable, that the change in the condition of the coal was a concomitant of the folding and upheaval of the rocks. The conversion, moreover, is most complete where the beds have been most disturbed; and there are tracts in Pennsylvania and Virginia, near the centre of the chain, where the coal is in a semi-bituminous state. Chemical analysis, likewise, has shown that a gradation from the most bituminous to the most anthracitic coal may be found in crossing the chain from north to south*. The associated shales, &c., of the disturbed regions exhibit no alterations.

It has also been supposed that the anthracite belonged to the transition, and the bituminous coal to the secondary period; but this belief, Mr. Lyell says, has been gradually abandoned, as the knowledge of the geological position and the fossil plants of the coal-districts have become better known. Both the anthracitic and the bituminous

* See papers by Prof. H. D. Rogers, Dr. Silliman, &c.

coal overlie the old red sandstone, and contain the same ferns, Sigillariæ, Stigmariæ, Asterophyllites, &c.; and they are as abundant and perfect in the anthracite as in the bituminous coal.

At the first point where Mr. Lyell, accompanied by Prof. Rogers, examined the Pottsville coal-measures, the strata are nearly vertical, being cut off by a great fault from the less inclined beds which form the northern prolongation of the measures. They present thirteen beds of anthracite, the lowest of which alternate with the uppermost strata of the coarse underlying conglomerate. The southern wall of an excavation from which the coal had been removed, and which wall occupied the place of the underclay, presented impressions of the stems and leaves of Stigmaria; and on the more solid and slaty beds of the opposite wall, or original roof, there were leaves of Pecopteris, reed-like impressions, and Calamites. In the slightly inclined northern continuation of the coal-measures, Mr. Lyell observed in the Peachmount vein, three miles north-east of Pottsville, a bed of anthracite eight feet thick, overlaid by the usual roof of grey grit, and underlaid by blue clay or shale with Stigmariæ. Impressions of ferns were likewise noticed in the coal itself. Only one instance was met with in the Pottsville coal-district, by Mr. Lyell and Prof. Rogers, of a Stigmaria, placed at right angles to the plane of stratification.

The Pottsville, or southern anthracitic coal-field of Pennsylvania was illustrated by a section resulting from the former labours of Prof. Rogers, under whose guidance Mr. Lyell examined the country. The following remarks may explain the general structure of the country; the names applied to the formations are not, however, those previously employed by the American geologists, but those suggested by Mr. Lyell, in conformity with the conclusions at which he arrived after his tour in New York, and a comparison of the strata of that state with their British equivalents. The contrast between the relative importance of most of the Silurian and Devonian groups in Pennsylvania and in New York, Mr. Lyell states, is very great, arising from a larger portion of sandstones and grits in the Pennsylvanian rocks. The section extends from north of Pottsville to the country ranging immediately south of Orwigsburg. To the south of the vertical coal-measures and the subjacent conglomerate there are displayed successively—1st, a vast series, composed of red shales 3000 feet thick, of grey sandstone 2400 feet thick, and of red sandstone 6000 feet thick, the whole being considered by Mr. Lyell as portions of the old red sandstone; and 2nd, of olive-coloured shale containing Devonian fossils. The dip of the strata is either nearly vertical or inverted. Still further south, and a short distance north of Orwigsburg, the olive-coloured shales are succeeded by very highly inclined or inverted beds of upper Silurian rocks flanking a protruded band of lower Silurian strata; and lastly, on the southern confines of the section is a trough of the Devonian olive-coloured shales resting on the upper Silurian strata.

Beautiful exhibitions of the underclay with its associated plant, and of the overlying roof with its distinct remains, were observed by Mr. Lyell and Prof. Rogers at Tamaqua, in the southern coal-field.

The thinning out of the grits and conglomerates of the west causes the beds of anthracite to be brought more nearly together in this district; and Mr. Lyell says, the decrease in the thickness of the intervening strata prepares the observer for the union of several of the seams still farther east, and for the enormous thickness of the anthracite at various places near the village of Mauch Chunk, or Bear Mount, particularly at the well-known Lehigh-Summit Mines. At this point a mass of anthracite forty feet thick, deducting three intercalated fire-clays and a fine thin vein of impure coal, is quarried in open day, a covering of forty feet of sandstone being entirely removed. In the south mine, where there is a sharp anticlinal fold in the coal, the Stigmaria-clay, four feet thick, was well seen, with nearly forty feet of coal above it and four below. In the Great mine Mr. Lyell observed the following section:—

Top, yellow quartzose grit.	
Coal, two or three inches of the uppermost part of the bed being in the state of dust, as if they had been crushed or rubbed by the yellow quartzose grit......	5 feet.
Blue fire-clay with Stigmariæ......................	15 inches.
Coal, including two or three seams of an impure slaty nature	25 feet.
Blue fire-clay with Stigmariæ......................	2 feet.
Coal, with an intervening layer of hard, bituminous slate	8 feet.

The anthracite, as in other parts of these coal-measures, often exhibits a texture exactly like that of charcoal; and frequently impressions of striated leaves, exactly resembling, as pointed out by Prof. Rogers, those of liliaceous plants, particularly the iris.

Mr. Lyell, accompanied by Prof. Rogers, afterwards examined the Room Run mines, on the Nesquahoning, where he saw a splendid exhibition of Stigmariæ in a bottom clay, one stem, about three inches in diameter, being no less than thirty-five feet in length. In the roof of slaty sandstone were impressions of Pecopteris, Glossopteris, and other ferns.

At Beaver Meadow, or the middle coal-field, a bed of anthracite is overlaid as well as underlaid by Stigmaria blue clay; the upper fire-clay, however, soon thins out, and is replaced by sandstone. No coal rested upon it, but Mr. Lyell observes that the carpeting of coal may not be always large enough to cover the flooring of fire-clay, or some change of circumstances or denudation may have interfered with the usual mode of deposition. Upon the whole, Mr. Lyell says, the accumulation of mud and Stigmariæ was, in Pennsylvania as in South Wales, the invariable forerunner of the circumstances attending the production of the coal-seams. The two extreme points at which he observed the Stigmaria-clay, Blossberg and Pottsville, are about 120 miles apart in a straight line, and the analogy of all the phænomena at those places, and still more on both sides of the Atlantic, is, he says, truly astonishing. In conclusion, Mr. Lyell states, that he had just received a letter from Mr. Logan, announcing the existence of the bottom clay, with Stigmariæ, in Nova Scotia; and that Mr. Logan had visited Mauch Chunk.

ON THE RIDGES, ELEVATED BEACHES INLAND CLIFFS AND BOULDER FORMATIONS OF THE CANADIAN LAKES AND VALLEY OF ST. LAWRENCE

ART. XI.—*On the Ridges, Elevated Beaches, Inland Cliffs and Boulder Formations of the Canadian Lakes and Valley of St. Lawrence;* by CHARLES LYELL, Esq., F. G. S., F. R. S., &c.*

AFTER adverting to a former paper on the recession of the Falls of Niagara, and the observations which he made jointly with Mr. Hall in the autumn of 1841, (see Proceed. Geol. Soc. Vol. III, p. 595,) Mr. Lyell gives an account of additional investigations made by him in June, 1842; in the course of which he found a fluviatile deposit similar to that of Goat Island, on the right bank of the Niagara, nearly four miles lower down than the great Falls. The fresh-water strata of sand and gravel here alluded to occur at the Whirlpool. They are horizontal, about forty feet thick, plentifully charged with shells of recent species, and are placed on the verge of the precipice overhanging the river. They are bounded on their inland side by a steep bank of boulder clay, which runs parallel to the course of the Niagara, marking the limit of the original channel of the river before the excavation of the great ravine. Another patch of sand, with fresh-water shells, was detected on the opposite or western side of the river, where the Muddy Run flows in, about one mile and a half above the Whirlpool. From the position of these strata it is inferred that the ancient bed of the river, somewhere below the Whirlpool, must have been three hundred feet higher than the present bed, so as to form a barrier to that body of fresh water, in which the various beds of fluviatile sand and gravel above-mentioned were accumulated. This barrier was removed when the cataract cut its way back to a point further south. The author also remarks, that the manner in which the fresh-water beds of the Whirlpool and Goat Island come into immediate contact with the subjacent Silurian limestone, no drift intervening, shows that the original valley of the Niagara was shaped out of limestone as well as drift. Hence he concludes that the rocks in the rapids above the present Falls had suffered great denudation while yet the Falls were at or below the Whirlpool.

Mr. Lyell thinks that the form of the ledge of rocks at the Devil's Hole, and of the precipice which there projects and faces down the river, proves the Falls to have been once at that point. An ancient gorge, filled with stratified drift, which breaks the continuity of the limestone on the left bank of the Niagara at the Whirlpool, was examined in detail by the author, and found to be connected with the val-

---

* Communicated to this Journal by the author, having been previously read before the Geological Society of London, and published in Vol. IV, No. 92, of their Proceedings.

ley of St. David's, about three miles to the northwest. This ancient valley appears to have been about two miles broad at one extremity, where it reaches the great escarpment at St. David's, and between two and three hundred yards wide at the other end, or at the Whirlpool. Its steep sides did not consist of single precipices, as in the ravine of Niagara, but of successive cliffs and ledges. After its denudation the valley appears to have been submerged and filled up with sand, gravel, and boulder clay, three hundred feet thick.

A description is next given of certain modern deposits, containing fresh-water shells, on the western borders of the Niagara, above the Falls, and in Grand Island, in order to show that the future recession of the Falls may expose patches of fluviatile sediment similar to those in and below Goat Island.

The author then passes to the general consideration of the boulder formation on the borders of Lakes Erie and Ontario, and in the valley of the St. Lawrence, as far down as Quebec. Marine shells were observed in this drift at Beauport, below Quebec, as first pointed out by Captain Bayfield, and also near the mouth of the Jacques Cartier river, and at Port Neuf and other places; also at Montreal, where they reach a height probably exceeding five hundred feet above the sea, the summit of Montreal mountain being seven hundred and sixty feet high, according to Bayfield's trigonometrical measurement, and the shells being supposed to be two hundred and forty feet below the summit. These shells, therefore, being more than three hundred feet above Lake Ontario, we may presume that the sea in which the drift was formed extended far over the territory bordering on that lake. The most southern point at which the author saw fossil shells belonging to the same group as those of Quebec was on the western and eastern shores of Lake Champlain, viz. at Port Kent and Burlington, in about lat. 44° 30′. Here, and wherever elsewhere the contact of the drift is seen with hard subjacent rocks, these rocks are smoothed, and furrowed on the surface, in the same manner as beneath the drift in northern Europe. The species of shells occurring in the drift, to which Mr. Lyell has made some additions, are not numerous, and are all, save one, known to exist, but are inhabitants, for the most part, of seas in higher latitudes. Many of them are the same as those occurring fossil at Uddevalla and other places in Scandinavia, and they imply the former prevalence of a colder climate when the drift originated. At Beauport there are large and far-transported boulders, both in beds which overlie and underlie these marine shells.

The author next describes the ridges of sand and gravel surrounding the great lakes, which are regarded by many as upraised beaches. He examined, in company with Mr. Hall, the "Lake ridge," as it is called, on the southern shore of Lake Ontario, and other similar ridges north of Toronto, which were formerly explored by Mr. Roy, (see Proceed. Geol. Soc. Vol. II, p. 537,) and which preserve a general parallelism to each other and to the neighboring coast. Some of these have been traced for more than one hundred miles continuously. They vary in height from ten to seventy feet, are often very narrow at their summit, and from fifty to two hundred yards broad at their base. Cross stratification is very commonly visible in the sand; they usually rest on clay of the boulder formation, and blocks of granite and other rocks from the north are occasionally lodged upon them. They are steeper on the side towards the lakes, and they usually have swamps and ponds on their inland side; they are higher for the most part and of larger dimensions than modern beaches. Several ridges, east and west of Cleveland in Ohio, on the southern shore of Lake Erie, were ascertained to have precisely the same characters. Mr. Lyell compares them all to the osars in Sweden, and conceives that, like them, they are not simply beaches which have been entirely thrown up by the waves above water, but that many of them have had their foundation in banks or bars of sand, such as those observed by Capt. Grey running parallel to the west coast of Australia, lat. 24° S., and by Mr. Darwin off Bahia Blanca and Pernambuco in Brazil, and by Mr. Whittlesey near Cleveland in Lake Erie. They are supposed to have been formed and upraised in succession, and to have become beaches as they emerged, and sometimes cliffs undermined by the waves. The transverse and oblique ramifications of some ridges are referred to the meeting of different currents and do not resemble simple beaches.

The base-lines of the ridges east and west of Cleveland, are not strictly horizontal according to Mr. Whittlesey, but incline five feet and sometimes more in a mile. Those near Toronto are said by Mr. Roy to preserve the same exact level for great distances, but Mr. Lyell does not conceive that our data are as yet sufficiently precise to enable us to determine the levels within a few feet at points distant several hundred miles from each other. No fossil shells have been obtained from these ridges, and the author concludes that most of them were formed beneath the sea or on the margin of marine sounds. Some of the less elevated ridges, however, may be of lacustrine origin, and due to oscillations in the level of the land since the great lakes existed, for unequal movements, analogous to those observed in Scandinavia, may have uplifted fresh-water strata above the barriers which divide Lake Michigan from the basin of the Mississippi, or Lake Erie from Ontario, or the waters of Ontario from the ocean. Considerable differences of level may have been produced in the ancient beds of these vast inland bodies of fresh water, while the modern deposit and the subjacent Silurian strata may to the eye appear perfectly horizontal.

The author then endeavors to trace the series of changes which have taken place in the region of Lakes Erie and Ontario, referring first to a period of emergence when lines of escarpment like that of Queenston, and when valleys like that of St. David's were excavated; secondly, to a period of submergence when those valleys and when the cavities of the present lake-basins were wholly or partially filled up with the marine boulder formation; and lastly, to the re-emergence of the land, during which rise the ridges before alluded to were produced, and the boulder formation partially denuded. He also endeavors to show, how during this last upheaval the different lakes may have been formed in succession, and that a channel of the sea must first have occupied the original valley of the Niagara, which was gradually converted into an estuary and then a river. The great Falls, when they first displayed themselves near Queenston, must have been of moderate height, and receded rapidly, because the limestone overlying the Niagara shale was of slight thickness at its northern termination. On the further retreat of the sea a second fall would be established over lower beds of hard limestone and sandstone previously protected by the water; and finally, a third fall would be caused over the ledge of hard quartzose sandstone which rests on the soft red marl, seen at the base of the river-cliff at Lewiston. These several falls would each recede further back than the other in proportion to the greater lapse of time during which the higher rocks were exposed before the successive emergence of the lower ones. Three falls of this kind are now seen descending, a continuation of the same rocks on the Genesee River at Rochester. Their union, in the case of the Niagara into a single fall, may have been brought about in the manner suggested by Mr. Hall, (Boston Journ. Nat. Hist., 1841,) by the increasing retardation of the highest cataract in proportion as the uppermost limestone thickened in its prolongation southwards, the lower falls meanwhile continuing to recede at an undiminished pace, having the same resistance to overcome as at first.

Mr. Lyell considers the time occupied by the recession of the Falls from the Whirlpool to be quite conjectural, but assigns a foot rather than a yard a year as a more probable estimate; thus he shows the Mastodon, found on the right bank near Goat Island, though associated with shells of recent species, to have claim to a very high antiquity, since it was buried in fluviatile sediment before the Falls had receded above the Whirlpool.

ON THE TERTIARY STRATA OF THE ISLAND OF MARTHA'S VINEYARD IN MASSACHUSETTS

Art. XII.—*On the Tertiary Strata of the Island of Martha's Vineyard in Massachusetts;* by Charles Lyell, Esq., V. P. G. S., &c.*

The most northern limit to which the tertiary strata bordering the Atlantic have been traced in the United States, is in Massachusetts, in Martha's Vineyard, lat. 41° 20 north, an island about twenty miles in length from east to west, and about ten from north to south, and rising to the height of between two and three hundred feet above the sea. The tertiary strata of this island are, for the most part, deeply buried beneath a mass of drift, in which lie huge erratic blocks of granite and other rocks, which appear to have come from the north, probably from the mountains of New Hampshire. The tertiary strata consist of white and green sands, a conglomerate, white, blue, yellow, and blood-red clays, and black layers of lignite, all inclined at a high angle to the northeast, and in some of their curves quite vertical. They are finely exposed near Chilmark on the southwest side of the island, and in the promontory of Gay Head at its southwestern extremity, where there is a vertical section of more than two hundred feet in height.

Attention was first called to this formation by Prof. Hitchcock in 1823, who appears to be the only American geologist who has examined them personally. He compared the beds at Gay Head to the plastic and London clays of Alum Bay in the Isle of Wight, to which, lithologically, they bear a striking resemblance, consisting in both cases of variously and brightly colored clays and sands with lignite, all incoherent and highly inclined. Various opinions, however, have been put forth as to the relative age of the Martha's Vineyard strata, which were assigned by Prof. Hitchcock, at a time when the tertiary formations of the United States were less known, to the eocene period, while Dr. Morton supposed them to be in part only tertiary, and that they rested on green-sand of the cretaceous period.

The section at Gay Head is continuous for four fifths of a mile; the beds dip to the northeast generally at an angle of from thirty five to fifty degrees, though in some places at seventy degrees. The clays predominate over the sands. In one place Mr. Lyell found a great fold in the beds, in which the same osseous conglomerate and associated beds of white sand, on the whole fifty feet thick, were so bent as to have twice a northeasterly, and once a southwesterly dip. In the yellowish and dark brown clay near the uppermost part of the section at Gay Head, and in the green-sand immediately resting upon it,† Mr Lyell

---

* From the Proceedings of the London Geological Society, Vol. IV, No. 92.

† Nos. 5 and 6 of Prof. Hitchcock's section.

found the teeth of a shark, that of a seal, vertebræ of Cetacea, crustacean remains, and casts of *Tellina* and *Mya*. These prevail at intervals through a thickness of nearly one hundred feet, and are followed by beds of sand and clay with lignite. Mr. Lyell found no remains in the red clays. Many rolled bones were found in the osseous conglomerate.

In the section at Chilmark similar strata to those at Gay Head occur, but the general dip is southwest. Some of the folds, however, give anticlinal dips to the northeast as well as the southwest, and there are many irregularities, the beds being sometimes vertical and twisted in every direction. Several faults are seen, and veins of iron-sand, which intersect the strata like narrow dykes, as if there had been cracks filled from above. One bed of osseous conglomerate at Chilmark, four yards in thickness, is vertical, and its strike is well seen to be north 25° east, so that the disturbances have evidently been so great that it would be difficult without more sections to determine positively the prevailing strike of these beds. The incumbent drift is very variable in thickness, and large erratics, from twenty to thirty feet in diameter, are seen resting on quartzose sand. The author saw no grounds for concluding that any cretaceous strata occur any where in the island, nor could he find any fossils which appeared to have been washed out of a cretaceous formation into the tertiary strata, as some have suggested.

Mr. Lyell proceeds to the consideration of the organic remains collected by himself in Martha's Vineyard.

Mammalia.—1. A tooth, identified by Prof. Owen as the canine tooth of a seal, of which the crown is fractured. It seems nearly allied to the modern *Cystophora proboscidea*.

2. A skull of a walrus, differing from the skulls of the existing species (*Trichecus rosmarus*, Linn.), with which it was compared by Prof. Owen, in having only six molars and two tusks, whereas those of the recent have four molars on each side, besides occasionally a rudimentary one. The front tusk is rounder than that of the recent walrus.

3. Vertebræ of *Cetacea*, some of which are referred by Prof. Owen to the Whalebone whales, and others to the Bottle-nosed (*Hyperoodon*).

Pisces.—Teeth of sharks resembling species from the Faluns of Touraine, viz. *Carcharias megaladon*, *Oxyrhina xiphodon*, *O. hastulis*, and *Lamna cuspidata*. With these were large teeth of two species of *Carcharias*, one resembling *C. productus*, a Maltese fossil. With the exception of the two last, Mr. Lyell found the same species in meiocene strata near Evergreen, on the right bank of James River in Virginia.

Crustacea.—A species considered by Mr. Adam White as probably belonging to the genus *Cyclograpsus*, or the closely allied *Sesarma* of Say, and another, decidedly a *Gegarcimus*.

Mollusca.—1. Casts of a *Tellina* allied to *T. biplicata*, a meiocene fossil, and of another near *T. lusoria*. 2. Cast of a *Cytherea* resembling *C. Sayana*, Conrad. 3. Three casts of a *Mya*, one of which bears close resemblance to *Mya truncata*.

Mr. Lyell concludes, from the various evidence here given, that the strata of Martha's Vineyard are meiocene. The numerous remains of Cetacea of the genera *Balæna* and *Hyperoodon* are adverse to the supposition of their being eocene, while such fossils abound in the meiocene beds of America. The other fossils all point to a similar conclusion.

ON THE GEOLOGICAL POSITION OF THE MASTODON GIGANTEUM AND ASSOCIATED FOSSIL REMAINS AT BIGBONE LICK, KENTUCKY

ART. XIII.—*On the Geological Position of the Mastodon giganteum and associated Fossil Remains at Bigbone Lick, Kentucky, and other localities in the United States and Canada;* by CHARLES LYELL, Esq. V. P. G. S. &c.*

WITH a view to ascertain the relations of the soil in which the bones of the Mastodon are found, to the drift or boulder formation, whether any important geographical or geological changes had taken place since they were imbedded, and what species of shells are associated with them, Mr. Lyell visited a number of places where they had been obtained. In this paper he gives the result of his researches.

The most celebrated locality visited was Bigbone Lick, in the northern part of Kentucky, distant about twenty five miles to the southwest of Cincinnati, situated on a small tributary of the river Ohio called Bigbone Creek, which winds for about seven miles below the Lick before joining the Ohio. A "lick" is a place where saline springs break out, generally among marshes and bogs, to which deer, buffaloes, and other wild animals resort to drink the brackish water and lick the salt in summer. The country around Bigbone Lick, and for a considerable distance on both banks of the Ohio, above and below it, is composed of blue argillaceous limestone and marl, constituting one of the oldest members of the transition or Silurian system. The strata are nearly horizontal and form flat table-lands intersected by numerous valleys in which alluvial gravel and silt occur; but there is no covering of drift in this region. The drift is abundant in the northern parts of Ohio and Indiana, but disappears almost entirely before we reach the Ohio.

Until lately herds of buffaloes were in the habit of frequenting the springs, and the paths made by them are still to be seen. Numbers of these animals have been mired in the bogs, and horses and cows have perished in like manner. Along with their remains are found innumer-

---

* From the Proceedings of the London Geological Society, Vol. IV, No. 92.

able bones of Mastodon, elephant, and other extinct quadrupeds, which must have visited these springs when the valley was in its present geographical conditon in almost every particular, and which must have been mired in them as existing quadrupeds are at present. The Mastodon remains are most numerous and belong to individuals of all ages. The mud is very deep, black, and soft. In places it is seen to rest upon the limestone, and at some points it swells up to the height of several feet above the general level of the plain and of the river. It is occasionally covered by a deposit of yellow clay or loam, resembling the silt of the Ohio, which is from ten to twenty feet thick, rising to that height above the creek and often terminating abruptly at its edges. This loam has all the appearance of having been deposited tranquilly on the surface of the morass and of having afterwards suffered denudation. The Mastodon and other quadrupeds have been mired before the deposition of the incumbent silt, for a considerable number of fossil bones have been found by digging through it. Accompanying the bones are fresh-water and land shells, most of which have been identified by Mr. Anthony with species now existing in the same region.

Mr. Lyell observes that the surface of the bog is extremely uneven, and accounts for it partly by the unequal distribution of the incumbent alluvium, which presses with a heavy weight on certain parts of the morass, from which other portions of the surface are entirely free. He also attributes it in part to the swelling of the bog where it is fully saturated with water near the springs.

The author is of opinion that the fossil remains of Bigbone Lick are much more modern than the deposition of the drift, which is not present in this district. But although the date of the imbedding of these mammalian fossil remains is so extremely modern, considered geologically, it is impossible to say how many thousand years may not have elapsed since the Mastodon and other lost species became extinct. They have been found at the depth of several feet from the surface, but we have no data for estimating the rate at which the boggy ground has increased in height, nor do we know how often during floods its upper portion has been swept away.

Ohio.—The Ohio River immediately above and below Cincinnati is bounded on its right bank by two terraces consisting of sand, gravel and loam, the lower terrace consisting of beds supposed to be much newer than those of the upper. In the gravelly beds of the higher terrace, teeth both of the Mastodon and elephant have been met with. Mr. Lyell was assured that a boulder of gneiss, twelve feet in diameter, was found resting on the upper terrace, about four miles north of Cincinnati, and that some fragments of granite had been found in a similar situation at

Cincinnati itself. These facts show that some large erratics have taken up their present position since the older alluvium of the Ohio valley was deposited. In travelling northwards from Cincinnati towards Cleveland, Mr. Lyell found the northern drift commence in partial patches twenty five miles from the former city and about five miles northeast of Lebanon, after which it continually increased in thickness as he proceeded towards Lake Erie.

New York: Niagara Falls.—In a former paper Mr. Lyell alluded to the position of the remains of Mastodon, twelve feet deep, in a freshwater formation on the right bank of the river Niagara at the Falls. He remarks that if we had not been able to prove that the cataract had receded nearly four miles since the origin of the fluviatile strata in question, we should have been unable to assign any considerable duration of time as having intervened between the inhumation of the Mastodon in marl full of existing shells and the present period. The general covering of drift between Lakes Erie and Ontario is considered to be of much higher antiquity than the gravel containing the bones of the Mastodon at the Falls.

Rochester.—In the suburbs of this city remains of the *Mastodon giganteum* were found associated with existing species of Mollusca in gravel and marl below peat.

Genesee.—Here remains of the *Mastodon giganteum* were found with existing shells in a small swamp, in a cavity of the boulder formation, so that the animal must have sunk after the period of the drift, when a shallow pond fed by springs was inhabited by the same species of freshwater Mollusca as now live on the spot.

Albany and Greene Counties.—Mr. Lyell examined, in company with Mr. Hall, two swamps west of the Hudson River, where the remains of Mastodon occurred in both places at a depth of four or five feet, precisely in such situation as would yield shell marl, and peat, with remains of existing animals in Scotland. Cattle have recently been mired in these swamps.

According to Mr. Hall the greatest elevation at which Mastodon bones have been found in the United States is at the town of Hinsdale, situated on a tributary of the river Alleghany in Cattaraugus County in the State of New York, where they occur at an elevation of fifteen hundred feet above the level of the sea.

Maryland.—In the museum at Baltimore, Mr. Lyell was shown the grinder of a Mastodon, distinct from *M. giganteum*, and which had been recognized and labelled by Mr. Charlesworth as *M. longirostris*, Kaup. It was found at the depth of fifteen feet from the surface in a bed of marl near Greensburg, in Caroline County, Maryland, and is considered by Mr. Lyell as a meiocene fossil.

Atlantic border.—Between the Appalachian Mountains and the Atlantic there is a wide extent of nearly horizontal tertiary strata, which at the base of the mountains are five hundred feet and upwards in height, but decline in level nearer the ocean, and at length give place to sandy plains and low islands skirting the coast, in which strata containing marine shells of recent species are met with, slightly elevated above the sea. Occasionally deposits formed in fresh-water swamps occur, below the mean level of the Atlantic or overflowed at high tide. In this district Mr. Nuttall discovered, on the Neuse fifteen miles below Newbern, in North Carolina, a large assemblage of mammalian bones, including those of the *Mastodon giganteum*, resting on a deposit containing marine shells of recent species. Mr. Conrad presented Mr. Lyell with the tooth of a horse covered with barnacles, from this locality. Professor Owen has examined it and could find no corresponding tooth of a recent species, but considers it as agreeing with the horse-tooth brought by Mr. Darwin from the north side of the Plata in Entre Rios, in South America.

South Carolina.—Remains of the Mastodon were found in digging the Santee Canal, in a spot where large quadrupeds might now sink into the soft boggy ground.

Georgia.—Bones of the Mastodon and Megatherium occur in this district in swamps formed upon a marine sand containing shells of species now inhabiting the neighboring sea.

Mr. Lyell in conclusion offers the following observations :—

1. That the extinct animals of Bigbone Lick and those of the Atlantic border in the Carolinas and in Georgia belong to the same group, the identical species of Mastodon and elephant being in both cases associated with the horse, and while we have the Mylodon and Megatherium in Georgia, the Megalonyx is stated by several authors to have been found at Bigbone Lick.

2. On both sides of the Appalachian chain, the fossil shells, whether land or fresh-water, accompanying the bones of Mastodons, agree with species of Mollusca now inhabiting the same regions.

3. Under similar circumstances Mr. Darwin found the Mastodon and horse in Entre Rios, near the Plata, and the Megatherium, Megalonyx and Mylodon, together with the horse, in Bahia Blanca in Patagonia; these South American remains being shown by their geological position to be of later date than certain marine newer pliocene, and post-pliocene strata. Mr. Darwin also ascertained that some extinct animals of the same group are more modern in Patagonia than the drift with erratics.

4. The extinct quadrupeds before alluded to in the United States lived after the deposition of the northern drift, and consequently the coldness of climate which probably coincided in date with the transportation of the drift, was not as some pretend the cause of their extinction.

ON THE UPRIGHT FOSSIL TREES FOUND AT DIFFERENT LEVELS IN THE COAL STRATA OF CUMBERLAND, NOVA SCOTIA

Art. XI.—*On the upright Fossil Trees found at different levels in the Coal Strata of Cumberland, Nova Scotia;* by Charles Lyell, Esq., F. G. S., F. R. S., &c.

[Communicated to this Journal by the author.]

The first notice of these fossil trees was published in 1829 by Mr. Richard Brown, in Haliburton's Nova Scotia, at which time the erect trunks are described as extending through one bed of sandstone twelve feet thick. Their fossilization was attributed by Mr. Brown to the inundation of the ground on which the forest stood. Mr. Lyell in 1842 saw similar upright trees at more than ten different levels, all placed at right angles to the planes of stratification, which are inclined at an angle of 24° to the S. S. W.

The fossil trees extend over a space of from two to three miles from N. to S., and according to Dr. Gesner, to more than twice that distance from E. to W. The containing strata resemble lithologically the English coal-measures, being composed of white and brown sandstones, bituminous shales, and clay with ironstone. There are about nineteen seams of coal, the most considerable being four feet thick. The place where these are best seen is called the South Joggins, where the cliffs are from one hundred and fifty to two hundred feet high, forming the southern shore of a branch of the Bay of Fundy, called Chignecto Bay. The action of tides, which rise sixty feet, exposes continually a fresh section, and every year different sets of trees are seen in the face of the cliffs.

The beds with which the coal and erect trees are associated are not interrupted by faults. They are more than two thousand feet thick and range for nearly two miles along the coast. Immediately below them are blue grits, used for grindstones, after which there is a break in the section for three miles, when there appear, near Minudie, beds of gypsum and limestone, and at that village a deep red sandstone, the whole having the same southerly dip as the coal at the Joggins, and being considered by Mr. Lyell as the older member of the carboniferous series.

Above the coal-bearing beds and stretching southward for many miles continuously along the shore, are grits and shales of prodigious thickness, with coal plants, but without vertical trees.

Mr. Lyell next describes in detail the position and structure of the upright trees at the South Joggins. He states that no part of the original tree is preserved except the bark, which is marked externally with irregular longitudinal ridges and furrows without any leaf-scars, precisely resembling in this respect the vertical trees found at Dixonfold, on the Bolton Railway, described by Messrs. Hawkshaw and Bowman. No traces of structure could be detected in the internal cylinder of the fossil trunks, which are now filled with sandstone and shale, through which fern leaves and other plants are occasionally scattered. Mr. Lyell saw seventeen vertical trees, varying in height from six to twenty feet, and from fourteen inches to four feet in diameter. The beds which inclose the fossil trees are usually separated from each other by masses of shale and sandstone, many yards in thickness.

The trunks of the trees, which are all broken off abruptly at the top, extend through different strata, but were never seen to penetrate a seam of coal however thin. They all end downwards, either in beds of coal or shale, no instance occurring of their termination in sandstone. Sometimes the strata of shale, sandstone and clay with which the fossil trunks have been filled are much more numerous than the beds which they traverse. In one case nine distinct deposits were seen in the interior of a tree, while only three occurred on the outside in the same vertical height.

Immediately above the uppermost coal-seams and vertical trees are two strata, probably of fresh-water origin, of black calcareo-bituminous shale, chiefly made up of compressed shells of two species of Modiola and two kinds of Cypris.

Stigmariæ are abundant in the clays and argillaceous sandstones, often with their leaves attached, and spreading regularly in all directions from the stem. The other plants dispersed through the shales and sandstones bear a striking resemblance to those of the European coal-fields. Among these are Pecopteris Conchitica, Neuropteris flexuosa? Calamites cannæformis, C. approximatus, C. Steinhaueri, and C. Nodosus, Sigillaria undulata and another species. The genera Lepidodendron and Sternbergia are also present. The same plants occur at Pictou, and at Sydney in Cape Breton, accompanied with Trigonocarpum, Asterophyllites, Sphænophyllum, and other well known coal-fossils.

The author then gives a brief description of a bed of erect Calamites, first discovered by Mr. J. Dawson, in the Pictou coal-field, about one hundred miles eastward of the Cumberland coal-measures before described. They occur at Dickson's mills, one mile and a quarter west of Pictou, in a bed of sandstone about ten feet thick. They all terminate downwards at the same level, where the sandstone rests on subjacent limestone, but the tops are broken off at different heights, and Mr. Dawson observed in the same bed a prostrate Lepidodendron with leaves and Lepidostrobi attached to its branches.

From the facts above enumerated Mr. Lyell draws the following conclusions:—

1. That the erect position of the trees, and their perpendicularity to the places of stratification, imply that a thickness of several thousand feet of coal-strata, now uniformly inclined at an angle of 24°, were deposited originally in a horizontal position.

2. There must have been repeated sinkings of the dry land to allow of the growth of more than ten forests of fossil trees one above the other, an inference which is borne out by the independent evidence afforded by the Stigmaria found in the under-clays beneath coal-seams in Nova Scotia, as first noticed in South Wales by Mr. Logan.

3. The correspondence in general characters of the erect trees of Nova Scotia with those found near Manchester, leads to the opinion that this tribe of plants may have been enabled by the strength of its large roots to withstand the power of waves and currents much more effectually than the Lepidodendra and Sigillariæ, which are more rarely found to retain a perpendicular position.

Lastly, it has been objected that if seams of pure coal were formed on the ground where the vegetables grew, they would not bear so precise a resemblance to ordinary subaqueous strata, but ought to undulate like the present surface of the dry land. In answer to this Mr. Lyell points to what were undoubtedly terrestrial surfaces at the South Joggins, now represented by coal-seams or layers of shale supporting erect trees, and yet these surfaces conform as correctly to the general planes of stratification as those of any other strata.

He also shows that such an absence of superficial inequalities and such a parallelism of successive surfaces of dry land, ought to be expected according to the theory of repeated subsidence, because sedimentary deposition would continually exert its levelling action on the district submerged.

ON THE COAL FORMATION OF NOVA SCOTIA

ART. XII.—*On the Coal Formation of Nova Scotia, and on the Age and Relative Position of the Gypsum and accompanying Marine Limestones;* by C. LYELL, Esq. F. G. S. &c. &c.

[Communicated to this Journal by the author.]

THE stratified rocks of Nova Scotia more ancient than the carboniferous consist chiefly of metamorphic clay-slate and quartzite, their strike being nearly east and west. Towards their northern limits these strata become less crystalline and contain fossils, some of which Mr. Lyell identifies with species of the upper Silurian group, or with the Hamilton group of the New York geologists.

The remaining fossiliferous rocks so far as they are yet known belong to the carboniferous group, and occupy extensive tracts in the northern part of the peninsula, resting unconformably on the preceding series. They may be divided into two principal formations, one of which comprises the productive coal-measures, agreeing precisely with those of Europe in lithological and palæontological character. The other consists chiefly of Red Sandstone and red marl, with subordinate beds of gypsum and marine limestone, but this series is also occasionally associated with coal-grits, shales, and thin seams of coal. A variety of opinions have been entertained respecting the true age of the last mentioned or gypsiferous formation, and it is the purport of this paper to show, first, that it belongs to the carboniferous group; secondly, that it occupies a lower position than the productive coal-measures. These last are of vast thickness in Nova Scotia, being largely developed in Cumberland County, and near Pictou, and recurring again at Sydney in Cape Breton. In all these places they contain shales, probably deposited in a fresh-water estuary, in which several species of Cypris and Modiola abound. The plants of these coal-measures belong to the genera Calamites, Sigillaria, Stigmaria, Lepidodendron, Pecopteris, Neuropteris, Sphenopteris, Nœggerathia, Palmacites, Sternbergia, Sphenophyllum, Asterophyllites, Trigonocarpum, with which are the trunks and wood of coniferous and other trees. Upon the whole nearly fifty species of plants have been detected, more than two-thirds of which are not distinguishable from European species, while the rest agree generically with fossils of the coal formation in Europe.

The internal cylindrical axis of petrified wood in the Stigmaria of Nova Scotia exhibits the same vascular structure, and the same scalariform vessels as the English specimens.

Mr. Lyell next describes the gypsiferous formation, especially the marine limestones of Windsor, Horton, the cliffs bounding the estuary of the Schubenacadie River, the district of Brookfield, and the cliffs at the bridge crossing the Debert River near Truro. Several species of corals and shells are common to all these localities, and recur in similar limestones in Cape Breton. In this assemblage of organic remains we find a Crustacean intermediate between the Trilobite and Limulus, Orthoceras, (two species,) Nautilus, Conularia, Encrinus and Cyathophyllum, besides some species of the carboniferous limestone of Europe, such as Euomphalus lævis, Pileopsis vetustus? Pecten plicatus, Isocardia unioniformis, Producta Martini, P. Scotica? Terebratula elongata, Fenestella membranacea? Ceriopora spongites, Goldf. For assistance in determining these the author has been chiefly indebted to M. De Verneuil.

The plants associated with these limestones consist of several species of Lepidodendron, Calamites, and others agreeing with carboniferous forms. With these Mr. Lyell found in Horton Bluff scales of a Ganoid fish, and in the ripple-marked sandstones of the same place Mr. Logan discovered footsteps which appear to Mr. Owen to belong to some unknown species of reptile, constituting the first indications of the reptilian class known in the carboniferous rocks. Several of the shells and corals of this group have been recognized by Messrs. Murchison and De Verneuil as identical with fossils of the gypsiferous deposit of Perm in Russia, and it had been successively proposed, (see Proceedings of the Geological Society, Vol. III, p. 712, and Mr. Murchison's Anniversary Address, Feb. 1843, Vol. IV, p. 125,) to refer these gypsiferous beds of Nova Scotia to the Trias, and to the period of the magnesian limestone. That they are more ancient than both these formations Mr. Lyell infers, not only from their fossils, but also from their occupying a lower position than the productive coal-measures of Nova Scotia and Cape Breton. In proof of this inferiority of position three sections are referred to; first, that of the coast of Cumberland, near Minudie, where beds of red sandstone, gypsum and limestone are seen dipping southwards, or in a direction which would carry them under the productive coal-measures of the South Joggins, which attain a thickness of several miles. Secondly, the section on the East River of Pictou, where the productive coal-measures of the Albion mines repose on a formation of red sandstone including beds of limestone, in which Mr. J. Dawson and the author found Producta Martini and other fossils common to the gypsiferous rocks of Windsor, &c. Some of these limestones are oolitic like those of Windsor, and gypsum occurs near the East River, fourteen miles south of Pictou, so situated as to lead to the presumption that it is an integral part of the inferior red sandstone group. Thirdly, in Cape Breton according to information supplied by Mr. Richard Brown, the gypsiferous formation occupies a considerable tract, consisting of red marl with gypsum and limestone. In specimens of the latter Mr. Lyell finds the same fossils as those of Windsor, &c. before mentioned. Near Sydney these gypsiferous strata pass beneath a formation of sandstone more than two thousand feet thick, upon which rest conformably the coal-measures of Sydney, dipping to the northeast or seaward, and having a thickness of two thousand feet.

To illustrate the gypsiferous formation the author gives a particular description of the cliffs bordering the Schubenacadie for a distance of fourteen miles from its mouth to Fort Ellis, which he examined in company with Mr. J. W. Dawson and Mr. Duncan. The rocks here consist in great part of soft red marls, with subordinate masses of crystalline gypsum and marine limestones; also three large masses of red sandstone, coal-grits, and shales. The strike of the beds, like that at Windsor, is nearly east and west, and there are numerous faults and flexures. The principal masses of gypsum do not appear to fill rents, but form regular parts of the stratified series, sometimes alternating with limestone and shale.

The author concludes by describing a newer and unconformable red sandstone, without fossils, which is seen to rest on the edges of the carboniferous strata on the Salmon River, six miles above Truro.

NOTES ON THE CRETACEOUS STRATA OF NEW JERSEY

3. *Notes on the* CRETACEOUS STRATA *of* NEW JERSEY, *and other Parts of the United States bordering the Atlantic.* By C. LYELL, Esq., M.A., F.R.S.

THE cretaceous and tertiary deposits of America, which intervene between the Alleghany mountains and the Atlantic, bear a great resemblance in mineral character to the sandy and argillaceous portion of the formations of the same age in the south-east of England. If all the white chalk, with its flints, together with the cherty beds of the green sand, were omitted, the remaining cretaceous strata in our island would consist of loose incoherent sand with green particles, red and highly ferruginous sandstones, white sands, and (in some places) beds of lignite; the overlying tertiary deposits, consisting of marls, clays, and variously coloured sands occasionally exhibiting green particles, like those of the green sand below the chalk; and as in the bottom of the London basin near Reading. Such, for the most part, is the succession of the beds in New Jersey; and, further south, in Maryland and Virginia, the Eocene strata are often as full of green particles as the cretaceous, so that they are only distinguishable by their fossils and relative position. Even the Miocene strata are sometimes, as in Virginia, of a blueish-green colour, and contain green particles of a similar kind. This fact alone of the identity in lithological character of the secondary and tertiary strata of the United States is calculated to put us on our guard against inferring that the green and ferruginous sands of New Jersey correspond in age to the lower rather than the upper part of the European cretaceous system. It is scarcely possible, on recognising so many of the common organic forms which are familiar to us in connection with the cretaceous rocks on this side of the Atlantic, and seeing them occur in beds which have the exact mineral type of the beds below the Gault, not to feel a strong inclination to regard them as the equivalents of our greensand, and to consider the white chalk as wanting. But when we dismiss from our minds, as we ought to do, the bias derived from the consideration of the mineral aspect of the beds, and compare the fossils of New Jersey with those derived from the European chalk, we find the agreement upon the whole to be far greater with the beds occurring in Europe above the Gault, than with those which are found below it. We are indebted to Dr. Morton for having pointed out, in 1834, the general agreement of the organic remains of the American and European cretaceous fossils, while, and at the same time, he and Mr. Conrad correctly observed that almost all the species were different. He divided the strata of New Jersey into the ferruginous sand, which he compared to our green sand formation, and the calcareous strata, which he identified with the white chalk of Europe. Prof. H. D. Rogers has since divided the New Jersey cretaceous beds into five formations, which are very useful, topographically considered, but which may be overlooked in the present paper, because only two of them, namely, those alluded to by Dr. Morton, have as yet yielded a sufficient number of fossils to entitle them to rank as palæontologically distinct.

In an excursion which I made in New Jersey, in September, 1841, in company with Mr. Conrad, we went first to Bristol, on the Delaware, next, by Bordentown, to New Egypt, and returned by the Timber Creek, recrossing the Delaware at Camden. On this occasion I had an opportunity of examining the strata of both these formations, and I collected nearly all the fossil species described by Dr. Morton, together with some few additional ones. I shall now, therefore, briefly notice these two deposits and their fossils, and consider them in reference to their European equivalents.

Although in this part of New Jersey there is no white chalk with flints, so characteristic of rocks of the same age in Europe, it is still impossible to glance at the fossils and not be convinced that Dr. Morton was right, as before hinted, when in 1834 he referred the New Jersey deposits to the European Cretaceous era, and remarked that the American species of shells were nearly all new or distinct from those before described, and yet very analogous to those of the chalk already known. Of the two well-marked subdivisions of the Cretaceous system the lower consists in great part of green sand and green marl, and was supposed by Dr. Morton, as already mentioned, to be the equivalent of the English green sand, while an upper or calcareous rock, composed chiefly of a soft straw-coloured limestone with corals, was thought to correspond with the white chalk of Europe. But after carefully comparing my collection, comprising about sixty species of shells, besides many corals and other remains, I have arrived at the conclusion that the whole of the New Jersey series agrees in its chronological relations with the European white chalk, or, to speak more precisely, with the formations ranging from the Gault to the Maestricht beds inclusive. Among the shells, in determining which I have been assisted by Professor E. Forbes, not more than five out of sixty seem to be quite identical with European species; but several others approach very near to, and may be the same as Europeans; and at least fifteen may be regarded as good geographical representations of well-known chalk fossils belonging for the most part to beds above the Gault in Europe. There are a few very peculiar forms among the American testacea, such as *Terebratula Sayii* Morton; and I found among the univalves a Bulla, but casts of the genus had previously been mentioned by Dr. Morton, and although not yet known in the European chalk a species occurs on the Continent in beds of the Jurassic system.

In the upper or straw-coloured limestones, I found on the banks of the Timber Creek, twelve miles south-east of Philadelphia, six species of corals* and several echinoderms, chiefly allied to *upper* cretaceous forms. The same calcareous formation also abounds in Foraminifera characteristic of the chalk, comprising, among others, the genera *Cristellaria*, *Rotalina*, and *Nodosaria*. Besides the shells there are also several remains of fishes, and of the series obtained by myself all those referred to the genus *Lamna* resembled species occurring in our chalk. They have been examined for me by Sir P. Egerton. One of them seems to approach very closely to *Lamna appendiculata*, and another comes very near to *Galeus pristodontus*; and indeed, if we may judge by so few specimens, seem identical. These are fossils of our upper chalk in Europe. There are also several forms of *Carcharias* not very unlike some tertiary species given me from the New Jersey chalk, several of which are figured by Dr. Morton; I will not dwell upon these however, since in Europe also there are many of the cretaceous Squalidæ which can scarcely, when the teeth alone are considered, be distinguished specifically from tertiary fossils.

There are three Saurian vertebræ in the New Jersey green sand in the collection of the Geological Society, which I have submitted to Mr. Owen's inspection. One of these, from the green sand of Mullica Hill, is the anterior dorsal vertebra of the Mosasaurus. Another is the posterior cervical vertebra of a Pliosaurus, a genus which Mr. Owen has constituted to include a portion of the Plesiosauri, and which approach still more nearly to the true Saurians. The vertebra in question resembles very closely that of *Pliosaurus brachydeirus* of the Kimmeridge clay. Until very lately, the Plesiosaurian type was not known higher in the series than in the Oolites; but it has now been shown to ascend to the chalk of Europe, so that its occurrence in the New Jersey strata is in strict accordance with European analogies. The third specimen (pre-

* These have been described by Mr. Lonsdale, and the description and figures will be given at the end of the present paper.

sented, I believe, by Professor H. D. Rogers) is labelled, "Woodstown, New Jersey;" a locality where those beds occur to which the great mass of shells before alluded to belong. It is a vertebra, penetrated by the green particles of the sand. Mr. Owen refers this to the dorsal vertebra of a crocodile of his *Procœlian* division, or those which, like the recent crocodiles, have the concavity in the forepart, and the convexity behind. This fact is important, as hitherto the Procœlian crocodiles in Europe have not been found in beds older than the eocene.

In concluding these remarks on the ferruginous and green sand formation of New Jersey, I may observe that the identification of four or five species out of sixty fossil shells with European cretaceous fossils would give an agreement of about seven per cent., which is by no means a small amount of correspondence, when we consider that the part of the United States above alluded to is distant between 3000 and 4000 miles from the chalk of Central and Northern Europe, and that there is more than 10° difference in the latitudes of the two districts compared, on the opposite sides of the Atlantic. It may doubtless be true, that the influence of temperature during the Cretaceous period was less powerful in limiting the range of species than it is now; and that the same forms prevailed more uniformly from India to Sweden, than they do at present. Nevertheless, the cretaceous fossils of Northern and Southern Europe differ sufficiently to show that the climate had then no small influence in causing distinct geographical provinces of species; and it seems natural that those species which are very abundant in Europe, such as *Belemnites mucronatus*, or those which have a great vertical range, such as *Pecten quinquecostatus*, should be the fossils found, if any, to recur in a distant part of the globe.

In the next place I proceed to give some account of the upper fossiliferous division of the New Jersey cretaceous deposit, which is for the most part arenaceous, but contains, in many places, layers of limestone and calcareous sand, with corals slightly aggregated together. It has been traced by Mr. Rogers to a distance of about 60 miles in a north-east and south-west direction, from Prosper Town to near Salem, having rarely a breadth of half a mile, and the thickness being from 6 to 20 feet. Its importance is derived, geologically speaking, from its fossils, and, in an economical point of view, from its affording the only lime procurable in this district. I saw the formation in question, on the banks of Timber Creek, a stream which flows into the Delaware, three miles below Philadelphia. The principal locality is twelve miles S. E. of Philadelphia, about a mile and a half south of the village of White Horse, in Gloucester County, New Jersey. Here a bed of soft calcareous stone, about 20 feet thick, is seen made up, in great part, of corals of the genera *Eschara*, *Escharina*, *Cellepora*, *Tubulipora*, and others*, together with the remains of echinoderms, such as *Cidaris* and *Spatangus*. It contains also some shells, as *Scalaria annu-*

* See the description of these corals by Mr. Lonsdale, in the Appendix.

lata, *Gastrochæna*, and *Teredo*, the whole indicating the sandy bottom of a shallow sea. I was so strongly reminded of the coralline crag of Sudbourn, and other places in Suffolk, when examining this rock, that I had some difficulty at first in persuading myself that it was not a tertiary deposit. It is, in a great part, a mass of white calcareous sand, more or less aggregated together, and the upper surface has been irregularly scooped out and rendered undulating, and is covered with a newer deposit of red clay and gravel, without fossils, the surface of which is even and level. This white sand and limestone pass downwards into light-green and ferruginous sand, with quartzose grains.

Near Hornerstown, I saw, on a branch of the Timber Creek, to which Mr. Conrad conducted me, a bed of this coralline aggregate, 8 feet thick, resting on the green sand or lower deposit before mentioned, with its characteristic fossils.

We have now to consider whether the calcareous or upper formation has been referred with propriety to the chalk. Mr. Forbes has examined the Echinoderms, and is of opinion that they are decidedly analogous to cretaceous forms. One of the species of *Spatangus* belongs to the same group as *S. subglobosus* of Goldfuss, a group which forms the genus *Holaster* of Agassiz, and which that naturalist regards as very characteristic of the upper part of the Cretaceous system.

One also of two species of Cidaris is allied to *C. vesiculosus*, and to other upper cretaceous species of Europe.

Dr. Morton had already observed, in regard to the corals, that some of the species resemble a Maestricht fossil, figured by Goldfuss; and the reader is referred to Mr. Lonsdale's comments on this subject in the Appendix.

The fossil called by Dr. Morton "*Belemnites ambiguus*," though probably not related to the Belemnite, is closely allied to a fossil which I have collected myself in the chalk of Sweden, associated with *Belemnites mucronatus*.

The last-mentioned, or upper of the two fossiliferous formations of New Jersey, has been called by Dr. Morton and Mr. Conrad the Medial Cretaceous, because there are others still higher in position in the Southern States, which they refer to the chalk period. One member of these, a white limestone, seen extensively on the Santee canal, and in other parts of South Carolina, as well as at Jacksonborough and Shell Bluff in Georgia, I have shown, in a former communication to the Society, to be Eocene tertiary. Another portion, called the Nummulite limestone of Alabama, I have not examined, and can therefore offer no opinion respecting it.

Upon the whole, the collection of fossils which I made in New Jersey confirms the principal conclusion to which Dr. Morton arrived, that there is a remarkable generic accordance between the fossil mollusca, corals, echinoderms, fish, and saurians of the cretaceous group, in New Jersey and in Europe. But the general analogy of the generic, and the identity of some specific, forms, which Mr. Forbes and Mr. Lonsdale have assisted me in comparing, has led me to refer all the fossiliferous formations of New Jersey to that part of the European series which ranges from the Maestricht beds to the gault inclusive.

North Carolina.

Of the same age are certain strata in North Carolina, at a place called Lewis's Creek near South Washington, forty miles north of Wilmington, and 340 geographical miles south-west of New Jersey, where I found *Belemnites mucronatus*, *Ostrea vesicularis*, *O. subspatulata* (a remarkable and new species figured in the Appendix), *Cellepora tubulata*, and other fossils.

The association of *Cellepora tubulata*, which abounds in the upper cretaceous formation of New Jersey at Timber Creek with *Belemnites mucronatus* in this locality of South Carolina, is important, as helping to show the near relation of the coralline limestone of New Jersey to the green sand containing Belemnites.

Georgia.

Some fossils have been communicated to me by Dr. Cotting, from Georgia, which make it probable that there are cretaceous strata there, lower than those of New Jersey; as among them are a *Pholadomya* and an *Ammonite;* both of which Mr. Forbes finds to be closely allied to certain Neocomien species from Neuchâtel.

In the collection of Mr. Conrad, from Alabama, I saw a species of Hippurite, derived from the cretaceous strata of that State, which I believe is the only example of any fossil of the Rudist family derived from the cretaceous rocks of North America. It affords another point of analogy between the cretaceous fauna co-existing on opposite sides of the Atlantic.

It is interesting to find, as the result of this investigation, that the marine fauna, whether vertebrate or invertebrate, testaceous or zoophytic, was divided at the remote epoch under consideration, as it is now, into distinct geographical provinces, although the geologist may every where recognise the cretaceous type, whether in Europe or America (and I might add India). This peculiar type exhibits the preponderating influence of a vast combination of circumstances prevailing at one period throughout the globe, circumstances dependent on the state of the physical geography, climate, and organic world, in the period immediately preceding, together with a variety of other conditions.

ON THE PROBABLE AGE AND ORIGIN OF A BED OF PLUMBAGO AND ANTHRACITE

2. *On the probable Age and Origin of a Bed of* PLUMBAGO *and* ANTHRACITE *occurring in mica-schist near Worcester, Massachussetts.* By C. LYELL, Esq., M.A., F.R.S., F.G.S., &c.

A BED of plumbago and impure anthracite described by Professor Hitchcock in his "Geology of Massachussetts" is found interstratified with mica-schist near Worcester, forty-five miles due west of Boston. It is about two feet in thickness, and has been made use of both as fuel and in the manufacture of lead-pencils. It is much mixed up with the associated rock, has the touch and somewhat of the lustre of plumbago, and gives a streak on paper. It is occasionally iridescent like coal, contains pyrites, which is also found in the associated clay slate and garnetiferous mica-schist,

* The accompanying plate exhibits the appearance of the fossil embedded in the rock, and partially cleared.

both of which are impregnated with carbonaceous matter. This plumbaginous rock presents numerous polished surfaces or slickensides, on some of which are delicate parallel striæ, which reminded me so strongly of the finely striated leaves common in coal, that I had first supposed them to be of the same nature; and was thus induced to search very diligently, but in vain, for vegetable impressions. In the old mine of plumbaginous anthracite about two miles to the north-east of Worcester, the accompanying clay-slate and mica-schist, (the latter containing garnets and veins of asbestus), dip towards the north at an angle of between 30 and 40 degrees, and a railway cutting east of Worcester, and to the south of the old mine above mentioned, has finely exposed to view the mica-slate and clay-slate with some layers of quartz inclosing a bed of similar plumbaginous anthracite; some of which has the iridescence of peacock coal. Professor Hitchcock has traced this group of strata in a north-easterly direction 50 miles, to the Merrimac river, and the beds are continued with the same strike, in a south-westerly direction, for many miles. In their course they exhibit a great variety of crystalline strata, and the mica-slate sometimes alternates with gneiss.

The schists including plumbago, at Worcester, as above mentioned, are separated from the anthracite occurring on the borders of Rhode Island and Massachussetts, by a district of gneiss and hornblende slate about thirty miles wide. The anthracite of those slates is impure and earthy, but has been worked for coal at Wrentham, Cumberland, Mansfield, and other places, where, in the accompanying carbonaceous and pyritiferous slates, I collected numerous impressions of the most common coal plants, such as *Pecopteris plumosa, Neuropteris flexuosa, Sphenophyllum, Calamites,* &c. This earthy anthracitic coal, as well as the accompanying slates, contain pyrites, as at Worcester; and the anthracite exhibits the same glazed surfaces and slickensides; but it does not soil the finger like that of Worcester, and its specific gravity has been shown by Dr. Jackson and Professor Hitchcock to be less than that of Worcester, but greater than that of the anthracite of Pennsylvania. There are layers and veins of quartz in the slates and micaceous sandstones forming the roof of this anthracite, affording another point of analogy between this series and the quartziferous rocks at Worcester. I have also seen numerous specimens from the anthracite and bituminous slate of the neighbouring district of Mansfield, in which the usual coal plants were imbedded; but the slate was more crystalline. I was presented by Professor Hitchcock with a specimen of distinct mica slate from this neighbourhood, in which rounded nodules of granite and quartz rock are included; and in the contiguous parts of Rhode Island a conglomerate belonging to what has been called the greywacké formation has been observed by Dr. C. T. Jackson (Survey of Rhode Island, p. 70.) to pass downwards into mica slate. The rocks of the coal measures now under consideration are accompanied by a red sandstone, which I examined at Attleborough, a few miles from Wrentham before mentioned. There is a conglomerate subordinate to it, and it may, with all probability, be referred to the old red sandstone which occurs beneath the coal in Pennsylvania; although, in the absence of fossils, the disturbed state of the strata, and the frequent concealment of their outcroppings by a thick covering of drift, it is usually difficult to determine the exact order of succession. It is, however, important to observe that the whole of this series, which Professor Hitchcock now inclines to refer to the coal and old red sandstone, was formerly called greywacké, and styled the transition formation, in consequence of the semi-metamorphic condition of several of the rocks. Their conversion into crystalline strata, in the immediate neighbourhood of masses of granite and syenite, is often complete. But besides this kind of alteration, resembling the effect of dykes and veins of intrusive igneous rocks, there is evidence here, as in the Alps of the Canton of Berne and elsewhere, of a more extensive and general change by chemical or plutonic action, affecting, with greater or less degrees of intensity, dense masses of stratified rocks.

Although many impressions of plants have been found in this anthracite formation, on the southern borders of Massachussetts and Rhode Island no traces of shells or corals have been discovered. In like manner we find an absence of all fossils except vegetable remains, in the anthracite coal district of Pennsylvania, and no fossils of any kind in the subjacent conglomerates and red sandstones.

The strata of conglomerate at Brooklyn, near Boston, and the greywacké slates and sandstones of that neighbourhood, some of which pass into metamorphic rocks, and in which no plants or other organic remains have as yet been found, are doubtless referable to the same carboniferous and Devonian formations as those above described.

After traversing this region in several directions, it appeared to me very probable that the stratified rocks, containing the plumbaginous anthracite of Worcester, consisted originally of similar sedimentary strata, which have been so altered by heat and other plutonic causes as to assume a crystalline and metamorphic texture, by which the grits and shales of the coal have been turned into quartzite, clay-slate, and mica-schist, and the anthracite into that state of carbon which is called plumbago or graphite.

The progressive debituminisation of the coal of the United States, as we proceed from Pittsburgh to the eastern and more disturbed axes of the Alleghany mountains, as pointed out by Professor H. D. Rogers, lend support to this conjecture.* In the Rhode Island anthracite, which is less combustible than that of Potsville, Pennsylvania, the change seems to have been carried farther; the volatile ingredients of the original coal having been still more completely expelled. In the impure plumbago of Worcester, we may have the last step in the series of transmutation, where only 3 per cent. of gaseous matter remains, where all traces of fossil plants

* See Appendix.

and vegetable structure have been obliterated, and where the lithological character of the accompanying sedimentary rocks has been entirely altered. I may remark that the Silurian and Devonian formations, which are so largely developed in the United States, yield no beds of coal or anthracite which could by metamorphosis be supposed to become turned into such a carbonaceous stratum as that of Worcester.

I shall conclude by observing that the difference of strike between the mica-schist containing plumbago at Worcester, and the nearest carboniferous rocks of Rhode Island and Massachussetts, affords no argument against the theory of both having belonged originally to the same group of sedimentary strata. In New England, and in Nova Scotia, the coal-measures frequently deviate widely from the same strike in continuous districts, and the direction of continuous anticlinal axes in the Alleghany mountains, composed throughout of similar silurian and carboniferous rocks, has been shown by Professors W. B. and H. D. Rogers, to vary more than 40° in different sections of that chain.

Notes on the Cretaceous Strata of New Jersey

3. *Notes on the* CRETACEOUS STRATA *of* NEW JERSEY, *and other Parts of the United States bordering the Atlantic.* By C. LYELL, Esq., M.A., F.R.S.

THE cretaceous and tertiary deposits of America, which intervene between the Alleghany mountains and the Atlantic, bear a great resemblance in mineral character to the sandy and argillaceous portion of the formations of the same age in the south-east of England. If all the white chalk, with its flints, together with the cherty beds of the green sand, were omitted, the remaining cretaceous strata in our island would consist of loose incoherent sand with green particles, red and highly ferruginous sandstones, white sands, and (in some places) beds of lignite; the overlying tertiary deposits, consisting of marls, clays, and variously coloured sands occasionally exhibiting green particles, like those of the green sand below the chalk; and as in the bottom of the London basin near Reading. Such, for the most part, is the succession of the beds in New Jersey; and, further south, in Maryland and Virginia, the Eocene strata are often as full of green particles as the cretaceous, so that they are only distinguishable by their fossils and relative position. Even the Miocene strata are sometimes, as in Virginia, of a blueish-green colour, and contain green particles of a similar kind. This fact alone of the identity in lithological character of the secondary and tertiary strata of the United States is calculated to put us on our guard against inferring that the green and ferruginous sands of New Jersey correspond in age to the lower rather than the upper part of the European cretaceous system. It is scarcely possible, on recognising so many of the common organic forms which are familiar to us in connection with the cretaceous rocks on this side of the Atlantic, and seeing them occur in beds which have the exact mineral type of the beds below the Gault, not to feel a strong inclination to regard them as the equivalents of our greensand, and to consider the white chalk as wanting. But when we dismiss from our minds, as we ought to do, the bias derived from the consideration of the mineral aspect of the beds, and compare the fossils of New Jersey with those derived from the European chalk, we find the agreement upon the whole to be far greater with the beds occurring in Europe above the Gault, than with those which are found below it. We are indebted to Dr. Morton for having pointed out, in 1834, the general agreement of the organic remains of the American and European cretaceous fossils, while, and at the same time, he and Mr. Conrad correctly observed that almost all the species were different. He divided the strata of New Jersey into the ferruginous sand, which he compared to our green sand formation, and the calcareous strata, which he identified with the white chalk of Europe. Prof. H. D. Rogers has since divided the New Jersey cretaceous beds into five formations, which are very useful, topographically considered, but which may be overlooked in the present paper, because only two of them, namely, those alluded to by Dr. Morton, have as yet yielded a sufficient number of fossils to entitle them to rank as palæontologically distinct.

In an excursion which I made in New Jersey, in September, 1841, in company with Mr. Conrad, we went first to Bristol, on the Delaware, next, by Bordentown, to New Egypt, and returned by the Timber Creek, recrossing the Delaware at Camden. On this occasion I had an opportunity of examining the strata of both these formations, and I collected nearly all the fossil species described by Dr. Morton, together with some few additional ones. I shall now, therefore, briefly notice these two deposits and their fossils, and consider them in reference to their European equivalents.

Although in this part of New Jersey there is no white chalk with flints, so characteristic of rocks of the same age in Europe, it is still impossible to glance at the fossils and not be convinced that Dr. Morton was right, as before hinted, when in 1834 he referred the New Jersey deposits to the European Cretaceous era, and remarked that the American species of shells were nearly all new or distinct from those before described, and yet very analogous to those of the chalk already known. Of the two well-marked subdivisions of the Cretaceous system the lower consists in great part of green sand and green marl, and was supposed by Dr. Morton, as already mentioned, to be the equivalent of the English green sand, while an upper or calcareous rock, composed chiefly of a soft straw-coloured limestone with corals, was thought to correspond with the white chalk of Europe. But after carefully comparing my collection, comprising about sixty species of shells, besides many corals and other remains, I have arrived at the conclusion that the whole of the New Jersey series agrees in its chronological relations with the European white chalk, or, to speak more precisely, with the formations ranging from the Gault to the Maestricht beds inclusive. Among the shells, in determining which I have been assisted by Professor E. Forbes, not more than five out of sixty seem to be quite identical with European species; but several others approach very near to, and may be the same as Europeans; and at least fifteen may be regarded as good geographical representations of well-known chalk fossils belonging for the most part to beds above the Gault in Europe. There are a few very peculiar forms among the American testacea, such as *Terebratula Sayii* Morton; and I found among the univalves a Bulla, but casts of the genus had previously been mentioned by Dr. Morton, and although not yet known in the European chalk a species occurs on the Continent in beds of the Jurassic system.

In the upper or straw-coloured limestones, I found on the banks of the Timber Creek, twelve miles south-east of Philadelphia, six species of corals* and several echinoderms, chiefly allied to *upper* cretaceous forms. The same calcareous formation also abounds in Foraminifera characteristic of the chalk, comprising, among others, the genera *Cristellaria*, *Rotalina*, and *Nodosaria*. Besides the shells there are also several remains of fishes, and of the series obtained by myself all those referred to the genus *Lamna* resembled species occurring in our chalk. They have been examined for me by Sir P. Egerton. One of them seems to approach very closely to *Lamna appendiculata*, and another comes very near to *Galeus pristodontus*; and indeed, if we may judge by so few specimens, seem identical. These are fossils of our upper chalk in Europe. There are also several forms of *Carcharias* not very unlike some tertiary species given me from the New Jersey chalk, several of which are figured by Dr. Morton; I will not dwell upon these however, since in Europe also there are many of the cretaceous Squalidæ which can scarcely, when the teeth alone are considered, be distinguished specifically from tertiary fossils.

There are three Saurian vertebræ in the New Jersey green sand in the collection of the Geological Society, which I have submitted to Mr. Owen's inspection. One of these, from the green sand of Mullica Hill, is the anterior dorsal vertebra of the Mosasaurus. Another is the posterior cervical vertebra of a Pliosaurus, a genus which Mr. Owen has constituted to include a portion of the Plesiosauri, and which approach still more nearly to the true Saurians. The vertebra in question resembles very closely that of *Pliosaurus brachydeirus* of the Kimmeridge clay. Until very lately, the Plesiosaurian type was not known higher in the series than in the Oolites; but it has now been shown to ascend to the chalk of Europe, so that its occurrence in the New Jersey strata is in strict accordance with European analogies. The third specimen (pre-

* These have been described by Mr. Lonsdale, and the description and figures will be given at the end of the present paper.

sented, I believe, by Professor H. D. Rogers) is labelled, "Woodstown, New Jersey;" a locality where those beds occur to which the great mass of shells before alluded to belong. It is a vertebra, penetrated by the green particles of the sand. Mr. Owen refers this to the dorsal vertebra of a crocodile of his *Procœlian* division, or those which, like the recent crocodiles, have the concavity in the forepart, and the convexity behind. This fact is important, as hitherto the Procœlian crocodiles in Europe have not been found in beds older than the eocene.

In concluding these remarks on the ferruginous and green sand formation of New Jersey, I may observe that the identification of four or five species out of sixty fossil shells with European cretaceous fossils would give an agreement of about seven per cent., which is by no means a small amount of correspondence, when we consider that the part of the United States above alluded to is distant between 3000 and 4000 miles from the chalk of Central and Northern Europe; and that there is more than 10° difference in the latitudes of the two districts compared, on the opposite sides of the Atlantic. It may doubtless be true, that the influence of temperature during the Cretaceous period was less powerful in limiting the range of species than it is now; and that the same forms prevailed more uniformly from India to Sweden, than they do at present. Nevertheless, the cretaceous fossils of Northern and Southern Europe differ sufficiently to show that the climate had then no small influence in causing distinct geographical provinces of species; and it seems natural that those species which are very abundant in Europe, such as *Belemnites mucronatus*, or those which have a great vertical range, such as *Pecten quinquecostatus*, should be the fossils found, if any, to recur in a distant part of the globe.

In the next place I proceed to give some account of the upper fossiliferous division of the New Jersey cretaceous deposit, which is for the most part arenaceous, but contains, in many places, layers of limestone and calcareous sand, with corals slightly aggregated together. It has been traced by Mr. Rogers to a distance of about 60 miles in a north-east and south-west direction, from Prosper Town to near Salem, having rarely a breadth of half a mile, and the thickness being from 6 to 20 feet. Its importance is derived, geologically speaking, from its fossils, and, in an economical point of view, from its affording the only lime procurable in this district. I saw the formation in question, on the banks of Timber Creek, a stream which flows into the Delaware, three miles below Philadelphia. The principal locality is twelve miles S. E. of Philadelphia, about a mile and a half south of the village of White Horse, in Gloucester County, New Jersey. Here a bed of soft calcareous stone, about 20 feet thick, is seen made up, in great part, of corals of the genera *Eschara*, *Escharina*, *Cellepora*, *Tubulipora*, and others*, together with the remains of echinoderms, such as *Cidaris* and *Spatangus*. It contains also some shells, as *Scalaria annu-*

* See the description of these corals by Mr. Lonsdale, in the Appendix.

lata, *Gastrochæna*, and *Teredo*, the whole indicating the sandy bottom of a shallow sea. I was so strongly reminded of the coralline crag of Sudbourn, and other places in Suffolk, when examining this rock, that I had some difficulty at first in persuading myself that it was not a tertiary deposit. It is, in a great part, a mass of white calcareous sand, more or less aggregated together, and the upper surface has been irregularly scooped out and rendered undulating, and is covered with a newer deposit of red clay and gravel, without fossils, the surface of which is even and level. This white sand and limestone pass downwards into light-green and ferruginous sand, with quartzose grains.

Near Hornerstown, I saw, on a branch of the Timber Creek, to which Mr. Conrad conducted me, a bed of this coralline aggregate, 8 feet thick, resting on the green sand or lower deposit before mentioned, with its characteristic fossils.

We have now to consider whether the calcareous or upper formation has been referred with propriety to the chalk. Mr. Forbes has examined the Echinoderms, and is of opinion that they are decidedly analogous to cretaceous forms. One of the species of *Spatangus* belongs to the same group as *S. subglobosus* of Goldfuss, a group which forms the genus *Holaster* of Agassiz, and which that naturalist regards as very characteristic of the upper part of the Cretaceous system.

One also of two species of Cidaris is allied to *C. vesiculosus*, and to other upper cretaceous species of Europe.

Dr. Morton had already observed, in regard to the corals, that some of the species resemble a Maestricht fossil, figured by Goldfuss; and the reader is referred to Mr. Lonsdale's comments on this subject in the Appendix.

The fossil called by Dr. Morton "*Belemnites ambiguus*," though probably not related to the Belemnite, is closely allied to a fossil which I have collected myself in the chalk of Sweden, associated with *Belemnites mucronatus*.

The last-mentioned, or upper of the two fossiliferous formations of New Jersey, has been called by Dr. Morton and Mr. Conrad the Medial Cretaceous, because there are others still higher in position in the Southern States, which they refer to the chalk period. One member of these, a white limestone, seen extensively on the Santee canal, and in other parts of South Carolina, as well as at Jacksonborough and Shell Bluff in Georgia, I have shown, in a former communication to the Society, to be Eocene tertiary. Another portion, called the Nummulite limestone of Alabama, I have not examined, and can therefore offer no opinion respecting it.

Upon the whole, the collection of fossils which I made in New Jersey confirms the principal conclusion to which Dr. Morton arrived, that there is a remarkable generic accordance between the fossil mollusca, corals, echinoderms, fish, and saurians of the cretaceous group, in New Jersey and in Europe. But the general analogy of the generic, and the identity of some specific, forms, which Mr. Forbes and Mr. Lonsdale have assisted me in comparing, has led me to refer all the fossiliferous formations of New Jersey to that part of the European series which ranges from the Maestricht beds to the gault inclusive.

North Carolina.

Of the same age are certain strata in North Carolina, at a place called Lewis's Creek near South Washington, forty miles north of Wilmington, and 340 geographical miles south-west of New Jersey, where I found *Belemnites mucronatus, Ostrea vesicularis, O. subspatulata* (a remarkable and new species figured in the Appendix), *Cellepora tubulata*, and other fossils.

The association of *Cellepora tubulata*, which abounds in the upper cretaceous formation of New Jersey at Timber Creek with *Belemnites mucronatus* in this locality of South Carolina, is important, as helping to show the near relation of the coralline limestone of New Jersey to the green sand containing Belemnites.

Georgia.

Some fossils have been communicated to me by Dr. Cotting, from Georgia, which make it probable that there are cretaceous strata there, lower than those of New Jersey; as among them are a *Pholadomya* and an *Ammonite;* both of which Mr. Forbes finds to be closely allied to certain Neocomien species from Neuchâtel.

In the collection of Mr. Conrad, from Alabama, I saw a species of Hippurite, derived from the cretaceous strata of that State, which I believe is the only example of any fossil of the Rudist family derived from the cretaceous rocks of North America. It affords another point of analogy between the cretaceous fauna co-existing on opposite sides of the Atlantic.

It is interesting to find, as the result of this investigation, that the marine fauna, whether vertebrate or invertebrate, testaceous or zoophytic, was divided at the remote epoch under consideration, as it is now, into distinct geographical provinces, although the geologist may every where recognise the cretaceous type, whether in Europe or America (and I might add India). This peculiar type exhibits the preponderating influence of a vast combination of circumstances prevailing at one period throughout the globe, circumstances dependent on the state of the physical geography, climate, and organic world, in the period immediately preceding, together with a variety of other conditions.

ON THE MIOCENE TERTIARY STRATA OF MARYLAND, VIRGINIA AND OF NORTH AND SOUTH CAROLINA

THE

QUARTERLY JOURNAL

OF

THE GEOLOGICAL SOCIETY OF LONDON.

I. PROCEEDINGS

OF

THE GEOLOGICAL SOCIETY.

February 26. 1845.

John Fowler, Esq. C. E., of Stockton-on-Tees, was elected a Fellow of this Society.

The following communications were read: —

1. *On the Miocene Tertiary Strata of* Maryland, Virginia, *and of* North *and* South Carolina. By Charles Lyell, Esq., M.A., F.R.S., &c.

Between the hilly country of the United States and the Atlantic there intervenes a low and nearly level region, occupied principally by beds of marl, clay, and sand of the cretaceous and tertiary formations. Maclure, in 1817, in his "Geology of the United States," laid down, with no small approach to accuracy, on a coloured map, the general limits of this great plain, and of the granitic district lying immediately to the westward. He also pointed out, that at the junction of these great geological regions almost all the great rivers descend suddenly by falls or rapids of moderate height, as the Delaware at Trenton, the Schuylkill near Philadelphia, the Potomac near Washington, the James River at Richmond, Virginia, the Savannah at Augusta in Georgia, and many others. At these points, therefore, the navigation is stopped, and a great many large cities have sprung up; so that the line which marks the western boundary of the tertiary, and the eastern of the granitic region, is at the same time one of peculiar geological, geographical, and political interest.

The general elevation of the great plain does not exceed a hundred feet, although it is sometimes considerably higher. Its width in the middle and southern States is very commonly from 100 to 150 miles. The tide, except in the more southern States, flows entirely across it, and the rivers intersecting it form large estuaries, which may have been due to the facility with which the incoherent materials of the cliffs were undermined and swept away, a process of waste still going on.

Throughout the greater part of the Atlantic plain, the cretaceous rocks, if present, are concealed by the overlying tertiary deposits, which consist chiefly of Miocene strata, extending from Delaware Bay to the Cape Fear River, and occupying portions of Delaware, Maryland, Virginia, and N. Carolina, an area about 400 miles long from north to south, and varying in breadth from 10 to 70 miles. There are, besides, some patches of the miocene formation in South Carolina and Georgia, where the Eocene, or older tertiary deposits, predominate almost exclusively, and where the limits of the miocene deposits have not yet been well ascertained.

I have endeavoured to show, in a former paper, read to the Society in February, 1843 *, that the fossils of Martha's Vineyard, an island off the coast of Massachusetts, especially the teeth of sharks, point to a near chronological connexion between the strata of that island and the miocene strata of Europe; but the evidence is far more complete and satisfactory in favour of a similarity of age between the deposits about to be described, which occur 350 miles to the south, or in Maryland and Virginia. In the last-mentioned States these formations extend over a wide area, and have been well described by Mr. Conrad, who identified them in age with the English crag, and called them "medial pliocene," and by Professors W. B. and H. D. Rogers, in their article in the Philos. Trans. of Philadelphia for 1836. These authors considered them entitled to the appellation of Miocene, as containing a proportion of recent shells corresponding to that known to characterise the miocene deposits of Europe.

The principal grounds which induce me to agree in this opinion, and to regard the beds of sand, clay, and marl now under consideration as corresponding in geological age with the crag of Suffolk, the faluns of the Loire, and other contemporaneous formations in Europe, are the following: —

First. On the banks of the James River and elsewhere I saw

* Proceed. of Geol. Soc., vol. iv. p. 31.

them resting on eocene deposits, or on sand and clay containing an assemblage of shells resembling those of the London and Paris basins.

2dly. The genera of shells, and the amount of species by which they are represented, agree for the most part with those which characterise the European miocene beds. Many of the most abundant species are allied, and some few are identical.

3dly. The proportion of fossil shells identified with recent species amounts to about 17 per cent, or about one-sixth of the whole, in 147 species collected by me, the recent species agreeing almost entirely with those now living in the neighbouring parts of the Atlantic.

4thly. There is the same mixture of shells of northern and southern forms as in the French faluns.

5thly. Of ten species of corals, all but one agree generically with those of the miocene beds of Europe, and two at least specifically. One of these is a *Lunulite*, the same as a fossil from the Suffolk crag and also from the faluns of Touraine; a second, *Anthophyllum lineatum*, is also common in the French faluns.

6thly. Among the remains of fish are several of the shark family, agreeing with species from the miocene beds of Europe, the faluns of Touraine, the molasse of Switzerland, and the tertiary formation of Malta, among which I may mention *Carcharias megalodon*, *C. productus*, *Lamna xiphodon*, *L. hastalis*, *L. cuspidata*. The ossicles also of the ears of fishes closely resemble those occurring in the Suffolk crag.

7thly. The absence of reptiles is another point of analogy, as also the presence of huge cetaceous remains.

The miocene deposits of the United States consist chiefly of incoherent sand and clay, and in this respect bear a strong resemblance to those of the same age in Europe. The associated siliceous sand imparts a sterile character to the space they occupy, but the soil has often been fertilised by the use of shell marl, derived by the agriculturist from parts of the same formation, a practice which Mr. Ruffin, of Petersburg, Virginia, editor of the Farmer's Register, has done much to promote. This use of the miocene marl and fossil shells affords another singular point of coincidence between the American strata, the English crag, and the faluns of the Loire, all of which furnish the same fertilising calcareous materials for improving light soils.

I began my examination of these middle tertiary deposits in the suburbs of Richmond, Virginia, where I saw in Shockoe Creek red clay and sand, from which I obtained *Artemis acetabulum*, and casts of miocene species of *Astarte* and *Mactra*, reposing on eocene marls with characteristic shells. Between the two formations, a remarkable bed of whitish and yellowish siliceous clay intervenes, from twelve to twenty-five feet thick, of an extremely fine texture. It affords a very sterile soil, and its course through the country is marked on the surface by a band of meagre vegetation, without trees or shrubs. It has been described by Professor

Wm. Rogers, in his State Report on the Geology of Virginia for 1840. When examined under the microscope, it is found to consist of an impalpable siliceous powder, derived from the cases of minute animalcules. Dr. Bailey has shown that the siliceous skeletons belong to several species of *Navicula*, *Gaillonella*, *Actinocyclas*, and other genera. The position of the infusorial earth, the bed being perfectly conformable to the miocene strata above, and the eocene below, is not decisive of its age; but I understand that the species are considered by Messrs. Tuomey, W. B. Rogers, and Bailey, as implying that it belongs to the miocene formation; and Dr. Mantell, who has examined them, informs me that many of them belong to living species.

On the right bank of the James River, at City Point, Virginia, about twenty miles below Richmond, in a cliff about thirty feet high, I observed the yellow and white miocene sands resting on dark green earth and marl of the eocene formation, just as the yellow sands of the crag rest on the blue London clay in some parts of the coast of Suffolk and Essex. Several miles below City Point, the overlying sandy deposit contains a bed of shelly marl, sometimes fifteen feet thick, as at Evergreen. Here I was much struck with the profusion of an *Astarte* (*A. undulata* Conrad), which resembles very closely, and may possibly be merely a variety of, one of the commonest and most characteristic fossils of the Suffolk crag, *A. bipartita*. The other shells also, of the genera *Natica*, *Fissurella*, *Artemis*, *Lucina*, *Chama*, *Pectunculus*, and *Pecten*, reminded me of shells of our crag and the French faluns, although the species are almost all distinct. The large *Venus tridacnoides*, however, is very peculiar, and the *Perna maxillata* is unlike any Suffolk or Touraine fossil, though closely allied to a miocene shell of the Mayence basin on the Rhine. A single coral is found plentifully at Evergreen, resembling an Astrea, and called by Mr. Lonsdale *Columnaria* (?) *sex-radiata*. It differs from the genus *Astrea*, as defined by Ehrenberg, in the stars not being subdivided. Large flattened masses of this coral, upwards of two feet wide, were lying on the beach, washed out of the marl. The teeth of sharks in the banks of the James River agree, some of them specifically, with those of the European miocene beds, and several cetaceous bones, analogous to fossils of the Suffolk crag and Touraine faluns, are frequently met with.

On the right bank or southern coast of the river, about a mile and a half south-west of Coggin's Point, a bluish green marl occurs, about sixteen feet thick and not stratified, in which, assisted by Mr. Ruffin, jun., I gathered in a short time more than thirty species of shells, beautifully preserved, of the genera before mentioned, as also *Oliva*, *Turritella*, *Teredo*, *Dentalium*, *Crassatella*, *Corbula*, *Panopœa*, *Cyprina*, *Tellina*, *Cardita*, *Ostrea*, and *Balanus*. The *Lucina divaricata*, precisely like that of Touraine, or the living West Indian variety, was not rare. This deposit was covered by a bed of reddish clay about six feet thick, without fossils. In many places a remarkable variety of bright green marl, somewhat of the colour of chlorite, alternates in this region with the yellow sand and white shelly marls, which more commonly characterise the miocene deposit.

Passing to the other side, or northern shore, of the estuary of the James River, I observed at the Grove Landing, near the old deserted village of Jamestown, and seven miles south of Williamsburg, Virginia, a range of cliff about forty feet high, in which greenish and yellowish sandy marls of the miocene formation are well exposed. At the top of the cliff is a bed of red earth, without fossils, about ten feet thick. Below this is a layer of shells almost entirely composed of *Chama congregata*, both valves in each individual being usually found united. Below is a mass of shells about twelve feet thick, consisting almost entirely of the genus *Pecten*, but associated with others, of which the *Astarte undulata* before mentioned is the most abundant. The Pectens are closely packed together, with only a slight quantity of accompanying earth. A geologist who should merely collect the shells which have fallen in great numbers on the beach would imagine that bivalves preponderate, to the exclusion of univalves, which, however, is not the case. The univalves are more readily destroyed by frost, rain, and the sun, and have therefore disappeared after exposure, but on digging into a fresh stratum, I met with shells of the genera *Conus*, *Oliva*, *Marginella*, *Fusus*, *Pyrula*, *Murex*, *Natica*, and others. I found no admixture of freshwater or land shells here or in any other locality.

At Burwell's Mill, near Williamsburg, I found a bed of marl about twelve feet thick, very rich in shells, so that I collected there more than seventy species, besides seven species of corals, mingled confusedly with the shells, in which last there appeared to be no order of arrangement, except that in one part, near the top, the *Chama congregata*, accompanied with a *Fissurella*, and a few others, predominated greatly. Among other signs of the slow accumulation of this mass of sandy marl and shells, I may mention that full-grown barnacles and corals are attached to several of the Pectens and other large shells. About one in five of the species of mollusca procured from this marl pit are not distinguishable from those now living in the neighbouring sea.

That remarkable variety of miocene marl which is of a bluish green colour, and which I have mentioned as occurring near Coggin's Point, is well exhibited in a valley near the town of Petersburg, Virginia, where it is used for fertilising the land. It rests on eocene strata, also containing green earth, but of a darker colour. The blue miocene marl is about sixteen feet thick, rich in shells, and is covered by yellow sand, and mottled red and grey clay, about fifty feet thick, with many other sandy strata still higher in the series, which may be observed near the summits of the hills bordering the valley.

All the miocene strata of the James River are horizontal, but the upper surface has been denuded very unevenly, and often with a deeply indented outline, after which the whole country has been reduced to one level by the deposition of an incumbent mass, consisting chiefly of red clay. The great width of estuaries such as that of the James and other rivers which have narrow channels in the higher and mountainous region, may probably be attributed to the facility with which, on the eastern coast of America, the action of the tides has removed the soft and incoherent tertiary deposit, when the land was gradually emerging from the sea. There is reason to believe, from what we know of the coast of S. Carolina and Georgia, that this emergence was accompanied by oscillations of level, which would greatly facilitate and prolong the period of the denuding operations.

North Carolina.

In the cliffs at Wilmington, North Carolina, resting on a calcareous eocene rock, are seen miocene shelly strata of the ordinary character, in which I collected about thirty species of shells. The recent species bore a larger proportion than usual to the extinct; but this may be owing to the circumstance that two-thirds of the whole consisted of marine bivalves, which is rather a larger proportion than that observed in the fossil miocene fauna, amounting to about 150 species, which I collected in Virginia. I have invariably found that the proportion of marine bivalves identical with the recent, whether in the English crag, the faluns of Touraine, or other tertiary formations, is greater than that of the marine univalves, a result in perfect harmony with the law of geographical distribution of living mollusca pointed out by M. Philippi, who has shown that the range of species in the marine bivalves is far more extensive than in the univalves. In Touraine I found the per-centage of recent species vary in the faluns as much as 10 per cent., and even more, in different places not very remote from each other; but when the number in each locality was considerable (150 species for example), this variation diminished. As, however, it would be very rash to assume that all the miocene deposits of the United States, especially in countries as far apart as Maryland and South Carolina, were of strictly contemporaneous origin, the fossil faunas of each region should be carefully distinguished, and considered separately.

Fossils of the Miocene Strata.

Mollusca.—Mr. Conrad, in his work on the fossils of the tertiary formations of the United States, Philadelphia, 1838, has described and figured many of the miocene shells, but unfortunately his work has never been completed. The same author has given a catalogue of 187 shells of the middle tertiary, or miocene formation, in the Appendix to Dr. Morton's "Organic Remains" of the Cretaceous Group. He has also described and figured some species in Silliman's Journal, vol. xli. p. 344. In his own cabinet, he has shown me more than 200 species of shells belonging to this formation, of which he considered about a sixth part to agree with species now living, and he supposed that a somewhat larger proportion would be identified, were our knowledge of the living mollusca more perfect.

I shall confine myself to remarks on those shells, amounting to 147 species exclusive of Balani, which I collected myself, and which I have carefully compared, with the assistance of Mr. G. B. Sowerby, with the tertiary shells of Europe, and with recent species.

The following table will show the miocene genera above alluded to, and the number of species in each: —

Miocene Mollusca of Virginia and South Carolina.

Marine Univalves (63 *species*).

Marginella	- 3.	Fulgur (Fusus)	- 3.	Turritella	- 5.
Voluta	- 2.	Cerithium	- 1.	Trochus -	- 2.
Oliva -	- 3.	Pleurotoma	- 6.	Margarita	- 1.
Conus -	- 2.	Cancellaria	- 3.	Solarium -	- 1.
Terebra	- 1.	Natica -	- 2.	Calyptræa	- 2.
Buccinum	- 7.	Sigaretus	- 1.	Crepidula	- 2.
Purpura	- 1.	Bulla -	- 2.	Fissurella	- 1.
Typhis	- 1.	Scalaria -	- 1.	Dentalium	- 1.
Fusus	- 7.	Eulima -	- 1.	Ditrupa -	- 1.

Marine Bivalves (84 *species*).

Solen -	- 1.	Mysia -	- 1.	Pectunculus	- 1.
Panopæa	- 2.	Lucina -	- 10.	Nucula -	- 2.
Myodora	- 1.	Sphærella	- 1.	Crenella -	- 1.
Mactra	- 7.	Astarte -	- 10.	Perna -	- 1.
Gnathodon	- 1.	Cytherea	- 1.	Lima -	- 1.
Crassatella	- 2.	Venus -	- 5.	Modiola -	- 1.
Amphidesma	- 5.	Artemis	- 1.	Pecten -	- 5.
Sphænia	- 1.	Isocardia	- 1.	Plicatula -	- 1.
Corbula	- 3.	Cardium	- 1.	Ostrea -	- 3.
Saxicava	- 2.	Chama -	- 1.	Anomia -	- 1.
Petricola	- 1.	Cardita -	- 2.	Orbicula -	- 1.
Tellina	- 2.	Arca -	- 4.		

Of the 147 species above enumerated, I have been able to identify the following 23 with recent species: —

American Miocene Shells identified with Recent Species.

Purpura lapillus Lin. Petersburg, Virginia.
Fusus cinereus Lin. Petersburg, Virginia.
Pyrula (*Fulgur*) *carica* Say. Petersburg, Williamsburg, &c.
P. canaliculata Say. Maryland.
Natica duplicata Say. Petersburg, &c.
N. heros Say. Maryland.
Calyptræa costata. Wilmington, North Carolina. (Syn. *Dispotæa ramosa* Conrad.)
Identical with a recent species which I have from Valparaiso.
Crepidula fornicata Lam. Williamsburg.
Dentalium dentalis. Wilmington.
Ditrupa gadus. Williamsburg, Maryland; and Petersburg, Virginia.

Solen ensis Lin. Williamsburg and Wilmington.
Panopæa Americana. Maryland and Petersburg.
Same as *P. Aldrovandi.*
Mactra lateralis Say. Petersburg.
Same as *M. similis.*
Lucina divaricata Lam. Williamsburg, Maryland; and Petersburg.
Identical with a West Indian variety.
L. anodonta Say. Evergreen, Virginia; and Wilmington, North Carolina.
L. squamosa. Petersburg.
L. contracta Say. Wilmington, North Carolina; and Williamsburg, Maryland.
It appears to be the same as *L. radula.*
Astarte lunulata Con. Williamsburg.
Identical with a recent shell from South Carolina.
Venus mercenaria Lam. Wilmington, North Carolina.
Nucula limatula Say. Petersburg.
N. proxima Say. Williamsburg.
Identical with recent specimens on the coast of Massachusetts.
Modiola glandula Totten. Petersburg.
Identical with a recent specimen from Massachusetts.
Pecten magellanicus Lam. Petersburg.

The above list would give a proportion of about 17 per cent. of recent species, or about one-sixth; but I have no doubt that if we possessed in London larger collections of the shells now inhabiting the American seas, I should be able to prove that a greater number of them agreed with fossils of the miocene strata in the United States. Thus I have not included *Anomia ephippium*, because the specific characters in this genus are so unsatisfactory; and I have not seen the recent shell which Mr. Conrad identifies with *Lucina crenulata;* nor have I been able to identify *Artemis acetabulum*, given as a recent species by Mr. Conrad, which does not agree with *A. concentrica* Lam., nor with the species which inhabits the Pacific coast of South America.

The general resemblance of the American fossils of this era with those of the Suffolk crag and faluns of the Loire was so great, that I was surprised, on a closer comparison, to find that I could only identify 9 species as common to both sides of the Atlantic, and that out of 147 American fossils, I could only find 13 species in the European miocene, which were so closely allied as to be entitled to be regarded as geographical representatives. A great part, therefore, of the analogy consists in the similarity of the genera.

List of Species common to the American and European Miocene Strata.

1. *Fusus rostratus* Dujardin.
   Found by me at Burwell's Mill, Williamsburg, Maryland; and agreeing with a common fossil of the faluns of Touraine.
2. *Purpura lapillus* Lin. Petersburg, Virginia.
   American variety: probably the same species as *P. crispata* of the red crag, Suffolk, some varieties of which agree with the European *P. lapillus.*
3. *Turritella plebeia* Say. Maryland.
   This agrees perfectly with an abundant fossil of the Touraine faluns, to which Dujardin has given the name, incorrectly perhaps, of *T. Linnæa.*

4. *Dentalium costatum.* Min. Con., Williamsburg.
This species agrees most perfectly with that of the Suffolk red and coralline crag. It is called *D. dentalis* by Conrad.

5. *Ditrupa gadus.*
I cannot distinguish this from a shell of the English crag.

6. *Lucina divaricata.* Petersburg, Va.
Identical with the Touraine variety.

7. *Lucina contracta* Say.
Seems undistinguishable from the recent *L. radula,* which is also found in the Suffolk crag.

8. *Perna maxillata* Lam.
Mr. Conrad has considered this the same as the fossil which occurs in the Mayence basin, which seems to agree well, though I have not perfect specimens to compare.

9. *Astarte undulata* Say.
I have some specimens of *A. bipartita* from the Suffolk crag, which agree perfectly with the American fossil, except that in the latter the sides of the hinge teeth are much more distinctly grooved. A few only of the English specimens exhibit a faint trace of this grooving. The species named by the Americans *A. vicina, A. arata, A. cuneiformis, A. obruta, A. perplana,* appear all to be varieties of the above species.

Geographical Representatives; or, American Miocene Species closely allied to European Miocene Fossils.

1. *Oliva,* n. sp. Wilmington.
Resembles *O. hispidula,* and may be considered as representing *O. Dufresnii* Bast., of the Touraine faluns.

2. *Voluta mutabilis.* Maryland and Petersburg, Va.
Decidedly represents *V. Lamberti* of the English crag and French faluns.

3. *Buccinum trivitatum* Say. Maryland and N. Carolina.
Closely resembles a common Touraine species, which Dujardin has called *B. elegans,* a name which cannot be retained. In the American fossil the outer lip is not thickened, as in that from Touraine.

4. *Cancellaria lunata.* Maryland.
It comes very near to a crag species in Mr. Wood's collection, and is allied to a Bordeaux species given me by M. Deshayes under the name of *C. contorta.*

5. *Trochus Audibarti* Bast. Williamsburg, Maryland.
The American shell is rather more acuminated than that of Bordeaux, but is so near that some may consider it a variety.

6. *Solarium,* new species. Williamsburg.
Nearly allied to *Solariella maculata,* Searles Wood, An. Nat. Hist. 9. t. 5. f. 7. from the coralline crag, Suffolk.

7. *Calyptræa sinensis?* Petersburg, Va.
This fossil nearly resembles, if it be not the same as, *C. sinensis,* found in the Suffolk crag.

8. *Panopæa reflexa* Say. Petersburg.
Nearly allied to *P. Faujasii,* crag, Suffolk.

9. *Corbula elevata* Con. Maryland.
Approaches near to the recent *C. nucleus,* a Suffolk crag species.

10. *Tellina lusoria.* Williamsburg.
Nearly allied to *T. donacina,* Suffolk crag.

11. *Astarte lirata* Conrad.
Nearly allied to *A. gracilis,* Coralline crag, Suffolk.

12. *Isocardia rustica.* Maryland.
Allied to *Isocardia cor,* Suffolk crag.

13. *Pecten Jeffersonius.* Petersburg.
Allied to *P. solarium* of the faluns of Touraine.

Miocene Species of Forms peculiar to America.

While the analogy of the American and European miocene shells is so striking, there are also some species of forms quite peculiar to America; *e. g.*:—

1. Oliva idonea.
2. Fusus quadricostatus *Say.*
3. F. sulcatus.
4. F. parilis *Con.*
5. *Pyrula* (Fulgur) *carica* Say.
6. *P. canaliculata* Say.
7. *Calyptræa costata* (C. ramosa *Conrad*).
8. Gnathodon Grayi.
9. Venus Tridacnoides.
10. *V. mercenaria* Lam.
11. *Modiola glandula* Totten.
12. *Pecten magellanicus* Lam.

The six species in italics in the above list are recent, and are confined to the western side of the Atlantic, and these characteristic forms imply that the beginning of the present geographical distribution of mollusca dates back to a period as remote as that of the miocene strata. It will be also seen that several of the 23 miocene shells identified with recent species in the list p. 419-20. are now inhabitants of the American side of the Atlantic.

On the other hand, when we examine the fossil shells of the European miocene strata, we find most of those which are identifiable with living species to belong to the British seas, to the Mediterranean, or to the coast of North Africa, and not to extend to the American side of the Atlantic, a fact which may be illustrated by the following list of species.

European Miocene Shells found living on the Eastern, and not inhabiting the Western, Side of the Atlantic.

	Touraine.	Crag.	Living.
1. Cypræa coccinella - -	+	+	British and Medit.
2. Erato cypreola - - -	+	+	British.
3. Terebra plicaria - - -	+		North Africa.
4. Buccinum mutabile - -	+		Mediterranean.
5. Columbella (Bucc.) Linnei	+		Mediterranean.
6. Cerithium lima - - -	+		Brit. and Mediterranean.
7. Eulima Cambessedesii -	+		Brit. and Mediterranean.
8. Trochus fanulum - -	+		Mediterranean.
9. Trochus striatus - -	+	+	Mediterranean.
10. Siliquaria anguina - -	+		Mediter. and Atlantic.
11. Fissurella neglecta - -	+	+	Mediterranean.
12. Petricola ochroleuca -	+		Mediter. and Ireland.
13. Tellina crassa - - -	+	+	British.
14. Lucina columbella - -	+		North Africa.
15. Arca lactea - - - -	+	+	British.

No one of the above shells is found fossil in the American miocene strata, and this fact strongly confirms the opinion before stated respecting the high antiquity of the separation of the existing molluscous faunæ, which we see may be traced back to a period when about four fifths of the species differed from those now living on the globe. The large *Pyrulæ* of the sub-genus *Fulgur*, and several other forms above enumerated, list, p. 419-20., would enable a conchologist, only acquainted with recent shells, to distinguish a set of American from a set of European miocene fossils. The genera *Pholadomya* and *Gnathodon*, and several others found in the miocene beds of the United States, would also enable the conchologist to recognise the American type.

It is worthy of remark, that the recent shells found in the American miocene beds are not only in about the same proportion to the extinct as is observed in the Suffolk crag or in the faluns of Touraine, but they also agree specifically in most cases with mollusca now inhabiting the neighbouring Atlantic. Now most of the recent miocene species of Touraine agree with species now living on the western coast of France or in the Mediterranean, and those of our crag are identifiable with species living in the British seas. This result appears to me to confirm the accuracy of the conchological determinations, for if any one of those palæontologists who are unwilling to believe that species pass from one geological period to another, should maintain that the living species are so numerous, and often resemble each other so closely, that false identifications may easily have arisen, I reply, that in that case, according to a fair calculation of chances, nine-tenths of the American miocene species said to be *recent* ought to have been identified with exotic species, instead of being found to agree with members of that very limited fauna at present known on the American shores. The same argument is clearly applicable to the identifications which have been made of fossil and recent shells in the European miocene formations. With the exception of one *Calyptræa*, I find no one of the miocene shells which I have identified with recent species to be identical with shells of the Pacific. The analogy is to an Atlantic, not a Pacific fauna.

Among other points of analogy between the miocene fossils on the banks of the James River and those of Touraine, I ought not to omit mentioning the multitude of large shells of the genus *Pecten*, as *P. Jeffersonius* and others, occurring near Williamsburg, (Maryland,) just as the large *Pecten Solarium* occurs in the faluns of the Loire near Doué. Nothing similar is observed in the eocene fossils, whether European or American. The same may be said of the large and conspicuous shells of the genus *Panopæa*, found in our Suffolk crag, and in the American miocene beds, but not in the older tertiary strata.

As it is very rare to meet with any land or freshwater shells in the red and coralline crag of Suffolk, so are they almost unknown as yet in the American strata of the same age. The *Gnathodon*, an estuary shell, has been found in several places, and I met with two odd valves of a *Cyræna*, well preserved in the miocene strata of Petersburg, Va. The species is rather larger than *C. consobrina* from the Nile. The *Buccinum quadratum* of Conrad, doubtless a marine shell, might easily be mistaken for a *Melanopsis*.

Even in Touraine, where the bones of mammalia are not uncommon in the marine beds, extremely few land and freshwater shells are met with.

Climate of the Miocene Shells. — Upon the whole, the shells of this formation in Virginia and Maryland resemble those of Touraine and Bordeaux more than the fossils of the Suffolk crag, as we might have expected from their nearer correspondence in latitude. Thus, for example, the genera *Conus*, *Oliva*, *Marginella*, and *Crassatella* (represented by large species), with other forms of warmer seas, which are wanting in our crag, are common to the miocene beds of Virginia, Maryland, Touraine, and Bordeaux. Yet when we consider that the shelly deposits on the James River, in the United States, are in the 37th degree of north latitude, while the French faluns are in the 47th, we are surprised to see so great an analogy in those characters by which the climate of a conchological fauna may be inferred. The American shells do not seem to indicate a more southern latitude by 10°, and the same remark applies equally when we compare the most southern beds of North Carolina with those of Bordeaux. We find in all these localities a great admixture of northern forms, as if the isothermal lines in the climate of the miocene period took a curve to the south when drawn from Europe to America, as they do now.

Polyparia. — The fossil corals which I observed in different miocene localities were usually few in number, both in species and individuals, with the exception of the *Columnaria* (?) *sex-radiata*, which resembles an Astrea, and is very conspicuous, from its size, in the shelly marls near the banks of the James River. The scarcity of these zoophytes may be a local accident, depending on the general rarity of calcareous deposits of this age in the United States. The following is a list of ten species collected by me, and of which a description will be communicated by Mr. Lonsdale in a subsequent report.*

Miocene Species.	Localities.
1. Columnaria? sex-radiata *Lonsdale*	Evergreen, on James River, Virginia; Petersburg, Virginia.
2. Anthophyllum lineatum *Lonsdale* (Caryophyllia lineata *Conrad*)	Petersburg and Williamsburg, Virginia.
3. Astrea hirtolamellata? *Michelin*	Williamsburg.
4. Heteropora? tortilis *Lonsdale*	Petersburg and Williamsburg, Virginia.
5. Cellepora informata *Lonsdale*	Petersburg, Virginia.
6. C. quadrangularis *Lonsdale*	Williamsburg, Evergreen, Virginia.
7. C. similis *Lonsdale*	Williamsburg, Virginia.
8. C. umbilicata *Lonsdale*	Williamsburg, Virginia.
9. Escharina tumidula *Lonsdale*	Petersburg, Virginia.
10. Lunulites denticulata *Conrad*	Williamsburg, Virginia.

Two species of the above list, namely *Anthophyllum lineatum*, identical with a Touraine species, and *Lunulites denticulata*, agreeing with one found in the Suffolk crag and faluns of Touraine, correspond with European fossils of the same age.

With respect to climate, Mr. Lonsdale regards this collection as indicating a temperature exceeding that of the Mediterranean,

* See this Report among the miscellaneous articles in the present Number of the Journal.

nearly corresponding with that of the faluns of Touraine, warmer than that implied by the polyparian fauna of the crag, and not so tropical as that of the Bordeaux tertiary beds.

Fossil Echinodermata.

In the shelly strata near Coggin's Point, on the James River, Virginia, I found a species of the *Spatangus* family, of the genus called by Agassiz *Amphidetus*, which also occurs in the English crag, and in the living fauna of Europe. I also found an *Echinus* at Williamsburg, and the spines of a second species. My friend, Professor E. Forbes, has favoured me with descriptions of these.

(*Family Spatangaceæ.*)

1. *Amphidetus Virginianus* Edw. Forbes.

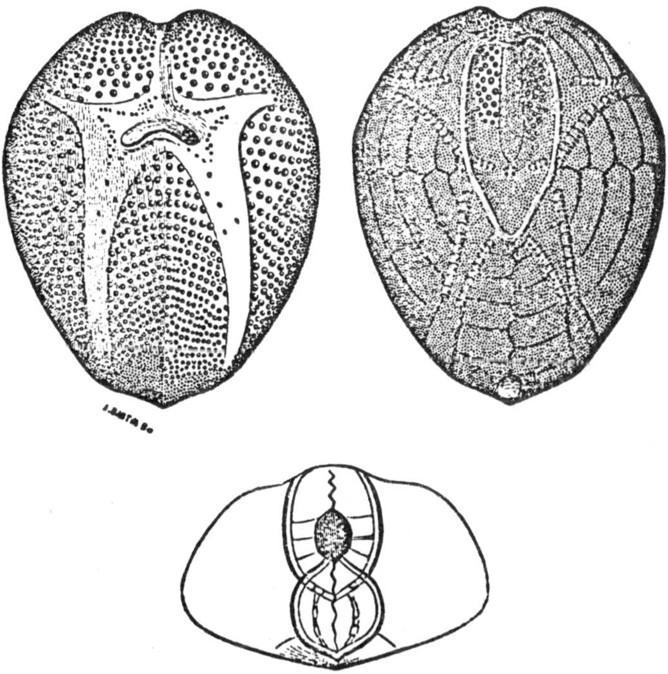

Amphidetus Virginianus (natural size).

Body broadly ovate; elevated and truncate posteriorly. Back oblique; dorsal impression lanceolate-scutate, area very slightly excavated : ambulacral spaces broad, triangular, depressed ; interambulacral spaces slightly convex. Anteal furrow broad, shallow ; sides slightly gibbous ; subanal impression broadly obcordate ; post-oral spinous space broadly lanceolate.

Dimensions of the smaller but more perfect specimens.

Lon. unc. $1\frac{11}{12}$; Lat. $1\frac{8}{12}$; Alt. 1.

Number of pairs of ambulacral pores.*

Ant. lat. dors. amb. 8 + 10.
Post. lat. dors. amb. 13 + 8.

Loc. Petersburg.

(*Family Cidarites.*)

2. *Echinus Ruffinii* Edw. Forbes.

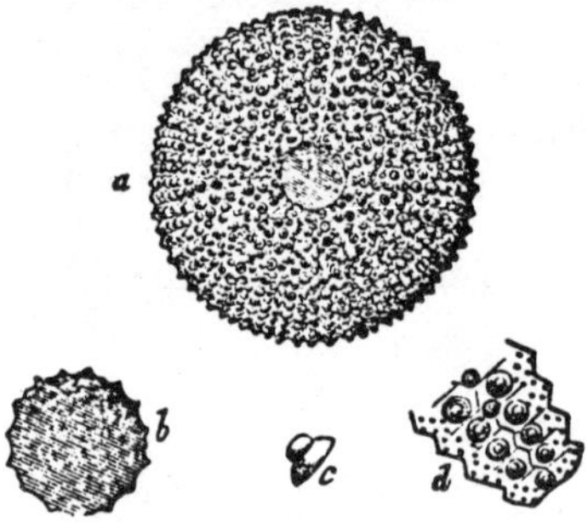

a, *Echinus Ruffinii*, viewed from above.
b, Ditto, mouth.
c, Ditto, a spinigerous tubercle.
d, Ditto, ambulacral plates and arrangement of pores.
(*a*, *b*, natural size; *c*, *d*, enlarged.)

Body subdepressed; ambulacral and interambulacral plates with several primary tubercles on each, closely ranged, having circles of secondary tubercles surrounding their (broad) bases; rows of pores very oblique, three pair of pores in each row, the uppermost distant from the other two. Beneath concave, mouth broad, widely notched, opposite each avenue. Anus narrow.

Lat. 1 unc.; Alt. $\frac{8}{12}$.; Lat. apert. $\frac{5}{12}$.

The primary tubercles, being very numerous and of almost equal size, give the testa a very granulated appearance.

Loc. Williamsburg, Virginia.

3. *Spines of an Echinus from James River.* Virginia.

Ridges fine, numerous; interstices apparently smooth, broader than the ridges.

Fish.

The remains of fish consist chiefly of teeth of the shark family, all belonging to existing genera.

1. *Carcharias megalodon.* Evergreen, James River.
2. *Lamna xiphodon.* Evergreen, precisely like specimens I have seen from Touraine.
3. *L. hastalis* (*Oxyrhina*). Evergreen, Virginia, like a specimen from Martha's Vineyard, and in the molasse of Switzerland.
4. *L. cuspidata.* James River, also common to the molasse.
5. *Carcharias productus.* Evergreen, also from the miocene of Malta.

Several other shark's teeth, and the ossicles of the ears of fishes, which I found in the miocene strata of Petersburg, Virginia, were very analogous to European miocene fossils.

* In another specimen the number was found to be —

Ant. lat. dors. amb. 8 + 13.
Post. lat. dors. amb. 11 + 11.

And four pair additional on each side of the ovarian holes.

Mammalia.

The vertebræ of whales, apparently similar to those before described as found in Martha's Vineyard, occur in the miocene beds of Petersburg and other places.

In the museum at Baltimore, the grinder of a Mastodon was shown me, evidently distinct from that of *M. giganteus.* It had been recognised by Mr. Charlesworth as the *M. longirostris* Kaup, and had been so ticketed by him. Professor Ducatel had the kindness to allow me to take the specimen to Philadelphia, where Dr. Harlan identified it with the *M. longirostris* by comparison with casts sent to him by Mr. Kaup. The tooth is narrower than in the common *Mastodon giganteus*, and the parallel rows of transverse mammillary processes, instead of being divided by distinct hollows, are connected together by projecting spurs.

I was informed that this tooth was found at the depth of 15 feet from the surface of a bed of marl, near Greensburgh, in Caroline County, Maryland. Though I have not visited the place, I have little doubt that this Mastodon was a miocene fossil, and it occurs in beds of the same age on the Rhine.

APPENDIX.—*Extract of a Letter from* W. LONSDALE, Esq. F.G.S., *to* Mr. LYELL, *on the Indications of Climate afforded by the Miocene Corals of Virginia.*

"July, 1845.

"From ten fossil species of Polyparia belonging to different families, it is difficult to form an opinion respecting the climate in which they were produced. Nevertheless, as recent lamelliferous corals are well known to be markedly distributed as respects climate, the following notice is submitted for consideration. The district in Virginia whence the miocene polyparia were procured being situated in about 37° of north latitude, is on the same parallel with the southern coast of the Mediterranean westward of Tunis, and the northern portion eastward of that point; and as very little appears to be known of the distribution of recent corals on the coasts of the United States, it is proposed to compare the nature of the fossil genera with that of those living in the Mediterranean; it is hoped that no great objection can be advanced against this comparison, or mode of inferring climate, as species are disregarded, and as the polyparia of the Bermudas and the West Indies agree generically with those of the Red Sea and Indian Ocean. The ten species belong to seven genera, *Columnaria* (?), *Astrea*, *Anthophyllum*, *Heteropora*, *Cellepora*, *Lepralia*, and *Lunulites*, the first three being lamelliferous Anthozoa; and though *Columnaria* is an extinct genus, yet it is believed that its requirements may be safely inferred to have been analogous to those of *Astrea*. *Heteropora* is also only found fossil; but being considered to be a true *Tubuliporidæa*, its habits may likewise be inferred: and the three last are well-known genera belonging to *Bryozoa*. The ascertained lamelliferous Anthozoa of the Mediterranean exhibit, when tabulated, a marked character, consisting almost wholly of simple or ramose groups, not numerous in genera or species, though abundant in specimens, and occasionally of ample growth, with very diminutive representatives either in number or size of the great *Meandrinæ*, *Madreporæ*, *Astreæ*, *Porites*, and numerous allied genera which swarm in the Red Sea, the Bermudas, the West Indies, and the Pacific and Indian Oceans. A perfect distinction, as a whole, may therefore be stated to exist between the Polyparia of the Mediterranean and those of southern seas; while in more northern regions there are representatives, very limited as to species, of the simple and ramose polypidoms of the Mediterranean.

"On attempting to compare the miocene fossil *Anthozoa* of Virginia with those of the Mediterranean, it will be found that the *Anthophyllum* has no generic representative, all the existing species, as restricted by Ehrenberg, belonging to the Red Sea and Indian Ocean; and if the fossil specimens be regarded as young, their dimensions are not inferior to those of *A. musicale* in the same stage of developement — that the *Astrea* has an equivalent in the *Astrea mediterranea* of M. Risso, or in a *Porites* discovered by Prof. E. Forbes; but the *Columnaria*, an abundant coral, exhibits dimensions allied to those of analogous genera of warmer seas. Again, as respects the *Heteropora*, *Celleporæ*, and *Escharina*, no satisfactory inference can be formed regarding climate, *Tubuliporideæ* and similar *Celleporideæ* being universally distributed; but *Lunulites* have been found only in the Mediterranean and more southern latitudes. The comparison, therefore, would lead to the inference, that a climate rather exceeding the one which prevails nearly throughout the Mediterranean existed in the region where these fossils lived; and it must be borne in mind, that, as in plants so in corals, size cannot be assumed as the sole indication of temperature, but the abundant developement of species of peculiar genera and families, whether the first be of large or small dimensions.

"It is desirable next to allude briefly to the relative geological position of the Virginia deposit as respects some European miocene accumulations, to the extent which each may appear to justify a comparison. The formations referred to are the crag of England, the faluns of Touraine, the deposits near Bordeaux and Dax; and the characters of the *Anthozoa* in each district will be again considered as the principal test of temperature. The crag, according to Mr. S. Wood's extensive list*, contains only four species of lamelliferous Polyparia, with two *Lunulites* and one *Orbitulites*, the first including a *Fungia*, a *Turbinolia*, and two corals, believed to be allied to genera living in the Mediterranean: one of the

* Annals of Nat. Hist. January, 1844.

Lunulites is also considered to be identical with the species found in Virginia. The Polyparia of the crag exhibit, therefore, a marked absence of the genera characteristic of high temperatures. The Touraine miocene strata contain, according to the fine series collected by Mr. Lyell, about nine species of *Anthozoa* and three of *Lunulites*, the former including an Anthophyllum, believed to be the species found in Virginia, two *Turbinoliæ* and two *Astreæ*, the whole indicating a somewhat greater temperature than that of the crag. The Bordeaux and Dax deposits include, according to M. Michelin's work on the fossil corals of France, eleven lamelliferous *Anthozoa* and two *Lunulites*, the former comprising a *Dendrophyllia*, believed to be identical with a Touraine species; six *Astreæ*, one *Gemmipora*, one *Porites*, and two *Madreporæ*, or an aggregate representative of a Red Sea list of polyparia. Lastly, the Turin polypidoms, for a knowledge of which I am wholly indebted to M. Michelin's work, present no less than 73 species of lamelliferous *Anthozoa*, a tabulated list of which will bear a detailed comparison, as respects genera, with similar summaries of Red Sea, or tropical polypidoms. From these data it may perhaps be inferred, that the American deposit was accumulated in a climate superior to that of the crag, possibly equal to that of the faluns of Touraine, but inferior to that of Bordeaux."

OBSERVATIONS ON THE WHITE LIMESTONE AND OTHER EOCENE OR OLDER TERTIARY FORMATIONS OF VIRGINIA, SOUTH CAROLINA AND GEORGIA

2. *Observations on the* WHITE LIMESTONE *and other* EOCENE *or Older Tertiary Formations of* VIRGINIA, SOUTH CAROLINA, and GEORGIA. By CHARLES LYELL, Esq., M.A., F.R.S., &c.

THE tertiary deposits occupying a lower position than the Miocene strata described in the last paper, were first referred by Mr. Conrad to the Eocene period, in the "Journal of the Academy of Natural Sciences" for 1830. Some of these strata have been observed by Mr. Conrad on the Potomac at Fort Washington in Maryland ("Fossil Tertiary Shells," p. 30.); but the most northern which I myself examined were in Virginia, at Richmond, at Petersburg, and at several points on the James River. The formation in this region consists in great part of greensand and marl, containing green earth, so precisely like that which characterises the cretaceous strata of New Jersey, that were it not for the distinctness of the fossil shells, it would be impossible in many places to separate these deposits by mere reference to their mineral composition.

Farther south, in N. and S. Carolina, and in Georgia, the eocene formation acquires a larger development and a new mineral type, consisting of highly calcareous white marl and white limestone, and passing upwards, especially in Georgia, into red and white clays, ferruginous sands, with associated layers of burrstone and siliceous rock. This calcareous form of the eocene rocks on the Santee River and elsewhere, had led some geologists to consider the solid limestones as an upper secondary or newer cretaceous formation, and as forming an intervening link between the secondary and tertiary formations; but after a careful examination of several localities, presently to be described, I found the white limestone to contain exclusively tertiary fossils, without any intermixture of species belonging to the true and unquestionable cretaceous rocks of New Jersey or Alabama. It appeared to me that there was the same chasm between the cretaceous and tertiary rocks in that part of America which I visited, as has been observed in Europe generally, and those organic remains which have been supposed to be common to the two formations in the United States, have been almost all referred by mistake to the older group, in consequence of part of the white limestone of S. Carolina, which is tertiary, having been erroneously referred to the cretaceous epoch.

The largest number of eocene shells found in a good state of preservation, are those of Claiborne, Alabama, where a large collection was made by Mr. Conrad, and descriptions and figures of them were published in 1832.* At the same period, Mr. Lea, of Philadelphia, received from a friend a fine collection of the same fossils from Alabama, and referred them also to the period of the London clay of England, and calcaire grossier of Paris. In his work intitled "Contributions to Geology," he gave figures and descriptions of more than 200 species, but unfortunately, in consequence of these two eminent naturalists having laboured simultaneously, and independently of each other, almost every shell received a distinct specific name. For a list of synonyms, I may refer the reader to the appendix to Dr. Morton's "Synopsis of Organic Remains of the Cretaceous Group, Philadelphia, 1834," drawn up by Mr. Conrad.

I shall now offer a few observations on several localities of the eocene strata which I visited, beginning with the most northern, in Virginia, and then proceeding southwards to N. and S. Carolina and Georgia.

Virginia.—Below Richmond, near Coggin's Point on the James River, the *Ostrea sellæformis* occurs in one of the uppermost of the eocene beds; and this fossil I afterwards found to be widely characteristic of the formation in S. Carolina and Georgia. At the same place, *Cardita planicosta,* so common in the London clay and Paris basin, is also found. It cannot be distinguished from one of the common varieties of the European shell, and is accompanied by an oyster very nearly allied to *O. bellovacina.* Professors W. B. and H. D. Rogers, have described and figured several of these fossils in the fifth and sixth volumes of the American Phil. Trans. for the years 1835 and 1839. Near Evergreen, on the right bank of the James River, twenty miles below Richmond, I found, in the eocene marl, a large piece of wood in a state of lignite, 7 feet long, and about 1 foot broad, bored by teredo.

* For some account of the Claiborne strata, see Conrad's "Tertiary Fossil Shells," 1842.

At Petersburg, 30 miles south of Richmond, I saw the eocene strata, containing several characteristic fossils, distinctly overlaid by a large mass of miocene shell-marl. At no other point is the older tertiary formation more remarkable for its lithological resemblance to the greensand of the cretaceous series.

North Carolina. — Very different is the aspect of the rocks near Wilmington, N. Carolina, where they consist in great part of a limestone containing siliceous pebbles, and casts of shells and corals, with many fishes' teeth. This limestone is less white and more compact than the white limestône of S. Carolina; on the shore at the town of Wilmington it is 12 feet thick, covered with a shelly miocene deposit 6 feet thick. It is quarried in the neighbourhood of the town, and burnt for lime. I obtained, besides corals, 31 species of shells, exclusive of balani, from this rock, almost all in the state of casts, and many of them, therefore, only capable of being named generically. Those in italics in the following list have been identified with species found elsewhere in eocene localities: —

List of Eocene Shells from the Limestone of Wilmington, N. Carolina.

Cypræa, identical with a cast from Shell Bluff, Georgia
Cypræa, two other species
Oliva Alabamensis
Oliva, allied to O. Laumontiana *Lam.* (fig. *a*)
Oliva
Voluta, two species
Conus
Strombus
Fusus
Buccinum, three species
Paludina, allied to P. Desnoyerii *Desh.* (fig. *c*).
Natica ætites
Turritella, two species
Vermetus
Infundibulum trochiforme
Crassatella, agreeing with a cast from Eutaw
Corbula
Lucina pandata
Cardium, agreeing with one from Shell Bluff
Cardita rotunda
Cucullæa
Arca, two species
Nucula magnifica
Pecten membranaceus Mort.
Terebratula Wilmingtonensis sp. n. (*Lyell & G. Sowerby*) (fig. *b*).

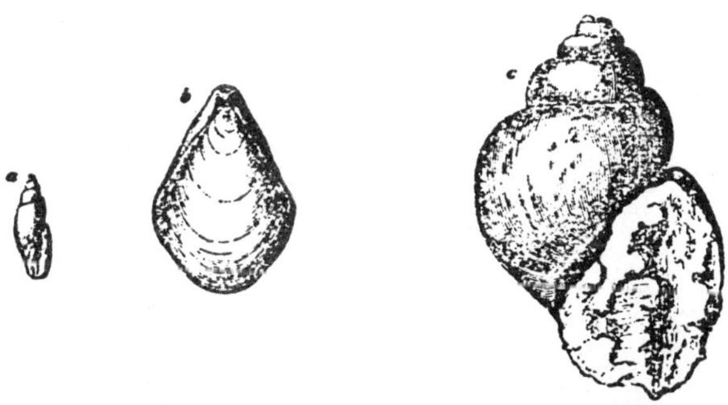

a. Oliva, (cast,) Eocene limestone, *Wilmington*, N. Carolina.
b. Terebratula Wilmingtonensis *Lyell & G. Sowerby.*
c. Paludina, (cast,) *Wilmington*, N. Carolina.

The two last-mentioned species of the above list, *Pecten membranaceus* and *Terebratula Wilmingtonensis*, alone preserve their shells, all the others being in the state of casts. The small Oliva (fig. *a.*), resembles in general form *O. laumontiana* Lam., or *O. nitidula* Lam., but is more slender than either. As it is only the cast of the inside, it cannot be fully described. It appears to have only two small folds in the columella, of which the anterior is the larger. The Paludina (fig. *c.*) is like *P. Desnoyerii* Desh. (a fossil of the white marl in the midst of the calcaire grossier), but it has six volutions, whereas the *P. Desnoyerii* has barely four and a half. In the Wilmington fossil, the spine is more acuminated and the volutions more distant, so that the suture must have been more distinct.

Terebratula Wilmingtonensis (Lyell and G. B. Sowerby), Wilmington, North Carolina. This shell resembles most nearly in general form *T. uva* Brod. (recent from the Gulf of Tehuantepec), and also approaches *T. bisinuata* Lam., a fossil of the Paris basin. The following are its characters: —

Terebratula, with an oblong, smooth shell, posteriorly acuminated, anterior margin nearly even, dorsal valve large, and posteriorly prominent.

I was informed that a species of Nautilus had been found in the Wilmington limestone. Among the *Polyparia* which I collected there, Mr. Lonsdale has observed the following species: —

1. Lunulites sexangula *Lons.*
2. Lunulites
3. *L. distans* Lons., also at Wantoot?
4. L. contigua *Lons.*
5. *Flabellum? cuneiforme Anthophyllum* of Conrad, also found at Eutaw and Cave Hall
6. *Dendrophyllia lævis* Lons., also at Shell Bluff
7. *Caryophyllia? subdichotoma* Lons., also at Shell Bluff
8. Eschara tubulata *Lons.*

As four out of these eight corals, those in italics, and ten out of the thirty-one shells, occur elsewhere in eocene localities, I consider the age of the Wilmington limestone, on which some doubts have been entertained, as set at rest. Among the teeth of sharks in the same rock I found, together with the usual eocene forms, a species of *Galeus*. The claws of crabs are also numerous.

I observed the same formation, and some of the same shells and corals, at Rocky Point, which is about twenty miles from Wilmington, on the N. E. branch of the Cape Fear River, where a similar conglomerate occurs, with green pebbles. At some points the rock is partly siliceous, and strikes fire with steel.

South Carolina.

From the low country near the level of the sea, at the mouth of Cooper River, to the junction of the Santee Canal, and from that point to Vance's Ferry on the Santee River, a calcareous formation of the eocene period occurs. At the first point where I saw it, in Dr. Ravenel's plantation called "the

Grove," near the mouth of the river, it appears in the form of a soft pulverulent limestone, in which two species of Scutella, (*S. macrophora* Ravenel, and another,) are very abundant. The soft limestone had been cut through to the depth of five feet in digging a canal, situated near "the Grove," about seventeen miles north of Charleston; its thickness here is unknown. I found in it *Pecten Lyelli* Lea, a Claiborne shell, and the upper valve of an oyster, which seems undistinguishable from *O. bellovacina;* also a species of Lucina, and a large Pecten allied to *P. pleuronectes;* also a species of Spatangus common to the limestone of the Santee canal.

At the Rock Landing, near the Grove, the white limestone is composed of triturated shells, and assumes a very hard and solid form. It contains fragments of Echinoderms, casts of shells, and corals (*Lunulite?*): it sometimes passes into an imperfect oolite.

Between the Grove and Vance's Ferry on the Santee River, a distance of about forty miles, is a continuous formation of white limestone, which I examined in company with Dr. Ravenel, first at Strawberry Ferry and Mulberry Landing, then on the banks of the Santee Canal, and afterwards at Wantoot and Eutaw. I then followed it in a north-westerly direction for twelve miles, by Cave Hall and Streeble's Mill, to near Halfway Swamp. On reaching Stoudenmire or Stout Creek, a tributary of the Santee, we found the limestone and marl to disappear beneath a newer deposit, also referable to the eocene period, of which I shall afterwards speak as the burrstone formation.

The soft limestone varies in hardness, passing frequently into a white marl, and resembling in texture some of the *craie tufau* of the Loire in France. It consists almost entirely of comminuted shells or corals, but it rarely exhibits any laminæ of deposition, and even where it attains a thickness of twenty or thirty feet, there would be a difficulty in determining whether it were horizontal, if a bed of oysters, *O. sellæformis,* like that at Vance's Ferry, did not occasionally occur. Notwithstanding its slight elevation above the sea, the Santee limestone cannot be less than 120 feet thick at Strawberry Ferry, being vertically exposed to the extent of 70 feet on the low banks and bottom of Cooper River, and to the height of 50 feet above these banks in the neighbouring hills. Its upper surface is very irregular, and is usually covered with sand, in which no shells have been found.

At Eutaw and other points, corals of the genera *Idmonea, Dendrophyllia, Flabellum, Tubulipora, Hippothoa, Farcimea, Vincularia, Eschara,* and others, occur, with a species of *Scalaria,* and other shells. These fossils, and the rock containing them, reminded me so much of the straw-coloured limestone of the cretaceous formation which I had seen on the banks of Timber Creek in New Jersey, that I do not wonder that some error has arisen in confounding the tertiary and secondary deposits of the Atlantic border. The species, however, prove, on closer inspection, to be different. This lithological resemblance led to the admission into

Dr. Morton's otherwise most accurate list of the cretaceous fossils of New Jersey, of the six following species, viz. *Balanus peregrinus*, *Pecten calvatus*, *P. membranosus*, *Terebratula lachryma*, *Conus gyratus*, *Scutella Lyelli*, and *Echinus inflatus* (see pl. 10., Morton's Synopsis), which came from the eocene beds of South Carolina, now under consideration, and led to a belief of the existence of a deposit intermediate between the chalk and tertiary strata, and containing fossils common to both.

One of the characteristic features of the region of tertiary white marl and limestone in South Carolina and Georgia, is the frequent occurrence of lime-sinks, or funnel-shaped cavities, arising from natural tunnels in the subjacent limestone, through some of which subterranean rivers flow. At Wantoot, there is one of these sinks in the limestone, and a spring issues from the rock so much above the temperature of the air during a frost as to send off clouds of steam.

At Çave Hall, two miles south of the Santee River, there is a cavern about twelve feet high at its opening, at the base of a precipice of limestone sixty feet perpendicular. Large beds of the *Ostrea sellæformis* occur in the limestone, which contains green particles in the lower strata. A stream is constantly flowing out of the mouth of the cave, and there is a line of sinks communicating with the underground river-course, in which the undermining process is continually going on. I was informed that a new "sink" had opened fifteen years ago within 110 yards of the mouth of the cave, and that a mule fell into the hollow while drawing the plough in the field above. Among other fossils from this place, I found in the limestone the tooth of *Myliobates*, and in the lower beds of calcareous greensand the same shells and corals as in the incumbent white limestone.

On reviewing the fossil Invertebrata which I collected from various localities in the Santee white limestone of South Carolina, I find many which will at once be recognised as species known to belong to the eocene formation of Claiborne and other places, among which I may mention *Trochus agglutinans*, *Pyrula inaurata* (*Fusus* Conrad), *Natica ætites*, *Dentalium*, same as one from Claiborne, *Lucina pandata*, *Lucina rotunda* Lea, *Lucina lapidosa*, *Crassatella* agreeing with a Wilmington (North Carolina) fossil, *Chama*, like one from Jacksonboro', Georgia, *Pecten calvatus*, *P. Lyelli* Lea, *Ostrea bellovacina*, *O. sellæformis*. Besides these we find *Terebratula lacryma*, *Ostrea Carolinensis*, *Pecten* allied to *P. pleuronectes* Lam., and shells of the genera *Nautilus*, *Voluta*, *Turritella*, *Scalaria*, *Vermetus*, *Lucina*, *Cytherea*, *Corbula*, *Cardium*, *Lima*, *Pecten*, *Ostrea*, *Terebratula*.

The Echinoderms are referable to the genera *Scutella* and *Cidaris*, and I met with several nautiliform foraminifera. The corals, already alluded to generically, will be described by Mr. Lonsdale.*

* See "Report on the Corals," &c. in the present Number of the Journal.

Burrstone Formation.

I have before mentioned that the white limestone of the Santee River, on being traced in a north-westerly direction, disappeared at Stoudenmire Creek, a tributary of the Santee, beneath a newer deposit of considerable thickness. The latter consists of slaty clays, quartzose sand, loam of a brick-red colour, and beds of siliceous burrstone, in some of which fossil sponges, having a coarse fibre, have been detected. Some of the clays break with a conchoidal fracture, and become stony when dried. One of the beds is extremely light, and resembles in appearance some kinds of calcareous tufa, but does not contain carbonate of lime. I at first supposed it to be of infusorial origin, but some practised observers have been unable with the microscope to discover infusorial cases. There were casts of shells in this rock, and in several of the associated strata, referable to the genera *Cypræa, Voluta, Natica, Trochus, Corbula, Mactra, Cardita, Cardium, Lucina, Nucula, Pectunculus, Pecten,* and *Serpula.* Among these, a *Corbula, Cardium,* and *Nucula,* seem to agree with Claiborne species; the rest did not agree with fossils from that locality, nor with those from the miocene beds of Virginia, but I was afterwards shown siliceous casts of *Ostrea sellæformis, Cytherea perovata,* and other eocene fossils, from strata of the same formation at Orangeburg and near Aikin in South Carolina. I believe, therefore, that the larger portion of the ferruginous sands, red clays, and white beds of kaolin (often miscalled chalk by the inhabitants of South Carolina and Georgia) belong to the Upper Eocene or Burrstone deposit.

At Aikin, fifteen miles S. E. of Augusta, and near the left bank of the Savannah River, the inclined plane of a railway has been cut through strata, 160 feet in thickness, consisting partly of earth and sand of a vermilion colour, and containing much oxide of iron; partly also of mottled clays, and white quartzose sand with masses of pure white kaolin. This compact kaolin appears fitted to make good porcelain ware. The globules of iron give a pisolitic appearance to some of the beds of quartzose sand. These beds at Aikin yielded no fossils, but I suppose them to be all referable to the Burrstone or Upper Eocene formation.

Georgia.

The same Burrstone formation is continuous from Aikin and Stoudenmire Creek, South Carolina, to Augusta in Georgia, and to the junction of the tertiary with the primary or hypogene rocks above that city. It must attain there in some places a thickness of more than 200 feet, and is very variable in its aspect and composition.

At a place called The Rocks, six miles west of Augusta, it consists of a highly micaceous quartzose grit and sand, having much the appearance of certain kinds of granite, and having been by some writers improperly termed gneiss. The mass exhibits occasionally a distinct cross-stratification, and pieces of compact kaolin, sometimes angular, are imbedded in it. In other places, as at Somerville, red, vermilion-coloured, and white clays, 180 feet thick, are seen. These rest in horizontal beds on the edges of highly inclined strata of chlorite schist and clay-slate, which are exposed to view at the rapids of the Savannah River, three miles above Augusta. On Ray's Creek, near this point, the old schists, much charged with iron, are seen to decompose into materials so like the red vermilion-coloured clays of the tertiary deposits, that they would be undistinguishable, were it not that the veins of quartz, which have not decomposed, still remain running through them. The quartz itself, when broken up, would furnish a white sand such as that found associated with the red clays, so that we have here a most satisfactory explanation of the derivative origin of a great part of the burrstone formation.

Savannah River.—I shall now describe several natural sections which are seen in the bluffs or cliffs bounding the alluvial plain of the Savannah River, in its course of about 250 miles between Augusta and the sea. The river has an average fall of about one foot per mile, or 250 feet between Augusta and the delta of the river. Like the Mississippi and all large rivers, which, in the flood season, are densely charged with sediment, the Savannah has its immediate banks higher than the plain intervening between them and the high grounds, which usually, at whatever distance from the river, present a steep cliff or "bluff" towards it.

Near Augusta, the Savannah cuts through the red clays and sands before mentioned. Forty miles below the city, a section from 120 to 150 feet high, and half a mile in extent, is observed in Shell Bluff in Georgia, on the right bank. Unfortunately, at the time of my visit the waters were high, and covered the bottom of the bluff. The lowest exposed portion of the cliff consisted of white pulverulent marl, derived chiefly from comminuted shells, which passed upwards into a solid limestone, sometimes concretionary, and containing numerous casts of shells; and above this was again seen pulverulent white marl. Still higher, the calcareous deposit becomes more sandy and clayey, and encloses a bed of huge oysters (*O. Georgiana* Conrad), which are found growing one upon the other, and have evidently not been drifted into their present place. The total thickness of these calcareous strata is about 80 feet, above which, beds of red loam and yellow sand, such as prevail at Aikin and Augusta before mentioned, and without fossils, are seen at the top of the cliff 40 feet or more in thickness.

After a diligent search of several days, I obtained casts of no less than thirty-nine species of shells from the limestone of Shell Bluff, twenty-four of which I have been able to identify either with eocene species, known to Mr. Conrad to occur at Claiborne, or to species found by me in other eocene localities, and I have no doubt that I could have identified more had my own collection from Claiborne been more complete.

Shells found at Shell Bluff agreeing with Species in other Eocene Localities.

1. Conus, like one from Jacksonboro', Georgia
2. Cypræa, like one from Wilmington, S. Carolina
3. Oliva Alabamensis *Conrad*
4. Pyrula, apparently agreeing with one from Claiborne
5. Voluta prisca (Turbinella prisca, *Con.*)
6. Trochus agglutinans
7. Melongena alveata, very common
8. Infundibulum trochiforme
9. Natica ætites *Con.*
10. Bulla, like one from Jacksonboro'
11. Crepidula lirata *Con.*
12. Dentalium thalloides *Con.*
13. Crassatella protexta *Con.*
14. Lucina pandata *Con.*
15. Lutraria lapidosa *Con.*
16. Cytherea Poulsoni *Con.*
17. C. perovata *Con.*
18. Cardita planicosta
19. C. rotunda *Con.*
20. Cardium, like one from Jacksonboro'
21. Nucula magnifica *Con.*
22. Chama, like one from Jacksonboro'
23. Ostrea sellæformis *Con.*
24. Pecten membranosus

The remaining shells, chiefly casts, which I collected, belong to the following genera: *Conus*, *Voluta*, *Turritella*, *Fusus*, *Cytherea*, *Lithodomus*, *Solenomya*, *Cardita*, *Cardium*, *Pectunculus* (two species), *Pecten*, *Arca*, *Ostrea*.

One of the species of Ostrea, *O. Georgiana*, has been supposed to agree with the large fossil so common in Touraine, and called by some *O. virginica*, but this identification is doubtful. The Shell Bluff species is very remarkable for its enormous thickness and length.

About nine miles below Shell Bluff, in the Long Reach, is a cliff about eighty feet high, called London Bluff, where the same shelly white calcareous beds again appear, covered with red clay and loam. The horizontal stratification is evident here, as in the cliff two miles below, where the large oysters are seen standing out in relief. Below this, at Stony Bluff in Burke County, near the borders of Scriven County, the calcareous beds have quite disappeared, and siliceous beds of the burrstone series are seen occupying the cliff, and resting upon brick-red and vermilion-coloured loam. This superposition is important, as concurring, with other facts, to show that the burrstone of this region with its eocene fossils is an integral part of that great red loam and quartzose sand formation, usually devoid of fossils, which occupies so large a space between the hypogene (or primary) region and the Atlantic. The quartzose and siliceous rock of Stony Bluff was, during the last war with Great Britain, quarried for millstones. It passes into a sandstone with distinct grains of quartz, and is full of cavities and geodes, partially filled up with crystals of quartz and agates. Portions of it are filled with spiculæ of fossil sponges, some of them in a decomposed state, and there are also seen in the same flints, when thin slices are cut and polished, minute flustriform corals and foraminiferous shells, which were detected by Mr. Bowerbank, who has had the kindness to examine the specimens for me microscopically. Loose flints of this formation, containing spines of echini and other fossils, are scattered over the surface of the country, like chalk-flints in England. At Millhaven, in Scriven County, about eight miles from Stony Bluff, and five in a direct line from the Savannah River, the siliceous beds crop out on the banks of a small stream. In the flints here I found casts of several shells of the genera *Valvata*, *Pecten*, and *Terebratula*, with a species of *Cidaris*. It was evident to me that this millstone belongs to the same formation of red loam and sand which extends almost uninterruptedly from Aikin and Augusta to this region, for I observed near Millhaven, where the deposit has been pierced through to the depth of 26 feet in wells, pieces of white kaolin embedded, like those before mentioned near Augusta.

Jacksonboro', Scriven County.

SECTION OF EOCENE STRATA ON THE RIGHT BANK OF BEAVERDAM CREEK, SCRIVEN COUNTY, GEORGIA.

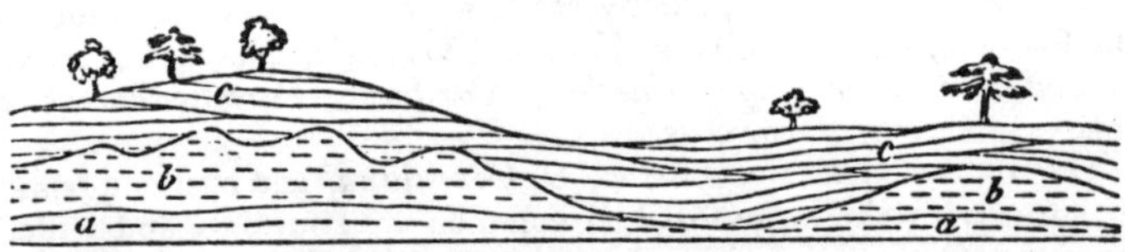

a. Limestone. *b.* White marl. *c.* Yellow and red sand and clay of the burr-stone formation.

About eight miles from the Savannah River, and one mile west of Jacksonboro', a limestone occurs, covered with sand, in the fork of Briar and Beaverdam Creeks, which has been quarried for lime. This limestone (*a*) passes upwards into marl (*b*), which has an undulating surface, as represented in the annexed section, and appears to have been denuded before the deposition of the incumbent sand and loam (*c*). The height of the cliff is about 25 feet, the thickness of the calcareous formation varying from 10 to 15 feet, and the yellow sand which rests upon it from 3 to 10 feet.

From this limestone of Jacksonboro' I obtained, besides corals and echinoderms, thirty species of shells, the larger portion of which were kindly presented to me by Colonel Jones of Millhaven. Those in italics agrée specifically with fossils from Claiborne, Alabama, or other eocene localities, which I have specified.

Fossil Shells of Jacksonboro', Scriven Co., Georgia.

Oliva Alabamensis
Voluta prisca
Conus, same as one from Shell Bluff
Rostellaria or Strombus (casts), four species
Fusus inauratus
Cerithium Georgianum *Lyell & G. Sowerby*, see fig.?
Melania, see fig.?
Paludina?

Natica ætites
Turritella
Bulla, same as one from Shell Bluff
Trochus agglutinans
Infundibulum trochiforme
Solarium canaliculatum
Dentalium
Crassatella
Lucina pandata
Cytherea
Cardium, like one from Shell Bluff
Chama, like one from Shell Bluff
Cardita
Lithodomus dactylus
Modiola
Mytilus
Avicula trigona ?
Pecten
Ostrea panda

c. Melania, cast.

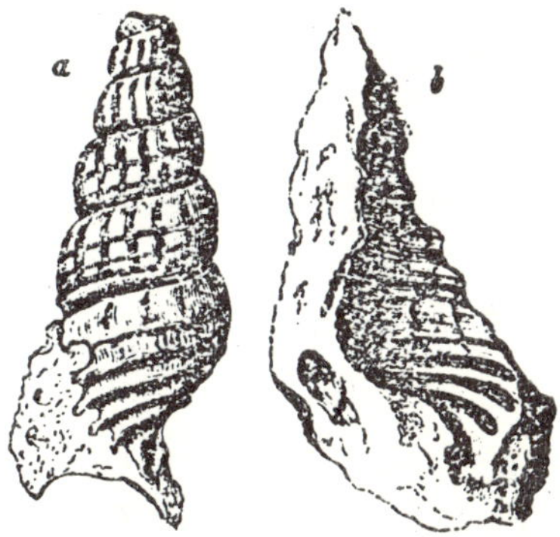

Cerithium Georgianum.
a. Nucleus. *b.* Cast of exterior.

The shell named *Cerithium Georgianum* in the above list (figs. *a*, *b*.), is very abundant, and closely resembles *C. lamellosum* Lam., a tertiary species of the Paris basin. Its characters are as follows:—

Cerithium, with an acuminated turreted shell, volutions 9 or 10, rounded, with rather obsolete and irregular longitudinal ribs, and with 5 or 6 transverse ridges, of which the three anterior are very prominent, and lamellose near the aperture.

Of the next shell in the list, named Melania (fig. *c*.), there is only a cast of the inside preserved, so that its external characters cannot be distinctly known.

Of the *Lithodomus dactylus* I obtained a beautiful cast, both of the exterior and interior, from the cavity of a fossil coral. It resembles the West Indian variety, and is an eocene species well known in the Paris basin. As, according to Philippi, it is one of the most cosmopolitic of living species, we have the less reason to be surprised at its great vertical range in the geological series.

A species of *Scutella* differing from those found by me in other places was common at Jacksonboro'. For the following description of it I am indebted to Professor E. Forbes, and it has been named, after Colonel Jones of Millhaven, *Scutella Jonesii* :—

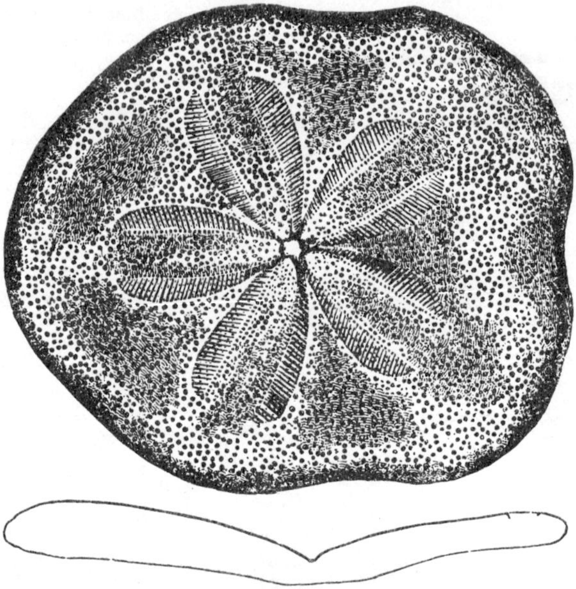

Scutella Jonesii Edward Forbes.

Body plane, shield-shaped, subpentangular, with sides and posterior margin undulated; angles obtuse.

Back centrally slightly convex, interambulacral spaces depressed; ambulacral spaces broad, somewhat convex, with parallel sides. *Avenues* petaloid, with their inner margins nearly straight. Pairs of pores in each avenue 37, united by oblique lines.

Oral surface concave, with five deep furrows radiating from the mouth to the margin.

Margin thick, rounded.

Lat. $2\frac{1}{12}$. lon. $2\frac{10}{12}$. crass. max. $\frac{4}{12}$.

This species appears to have been marked with spots.

Locality.—Jacksonboro'.

I have no doubt that all the sandy soil on which the long-leaved pitch pine grows in the neighbourhood of Jacksonboro' belongs to the burrstone formation, consisting of sand, ferruginous sandstone, and red loam, and although it is rare to find the limestone and marl exposed to view, I believe it to be everywhere subjacent. For we meet not unfrequently in Scriven, and several of the adjoining counties, with lime-sinks, or deep depressions, more or less circular in shape, in the vertical walls of which we observe sections of the horizontal beds of sand and mottled red and yellow loam and clay. As the water does not stand in these sinks, there is evidently a subterranean drainage, by which the loose sand has been carried down, and the surface undermined, as before described at Cave Hall. I saw several lime-sinks near Jacksonboro', and one about 16 miles south of Millhaven, on the east side of the road to Savannah, at Reeve's mill. It was 60 paces in circumference, and 80 feet in depth, and the beds gone through consisted of yellow and deep-red sand, in some parts ferruginous, with beds of mottled red and white steatitic clay.

All the bluffs which I examined on the Savannah River below Briar Creek belonged to the beds above the limestone, and are referable for the most part, if not entirely, to the burrstone formation. I observed several sections in the Long Reach in Scriven's County, where the red loam and yellow sand is conspicuous; and there is a fine section in Hudson's Reach, at a place called Tiger-Leap, where beds of fuller's earth occur. A few hundred yards below Tiger-Leap, where a small creek or brook enters on the right bank of the Savannah River, I found in some of the white clays impressions of *Mactra*, *Pecten*, and *Cardita*, with fragments of fishes' teeth, particularly of the genus *Myliobates*, several of the genus *Lamna*, and one of the genus *Galeus*. These bluffs of loam, clay, and sand are often 80 feet in height; and after passing Scriven I found, in the county of Effingham, similar sections, as at Sister's Ferry and Ebenezer. In the section at Sister's Ferry there is not only the brick-red loam, and the red and grey clay and sand, but layers of steatitic clay, which, although soft when moist, become hard and acquire a conchoidal fracture when dried.

On the whole it appears, from the information I obtained, that the less elevated part of South Carolina and Georgia, intervening between the mountains and the Atlantic, has a foundation of cretaceous rocks, containing *Belemnites*, *Exogyra*, and other fossils, above which are, first, eocene limestones and marls, and, secondly, the burrstone formation, with its red loam, mottled clays, and yellow sand. I am informed by Mr. Vanuxem that a tertiary lignite formation is sometimes interposed between the cretaceous beds and the eocene limestone; but I had no opportunity of verifying this fact in the sections which I saw, partly, I believe, owing to the swollen state of the rivers at the time of my visit. The remarkable difference of the fossils found in the eocene limestone at different points may lead some to the suspicion that there exists in this country a considerable succession of minor divisions of the eocene period, but I am inclined to ascribe the circumstance principally to two causes: first, that the number procured in each place is small, and therefore represents a mere fraction of the entire fauna of the period under consideration; and, secondly, that we have not yet any great eocene collection from any part of the United States. If we had 1000 shells from Alabama instead of little more than 200 (those, namely, which have been found at Claiborne), we should be able to form a more correct opinion respecting the mutual relations of the strata at distinct points, such as Shell Bluff, Jacksonboro', Eutaw, the Santee canal, and Wilmington, North Carolina.

The difficulty of classifying the tertiary strata of the southern states arises mainly from the wide extent of red and white clays, and siliceous sand, without fossils. The sterile sands which form the soil of the pine barrens in the lower plains of Virginia and North Carolina appear to belong to the miocene period, while those of a large part of South Carolina and Georgia are eocene. Some of the red ochreous and vermilion clays, as, for example, those of Martha's Vineyard and at Richmond, Virginia, are miocene. Others of similar character in South Carolina and Georgia, as at Orangeburg, Aikin, Stony Bluff, and Millhaven, belong to the burrstone formation, which is of the eocene period.

The species of eocene shells common to the United States and Europe appears to be very small. I have in my cabinet eighty-five species, in a good state of preservation, from Claiborne, Alabama, presented to me by Mr. Conrad; and I procured from the various localities already enumerated in this paper about forty species which I could not identify with the above, or with any which I have seen from Claiborne. Out of these 125 species I have been able to identify the following seven only with European eocene shells: namely, *Bonellia terebellata*, *Trochus agglutinans*, *Solarium canaliculatum*, *Infundibulum trochiforme*, *Cardita planicosta*, *Lithodomus dactylus*, *Ostrea bellovacina*. The proportion, therefore, of species common to Europe and the United States scarcely exceeds five per cent., and the proportion of species now living and identical with the American eocene shells appears to be still smaller. In regard to geographical representations, I found at least one fourth of the species to be very closely allied to European eocene fossils, while another fourth presented forms differing greatly from any species procured from the eocene strata of Europe, although belonging to genera which are abundantly represented in these formations.

[ON THE EOCENE OF ALABAMA AND GEORGIA]

POSTSCRIPT.

1. Since the body of this number of the Journal was printed, we have received an interesting letter from CHARLES LYELL, Esq., dated Feb. 4th, at Claiborne, (Ala.) addressed to B. Silliman, Jr., from which we make the following extracts:

"As you expressed a wish to receive as early a notice as possible, of some of the results of my geological explorings in the South, I will send you a brief statement of several conclusions to which I have come since I entered Georgia and Alabama.

"1st. I have visited some of the principal localities where the bones of the gigantic Cetacean (the Zeuglodon) have been discovered in Clarke County, Ala., in the fork of the rivers Alabama and Tombeckbee, and find the geological position of the bones to be every where the same, namely, in the white tertiary limestone of the Eocene period, corresponding in age to that of the Santee River, in South Carolina, or of Burke County, in Georgia, or that of the upper part of the celebrated bluff of Claiborne, in Alabama.

"2d. The beds in which the remains of the Zeuglodon occur, are above the level of those Claiborne deposits, so well studied and described by Mr. Conrad in 1832–3, as containing several hundred perfectly preserved species of lower tertiary shells.

"3d. Part of the head of the Zeuglodon and vertebræ, extending to a length of 30 feet, were procured by Mr. Koch in 1845, at a place which I visited, 4½ miles southwest from Clarksville, Ala., in company with Mr. Pickett, who assisted in the exhumation made by Mr. Koch. But the main body of the vertebræ, (as I learn from the same gentleman, and other persons,) which entered into the skeleton exhibited in the United States in 1845, under the name of Hydrarchos, were procured in Washington County, Alabama, fifteen miles distant, in a direct line, from the place where the head was discovered.*

"4th. I have information of more than forty other localities, where separate bones of the same huge animal have been met with, in Clarke and Washington counties, all in such perfect condition, and each so bulky and so distant from the rest, that I can hardly doubt their having belonged to as many distinct individuals.

* Another correspondent (S. G. Houston, Esq.) gave us a confirmation of Mr. Lyell's statement, under date of Washington County, Ala., Nov. 1st, 1845. Mr. H. says that the fossils were found, a bone here and a bone there, scattered over a space of 25 miles by 10 or 12.—*Eds.*

"5th. The age of the nummulite limestone of Alabama, seems to have been involved in considerable obscurity, it having been commonly referred to the secondary period, or considered as an upper cretaceous formation, or as possibly intermediate in age between the tertiary and secondary strata. After visiting the neighborhood of Suggsville, Macon, and Clarksville, Ala., I am persuaded that it is a tertiary limestone newer than all the beds of the well known Claiborne bluff, and lying immediately above the calcareous rock in which the Zeuglodon occurs. It is therefore by no means in the lowest part of the Eocene system of the South. Hitherto I have always met with it in hills considerably higher above the sea, than the top of the bluff at Claiborne, a circumstance which, in a region where the stratification is horizontal, would imply a newer deposit, even if the sections near Suggsville and Clarksville, showing actual superposition, had been less satisfactory. I did not meet with the limestone in question in the bluff at Claiborne, which I have no doubt is owing to the fact that the calcareous strata are cut off at the top, before they extend upwards into the nummulitic beds.

"6th. The conclusion last stated, will make it necessary in future to omit from the list of cretaceous fossils, such species of shells as have been referred to the secondary formations, simply on the ground of their occurrence in the nummulitic limestone. The fossils which I met with most abundantly in the latter, were the Nummulites Mantelli, (Morton,) and Pecten Poulsoni. I also found in the lower part, Pecten perplanus, Ostræa panda, and Plagiostoma dumosum, a large species of Cypræa, a Lunulite, and other corals.

"7th. The shale of the tertiary white limestone country of Alabama, is covered, like that of the Savannah River in Georgia, with river deposits of red clay, with quartz gravel, fine yellow, white, and pink sands, white porcelain earth, and occasionally chert and clay with sand intermixed. In many places in Georgia, where the buhr-stone occurs in this series, I have found in it distinct casts of Eocene testacea and corals, and I have shown, in the Proceedings of the Geological Society of London, the manner of its superposition to the subjacent calcareous beds in Scriven County, Georgia, which is equally applicable at Claiborne and in Clarke County, Alabama. Whether this upper deposit in Alabama be Eocene, and identical in age with that of Georgia, as I suspect, I am unable as yet positively to prove by the evidence of organic remains, although I have met with some faint impressions of shells and silicified wood in Clarke County.

"8th. The underlying Eocene or white limestone series, has a very undulating surface, on which the newer beds, many hundred feet thick, were deposited. The inequalities having been thus removed, and the hollows filled up, the existing valleys were excavated in both deposits. From this disposition of the beds, it happens that a deep cut in one place may pass exclusively through the limestone, and then at another, in the immediate neighborhood, through the incumbent newer strata. At Claiborne itself the bluff at the new landing consists of the calcareous formation about 150 feet thick, with a capping of 20 feet of the newer deposit, while about a mile lower down in the same bluff the appearances are reversed, and the upper formation occupies more than 100 feet of the higher part of the bluff, the calcareous beds cropping out from beneath. The section given by me in the Quarterly Journal of the Goological Society, No. 4, p. 438, in reference to the Jacksonboro' beds in Georgia, will enable you to understand this arrangement."

COAL FIELD
OF
TUSCALOOSA, ALABAMA

Art. IV.—*Coal Field of Tuscaloosa, Alabama*,—being an extract of a letter to Prof. Silliman, from Charles Lyell, Esq., dated Mobile, Alabama, Feb. 19th, 1846.

Since I wrote to you last, I have been engaged in the study of the carboniferous rocks and coal of Alabama, respecting the existence and extent of which, I was unable to obtain any accurate information, when I compiled the small geological map of the United States that accompanied my Travels. Mr. Conrad had mentioned to me the existence of a bed of coal, the geological age of which he had not ascertained, above Tuscaloosa, and I learnt from several persons on my way south, as I passed through Georgia, that the city of Mobile was supplied with bituminous coal, for fuel and gas, brought down the Tombecbee River, by a navigation of more than 350 miles. When I heard that it was procured not far from the northern outcrop of the newer formations, whether cretaceous or tertiary, that it occurred very near the falls of the Warrior River, the principal tributary of the Tombecbee, and that it was used for the manufacture of gas at Mobile, I suspected that it might be related to the Richmond coal, which is bituminous, situated near the falls of the James River, and placed nearer the Atlantic than the coal of the Allegheny range, than which it has also been shown to be newer in date, by Prof. W. B. Rogers.

In order to determine, if possible, its chronological relations, and obtain information respecting its extent, I ascended the Tombecbee, from Mobile to Tuscaloosa, where, at the University, I found Prof. Brumby, who had examined with considerable care the geographical boundaries of the productive coal measures, and the structure of the region. With him, I made an excursion to some of the pits, or rather open quarries of coal, where the edges of the beds of several seams have been dug into by different individuals, entirely ignorant of mining operations, but with no small success, the quality being good at the points of natural outcrop. I found the coal seams every where covered with beds of ordinary black carbonaceous shale, full of impressions of more than one species of Calamite, with ferns of the genera Pecopteris and Neuropteris, and impressions of Sigillaria and Lepidodendron. In some of the beds, Stigmaria has also been met with not unfrequently. I recognize a specific identity between several of the most common of these coal plants and those of Europe, Ohio, and Pennsylvania, and I observe that they are completely different from the vegetable remains that are most abundant and characteristic in the newer or Virginian coal field, near Richmond, which I had lately an opportunity of examining on my way South. The strike also of the coal beds of Alabama, on the Warrior River, where I saw them, and those of a coal field immediately to the east, examined by Prof. Brumby, is northeast and southwest, agreeing with the general direction of the Allegheny Mountains, of which, geologically speaking, they are evidently a southern prolongation. They are in fact portions of the great Apalachian coal field, exhibiting all the same mineral and palæontological characters, the beds having been bent into anticlinal and synclinal ridges, similar to those of the Alleghenies, with corresponding dips to northwest and southeast.

Alabama is the first state, as we proceed from the northeast, where we can pass from the Atlantic plain, occupied by tertiary and cretaceous strata, to the older rocks of the Apalachian region, without first crossing an intervening zone of the hypogene or granitic rocks. In the eastern part of the State, the crystalline formations still appear, as at Wetumpka, where, as I learn from Prof. Brumby, the lower cretaceous rocks rest upon mica schist. But on going westward to the Cahawba River, and still farther west to the Warrior, at Tuscaloosa, we find the carboniferous strata concealing the granitic rocks, or coming into direct contact with the cretaceous. At the falls of the Warrior, at Tuscaloosa, beds of gravel containing well rounded pebbles of quartz, are seen to rest unconformably on the gray micaceous sandstone of the coal, which is full of Calamites and impressions of Lepidodendron and Sigillaria. I should have felt very doubtful respecting the true age of the overlying unconsolidated gravel and sand, in which no organic remains have yet been discovered, in that neighborhood, if I had not carefully examined the position of similar beds of loose gravel 30 feet thick, at Montgomery, 100 miles southeast, on the Alabama River, as also at spots between Montgomery and Wetumpka. In nodules of an impure limestone, in a loam overlying that gravel, and almost in contact, I found casts of Inoceramus, Rostellaria arenarum, and other common chalk fossils. They are well seen at the bluff, in the suburbs of Montgomery, and there are beds of gravel, and various colored sands and clays below them; the general dip of the cretaceous strata being to the south, which brings up the edges of certain calcareous marls, newer than the gravel, but of the same epoch, producing prairie soils, often destitute of natural wood, which run for a great distance east and west, or in the line of their strike, crossing the Alabama several miles south of Montgomery, and being again seen in their western prolongation on the Tombecbee, at Arcola and other places. It is due to Mr. Conrad, to state that in an outline map which he presented to me in 1842, with permission to publish any part of it in my general map, he had traced out the course of the lower cretaceous beds; especially their northern boundary in Alabama, with great approach to accuracy, from observations made by him, and information obtained during his tour in 1833.

The members of the carboniferous series, which I examined in an excursion with Prof. Brumby, within a distance of between

30 to 40 miles northeast of Tuscaloosa, may be said to consist of three subdivisions: 1st, the productive coal measures; 2d, quartzose grit, used for millstones, with schistose sandstones and slates; 3d, limestone and chert. The first of these, or the productive coal measures, consist of the usual white sandstones and grits, so common in all coal fields, with greenish and gray micaceous sandstones, and some ripple-marked slabs of yellow sandstone, with Calamites. Associated with these, are shales and clays, with seams of coal, the thickest of those which I saw being about four feet, but a ten foot seam has been discovered sixty miles farther to the north. I saw a thickness of many hundred feet of this formation, but have no data for estimating its entire vertical extent. To the above succeeds, in the descending order, a great deposit of quartzose grit, used for millstones, which resembles in position and mineral character, the millstone grit of the north of England, and perhaps occupies the place of the fundamental conglomerate of the great Apalachian coal field. It passes downward into thinly laminated sandstones, and dark slates of small thickness. Under this group lies a formation of limestone, usually blue, with white veins of carbonate of lime, and with much chert and hornstone intermixed. In the pure limestone, which is fetid, when broken, no fossils have been found, but in some of the associated siliceous beds, Prof. Brumby has obtained Encrinites, Producta, Orthis, and one or two corals, which have not yet been determined, but the genera which I saw, are such as might belong to the carboniferous limestone. In some of the inferior beds of limestone, there is a great mixture of iron, and throughout the range of this formation, there has been traced an enormous mass of brown hematite, which seemed to me to constitute, where I examined it at Murphy's, 30 miles from Tuscaloosa, a regular bed, rather than a vein. From the abundance, accessibility and richness of this ore, its proximity to the coal field and to the navigation of the Tombecbee River, I can hardly doubt that, like the coal itself, it is destined, at no distant day, to be a source of great mineral wealth to Alabama. I may also mention that large specimens of pure galena, obtained from this limestone formation, have been shown me by Prof. Brumby, who found them in the counties of Tuscaloosa, Bibb, Shelby, Jefferson, and Blount, as also in the Tennessee valley, in Alabama;

but no search has yet been made by miners to ascertain whether the lead could be worked to advantage.

It would have been impossible for me, during my short visit to this region, to form more than a conjectural opinion respecting its structure, still less could I have determined the geographical area of the Alabama coal fields, had not these subjects been previously studied, with great care and scientific ability, by Prof. Brumby. Of the extent of the coal, he published a brief account in Barnard's Almanac for 1838, and communicated the same to Dr. Silliman. It appears that south of the waters flowing to the Tennessee, there are two parallel coal fields—those of the Warrior and Cahawba rivers. The western of the two, that of the Warrior, has a length of more than 90 miles from northeast to southwest, and a breadth of from 10 to 30 miles. It extends through the counties of Tuscaloosa, Walker, Jefferson and Blount, on both sides of the Warrior River, and its several branches. Throughout all this area, seams of bituminous coal crop out, the number and thickness of which it has been impossible as yet to determine, as no regular survey has been commenced of this State. That there must, however, be many distinct seams, is evident, as I saw several different ones between 11 and 20 miles of Tuscaloosa, which had been worked with profit, and considerably farther to the northeast, three have been seen superimposed upon each other, the largest and lowest appearing in the bed of the Warrior River, and being no less than 10 feet thick.

The more eastern coal field, or that of the Cahawba, is nearly of equal length and breadth, terminating southwards at Centreville, where it meets the lower cretaceous beds, and extending from thence through Bibb, Shelby, Jefferson and St. Clair counties, to the sources of the Cahawba River. In this, also, numerous beds of coal of good quality have been detected, and some of them worked to a small extent.

From what I saw of the structure of the country between Tuscaloosa and Rook's valley, and from the facts communicated to me by Prof. Brumby, I have no doubt that a section passing from northwest to southeast, from the coal district of the Cahawba to that of the Warrior River, would present an additional axis along the line of the water-shed between the two rivers, where the beds are highly inclined, and occasionally vertical, and the productive coal measures occur on both sides, their strata having a slight dip, and being in some places nearly horizontal. These views will best be explained by the annexed diagram.

Coal Fields of Alabama.

References.—*a*, Warrior river; *b*, Rook's and Jones' valley; *c*, Cahawba river. 1, Coal measures; 2, Grit; 3, Limestone and Chert; 4, Grit; 5, Coal measures. Length of section, 50 miles.

On the northern confines of the State of Alabama, there is a third coal field, that of the Tennessee valley, which is separated from the two former by a broad but low chain of mountains, which run nearly east and west, and intervene between the Tennessee and the sources of the Warrior and Cahawba rivers. These mountains, according to Prof. Brumby, consist of strata, like those before described as occurring immediately below the productive coal measures between the Warrior and Cahawba rivers, and consisting chiefly of quartzose grit and limestone. I could not ascertain whether this coal of northern Alabama, which runs about 100 miles east and west, along the Tennessee River, is actually continuous with the southeastern coal measures of the Apalachians, in the State of Tennessee, but it seems highly probable that they are united.

A careful comparison of the fossil plants of the Alabama coal field, which I have collected myself, or which have been liberally presented to me by Prof. Brumby, with those of the more northern coal fields, and of Europe, will be a subject of peculiar interest, as they are situated in lat. 33° 10′ north, and constitute, if I mistake not, the extreme southern limit to which the peculiar vegetation of the ancient carboniferous era has yet been traced, (in the northern hemisphere,) whether on the western or eastern side of the Atlantic.

ON THE EVIDENCE OF FOSSIL FOOTPRINTS OF A QUADRUPED ALLIED TO THE CHEIROTHERIUM

Art. III.—*On the Evidence of Fossil Footprints of a Quadruped allied to the Cheirotherium, in the Coal Strata of Pennsylvania;* by Charles Lyell, Esq., F. R. S., F. G. S., &c.

(Communicated by the Author.)

On my way from Pittsburg to Philadelphia, I visited the rocks in which Dr. King, of Greensburg, first discovered, in 1844, the footmarks supposed to have been made by a large reptile, and which stand out in relief from the lower surface of slabs of sandstone, resting on a thin layer of fine clay. Having visited the quarry and enquired into all the circumstances of the discovery, having seen almost every specimen hitherto obtained, and considered the relations of the sandstone with the rocks lying above and below it, I have no hesitation in declaring my conviction, first, that the footprints are genuine, and were made by a quadruped nearly allied to the Cheirotherium of Europe, if not the same; secondly, that they occur in the middle of the carboniferous formation, having both above and below them seams of coal and strata containing Lepidodendron, Sigillaria, Stigmaria, Calamites, &c.

Great praise is due to Dr. King for the exertions which he has made in bringing these fossils to light, and duly appreciating from the first their extraordinary importance and value. A young man, engaged in an extensive medical practice in a remote town, acting under every discouragement which want of sympathy in those immediately around him, and of access to scientific books or museums could produce, he has persevered and succeeded in forming a fine collection of the fossil tracks, and become acquainted with the relations and organic remains of the contiguous strata.*

I shall begin by describing the first locality to which the attention of Dr. King was called in 1844, a stone quarry about five miles S. E. of Greensburg in Westmoreland county, Pennsylvania. The slabs of argillaceous sandstone are extracted here for paving, but the excavation begun in the bank of a small stream was soon desisted from in consequence of the increasing thickness of the

* Three papers have appeared in this Journal from Dr. King. The first describes and figures the footmarks from five miles S. E. of Greensburg, vol. xlviii, p. 343; the second, those of Derry, vol. xlix, p. 216; and lastly, the supposed hoofed-prints at Connelsville, second series, vol. i, p. 268.

overlying useless shale. Between the slabs of stone, which are a few inches thick, are thin parting layers of a fine unctuous clay well fitted to receive and retain faithful impressions of the feet of animals. About twenty three footsteps have been observed, some more and others less distinct. One of these was seen by Dr. King, to be impressed on the upper surface of a layer of clay, but the specimen was unfortunately left exposed to the weather and destroyed. The other Cheirotherium footsteps stand out in relief from the under surface of a bed of sandstone, and are accompanied by small mud-veins which have been described as fucoids for which they were mistaken. They are in fact casts of those cracks of various sizes which are produced by the drying and shrinking of mud. As the footprints are traversed and occasionally distorted by these veins, it is evident that the shrinkage took place after the animal had walked over the mud. There is every where seen a double row of tracks, which occur in pairs, each pair consisting of a hind and fore foot, and each being at nearly equal distances from the next pair. The toes, on each of these parallel rows, turn, the one set to the right, the other to the left.

The geological position of this sandstone is perfectly clear. It is situated in the midst of the Apalachian coal field, and there is no formation to be seen, except the carboniferous, for a considerable distance in any direction. The beds are gently inclined to the S. E., and occur on the side of one of those troughs which intervene between two anticlinal folds or ridges which are not only found in the principal chain of the Alleghany mountains, but also, as in this instance, in the lower and less hilly country immediately west of the mountains. The sandstone and shale containing the footsteps, crop out from beneath the main or Pittsburg seam of coal, which is about nine feet thick, and has been worked at its outcrop in the immediate neighborhood. This coal is less than a hundred feet above the sandstone, and there are several other seams of workable coal which lie at lower levels; and impressions of Lepidodendron, Sigillaria, Stigmaria, Calamites, ferns, and other carboniferous plants, have been found both above and below the level of these reptilian footsteps.

The second locality which I visited, where the supposed fossil footprints of birds and of quadrupeds resembling dogs, were observed by Dr. King, is situated within a mile of the town of

Derry in Westmoreland county, about fourteen miles north of Greensburg, (see this Journal, vol. xlix, p. 216.) The markings here are all upon the surface of a white coal grit or sandstone, forming a bare ledge of rock, no vegetation, scarcely even any mosses or lichens having grown on the pure quartzose stone. The ledge is about thirty five feet in its longest diameter, and about thirty two in breadth. It projects above the general level about three feet at one end, and slopes down gradually but irregularly till it is only an inch or two high at the lower extremity. There are several other similar but much smaller ledges of bare sandstone within a very short distance, on which there are no footprints. The large ledge is covered with distinct representations, not only of the tracks of birds and dogs, but also of other animals. Those of birds in particular, are cut sharply and deeply, and there is one long series of steps in succession, as if a bird had walked from the lower to the upper end of the ledge, and then over the end, so as to leave on a steep slope inclined to the horizon at an angle of 22°, a clear and vivid imprint. I found it quite impossible to imagine the layers of sand when in a soft state or capable of receiving footprints, to have been ever so placed, as to admit of such an impression being made, and it is obvious that after the loose sand had consolidated and the ledge had acquired its present form by denudation, no bird could have left the slightest trace of its passage over so hard a rock. I may state indeed that I have never seen in Germany, England, or America, any impressions of the tracks of birds or quadrupeds indented on so coarse a sandstone as that of Derry, although casts in relief have frequently been taken in such a matrix, the sand having been poured into the hollows made in the soft mud on which the animals had trodden.

Another serious objection to the supposed geological antiquity of the footprints, presented itself to me on my first view of this ledge near Derry. The impressions are most of them extremely sharp, although the sandstone has not only been exposed for ages to the weather, but evidently acted upon by currents of water, which have shaped out channels and cavities of various depths, and some perfectly round, deep, and regularly formed pot-holes, one of them eighteen inches deep and more than a foot in diameter. I counted on this single ledge no less than nineteen pot holes, between some of which, part of the principal series of bird tracks passes.

Another objection to the genuineness of the footprints, arises from the unevenness of the surface of the rock, on which, as the successive layers lie nearly horizontal (which is shown by sections exhibited in the vertical walls of the pot-holes,) distinct planes of stratification are laid open. The only way of imagining how so great a multitude of tracks could be exposed superficially on many different planes, would be to suppose that each superimposed layer was originally covered with these markings, all the tracks turning in one direction, so that wherever the excavating power of water cut into a subjacent layer, it laid open to view some new tracks. But this hypothesis is so far from being borne out by the facts, that after cutting into the stone in more than twenty places and removing small slabs with imprints upon them, Dr. King has never been able to detect a single indication of a footprint in the rock below.

I have lately seen good imitations of the tracks of birds on a slab of limestone sculptured by Indians, at New Harmony, in the museum of Dr. D. D. Owen. It was brought from St. Louis in the state of Missouri. The different joints of the toes were indicated as in the Derry stone. In the immediate neighborhood of the ledge of Derry there are numerous graves of Indians, and the same place is known to have lain in the line of one of their paths leading from the Alleghany mountains to the west.*

It seems indeed, in the highest degree probable, that the aborigines of North America, whose skill in following the trail of all kinds of game is so well attested, and who are known to have cut in some places on the rocks, rude imitations of the forms of animals, should sometimes have employed, as the symbols of the birds or quadrupeds which they hunted, a copy of these footprints with which they were so familiar. But the true origin of these markings on the Derry sandstone, may, I think, be considered as set at rest by observations lately made in another locality

* The supposed *human footprints* in solid limestone, first discovered and figured by Mr. Schoolcraft in 1822, and considered by him to be genuine fossils, (Am. Jour. Sci., vol. v, p. 223,) will not soon be forgotten. These curious intaglios remained for twenty years unexplained, until Dr. Owen, in 1822, (Amer. Jour. of Science, vol. xliii, p. 14,) demonstrated to the satisfaction of all, their artificial origin, and attributed them (justly no doubt) to the ancient inhabitants of America. We refer the reader to his most interesting paper for a full detail of the evidence, with figures of the acompanying fossils, and some of the instruments which may have been used in the process of sculpture.—*Eds.*

in Fayette county, near Connelsville, (Second Series, vol. i, p. 268,) nineteen miles from Greensburg, and about thirty from Derry.

Dr. King and the Rev. Mr. Hackey have visited this place, and they inform me that on the surface of a hard ledge of sandstone exposed there, some unquestioned Indian hieroglyphics with the representation of two human heads and of a serpent, are to be seen. These are accompanied with tracks of birds and of hoofed quadrupeds, and lastly of a footprint resembling exactly that of a dog or wolf, and identical with some of the most common tracks on the stone at Derry.

I have now only to express my thanks to Dr. King for the assistance which he gave me in prosecuting these inquiries, in which his sole desire was to arrive at the truth, and to correct any mistakes into which he might have fallen. He now agrees with me in regarding those imprints only as genuine and fossil, which occur in Unity township, five miles from Greensburg. They consist, as I before stated, of the tracks of a large reptilian quadruped, in a sandstone in the middle of the carboniferous series, a fact so full of novelty and interest, that when we reflect on its importance, all disappointment in the abandonment of the spurious footprints is forgotten. In no formation have so many excavations affording facilities of palæontological researches in Europe and America been made, as in the carboniferous. No other group of strata has yielded such unequivocal evidence of the presence on the spot, of dry land, or of its existence in the immediate vicinity; and yet here in Pennsylvania, for the first time we meet with evidence of the existence of air-breathing quadrupeds capable of roaming in those forests, where the Sigillaria, Lepidodendron, Caulopteris, Calamites, Ferns, and other plants, flourished. Few geologists will now think it safe to assume, that no other species or genus or even class of terrestrial vertebrata existed at that period.

In conclusion, it may be proper for me to state, that having formerly examined, in company with Prof. Hitchcock, the Ornithichnites of the valley of the Connecticut river, I have by no means been led to doubt the genuineness of those imprints from what I saw at Derry. The fossil markings on the Connecticut are not superficial, but found under circumstances in all respects most strictly analogous to those which have led me to believe in the reality of the Cheirotherium of the coal strata near Greensburg.

Philadelphia, April 26, 1846.

OBSERVATIONS ON THE FOSSIL PLANTS OF THE COAL FIELDS OF TUSCALOOSA, ALABAMA

Art. XXIV.—*Observations on the Fossil Plants, of the Coal Field of Tuscaloosa, Alabama;* by C. Lyell, Esq., with a description of some species by C. T. F. Bunbury, Esq., F. G. S.

In a former number of the American Journal,‡ I described the geographical position of the coal fields of Alabama, and stated that the carboniferous strata of the Warrior river extend southwards to the town of Tuscaloosa in the neighborhood of which, aided by Professor Brumby, (who had already made some progress in the investigation,) I succeeded in collecting impressions of Sigillaria, Stigmaria, Lepidodendron, Calamites, Neuropteris and several other ferns. I also stated that I recognized a specific identity between several of these fossils, and some of the most abundant coal plants of Pennsylvania and Europe. On my return to England in June, 1846, I submitted the specimens to my friend C. T. F. Bunbury, who immediately compared them with the best published plates and descriptions, and with European fossils in the cabinets of the Geological Society of London. The result of his examination has fully confirmed the conclusion to

* The first specimens of this fossil were found by H. A. Prout, M. D., of St. Louis, sometime previous to our discovery. Most of the specimens yet obtained, are compressed; but Dr. P. has one in his cabinet which is detached, and in the original spheroidal form.

† For want of the necessary works of reference, we are unable, at present, to decide whether these genera, belong to undescribed species.

‡ Second Series, i, 371.

which I had arrived, and it will be seen in the sequel that notwithstanding the loss of several ferns and Sigillariæ which were obtained in a disintegrating matrix near the outcrop of the strata, where the shales are changed into soft, pale, laminated clay, this botanist has been able to detect no less than sixteen forms, of which the following is an enumeration.—1. Sphenopteris latifolia, *Ad. Brongn.* 2. S. Dubuissoni? *Ad. Brongn.* 3. Sphenopteris, allied to the last, perhaps a variety of the same. 4. Neuropteris tenuifolia, *Ad. Brongn.* 5. Neuropteris Grangeri, or N. gigantea? 6. Calamites cannæformis. 7. Calamites, obscure specimen allied to the foregoing. 8. Lepidodendron elegans. 9. Lepidodendron allied to L. dilatatum, *Foss. Flora.* 10. Lepidophyllum? 11. Sigillaria, decorticated. 12. Stigmaria ficoides. 13. Poacites? 14. Bechera tenuis, n. sp., very nearly allied to B. grandis, *Foss. Flora.* 15. Asterophyllites? flaccida. 16. Phyllites, resembling the leaf of Sparganium or Eriocaulon.

The Palæontologist will perceive at once that no less than half of the specimens in the above list, agree with well-known European fossils of the old carboniferous formation, and the rest belong to genera which are common in our coal measures, and may perhaps agree with European fossils when procured in a better state of preservation. The leaves however resembling Sparganium, or still more closely some of the Junci such as Eriocaulon, and which are very abundant, appear to Mr. Bunbury to be new, (see fig. p. 232.)

The three species of Sphenopteris mentioned in the above list all differ from any ferns which I met with in my travels in North America in 1841-2, but one of them S. latifolia is a common Northumberland species found at Newcastle. Neuropteris tenuifolia is also met with at Newcastle, and N. Grangeri occurs in Staffordshire, as also at Zanesville, Ohio. Calamites cannæformis is one of the most abundant forms both in Europe, the United States, and Nova Scotia, and the same may be said of Lepidodendron elegans, the Alabama variety being that commonly met with at Sidney, Cape Breton. Lepidodendron dilatatum is a Northumberland fossil. Lepidopyllum, probably the leaves of the preceding genus, is also a form frequently met with in the British and other European coal strata. The same is true of Stigmaria ficoides, and the genera Stigmaria and Sigillaria are as yet exclusively characteristic of the carboniferous group. Lastly, Bechera tenuis has been found in Coalbrook Dale, Shropshire.

When we recollect that the Tuscaloosa coal field is situated in lat. 33° 10′ North, and that so many of the species now first examined agree with fossils occurring more than twenty degrees further north in the British Isles, and at a distance of five thousand miles, with a broad ocean intervening, we cannot but be struck with this wide diffusion of a Flora, so singularly uniform in its character. The phenomena seem to imply at the remote period in question the existence, first, of continuous land across the space now occupied by the Atlantic, or at least of a chain of islands in that region; and secondly, a remarkably equable climate; for Dr. Hooker, author of "the Antarctic Flora," has lately observed that no existing Flora, viewed as a whole, preserves so uniform a character for so great a distance in a north and south direction, as that extending from the south of the island of Chiloe to Cape Horn, a range of twelve or thirteen degrees of latitude. This wide distribution occurs where the climate in winter and summer, and throughout the whole year, is peculiarly equable.

Alabama is the most southern spot in the northern hemisphere where the ancient carboniferous flora has yet been studied; geologists therefore will rejoice to hear that Prof. Brumby is fully alive to the importance of a more full investigation of the plants of that country, of which he will soon, it is hoped, have it in his power to form a considerable collection.

NOTICE ON THE COAL-FIELDS OF ALABAMA

3. *Notice on the* COAL-FIELDS *of* ALABAMA; *being an extract from a Letter to the* PRESIDENT *from* CHARLES LYELL, Esq., F.R.S., *dated Tuscaloosa, Alabama, 15th February* 1846.

SINCE my arrival in Alabama I have devoted part of my time to the investigation of the carboniferous rocks, and have obtained information respecting some coal-fields, the very existence of which in this State was unknown to me in 1844, when I compiled the Map of the Geology of the United States, published in my 'Travels.' On my way southwards, I had learnt from several persons in Georgia that the city of Mobile was supplied with bituminous coal, brought down the

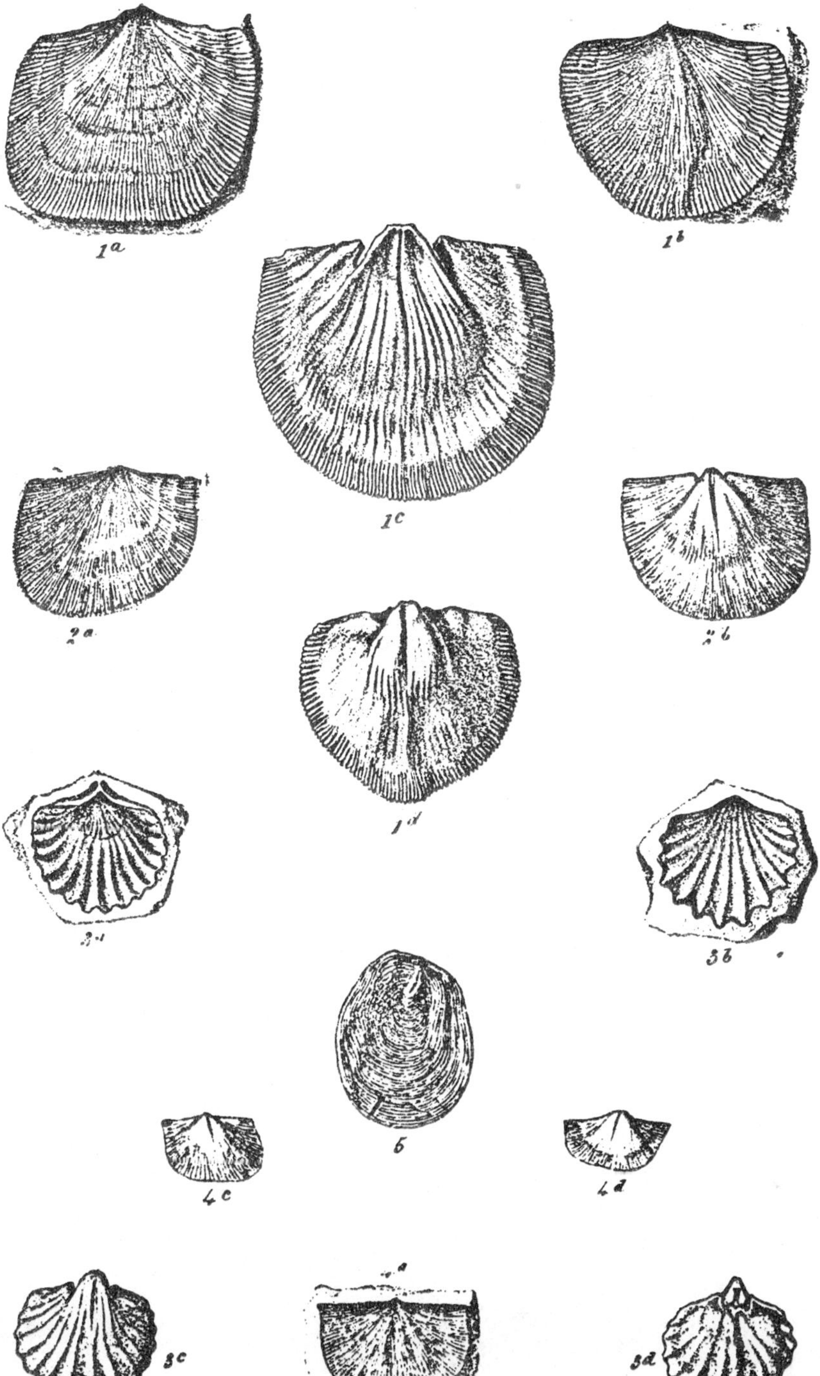

G.B. Sowerby Junr. ad lap. delt.

Reeve imp.

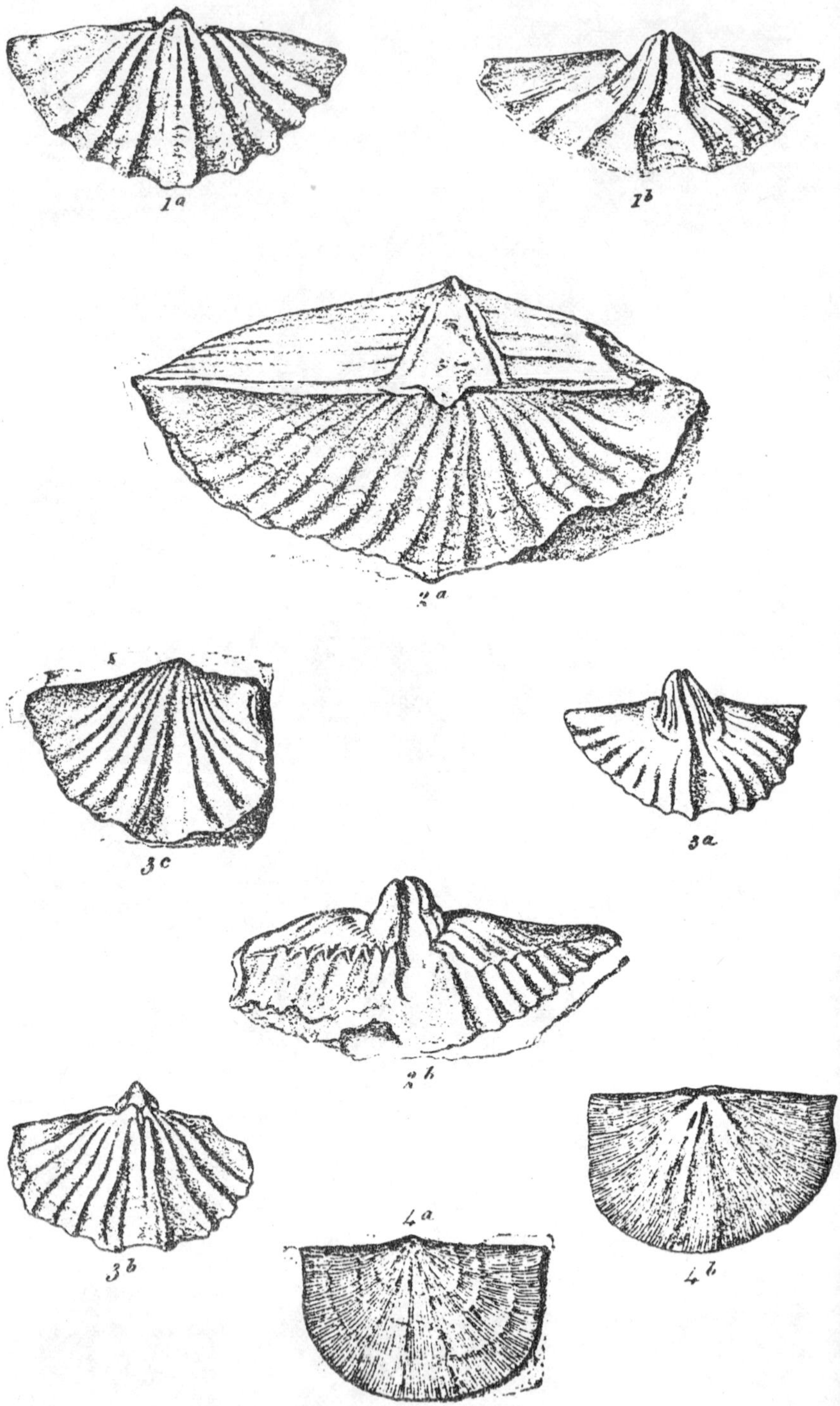

G. B. Sowerby Junr. ad. lap. delt.

Reeve imp

Tombecbee River from Tuscaloosa, a navigation of about 400 miles. This coal is procured from the neighbourhood of Tuscaloosa, a place situated near the centre of Alabama, and more than a hundred miles further south in a direct line than the southern limit which I had assigned to the Appalachian coal-field, which I supposed to terminate near the great bend of the Tennessee River.

The fact of coal occurring near Tuscaloosa had been previously mentioned to me by Mr. Conrad, but he was uncertain respecting its age; and the circumstance of its occurrence near the Falls of the river, not far from the northern outcrop of the cretaceous strata, together with the fact of its quality being preferred to all other coal for the manufacture of gas at Mobile, made me suspect that it might prove to be of the age of the Richmond coal, which is also bituminous, situated near the Falls of the James River, and which, as Professor W. B. Rogers has pointed out, is newer than the ancient carboniferous series.

In order to determine on the spot the question in regard to its age, I ascended the Tombecbee from Mobile to Tuscaloosa, where at the University I found Professor Brumby, who had examined with considerable care the geographical boundaries of the productive coal-measures and the structure of the region. With him I made an excursion to some of the pits, or rather open quarries of coal, where the edges of the beds of several seams have been dug into by different individuals entirely ignorant of mining operations, but with no small success, the quality being good at the point of the natural outcrop. I found the coal-seams covered everywhere with beds of ordinary black carbonaceous slate full of impressions of more than one species of Calamite, with ferns of the genera *Sphenopteris* and *Neuropteris*, and impressions of *Sigillaria* and *Lepidodendron*. In some of the beds *Stigmaria* has also been met with not unfrequently, and I recognise a specific identity between several of the most common of these coal plants, and those which I formerly obtained from the mines of Ohio and Nova Scotia. I also observed the complete difference between these fossil plants and those most characteristic of the newer or Virginian coal-field near Richmond, which I lately had an opportunity of examining on my way south. The strike of the coal-beds in Alabama is also, where I have seen them, north-east and south-west, agreeing with the general direction of the Allegany Mountains, of which, geologically speaking, they are evidently a southern prolongation. They are, in fact, portions of the great Appalachian coal-field, with all the same mineral and palæontological characters, the beds having been bent into similar ridges to those of the Alleganies, with corresponding dips to the north-west and south-east; and we have no reason to suppose that Tuscaloosa, in lat. 33° 10′ south, is the extreme southern limit of the formation, for the carboniferous strata are merely concealed from observation south of this point by the lower gravelly and sandy beds of the cretaceous group which extends to Tuscaloosa.

In the eastern part of Alabama, a zone of hypogene or granitic rocks separates the tertiary and cretaceous strata of the Atlantic plain from the older formations of the Allegany Mountains, as is usual along the eastern border of the United States. This granitic formation still appears at Wetumpka on the Alabama River, where the lower cretaceous rocks, as I learn from Professor Brumby, rest on mica schist; but on reaching the Cahawba further west, and the Warrior River at Tuscaloosa, we find the carboniferous strata concealing the granite, and coming, as before stated, into direct contact with the cretaceous rocks, which I have seen resting unconformably upon the coal at the Falls at Tuscaloosa. At this junction the cretaceous group consists of beds of quartzose gravel, such as I have seen intimately connected with cretaceous fossils at Montgomery on the Alabama River. At Tuscaloosa the underlying gray, micaceous sandstone of the coal is full of Calamites and impressions of Lepidodendron and Sigillaria, which I beheld with no small interest, as they constitute, if I mistake not, the extreme southern limit to which the peculiar vegetation of the ancient carboniferous æra has yet been traced, whether on the western or eastern side of the Atlantic. It is due to Mr. Conrad to state, that in an outline map of the Geology of Alabama made during his tour in 1833, and which he presented to me in 1842, with permission to publish any part of it, I find the northern boundary of the lower cretaceous deposits in their course through the States marked out with considerable approach to accuracy.

The several members of the carboniferous series which I have seen while in company with Professor Brumby, within a distance of between thirty and forty miles north-east of Tuscaloosa, consist, first, of productive coal-measures, containing the usual white quartzose sandstones and grits, with greenish and yellowish sandstones, some of which are thinly laminated and ripple-marked, and contain Calamites. These form the highest beds, and below them shales and clays predominate with several subordinate seams of coal, from three to upwards of four feet thick. Below these beds, of which I saw a thickness of many hundred feet, there lies, secondly, a great deposit of quartzose grit, of which millstones are sometimes made, and which reminded me of our millstone-grit, and of the fundamental conglomerate of the Appalachian coal-field. It passes downwards into thinly laminated sandstones and dark slates of small thickness. This group is succeeded (3) by fetid limestones, with chert and hornstone, usually without fossils, but in some of the siliceous beds of which, casts of Encrinites, Producta, Orthis and several corals abound. This inferior formation, which may perhaps belong entirely to the carboniferous series, also contains still lower down a limestone charged with iron; and an enormous mass of brown hematite appears to constitute a regular bed, and not a vein, and to be destined one day, like the coal, to be a source of great mineral wealth to Alabama.

It would have been impossible for me, during my short visit, to form more than a conjectural opinion respecting the structure of this coal-field, still less to determine its geographical area, had not these subjects been previously studied with great care and scientific ability by Mr. Brumby. Of the extent of the coal in Alabama he published a brief account in 1838 in Barnard's Almanac, and communicated the same to Dr. Silliman; and from the observations which we have lately made together, and from his notes and information, it may be inferred that a section from the north-west to the south-east, passing through the basins of the Warrior and Cahawba rivers, would present an anticlinal axis along the line of the water-shed between the two rivers, in the middle of which the beds are highly inclined, and often vertical, while on both sides the productive coal-measures occur in separate basins, their strata having a slight dip, and being in many places nearly horizontal. These views will best be explained by the annexed section.

SECTION across the ALABAMA COAL-FIELDS.

Length of section 50 miles.

In regard to the most western of the two coal-fields, or that on the Warrior River (the principal tributary of the Tombecbee), it has been found by Professor Brumby to be no less than ninety miles long from north-east to south-west, with a breadth of from ten to thirty miles, extending through the counties of Tuscaloosa, Walker, Jefferson and Blount, on both sides of the Warrior River and its several branches. Throughout all this area, seams of bituminous coal crop out; but the number and thickness of these it has as yet been impossible to ascertain, no survey having been made. They must however be numerous, for I saw several separated by a slight thickness of intervening strata at points between eleven and twenty miles north-east of Tuscaloosa; and I am informed, that in one place in the bed and banks of the Warrior River three seams are exposed to view, one above the other, the lowest and largest being ten feet in thickness.

The more eastern coal-field, or that of the Cahawba, is nearly of equal length and breadth, terminating southwards at Centreville, where it meets the lower cretaceous beds, and extending from thence through Bibb, Shelby, Jefferson and St. Clair counties, to the source of the Cahawba River. In this also numerous beds of coal of good quality have been found, and worked to slight depths.

A third coal-field on the northern confines of the State of Alabama, is that of the Tennessee Valley. It is separated from the two former by a broad but low chain of mountains, running nearly east and west, which intervenes between the Tennessee and the sources of the Warrior and Cahawba rivers. These mountains, according to Professor Brumby, consist of strata older than the productive coal-measures, and similar to those seen by me in Rook's Valley.

The coal on the Tennessee, above alluded to, may perhaps be continuous with that of the great Appalachian coal-field. I hope hereafter to be enabled to give a more full account of the fossil plants of these Alabama coal-fields, a comparison of which, since they form the extreme southern limit of the carboniferous flora, with those of the north, will deserve particular attention.

London, June 23, 1846.—The above observations were written at Tuscaloosa in February last, and sent from thence to the Geological Society. On my return to England I found the paper already in type, and about to be printed off, but an opportunity having been offered me of adding this note, I submitted the specimens to my friend Mr. Charles J. F. Bunbury, F.G.S., who immediately compared them with published plates and the fossil plants in the Society's collection. The result of his examination confirms the conclusions to which I had arrived, and his specific identification of several of the Alabama remains with well-known European fossils is highly interesting. Although the decomposed state of the matrix in which the plants were obtained, near the outcrop of the strata where the shales are changed into soft, pale, laminated clays, has occasioned the loss of some of the Ferns and Sigillariæ, Mr. Bunbury has nevertheless been able to give a list and description of the following sixteen forms:—1. *Sphenopteris latifolia*, Ad. Brongn. 2. *S. Dubuissoni*? Ad. Brongn. 3. *Sphenopteris*, allied to the last, perhaps a variety of the same. 4. *Neuropteris tenuifolia*, Ad. Brongn. 5. *Neuropteris Grangeri*, or *N. gigantea*? 6. *Calamites cannæformis*. 7. *Calamite*, obscure specimen allied to the foregoing. 8. *Lepidodendron elegans*, var.? 9. *Lepidodendron* allied to *L. dilatatum*, Foss. Flora. 10. *Lepidophyllum*? 11. *Sigillaria*, decorticated. 12. *Stigmaria ficoides*. 13. *Poacites*? 14. *Bechera tenuis*, n. sp., very nearly allied to *B. grandis*, Foss. Flora. 15. *Asterophyllites*? *flaccida*. 16. *Phyllites*, resembling the leaf of *Sparganium*.

The palæontologist will perceive at once that no less than half of the species in the above list agree with well-known European fossils of the old carboniferous formation, and the rest belong to genera which are common in our coal-measures, and may perhaps agree with European fossils when procured in a better state of preservation. The leaves resembling Sparganium, however, which are very abundant, appear to Mr. Bunbury to be new.

When we recollect that the Tuscaloosa coal is found in lat. 33° 10′ north, and that several of its most common fossils are of the same species as those of Northumberland in lat. 54° north, at a distance of nearly 5000 miles (the broad Atlantic Ocean now intervening), we cannot but be struck with this new proof of the wide extension of a uniform flora in the Carboniferous period, especially as Alabama is situated much farther south than any region in which this ancient flora had previously been studied, whether in Europe or North America. Geologists will therefore rejoice to learn that Professor Brumby is fully alive to the importance of a more full investigation of the plants of the country, of which he will soon, it is hoped, have it in his power to form a large collection.

C. L.

ON THE NEWER DEPOSITS OF THE SOUTHERN STATES OF NORTH AMERICA

2. *On the Newer Deposits of the* SOUTHERN STATES *of* NORTH AMERICA. By CHARLES LYELL, Esq., F.G.S.

IN a communication formerly made, I gave a short notice on the Alabama coal-fields; I now propose to say something, but very briefly, of the newer deposits of the South, the post-pliocene and tertiary, so far as I have examined them in my present tour.

First, in regard to the post-pliocene occupying the coast and the lowest lands for some leagues inland in Georgia, the marine shells contained in deposits of clay and sand differ in no way from those of the adjoining sea. They are found arranged in groups, littoral and pelagic, as now on or near the shore. Such is observed to be the case in the island of Skiddaway, which is part of the delta of the Savannah river, and in the country between the mouth of the Alatamaha and the Turtle river in Glynn county, Georgia. These marine sandy deposits resemble exactly those which must be now forming off the neighbouring coast, and are covered in some spots by dark-coloured clay containing the remains of quadrupeds of extinct species, the *Megatherium*, *Mylodon*, *Mastodon giganteus*, *Elephas primigenius*, the Horse and others, with a large Chelonian*. I visited numerous localities on the Brunswick canal, near Darien, in company with Mr. Hamilton Couper, to whom we owe the discovery and careful exhumation of these remains, nearly all of which he has munificently presented to several public museums in the United States, especially those of Washington and Philadelphia. Everywhere I observed that there is no intermixture of the deposit containing the land animals with the subjacent bed containing the marine shells, and the mammalian remains appear to me to have been deposited in one of the arms of the ancient delta of the Alatamaha, when the relative level of land and sea was not the same, and yet differed but slightly from the present.

Mr. Couper's collection of the fossil shells underlying the bones comprises a large proportion of all the species which have been found on the coast of South Carolina, Georgia and Alabama. I have myself seen most of these fossils *in situ*, near Savannah or in the excavations made for the Brunswick canal, and they establish the fact that the ocean was inhabited before the time of the extinct mammalia by precisely the same conchological fauna as now: nor ought this conclusion to surprise us, as it is quite in harmony with that which I formerly deduced from the association in all parts of North America, from Canada to South Carolina, of the bones of the Mastodon and other lost quadrupeds with freshwater and land shells, not differing specifically from those of the adjoining rivers, lakes and forests. Such facts seem distinctly to imply that the temperature both of the atmosphere and the ocean has not materially changed since the time of the lost quadrupeds.

Secondly, on the coast of Georgia I examined, with Mr. H. Couper, the position of those submerged trunks or stools of the cypress (*Cupressus disticha*) which Bartram formerly alluded to in his 'Travels,' as showing that the level of the land on this coast had changed relatively to the sea in modern times. These trees can only grow in freshwater swamps, and there they are not permanently submerged at their base even in fresh water. At the mouth of the Alatamaha they are now found below the level of high tide in the salt marshes, with a deposit of mud over them. They also occur all with their upper surfaces cut off at the same level, and covered with several feet of alluvial matter in swamps higher up than the line of

* The *Hippopotamus* and *Sus* were formerly enumerated by Mr. Cooper and Dr. Harlan in their list of the fossil genera met with in digging the Brunswick canal. But Prof. Owen, in a communication recently sent to the Academy of Sciences at Philadelphia, has shown that the tusk referred to the Hippopotamus belongs to the Mastodon, so that there is now no evidence of a fossil Hippopotamus in America. The jaw to which the name of *Sus americana* had been applied exhibits a new type of organization, making some approach both to Lophiodon and Toxodon, but differing from either. For this fossil Mr. Owen has proposed the name of *Harlanus americanus*.

brackish water. They are so placed as to show a vertical depression of the land to the amount of at least four feet. But a full explanation of the curious appearances presented by them would require diagrams and a more detailed description than I have now time to send.

Thirdly, on the shores of the Bay of Mobile which opens into the Gulf of Mexico, I first met with inland deposits of that bivalve shell called *Gnathodon cuneatus*, an inhabitant of brackish water, but now met with in banks of sand extending several miles above the influence of the salt water, and rising three or four feet above the height of the present tides. Mr. Conrad first directed my attention to this remarkable formation, and I was taken to see it at several places by the Rev. Mr. Hamilton at the mouth of the Alabama river and in the suburbs of the city of Mobile. In neither of these localities, where large and small individuals of the Gnathodon form dense, shelly masses, could I detect any intermixture of other shells; yet on examining the mud of the adjoining sea-shore at low water, about a mile west of the embouchure of the Alabama, I met not only with the Gnathodon abounding in a living state in the mud, but also with a Nerita, and occasionally shells of *Cyrena carolinensis*. I tasted the water here and found it perfectly fresh, but was assured that this was owing to the wind blowing off the shore, and that the same testacea cannot live in water permanently fresh. The Gnathodon is not known to exist at present at the mouth of any American river farther north than the Alabama. The accumulation of fossils is far too considerable and extensive to be referred to the Indians, who are supposed by some observers to have used them for food, and to have left the shells scattered over the ground. On the other hand, the facts above stated may perhaps be explained without assuming that there has been an upheaval of the coast since the brackish waters of the Gulf of Mexico were inhabited by the Gnathodon. It is possible that the action of the waves on the shore may have thrown up these shells in the form of sea-beaches.

Fourthly, I have not fallen in with any deposits of miocene shells in the course of my present tour in Georgia and Alabama. The most southern point where I know them to occur is South Carolina. Dr. Gibbes of Charleston showed me a large number of miocene fossils obtained by him and Mr. Tuomey on the banks of Goose Creek, near the mouth of the Cooper river, at a place not visited by me in 1842.

Fifthly, the eocene formations are of great thickness, and occupy an extensive area in Georgia and Alabama. I formerly described in the Society's 'Proceedings' and 'Journal' the two formations of this age which I saw in 1841–42 in South Carolina and Georgia, the one chiefly calcareous, consisting of white limestone and marl, the other the burr-stone formation, composed principally of clay and sand; and I showed in a diagram representing a natural section at Jacksonboro', Georgia (Journal of the Geol. Society, vol. i. p. 438), the manner in which these different members of the eocene series, both horizontally stratified, rest one upon the other. My examination of the cuttings recently made for a railway about 200 miles in length between Savannah and Macon in Georgia, in a line parallel to the section which I formerly observed in the bluffs of the Savannah river, has confirmed the opinions before expressed by me, namely, 1st, that the clays and sands containing the burr-stone are eocene, as shown by casts of testacea and corals; 2ndly, that they overlie the calcareous eocene limestone and white marl. In like manner I have found throughout Alabama that the lower tertiary deposits in which the calcareous strata predominate are horizontal, with a very undulating or uneven surface, and have been covered by newer beds, also horizontal, of blood-red clay, silt, quartz gravel, white porcelain earth, yellow ochreous clay, and white, pink and yellow siliceous sands and ferruginous sandstone. Chert occurs in some places in this upper deposit, which is so barren of organic remains in Alabama, that I could only infer its eocene date from the analogy of Georgia. The older eocene limestone and marl must have had a very uneven surface, shaped into hills and valleys, often bounded by steep precipices, before the incumbent clay and sand, which is occasionally from 300 to 400 feet thick, was thrown down. When the original inequalities had been for the most part removed by the deposition of the argillaceous and sandy beds, the existing ravines and valleys were excavated at the expense of both formations, and hence it occasionally happens that two closely adjoining sections, each on the same level and of the same height, exhibit a distinct series of horizontal beds. This diversity is exemplified in different parts of the bluff at Claiborne on the Alabama river, which has become celebrated for the great number of perfect fossil shells obtained from it by Messrs. Conrad and Lea. At the new landing, the perpendicular precipice exhibits more than 150 feet of the calcareous formation, chiefly composed of white limestone and marl, while the red clay and sand appear only at the top of the cliff about twenty feet thick. But this upper formation, about a mile lower down in the same bluff at Claiborne, is more than 100 feet thick, composed of sand and clay without fossils, a small portion only of the calcareous beds cropping out from beneath.

Sixthly, owing to the extent and thickness of the overlying clay and sand, it is often impossible to obtain a clear section of the various subdivisions of the eocene white limestone of Alabama. To this cause I attribute the obscurity in which the true age of the nummulite limestone of Alabama has hitherto been involved. It has been considered sometimes as an upper cretaceous group, sometimes as intermediate in age between the tertiary and secondary series. After visiting Claiborne and the country on the other side of the Alabama river in the fork of that river and the Tombecbee, passing by Suggsville, Macon, Clarksville and Creagh's, all in Clarke county, I am persuaded that the nummulite limestone is an eocene rock, newer than all the beds of the well-known Claiborne bluff. It is in fact more modern than the sandy deposit from which the eocene shells described in the publications of Messrs. Conrad and

Lea were derived. The annexed section will explain my view of the structure of this country better than a detailed description.

SECTION showing the position of the NUMMULITIC LIMESTONE near CLAIBORNE.

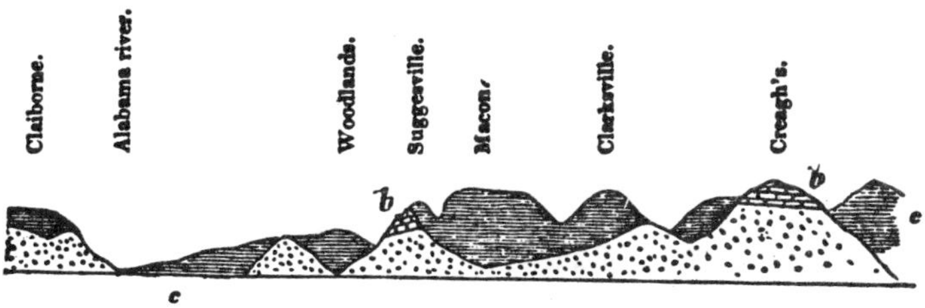

a. White eocene limestone, marl, &c.
b. Nummulitic limestone (eocene).
c. Blood-red clay, yellow sand, &c. (eocene?)

I am informed by Captain Bingham the engineer, that some of the hills between Clarksville and Creagh's, on the tops of which I observed the nummulite limestone, are about 400 feet higher than the top of the bluff at Claiborne.

Seventhly, the conclusion last-stated will make it necessary in future to omit from all lists of American cretaceous fossils such species of shells as have been considered secondary, simply on the ground of their occurrence in the nummulite limestone of Alabama. The fossils which I have met with most abundantly in the latter are the *Nummulites Mantelli* and *Pecten Poulsoni*. I have however many others, especially casts of shells and corals, which I shall hereafter describe. The nummulitic formation is from 50 to 100 feet in thickness, consisting for the most part of a soft cream-coloured stone, hardening on exposure, while in other beds it is highly indurated. In its lower portion, where it joins the ordinary white limestone, I found *Pecten perplanus*, *Ostrea panda* and *Plagiostoma dumosum*. Lunulites and other corals also occur in this rock. The recognition of the true place of this limestone in the series will remove almost every fossil from the list of those species which were still supposed to be common to the cretaceous and tertiary groups of the United States.

Eighthly, I have visited several of the principal localities where the bones of that gigantic cetacean, called Basilosaurus by Dr. Harlan and Zeuglodon by Mr. Owen, have been discovered in Clarke county before-mentioned. These bones are everywhere in the same geological position, namely in the eocene white limestone, below the level of the nummulitic rock and above the beds which contain the greater number of perfectly preserved eocene shells, such as *Cardita planicosta* and others.

Ninthly, on Creagh's plantation and about four miles and a half S.W. from Clarksville, I visited the spot where Mr. Koch procured in 1845 the head and part of the vertebral column, about thirty feet in length, of the Zeuglodon, called by him Hydrarchos. I was accompanied by Mr. William Pickett, a gentleman who also assisted

Mr. Koch in the exhumation of those bones. From him and other persons I learn that the main body of the vertebræ which entered into the skeleton exhibited in the United States by Mr. Koch in 1845 under the name of Hydrarchos were procured in Washington county, Alabama, at a place fifteen miles distant in a direct line from the locality where the head was dug up.

I have information of about forty other places where separate bones of this huge animal have been met with in Clarke and Washington counties alone. From several of these I have seen the bones, and they are in so perfect a condition and so bulky, so distant the one from the other, and must have been so difficult to transport; that I can hardly doubt their having belonged to forty distinct individuals.

ON FOOT-MARKS DISCOVERED IN THE COAL-MEASURES OF PENNSYLVANIA

2. *On* Foot-marks *discovered in the* Coal-measures *of* Pennsylvania. By Charles Lyell, Esq., F.G.S.

I intended to draw up a paper on what I have learned and observed respecting the proofs alleged to have been found of the existence of mammalia, birds and reptiles in the Pennsylvania coal-field; but I cannot do this until I receive my specimens, and have had more time to make sections and maps, in order to do the subject justice. I must therefore content myself with this brief notice. The discoveries of the impressions to which I shall refer were made by Dr. King, a physician of this place, whom I have seen, and who has been most anxious to facilitate my investigations, and to give me every help in arriving at the truth, which was his only thought,

* The skull is mutilated in the specimen on which this species was founded.

whether I confirmed or invalidated his conclusions. I shall first speak of the reptilian foot-marks.

The quarry where they were found is about five miles S.E. of Greensburg, in the county of Westmoreland, Pennsylvania. The stone quarried is a sandstone which rises up from beneath the main or Pittsburg seam of coal, which has been worked at its outcrop in the immediate neighbourhood*. There are several other seams of coal, which lie at lower levels, and impressions of Lepidodendron, Sigillaria, Stigmaria, Calamites, Ferns, and other plants have been found in the beds which lie above and below the sandstone containing the foot-marks. The slabs of sandstone, which are extracted in the quarry for paving, are separated by parting layers of a fine unctuous clay, such as would be admirably fitted to receive the most delicate and faithful impressions of the feet of animals treading upon it.

One of these Cheirotherian impressions was observed by Dr. King imprinted on the upper surface of one of the layers of clay or shale before-mentioned, but the specimen was unfortunately left exposed, and was destroyed by the weather. Twenty-two other footsteps were discovered on the under sides of the slabs of sandstone, standing out in relief from a surface which also exhibits those large and small mud-veins, which are the casts of cracks produced by the drying and shrinking of the mud on which the animal walked. The shrinkage took place after the foot-marks were made, so that these prints were traversed and slightly distorted by the cracks. The casts of these shrinkage cracks were mistaken for Fucoids.

The Cheirotherian tracks occur in pairs, each pair consisting of a hind- and fore-foot. There are two rows of these, which are parallel, or have been formed, the one by the right fore- and hind-feet, the other by the left; the toes turning one set to the right and the others to the left, and the distances between the successive footsteps being about the same throughout.

I shall now advert to the supposed foot-prints of birds, and of quadrupeds resembling dogs, cloven-footed and other animals.

The principal place where these imprints were observed, and which I visited, is about a mile distant from the town of Derry in Westmoreland county, Pennsylvania. They were not found beneath the soil, nor on the under surface of solid strata, like the Chirotherian marks at Greensburg, but consisted of impressions or incisions on the upper surface of a denuded ledge of white coal-grit or sandstone, which has been exposed for ages, and so acted upon by water, that channels and cavities have been shaped out, and numerous deep pot-holes, some of them more than a foot in diameter and eighteen inches deep, excavated in it. To suppose a series of deep and sharp foot-prints of birds to have been retained unblunted by weathering, or by the currents of water which denuded the sandstone, would be sufficiently difficult; but still more to account for the fact, that the last of the imprints is on the steep end of the ledge, inclined at an

* "The Pittsburg seam is 10 feet thick, and has been ascertained to occupy an area of about 14,000 square miles."—*Lyell's Travels in America*, vol. ii. p. 27.

angle of 22°, so that the bird, after walking on the planes of stratification, which are nearly horizontal, must have gone over the edges also. Loose sand, capable of receiving an impression, could never have acquired the present outline in a ledge of rock; and, on the other hand, sand consolidated into stone, as this evidently was when denuded, could never have been marked subsequently by any known agency but the hand of man, in imitation of the feet of birds and quadrupeds.

In order to explain the great number of the foot-prints appearing on so uneven a surface, and where several different stratified layers of sandstone have been cut into, we should have to imagine that there were originally a multitude of impressions on each superimposed layer, all nearly in one direction; that wherever the denuding action of water had happened to cut into any inferior layer, it always laid open to view new tracks. But this hypothesis is so far from being borne out by the facts, that after cutting into the stone in more than twenty places, and removing small slabs with imprints on them, Dr. King has never been able to detect a single indication of a foot-mark in the rock below.

One of the dog-like foot-prints on this Derry stone agrees with one of those which are found in another locality, in Fayette county, near Connelsville, nineteen miles from Greensburg, and about thirty from Derry. Dr. King and the Rev. Mr. Hackey, who have visited that place, inform me that unquestioned Indian hieroglyphics, and the representation of a serpent and two human heads, are there to be seen with bird tracks and those of hoofed quadrupeds. I have seen good imitations of the foot-prints of birds brought to Indiana by Dr. David Dale Owen from St. Louis, Missouri, on slabs of limestone sculptured by Indians; nor can anything be more probable, than that the aboriginal inhabitants of North America, who are so accustomed to trace and follow the trail of every kind of game, should occasionally employ as symbols of birds and quadrupeds, an imitation of the foot-prints which they leave on soft mud and sand. As Dr. King agrees with me in abandoning as spurious all the imprints except those of the large reptile, I will not dwell longer on the refutation of the evidence; but I may mention that there are numerous graves of Indians near the sculptured sandstone of Derry, and it is known to have lain in the line of one of their principal paths leading from the Alleghany mountains to the west.

The rectification of a few mistakes, with which some of the first observations in this new field have been connected, is a matter of small interest and moment compared to the important truth which has been brought to light through Dr. King's exertions;—that the land on which forests of Sigillaria and Lepidodendron grew, gave support also to large air-breathing quadrupeds. Few geologists will now be prepared to believe that this single species or genus of reptiles, or that one class only of vertebrated animals, had possession of the islands and continents on which so widely-extended and magnificent a vegetation flourished.

It may be as well for me to add, that the rejection of the supposed Ornithichnites of Derry as spurious has by no means led me to doubt the genuineness of those that I formerly examined in the red sandstone of the Connecticut valley, and which have been so fully described by Professor Hitchcock. On the contrary, those fossil foot-prints were found under circumstances in all respects analogous to those which lead me to believe in the reality of the Cheirotherium which has been discovered in the coal strata near Greensburg. These reptilian tracks occur in one locality only; no others have yet been found in the same place, nor under similar circumstances elsewhere.

ON THE STRUCTURE
AND PROBABLE AGE
OF THE COAL-FIELD
OF THE JAMES RIVER
NEAR RICHMOND, VIRGINIA

1. *On the Structure and Probable Age of the* Coal-Field *of the* James River, *near* Richmond, Virginia. By Charles Lyell, Esq., V.P.G.S.

Contents:—*Geological structure of the Coal-field*, p. 261.—*Vertical Calamites*, p. 262.—*Thickness of Coal*, p. 263.—*Sections*, p. 264.—*Organic structure and mineral composition of the Coal*, p. 268.—*Analyses*, p. 269.—*Beds of Coal altered by Trap*, p. 270.—*Natural Coke*, p. 272.—*Position of Trap*, p. 273.—*Age of the Coal as determined by organic remains*, p. 274.—*Shells*, p. 274.—*Fossil fish*, p. 275.—*Whether the strata are of freshwater origin*, p. 278.—*Fossil plants*, p. 278.—*The Coal-measures probably of the age of the Inferior Oolite and Lias*, p. 279.

Geological Structure of the Coal-field.

There are two regions in the state of Virginia (a country about equal in area to the whole of England proper) in which productive coal-measures occur. In one of these, which may be called the Western Coal-field, the strata belong to the ancient carboniferous group, characterized by fossil plants of the same genera, and to a great extent the same species, as those found in the ancient coal-measures of Europe. These strata form an integral part of the system of rocks constituting the Appalachian chain, and have been disturbed by the same movements, and subsequently exposed to the same denuding action. Another coal-field, wholly disconnected from the above in its geographical and geological relations, is found to the east of the Appalachian mountains, in the middle of that granitic region sometimes called the Atlantic Slope. This district of granite, gneiss, hornblende, schist, and other hypogene rocks, intervenes between the palæozoic rocks of the Alleghanies and the low country bordering the Atlantic, which is composed of cretaceous and tertiary deposits. In consequence of the isolated position of these eastern coal-beds, the lowest of which rest immediately on the fundamental granite, while the uppermost are not covered by any overlying fossiliferous formations, we have scarcely any means of determining their relative age, except by the characters of their included organic remains. The study of these induced Prof. W. B. Rogers, in his memoir published in 1842 (Trans. of American Geologists, p. 298), to declare his opinion that this coal was of newer date than that of the Appalachians, and was about the age of the oolite, a conclusion which, after a careful examination of the evidence on the spot, and of all the organic remains which I could collect, appears to me to come very near the truth. The only doubt that can be entertained on the subject in the present state of our knowledge is, whether we should refer the strata to the triassic or oolitic period; and if we incline with Prof. W. B. Rogers to the latter opinion, these rocks must then be considered as the only ones hitherto known in all Canada and the United States, which have been proved by their organic remains to be of contemporaneous origin with the oolitic or jurassic formation of Europe.

Before I discuss the question of the relative age of the strata, I shall proceed to describe briefly the geographical extent, position, and geological structure of the coal-field. The tract of country occupied by the crystalline or hypogene rocks which runs parallel to the Alleghany mountains, and on their eastern side, is in this part of Virginia about seventy miles broad; in the midst of this space the coal-field occurs in a depression of the granitic and other hypogene rocks on which the coal rests, and by which it is surrounded along its outcrop. The length of the coal-field from north to south is about twenty-six miles, and its breadth varies from four to twelve miles. It extends over portions of six counties, from Amelia the most southern, through parts of Chesterfield, Powhatan, Henrico, Goochland and Hanover. The James River flows through the middle of it, about fifteen miles from its northern extremity, while the Appomattox River traverses it near its southern borders; on its eastern side it is distant about thirteen miles from the city of Richmond, the capital of the State of Virginia: it occupies an elliptical area, the beds lying in a trough, the lowest of them usually highly inclined where they crop out along the margin of the basin, while the strata higher in the series, which appear in the central parts of the basin, are very nearly horizontal. The general strike is about N.N.E. and S.S.W., while that of the nearest ridges of the Appalachian chain is about N.E. and S.W.

A great portion of these coal-measures consists of quartzose sandstone and coarse grit, some of the beds in the lower part of the series resembling granite or syenite, being entirely composed of the detritus of the neighbouring granitic and syenitic rocks. Dark carbonaceous shales and clays, occasionally charged with iron ores, abound in the proximity of the coal seams, and numerous impressions of plants, chiefly ferns and Zamites, are met with in shales, together with flattened and prostrate stems of Calamites and Equisetum. These last however, the Calamites and Equisetum, are very commonly met with in a vertical position, more or less compressed perpendicularly. That the greater number of Calamites standing erect in the beds above and between the seams or beds of coal which I saw at points many miles distant from each other, have grown in the places where they are now buried in sand and mud, I entertain no doubt. This fact would imply the gradual accumulation of the coal-measures during a slow and repeated subsidence of the whole region*.

The coal-seams have hitherto been all found at or near the bottom of the series, and the plants in beds below or between them, or immediately overlying. One or two species of shells (Posidonomya?) also occur in the same part of the series, at a small height above the coal-seams, and above these a great number of fossil fish chiefly referable to two nearly allied species of a genus very distinct from any ichthyolite

* It is worthy of remark, that the *Equisetum columnare*, the same species as that found in the Virginian coal-field, also occurs in an upright position, and extends over a wide area in those oolitic strata near Whitby in Yorkshire, which are supposed, from the general evidence of the fossil plants, to be the equivalents in age of these secondary coal-measures of America.

hitherto discovered elsewhere. Above these fossiliferous beds, which probably never exceed 400 or 500 feet in thickness, a great succession of grits, sandstone and shales, of unknown depth, occur. They have yielded no coal, nor as yet any organic remains, and no speculator has been bold enough to sink a shaft through them, as it is feared that toward the central parts of the basin they might have to pass through 2000 or 2500 feet of sterile measures before reaching the fundamental coal-seams.

Section showing the Geological Position of the James River, or East Virginian Coal-Field.

Fig. 1.

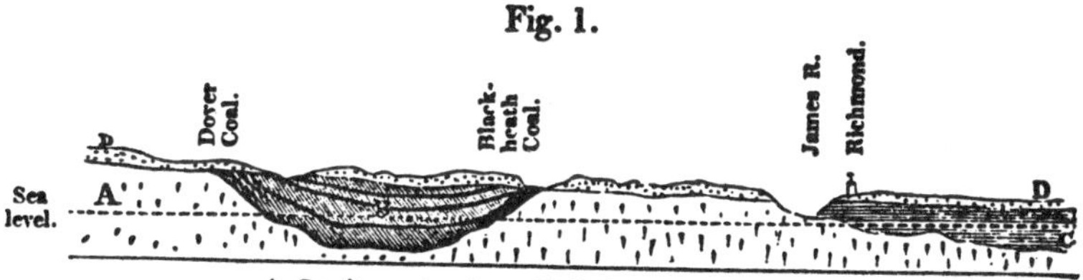

A. Granite, gneiss, &c. B. Coal-measures.
C. Tertiary strata. D. Drift or *ancient alluvium.*

The annexed ideal section (1) will show the manner in which I suppose the coal-field to be placed in a hollow in the granitic rocks, the whole country having suffered by great denudation, and the surface having been planed off almost uniformly, and at the same time overspread by a deep covering of gravel and red and yellow clay, concealing the subjacent formation from view, so that the structure of the region could not be made out without difficulty but for the artificial excavations. It will be seen by the section that the tertiary strata appear at Richmond, about thirteen miles from the eastern outcrop of the coal, and they continue to occupy the lower country between that city and the Atlantic.

It has been stated, that the only beds of coal hitherto discovered lie in the lower part of the coal-measures, and that consequently they crop out all round the margin of the basin. As the dip is usually at a considerable angle, vertical shafts from 400 to 800 feet deep are required to reach the coal at the distance of a few hundred yards inside the edge of the basin; it is only therefore along a very narrow band of country that the coal comes up to the surface naturally, and even here it is for the most part buried under superficial red and yellow clay with sand and gravel, the whole often 30 or 40 feet thick. I saw no erratic blocks or far-transported materials in the detritus, but it is so continuous as to conceal the junction of the carboniferous and granitic rocks; and I felt some curiosity to learn how the existence of coal in such a country had been originally suspected. It appears that in Chesterfield, the county where the richest seams occur, the coal enters very largely into the composition of the overlying gravel and sand, and at the old Blackheath pits, which were first worked between the years 1785 and 1790, a fallen tree displayed fragments of coal in the soil adhering to its roots, which led to excavations, and the discovery, at the depth of a few yards, of a rich bed of bituminous coal, no less than 30 feet thick, of excellent quality. Being so near the surface, it was for a time worked with great profit; but very exaggerated notions were formed, both here and in the vicinity, of the real thickness of the main seam, due allowance not being made at first for the dip of the beds through which the shafts passed down obliquely to the planes of stratification. In the valley of a small tributary of the James River, about two miles north of Blackheath, I saw, exposed in a natural section of drift, a bed of rubbly coal, which might easily have betrayed the existence of the seam from the waste of which it had been derived.

It has been usually stated in reports on this coal-field, that north of the James River the coal is divided into two or three seams, while to the south of the same river it forms one huge seam from 30 to 40 feet thick. Such however is not the case, the coal being almost everywhere separated into three distinct beds, and sometimes more, and this division being observable quite as much in the southern as in the northern half of the trough. The uppermost seam is almost always the most important. Thus in the Clover-hill pits, near the Appomattox River, towards the southern extremity of the basin, I found five seams of coal, all as usual at the margin of the basin. Here, about 160 feet of grit and shale intervene between the lowest of the five seams and the fundamental granite and hornblende schist. Of this section I shall have to say more in the sequel. In all the localities I visited, the thickest bed of coal or the main seam is the uppermost, except at Clover-hill just mentioned, where a thin layer of coal is met with still higher; but it is not continuous for any distance. Along the western margin of the basin at Dover, for example, I observed the same division of the coal into two or more seams; nor am I aware of any place, except Blackheath and the adjoining parts of Chesterfield county, where all the seams appear united as it were into one, of unusual strength, from 30 to 40 feet thick,—to explain which we may perhaps adopt the theory of Mr. Bowman, and suppose the vegetation to which the coal is due to have grown uninterruptedly in this spot, while elsewhere it was frequently arrested for a time in its progress by subsidence, which submerged certain areas and permitted the deposition of mud and sand to take place upon the materials of the future coal. This unity however of the coal-seam in the Blackheath region is of small extent, and even in the purest mass there occur thin intercalated layers of shale and sandstone with pyrites, to separate which from the coal costs the miner no small labour. These "partings" of sedimentary matter, which in some spots swell out rapidly from a few inches to a thickness of several feet, or even yards, make the great seam of Blackheath much less of an *exceptional* case in a geological point of view than might at first be supposed. The inconsiderable distance to which the undivided state of the seam extends, is shown by the section obtained only a quarter of a mile to the south of Blackheath in the Midlothian pits, where a shaft sunk to the depth of 774 feet passed through the undermentioned beds.

Section at the Midlothian Pit, half a mile south of Blackheath, on the eastern outcrop of the Coal.

	Feet.	Inches.
Sandstone and shale	570	0
Slate with Calamites	1	6
Sandstone and shale	43	10
Sandstone with Calamites	8	0
Sandstone and slaty shale	48	0
Slate and long vegetable stems	2	6
Sandstone	6	6
Slate with Calamites, numerous	5	6
Sandstone	14	0
Black rock (carbonaceous?)	13	0
Slate	5	0
Main coal	36	0
Sandstone, not laminated	5	0
Slate	4	0
Coal	1	0
Slate	3	0
Sandstone or grit	7	0
	773	10

Granite, depth unknown.

It will be seen that here, within a quarter of a mile of the Blackheath pit, there are two seams of coal—one, the uppermost, more than 30 feet thick, and the other separated from it by 9 feet of intervening shale and sandstone about one foot thick; this lowest seam being within four feet of the fundamental granite, and parted from it by beds of grit and shale. Some deduction must be made from all the estimated thicknesses of the beds as above stated, the inclinations of the beds cut through by the vertical shaft being at an angle of 20°. After making this deduction, however, the magnitude and persistency of this seam, from 30 to 40 feet in thickness, consisting of as pure a mass of rich bituminous coal as can perhaps be found in the world, even in the old carboniferous formations, are truly remarkable, especially when we take into account its geological age. I was indeed not a little surprised when I first arrived at Blackheath and descended a shaft 800 feet deep, to find myself in a chamber more than 40 feet high, caused by the removal of the coal. Timber props of great strength are required to support the roof, and although the use of wood is lavish here as everywhere in the United States, the props are seen to bend under the incumbent weight. The great seam in some places at Blackheath is seen actually to touch the fundamental granite, or is parted from it only by an inch or two of shale. The unevenness of the granite floor is extremely great, as represented in the annexed drawing, Fig. 2*; and the manner in which, in these *troubles* as they are called, the coal is squeezed out at one point and made to swell and thicken at another, B, must be referred to movements of the rocks

* For this and several other documents relating to this coal-field, and for numerous organic remains, I am indebted to the kindness and exertions of my friend Mr. A. F. D. Gifford, under whose superintendence I found the Blackheath mines. I have also to acknowledge the liberality of Dr. Samuel H. Royal of Blackheath, who presented me with fossil fish, and plants of great rarity and beauty.

Junction of Coal and Granite, Blackheath Mines.

Fig. 2.

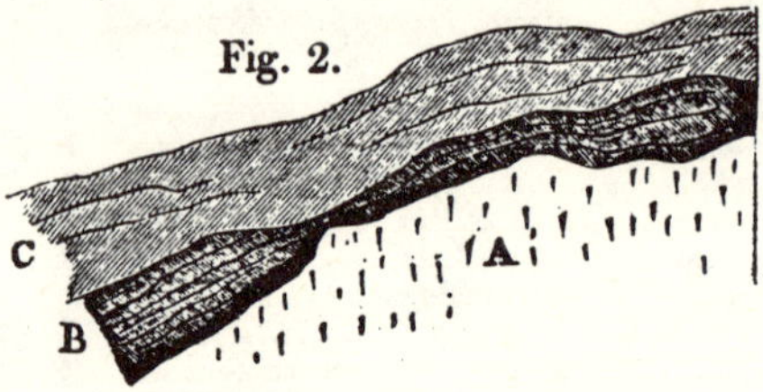

C. Sandstone and shale. B. Coal. A. Granite.

and the forcing of the granite against the coal, the distinct layers of which are often cut off abruptly one after the other by the granite in contact. The dips of the strata of coal in the neighbourhood of these sudden thickenings and swellings of the seams vary from 20° to 70°, and at one point attained an angle of 84°. Polished surfaces or *slickensides*, with their planes turned in various directions, also bear testimony to the sliding and friction which the beds have undergone. Some of the alleged varieties in the thickness of the great seam at short distances may be explained by sudden changes in the angle of dip, which, as before stated, might cause an exaggerated estimate wherever an inclined stratum is cut through vertically in vertical shafts; but it appears that where the coal is nearly horizontal, it often is far from uniform in strength at points a few hundred yards apart. The beds which I actually observed at Blackheath and the neighbourhood varied in their inclination from 8° to 42°, the dip being always westerly.

The great uncertainty of the depths at which the coal can be won in sinking shafts at known distances from the natural outcrop or from previous workings, proves the irregularity of the dip, and implies in some places sudden flexures, or more probably vertical faults. An example of one of these occurs in comparing the sections exposed in the Creek and White Chimney pits. A shaft at the former or Creek mine being sunk 400 feet deep to the main seam, the coal was followed for 15 yards westward in the direction of the dip at an angle of 20°. The White Chimney or Old Midlothian shaft was then sunk 500 yards westward of the former, and the same great bed of coal, with the granite below it, was entered at the depth of 360 feet. If the dip of about one in three continued westward, in this instance, an upthrow of more than 100 yards would be required to account for the position of the coal so near the surface in the more westerly pit.

About five miles north of the Midlothian shaft last mentioned, at Duval's pit, which is also on the eastern outcrop, the coal-measures are seen near the granite to be arranged in a small trough extending only a few hundred yards east and west. At this place there are three beds of coal, which in the deepest part of the small basin are only 200 feet from the surface. Again, at the northern extremity of the whole coal-field, at the Deep-run pits, there is another example of a small basin only a mile and a half in its longest diameter, from N.N.E. to S.S.W., and half a mile in breadth. The coal there is similar in quality to that of Blackheath, and I found the accompanying shales full of that splendid fern, the *Tæniopteris magnifolia*, spread out between the laminæ, as in the shales of Clover-hill at the southern end of the coal-field. The same Calamites, Equisetum and Zamites also occur there, and in nearly every locality, though some species of ferns have only as yet been met with at one place, as for example the *Filicites fimbriatus* (Bunbury) and *Pecopteris bullata* (Bunbury), which I found at the Deep-run mines just mentioned, and *Neuropteris linnææfolia* found at Blackheath. It is the opinion of some of the miners, that where several seams of coal occur, they would form, if united, a single seam, about equal in the aggregate to the average thickness of the great single bed of coal in the Blackheath region; but it seemed to me that the strength of the beds in most of the workings would never approach in importance the single main seam in the Blackheath and Midlothian pits—and here also the seam is sometimes no less than 30 feet thick where other beds of coal intervene between it and the granite. This may be seen about two miles north of the Engine pit, at the place where the coal was first obtained in 1785, in the valley of a small tributary of the James River. The main seam is there 30 feet thick at its outcrop, and in the coal-measures separating it from the granite, consisting of grit and shale 200 feet thick, are seen two beds of coal, the uppermost three feet and the lowest about one foot thick.

It was before stated (page 263), that there have been no borings for coal in the central parts of the coal-field; the only section I could obtain of the beds far from the margin of the basin is that afforded by the cuttings along the Richmond and Lynchbury canal, and these are only a few yards in depth. The beds here consist of shale and quartzose grit, one undivided stratum of which, ten feet thick, was very conspicuous.

Wherever I observed these strata closely, their dip was about 10° south-east; but on reaching the western outcrop of the coal at Dover, only three-quarters of a mile from where the beds are so slightly inclined on the canal, I found the coal-measures dipping at an angle of no less than 50° east, or opposite to that prevailing at Blackheath on the other side of the trough. The seams of coal at Dover correspond very closely at their outcrop near the granite with those before mentioned, where the western dip prevails, the main seam at Dover being 16 feet thick, and there being two other thinner ones below. Only a mile or two south of the open shaft at Dover, where I saw the beds dipping at an angle of 50°, I observed the strata near the granite to be almost horizontal. The same may be said of the strata three miles south of the Powhatan pits, near their junction with the crystalline rocks on the eastern outcrop. I was told also of instances of the strata dipping towards the granite, as for example three miles south-east of the Midlothian pits before mentioned. In short, the disturbances have been so great, that although the general structure of the coal-field is that of a basin or trough, as before stated, the deviations from this form are numerous.

Organic Structure and Mineral Composition of the Coal.

The coal of Eastern Virginia, although derived from a different vegetation from that of the ancient carboniferous period, resembles very closely the older coal in structure, appearance and composition. That of the Blackheath mine has usually a highly resinous lustre and conchoidal fracture, and always contains at least as large a proportion of gaseous or volatile ingredients (hydrogen, oxygen and nitrogen) as the coal of the palæozoic rocks of the United States.

The coal is also divided into horizontal layers of slight thickness parallel to the planes of stratification, as in the older kind of coal. Sometimes these layers consist alternately of highly crystalline and resinous coal with a bright lustre, and of other portions exactly resembling charcoal in appearance. The same is observed in some of our Welsh coal, where the charcoal is called "mother coal."

My friend Dr. Hooker has had the kindness to examine for me some specimens of this Virginian charcoal, which I procured at the Blackheath mines, and others from those of Clover-hill before alluded to, and he finds vegetable structure in both, but appearing in each locality to belong to a different species of plant. At first he thought they might be referable to ferns, but abandoned that opinion from the total absence of cellular and scalariform tissues. The prominent glands of the fibres are much more minute than the glands of coniferous tissue, whilst the large perforated tubes are foreign to that order. They depart still more widely from Zamia, and do not indeed present any obvious affinity with any existing natural order. "Both are very opake, much crushed, and so fragile that it is difficult to obtain fragments fit for microscopic investigation. They principally consist of a mass of parallel fibres or elongated cells, amongst which occur very large tubes whose walls are pierced with circular, or longitudinally or transversely elongated holes, either scattered or placed very close together.

Vegetable Structure of Mineral Charcoal from Clover-hill Mines, Virginia.

Fig. 3.

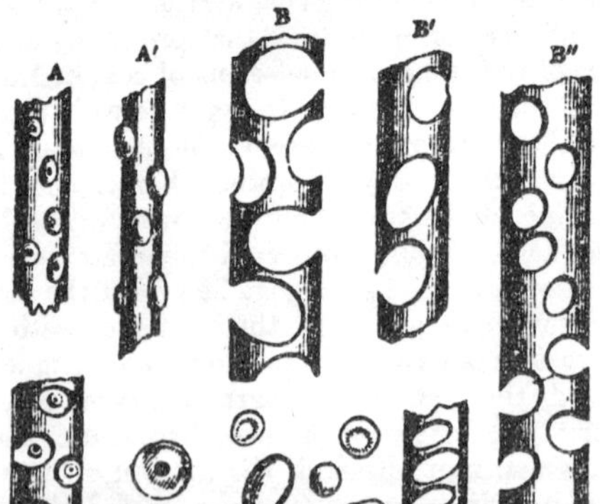

"In this (Fig. 3.) the fibres (A) are larger in diameter, covered with prominent dots or glands, each often having a dark spot in the centre. The perforations in the tubes (B) are often filled up with circular discs (as at C), which have a central dark spot (D), or are transparent throughout (E). The perforations of some of the tubes are so closely placed and so much elongated transversely (F), that the tube resembles the scalariform ducts of a fern.

Blackheath Specimen (Creek Mines).

Fig. 4.

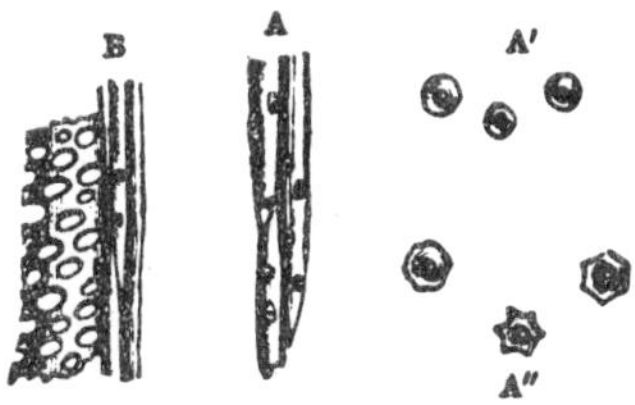

"The fibres of this specimen, fig. 4. (A), are smaller and more closely packed, covered with more numerous prominent dots or glands; the latter have almost uniformly a central spot (A'), are extremely minute, very caducous, and are often star-shaped (A''), or of various outline. The tubes are very rarely seen, and have many series of generally transversely elongated perforations (B)."

Next to the ferns, the most abundant vegetable remains in the coal-measures under consideration are two species of Calamites and the *Equisetum columnare,* especially the former, and the bark of the Calamites is usually converted into coal. It is therefore natural to conjecture, that the thin layers resembling charcoal may have been derived from Calamites, a plant of which the botanical structure is still unknown. Dr. Hooker, in reply to this suggestion, tells me that some fossil charcoal in the Museum of the Geological Society, from the English coal formation, supposed to be derived from Calamites, is very like that of the Virginia coal above described.

The cities of New York and Philadelphia have for many years supplied themselves with coal from the Blackheath mines, for the manufacture of gas for lighting their streets and houses. The annual quantity taken by Philadelphia alone has of late years amounted to 10,000 tons. It will appear from the annexed table of analyses of specimens of coal, which Mr. P. H. Henry has had the kindness to make for me, that the proportion of volatile ingredients—hydrogen, nitrogen and oxygen—in this newer coal of the James River comes exceedingly close to that found in the older coal of America, which it so much resembles in aspect and structure. The specimen from Tuscaloosa in Alabama was from the old coal formation described by me in the Journal of this Society for 1846, vol. ii. page 278, in which are found precisely the same plants as in the coal of the Appalachians, of Nova Scotia, and of the north of England. The other three specimens, from

Analysis of American Coals and Coke, dried at 250° Fahr. Moisture 1 to 1½ per cent.

	Coal and Coke of the Newer Coal-field of Eastern Virginia.				Ancient Coal of Alabama.
	Coal, Clover-hill.	Coal, Blackheath.	Coal, Deep-run.	Coke—Townes and Powell's.	Coal, Tuscaloosa.
Carbon	76·49	80·38	82·90	86·54	80·96
Hydrogen ...	5·23	4·06	4·77	4·23	5·13
Oxygen and Nitrogen	8·41	6·19	5·97	4·53	7·83
Ash............	9·87	9·35	6·36	4·70	6·08
	100·00	100·00	100·00	100·00	100·00

Clover-hill, Blackheath, and the Deep-run mines respectively, represent the southern, middle, and northern parts of the coal-field treated of in this memoir. On comparing this analysis of the Clover-hill coal as here given, with Dr. Percy's analysis of the older coal from Pomeroy on the Ohio (see this Journal, vol. i. p. 199, May 1844), we cannot but be struck with the identity of the results, as far as relates to the proportions of carbon and hydrogen. Thus the ancient coal of Pomeroy gave, carbon 76·70, hydrogen 5·67, in which it will be seen that the figures are the same, except the decimals, as in the analysis of the newer coal from Clover-hill. Nearly the same remarks are applicable when we compare the Blackheath coal and that of Tuscaloosa in the annexed table.

That there should be a disengagement of inflammable gas from this coal, containing so much volatile matter, was to be expected; accordingly, in spite of the usual precautions taken to prevent accidents, several fatal explosions have occurred; one of these happened at Blackheath in the old pit in 1839, by which forty-five negroes and two white overseers lost their lives. They were killed, not by choke damp, but by the combustion of carburetted hydrogen, attributed by the miners to the spontaneous ignition of coal mixed with decomposing pyrites. The most approved methods of ventilating the mines had been introduced, but another terrible accident occurred so late as the year 1844, at the depth of more than 900 feet, at Blackheath, occasioned by the leaking out of gas from some deserted works, which had been ineffectually dammed off from the new galleries.

On Beds of Coke and Altered Coal, and the Action of Intrusive Igneous Rocks.

Although rocks of volcanic origin do not form a prominent feature in the geological structure of this coal-field, as they do in the region of the new red sandstone of Connecticut and New Jersey, they are not wanting, and I shall next describe the manner of their interference with the strata and the changes they cause at and near their contact.

At the southern extremity of the coal-field at Clover-hill before alluded to, four miles north of the Appomattox River on "Johnson's

Tract," a dike of greenstone is seen about 20 feet thick, running W. 10° S.; the coal-measures crop out here near the granite at various angles between 10° and 45°. Near the locality last mentioned, I obtained at the Beaver mine a section passing through five seams of coal: see Section Fig. 5. The fundamental rocks consisted of granite

Section at Clover-hill, Virginia.

Fig. 5.

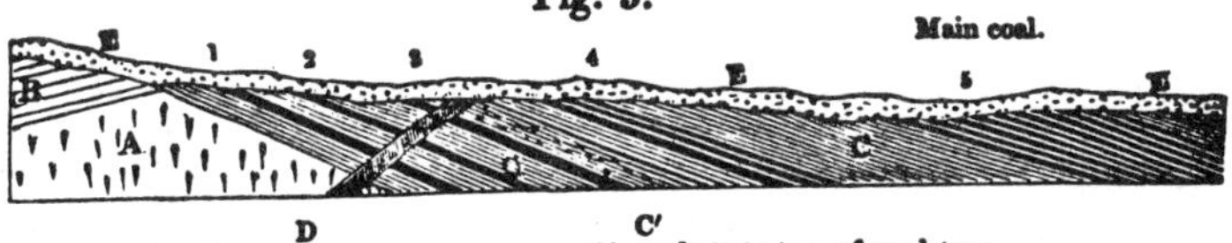

A. Granite.
B. Hornblende schist.
C. Coal-measures.
D. Dike of greenstone.
E. Drift, or ancient alluvium.
C'. 10 feet stratum of sandstone.
1, 2, 3. First, second, and third coal-seams.
4. Four foot coal.
5. Main seam of coal, 10 to 12 feet thick.

and hornblende schist, on which repose coal-measures, chiefly of white sandstone and shale 500 feet thick with three seams of coal, the uppermost of which was cut through by the dike of greenstone, the coal, as the miners expressed it, having been burnt up by it and altered into a kind of coke. On the altered coal No. 3 rests a solid undivided stratum of sandstone 10 feet thick, and on this a seam of coal 4 feet thick, to which succeeds slaty shale 30 feet thick, and on this rests the fifth seam or main coal, 10 to 12 feet thick. Then follow strata of sandstone, shale and other rocks without coal for a thickness of about 160 feet, exposed continuously. The next point to the north where I saw greenstone in connexion with the coal-measures was in Sallé's Tract, within a mile-and-a-half of the James River and near the United States' Arsenal, about sixteen miles north of Clover-hill pits last mentioned. In this region some of the coal is affected and more or less deprived of its bituminous qualities, although not in contact with the igneous rock, which however I traced to within 120 feet of the altered coal, and it may approach much nearer.

Not far from the same locality, in "Burfoot's Tract," about three-quarters of a mile from the Arsenal on the James River, the principal coal-seam was described to me as damaged by the near approach to it of the whin and a kind of coke produced. The mass of trap was visible, a greenstone of the ordinary character 28 feet thick, and near it a brecciated rock of hardened shale varied by patches of carbonaceous matter resembling impure coke in appearance. The locality, like all the others, is near the junction of the coal-measures and the granite. Before visiting the Richmond or East Virginia coal-field, I had been asked by some geologists, whether it was not a singular phænomenon that some upper beds of coal had been deprived of all their volatile matter, while others below remained perfectly unaltered and bituminous? The explanation appears to be simply this: that the greenstone, although intrusive, has often here, as is so common elsewhere, made its way between the strata like a conformable deposit, and has driven out the gaseous matter from the upper coal, while its influence has often not extended to lower seams. In fact, at Duval's and farther north, at Townes and Powell's pits, the trap overlies the coal, and forms a bed having the same dip as the coal-bearing strata.

At the first-mentioned of the above places (Duval's pit, on the south bank of the James River), a shaft, open at the time of my visit, had passed through—first, 105 feet of sandstone and shale, then the greenstone reduced to a thickness of 6 feet and parallel to the strata. Below the trap, strata 25 feet thick, similar to those above, were met with, and then the main seam of coal 10 feet thick, of which the quality was not affected as in some adjacent mines, owing probably to the thinness of the greenstone and the interposition of the other rocks. Below the coal the miners entered black slaty rocks 35 feet thick with numerous prostrate and upright Calamites, and then reached a second bed of coal 3 feet thick of excellent quality resting on other carbonaceous slates with Calamites. Next came the bottom coal 6 feet thick, also unchanged: some other thin partings of coal between this bed and the fundamental granite, show how various are the modifications under which the seams of coal present themselves,—a diversity probably connected with the variations in the amount of the subsidence to which the erect position of the Calamites observed at intervals over so wide an area clearly points.

Coke of Townes and Powell's Pits.

The most remarkable example of natural coke met with in this or any other coal-field known to me, occurs at Townes and Powell's pits at Edge-hill, a locality between five and six miles north of the James River and ten north of Blackheath, being also on the eastern outcrop of the basin and within 500 yards of the granite. A large quantity of coke, used for furnaces and other economical purposes, is here extracted from a bed about 8 feet thick, below which at greater depth are two beds of imperfect coal. The measures passed through above the 8-feet-bed of coke are 110 feet thick, including a conformable bed of blue basalt 16 feet thick. The shale immediately below the trap was white for 11 feet, and then 25 feet of dark leafy shale succeeded, below which came the bed of coke resting on white shale, and lower down coal-measures with two seams of inferior coal, each about 4 or 5 feet thick. The shale, 47 feet thick, interposed between the basalt and the coke, exhibits so many polished surfaces or slickensides, and is so much jointed and cracked, and in some places disturbed and tilted (in one area of 12 feet the beds being vertical), that we may probably attribute the change from coal to coke, not so much to the heating agency of the intrusive basalt, as to its mechanical effect in breaking up the integrity of the beds, and rendering them permeable to water, or the gases of decomposing coal. In some places in the same district where the upper part of a seam is coke, the lower is coal, and there is sometimes a gradation from the one to the other, and sometimes, I was told, a somewhat abrupt separation. The general dip of the strata in this locality is W.S.W. Calamites are abundant in the beds associated with the coke.

I observed that some of the basalt of the neighbourhood assumed a very regular concretionary and spheroidal structure on exposure to the air. About 200 yards north of the great coke-pit, the trap, 14 feet thick, is found within 26 feet of the surface. At the distance of about two miles south of Townes and Powell's, at Crouch's pits, 300 feet deep, where no trap occurs, the coal is bituminous and unchanged.

Analysis of the Coke.—It will be seen by the table before given (p. 270), that Mr. P. H. Henry, who has very carefully analysed this coke, finds it to contain, carbon 86·54, hydrogen 4·23, oxygen and nitrogen 4·53, ash 4·70. He informs me that, according to Mr. Regnault, Welsh anthracite contains only 3·3 per cent. of hydrogen, and it appears in the paper before cited (Quart. Journ. Geol. Soc. vol. i. p. 199), that Dr. Percy found only 2·45 per cent. of hydrogen in the Pennsylvanian anthracite, which I submitted to his examination, from the Mauch Chunk mines. In the Frostburg coal (see same paper), the total quantity of volatile ingredients (hydrogen, oxygen and nitrogen) does not greatly exceed that detected in the coke of Edge-hill, or Townes and Powell's, although the latter is so much less inflammable.

Mr. Mitchell, in his 'Geology of North Carolina' (1842), p. 133, observes, that the theory of the fluviatile origin of the sandstone of North Carolina (which, like that of East Virginia, contains coal) is open to this objection: "That it takes no account of the trap, which is not only imbedded in it, but traverses the sandstone in a thousand different directions. The same trap (he adds) is not met with in the slates and granite which lie adjacent to the sandstone, and the area of a river or estuary could not have been liable throughout its whole extent in length and breadth to injections and eruptions of trap, whilst nothing of the kind occurred upon its banks." The same might be said of the red sandstone districts of the valley of the Connecticut and of New Jersey, for there also we no sooner pass the limits of the sedimentary strata and enter on the hypogene region, than all the conspicuous masses of trap cease to show themselves. But the explanation of this phænomenon has probably occurred to the mind of every experienced geologist. The same greenstone which alters the coal at Clover-hill, above alluded to, crosses in the form of a narrow dike through the adjacent hornblende schist; and Mr. Percival of Newhaven pointed out to me in 1842 several trap dikes in Connecticut traversing the mica schist which forms the western border of the "new red sandstone" district of that State. I saw other dikes to the eastward of the same region, east of Newhaven; and if they are not everywhere traceable, it arises from the wide extent and depth of the "drift" or ancient alluvium which everywhere overspreads the hypogene formations. The apparently abrupt cessation of the volcanic rocks, when we enter the granitic area, arises from the more compact and crystalline form which they assume where they have been consolidated in fissures which were deep-sited originally, before the denudation of the country exposed the hypogene formations to view. Under such circumstances the narrow dikes of more modern igneous rocks are scarcely distinguishable from some members of the older and accompanying plutonic family, and they nowhere expand into dense conformable masses, nor form amygdaloids nor trap tuffs. Hence they are easily overlooked in a cursory survey, although in reality very numerous, and often persistent for great distances.

On the Geological Age of the Coal-Fields of Eastern Virginia, as determined by Fossil Remains.

I shall now consider the chronological relation of the coal-field, in determining which we have no direct evidence from superposition. The only other strata occupying a similar position in the hypogene region on the Atlantic Slope, with the exception of some of a similar character near Raleigh in North Carolina, are those commonly called the new red sandstone of Massachussets, Connecticut, and New Jersey. In part of New Jersey the strata of that red sandstone are unconformable to a part of the palæozoic series of the Appalachian system, and are therefore subsequent in origin to the movements which gave to the old carboniferous and other still more ancient groups of the Alleghany mountains, their present strike, dip, and flexures. The red sandstone therefore being posterior in date to the Appalachian coal strata, the next question is, whether the coal-field near Richmond is of the same age as the red sandstone, or of newer or older date? The difficulty of replying to the inquiry consists in this: that most of the fossils of the so-called new red sandstone are fish, or the foot-tracks of birds, without any plants, while those of the coal near Richmond are almost exclusively confined to plants, so that we have scarcely as yet any satisfactory terms of comparison in the same family of organic beings common to the two. Before alluding to the plants, I shall say something of the small number of shells and ichthyolites which I met with in the coal-field near Richmond.

Shells of the Coal-Field.

Fig. 6.

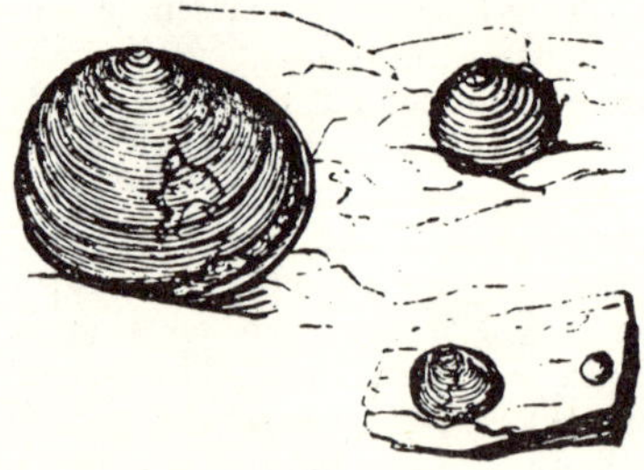

In the carbonaceous shales associated with the main seam, and usually not far above, a great number of minute flattened bivalves are often observed in some places, as at Harden's pits, north of Blackheath (see *a*, fig. 6). They resemble Cyclas in outline; they are thin and compressed, have a horny texture, oval and inequilateral, with a surface concentrically furrowed and somewhat wrinkled. They resemble Posidonomya, and more than any other the *P. minuta*. They occur in such immense numbers as to divide the shale-like plates of mica into very thin laminæ. Every fresh surface exhibits a layer of them. In their gregarious character they resemble the genus Cypris as it occurs in the cypriferous marls of Auvergne.

With them is another smaller and more convex bivalve shell which resembles a young Astarte, but may perhaps be the young of the preceding (see *b*, fig. 6). The same shell was also presented to me by Dr. Werth, from the Creek mines south of Blackheath, also in black shale.

At the Deep-run pits I saw traces of the same shell, but in none of these places in a perfect state: the genus Posidonomya in Europe is considered to be characteristic of the triassic period, in which the *P. minuta* occurs in England (see fig. 4. plate 28. vol. v. Geol. Trans. 2nd series). I have stated, that the manner in which these extremely thin shells occur by myriads, like plates of mica between the partings of the shale, reminded me of the exuviæ of the Cypris in the freshwater marls of Auvergne. Mr. Morris has shown me a recent bivalve crustacean from Bahia Blanca in South America, having a delicate, crumpled or corrugated horny shell, bearing so great a resemblance to the Posidonomya, that I cannot help sharing his doubts whether the latter genus may not be allied to Cypris rather than to any genus of the true Mollusca, *i. e.* to Avicula.

Fossil Fish.

After much careful search, and the indefatigable exertions of my friend Mr. A. F. D. Gifford, residing at the Blackheath mines, I was unable to procure the remains of more than three species of fossil fish, most of which, and the only perfect ones, are referable to a new genus of the homocercal class. On seeing these ichthyolites (September 1846), M. Agassiz, ignorant of their locality, at once suggested, from the analogy of European forms, that they were of the age of the lias; as being all homocercal, and the large smooth scales of one species belonging to the genus Tetragonolepis. This opinion would accord with the inference previously deduced from the fossil plants by Professor W. B. Rogers (Trans. Assoc. American Geol. 1840–42, p. 298).

As all the fish hitherto obtained from the sandstone (commonly called the "new red") of the Connecticut River, four or five species in number, are heterocercal, this would indicate a distinct and probably a more modern date for the Virginian coal-bearing strata. Could we therefore feel sure that the beds containing ichthyolites and the foot-prints of birds in the valley of the Connecticut were triassic, it would afford strong ground for presuming the oolitic age of the Virginian coal strata. But I conceive the question is still open, whether the Connecticut sandstone be of Permian or triassic date.

I shall now offer a description of the Virginian ichthyolites, drawn up chiefly from notes supplied me by Prof. Agassiz and Sir Philip Egerton.

Dictyopyge macrura, Pl. VIII. This species is the most abundant, occurring in strata of argillaceous sandstone about 150 feet above the main seam of coal in Chesterfield county. It is the same one which was described by Mr. W. C. Redfield, under the name of *Catopterus macrurus* (Silliman's American Journal of Science, 1841, vol. xli. p. 27). The following are the characters given by him :—

Catopterus macrurus, Large-finned Catopterus. This beautiful species is distinguished by its broad and flowing fins, of which the anal is so extended as to be nearly joined by the caudal fin. The latter is finely extended. The length of this species is from four to five inches; its width from one and one-eighth to one and three-eighths of an inch. The margins of the fins are remarkably fine and beautiful. The posterior margin of the scales seems to be curled slightly outward, giving the surface a somewhat roughened appearance.

The genus Catopterus was instituted by Mr. Redfield for certain species of heterocercal fish from the Connecticut red sandstone, and the species under consideration having, as before stated, a homocercal tail, cannot be comprehended in it. Sir P. Egerton also remarks, that the dorsal fin is more strictly opposite to the anal than in *Catopterus Redfieldi*, which in this respect approaches nearest to the species under consideration; the tail, according to Agassiz and Sir P. Egerton, bearing the greatest resemblance to that of the genus Pholidophorus. The numerous and uniform articulations give to its large anal fin a net-like appearance, which has led Sir Philip to suggest the generic name of *Dictyopyge*—from *δικτυον, net*, and *πυγη, anus*. The specific name of Mr. Redfield must of course be retained.

I procured several specimens of this ichthyolite, the most perfect of which here figured (Pl. VIII.) is exactly 6 inches long from the mouth to the extremity of the tail, and 1¼ inch broad, but some were a fifth shorter. I do not observe the curled appearance mentioned by Mr. Redfield, in the posterior margin of the scales, which may have arisen perhaps from their being in a disturbed or injured state. The somewhat lengthened shape of the scales of the lateral line, pointed out to me by Mr. Dinkel the artist, causes a marked dividing line between the scales above and below it. In figure *e*, Pl. VIII., the peculiar shape of these scales of the lateral line will be seen, some of them having the opening of the tube outwards, others not, and the projection of the tube in the anterior part of each scale fitting into a notch in the scale immediately preceding.

f, Pl. VIII., shows the shape of the scales immediately above and immediately below the lateral line, in which there is a small notch or incision not exhibited in the remoter scales.

b, Pl. VIII., represents the large anal fin, also seen more fully developed in fig. *b*, Pl. IX.

c, Pl. VIII., shows a faint impression of the pectoral fin.

d, Pl. VIII., shows part of the ventral fin and an impression of the rest.

Plate IX. fig. 1. represents the posterior part of *Dictyopyge macrura*, obtained in the same slab as the specimen figured in Pl. VIII.; *d* is the ventral, and *b* the anal fin.

I may also remark, 1st, The opercular and frontal bones of the head are covered with extremely minute scales. 2ndly, The raylets on the anterior margin of the caudal fin alluded to in Mr. Redfield's description are stronger and broader in the lower lobe of the tail than in the upper. 3rdly, The insertion of the anterior ray of the dorsal fin begins nearly opposite the middle of the anal fin, and in consequence of the great extension of the latter towards the tail, it continues to be opposite the whole length of the dorsal. The pectoral and ventral fins are rarely to be seen, but have left impressions in some specimens.

Second species of the genus Dictyopyge.

Portions apparently belonging to a second species of the same genus occur in some of the slabs of stone from Chesterfield county, the scales being so much larger in proportion to the size of the individual, that in corresponding parts of the body, five scales occupy the place of seven in the preceding species.

Tetragonolepis.

Pl. IX. fig. 2. represents nearly the whole side, with part of the lateral line near the head, of a specimen from Blackheath: natural size.

Although a few fragments only of this ichthyolite were obtained, they were recognized at once by M. Agassiz as belonging to the distinct genus Tetragonolepis, and to resemble the large liassic species with smooth scales. A scale from the lateral line, magnified about four times (*a*, fig. 2. Pl. IX.), shows the manner in which the tube in some of these scales opens outwards. Each scale in the lateral line has also a notch posteriorly.

I have stated that all the fish of the red sandstone of Connecticut differ from those of Virginia. A fine collection of the former was made by Prof. B. Silliman, jun., and myself, at Durham, Connecticut, in 1842, which M. Agassiz and Sir P. Egerton have examined for me. They belong to three species:—1st, *Palæoniscus fultus*, Agassiz (*Ischypterus fultus*, Egerton), of various sizes and ages, some of the varieties comprising the *Palæoniscus macropterus* of Redfield, which is considered by M. Agassiz to be the same. Sir P. Egerton has proposed for Palæoniscus the generic name of *Ischypterus*, from *ισχυς*, *strength*, and *πτερον*, *fin*, from the great size and strength of the fulcral rays of the dorsal fin. They differ from Palæoniscus, as Mr. Redfield first pointed out, by having the vertebral column prolonged to a more limited extent into the upper lobe of the tail, or, in the language of M. Agassiz, they are less heterocercal. The teeth also, according to Sir P. Egerton, who in 1844 examined for me a fine series of specimens from Durham, Connecticut, differ from those of Palæoniscus in being strong and conical. The oral aperture is smaller and the scales smoother than in the Palæonisci of the magnesian limestone or Permian beds of Europe, in which the scales are striated and serrated on the posterior margins. Some of the *carboniferous* Palæonisci of Europe resemble the American fish in having similar smooth scales.

2ndly, *Catopterus gracilis.*

3rdly, *Catopterus Redfieldi*, a broader fish than the preceding, and with scales not so long in proportion to their depth.

These three, and all others found in the same (new red sandstone?) formation, are heterocercal. Sir P. Egerton is of opinion that far too much importance has been attached by some geologists to the circumstance of the vertebral column not being prolonged so far into the upper lobe of the tail in these genera, Ischypterus and Catopterus, as it is in true Palæonisci, when they have insisted upon it as an argument in favour of the red sandstone of Connecticut being less ancient than the palæozoic period. For at Autun, in France, we find the genus Ischypterus accompanying the true Palæonisci with fossil plants specifically the same as those of the old coal.

It would be rash to attempt to settle the chronological relations of any group of strata by the presence of some one peculiar generic form of fish; but if we are to reason on the occurrence of Ischypterus in the present state of our knowledge, it must be in favour of identifying the red sandstone of Connecticut with the beds which belong to the uppermost members of the old coal of Europe, as at Autun the coal-measures are almost passing upwards into Permian. The Irish *Palæoniscus catopterus* of Roan (or Rhone) Hill, Tyrone, referred by Captain Portlock to the trias, is a true Palæoniscus, and not allied generically either to the Ischypterus of Egerton or the Catopterus of Redfield.

Whether the Coal-Field of the James River be of marine or freshwater origin.

I know of no character in any of the fossils of this coal-field which is inconsistent with the hypothesis that all the strata may have been deposited in a lake, estuary or river-delta. The plants are all terrestrial, and the occurrence of erect Calamites and Equiseta at various heights in the series, and at points widely distant, implies, according to my view, that they were not drifted, but grew on the area where we now find them. Hence I infer the constant proximity of land, the drainage of which may have supplied a body of fresh water. Of the genus Posidonomya too little is as yet known to permit us to reason on its habits. In regard to the fish (omitting the new genus), I am informed by Sir P. Egerton that Tetragonolepis is accompanied in the British lias by Lepidotus, which last genus abounds in the freshwater strata of the Wealden in England.

The Coal-Field probably of the age of the Inferior Oolite and Lias.

It will appear from the subjoined account, drawn up by Mr. Charles J. F. Bunbury, of the fossil plants procured by me from the coal-field, that they belong to fifteen forms, and that they bear upon the whole a very decided analogy to the oolitic and triassic Flora of Europe. The *Calamites arenaceus*, one of the most abundant plants, has as yet been traced up no higher than the trias in Europe*.

* Mr. Prestwich has mentioned the plant as occurring in the old coal of Coalbrookdale (Geol. Trans. 2nd series, vol. v. p. 488), but Mr. Morris informs me, after inspecting the specimens, that this was a mistake for *Calamites Suckowii*.

The *Equisetum columnare* is said to be common to the trias and oolite in Europe. Among the ferns, *Pecopteris Whitbiensis*, which I procured, seems to have been correctly identified by Mr. W. B. Rogers, and is one of the commonest species of the Yorkshire oolite. The large Tæniopteris is also allied to some of the Yorkshire oolite ferns. The genus Zamites is represented by many species in the oolitic rocks of Europe, but is also triassic.

Fig. 7.

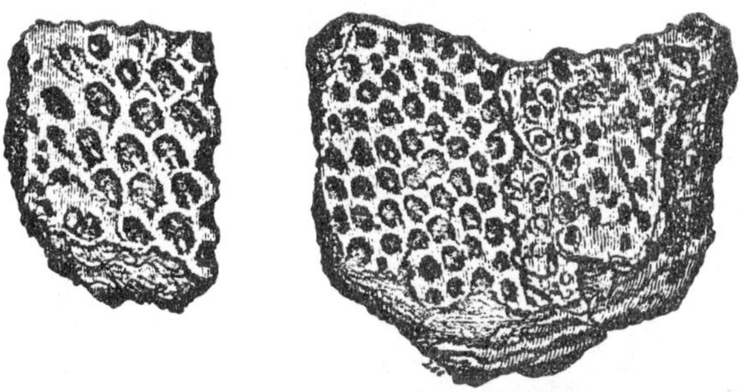

The fossil, somewhat like a Stigmaria—see fig. 7 (No. 13. of Mr. Bunbury's description of plants)—which Dr. Hooker has had the kindness to figure for me, has more of the aspect of a plant from the old coal than any other, but we have not as yet sufficient knowledge of its true relations to found any argument upon it. I may remark also, that if some of the plants and the shell called Posidonomya should indicate a formation rather older than the Whitby oolite, they may still belong to part of the great jurassic period; for the Whitby coal occurs in, or immediately below, the great oolite, which is succeeded in descending order by the inferior oolite and the lias. Now as we have not in Europe any coal-fields, nor any large development of a fossil flora in these older subdivisions of the jurassic group, the Virginian strata may correspond in age either to the inferior oolite or the lias, and for that reason may have more of a triassic character in their organic remains than the coal strata of Whitby in Yorkshire*. The occurrence of the large smooth scales of a fish of the genus Tetragonolepis, so characteristic of the lias, favours this view. It has been well remarked by Prof. W. B. Rogers, that the few plants discovered by Captain Grant in connexion with the oolite-coal series in Cutch, resemble very closely those of the East Virginian coal-measures†. There can be no doubt that the Indian fossils belong to

* Some fossil plants figured and described by Professor Hitchcock, from "the new red sandstone formation of Connecticut and Massachussetts" (Trans. of Amer. Geol. 1840–42, p. 294 and pl. 13.) do not enable us to decide the age of that formation, or to compare it with that of Virginia. The supposed Tæniopteris, fig. 8, cannot belong to that genus, the striations of the leaf not being at right angles to the midrib, and the entire portion represented being of the same diameter throughout. The plant (fig. 3 and 4) is probably related to the genus Lycopodites, which is also found in the Virginian coal-field; if, as conjectured, it be Voltzia, this genus is common to the magnesian limestone (Permian) and trias of Europe.

† Grant, Trans. Geol. Soc. Lond. vol. v. 2nd series.

the oolitic period, from the character of the numerous marine shells found in some of the associated strata.

Upon the whole, therefore, I arrive at the same result as Prof. W. B. Rogers, namely, that the coal-field of Eastern Virginia agrees in age with the lower members of the oolitic or jurassic group of Europe, —a conclusion which is important, because these strata constitute at present the only representatives of this group hitherto discovered in the United States of North America. If future researches should require any modification of this opinion, we may then expect that the trias will be the group to which the American formation will be referable.

In conclusion I may observe, that I was much struck with the general similarity of this more modern coal-field, and one of palæozoic date near St. Etienne in France, which also rests on granitic rocks, from the detritus of which its coarse grits and sandstone are composed. In both these coal-fields upright Calamites abound; fossil plants are met with in both, almost to the exclusion of all other classes of organic remains, shells especially being absent. The character of the coal is similar, but in the richness and thickness of the seams, the Virginian formation is pre-eminent. When we behold phænomena so identical repeated at an epoch so much more modern in the earth's history, and at a time when a very distinct vegetation had been established, we may derive from the fact a useful caution in regard to many popular generalizations respecting a peculiar state of the globe during the remoter of the two periods alluded to. Some geologists, for example, have supposed an atmosphere densely charged with carbonic acid to be necessary to explain the origin of coal—an atmosphere unfit for the existence of air-breathing vertebrate animals; but this theory they will hardly be prepared to extend to so modern an epoch as the oolitic or triassic.

DESCRIPTION OF THE PLATES.

Plate VIII. Figure of *Dictyopyge macrura*, size of nature, from the oolitic coal-field of Eastern Virginia. This specimen was obtained from the Blackheath mines in Chesterfield county, Virginia. None of the parts here represented have been restored. See p. 275.

b. The anal fin, which is not so perfect as in some other specimens from the same region.

c. A faint impression of the pectoral fin.

d. Part of the ventral fin and an impression of the remainder.

e. Scale from the lateral line.

f. Shape of the scales from immediately above and immediately below the lateral line. See p. 276.

Plate IX. fig. 1. Posterior part of the same fish, *D. macrura*, from the same slab of stone.

d. Ventral fin.

b. Anal fin.

fig. 2. *Tetragonolepis*: portion of a fossil fish from the same strata, Blackheath, Virginia. See p. 277.

a. Scale of the lateral line.

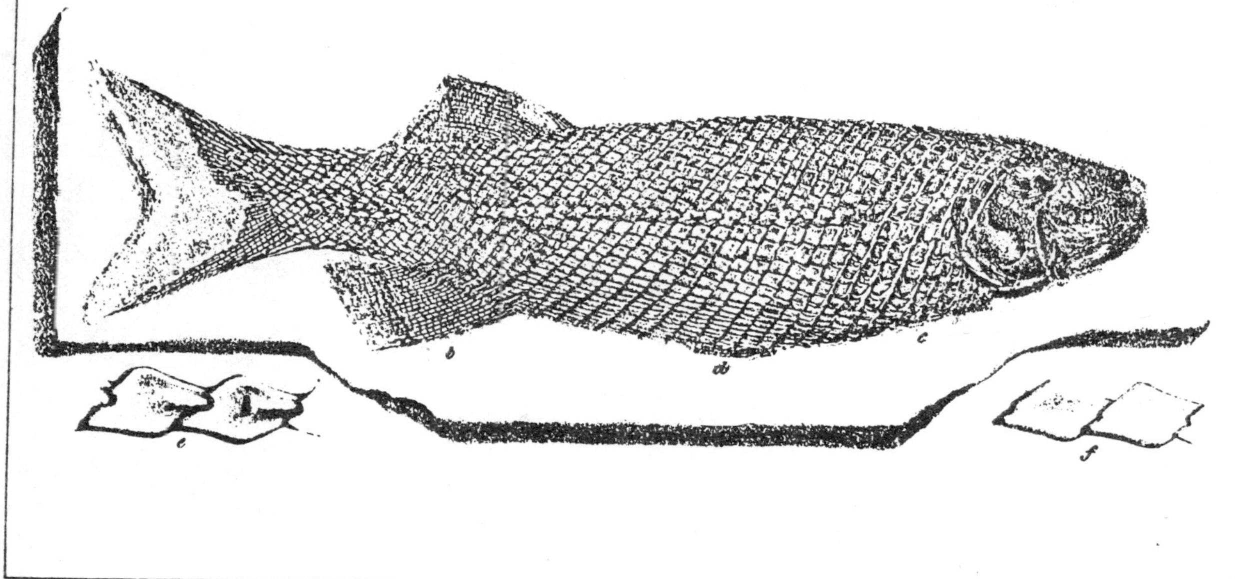

Jos: Dinkel del.

Hullmandel & Walton Lithographers.

DICTYOPYGE MACRURA BLACKHEATH, VIRGINIA

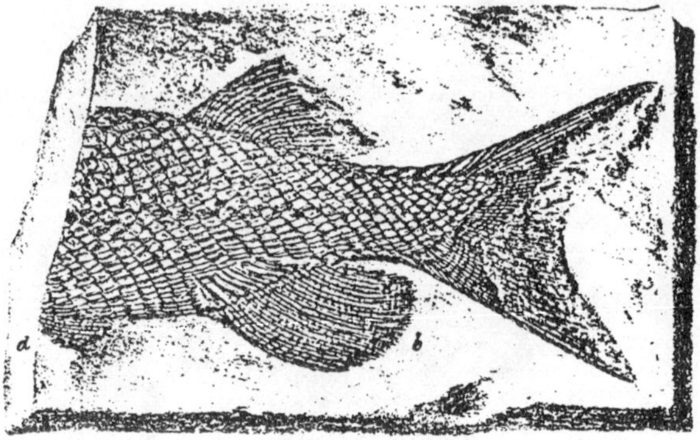

Fig. 1.

DICTYOPYGE MACRURA.

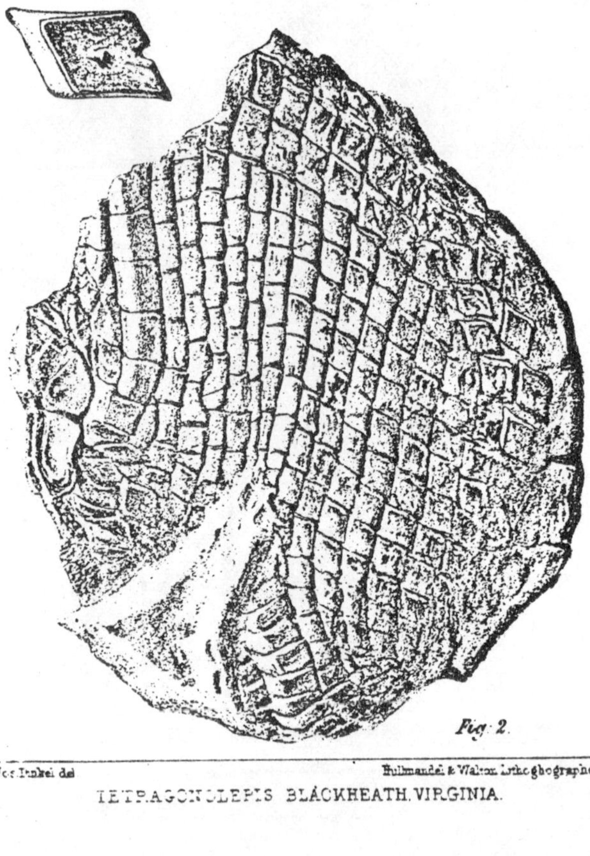

Fig: 2.

Jos. Dinkel del   Hulmandel & Walton Lithographers

TETRAGONOLEPIS BLACKHEATH, VIRGINIA.

ON THE DELTA AND ALLUVIAL DEPOSITS OF THE MISSISSIPPI

Art. IV.—*On the Delta and Alluvial Deposits of the Mississippi, and other points in the Geology of North America, observed in the years* 1845, 1846; by C. Lyell.*

The delta of the Mississippi may be defined as that part of the great alluvial plain which lies below, or to the south of the branching off of the highest arm of the river, called the Atchafalaya. This delta is about 13,600 square miles in area, and elevated from a few inches to ten feet above the level of the sea. The greater part of it protrudes into the Gulf of Mexico beyond the general coast line. The level plain to the north, as far as Cape Girardeau in Missouri above the junction of the Ohio, is of the same character, including, according to Mr. Forshey, an area of about 16,000 square miles, and is, therefore, larger than the delta. It is very variable in width from east to west, being near its northern extremity, or at the mouth of the Ohio, 50 miles wide; at Memphis 30; at the mouth of the White River 80, and contracting again farther south, at Grand Gulf, to 33 miles. The delta and alluvial plain rise by so gradual a slope from the sea as to attain, at the junction of the Ohio, (a distance of 800 miles by the river,) an elevation of only two hundred

* Abstract of a discourse before the British Association, Sept. 14, 1846, from the Athenæum, for September 26.

feet above the Gulf of Mexico. Mr. Lyell first described the low mud banks covered with reeds at the mouths of the Mississippi, and the pilot-station called the Balize; then passed to the quantity of drift wood choking up the bayous, or channels, intersecting the banks; and, lastly, enlarged on the long narrow promontory formed by the great river and its banks between New Orleans and the Balize. The advance of this singular tongue of land has been generally supposed to have been very rapid, but Mr. Lyell and Dr. Carpenter, who accompanied him, arrived at an opposite conclusion. After comparing the present state of this region with the map published by Charlevoix, 120 years ago, they doubt whether the land has, on the whole, gained more than a mile in the course of a century. A large excavation, eighteen feet deep, made for the gas works at New Orleans, and still in progress, March, 1846, shows that much of the soil there consists of fine clay or mud, containing innumerable stools of trees, buried at various levels in an erect position, with their roots attached, implying the former existence there of fresh-water swamps covered with trees, over which the sediment of the Mississippi was spread during inundations, so as slowly to raise the level of the ground. As the site of the excavation is now about nine feet above the sea, the lowest of these upright trees imply that the region where they grew has sunk down about nine feet below the sea level. The exposure, also, in the vertical banks of the Mississippi at low water for hundreds of miles above the head of the delta, of the stumps of trees buried with their roots in their natural position, three tiers being occasionally seen one above the other, shows that the river in its wanderings has opened a channel through ancient morasses, where trees once grew, and where alluvial matter gradually accumulated. The old deserted beds, also, of the river, with banks raised fifteen feet above the adjoining low grounds, bear testimony to the frequent shifting of the place of the main stream; and the like inference may be drawn from the occurrence, here and there, of crescent-shaped lakes, each many miles in length and half a mile or more in breadth, which have once constituted great curves or bends of the river, but are now often far distant from it. The Mississippi, by the constant undermining of its banks, checks the rise of large commercial towns on its borders, and causes a singular contrast between the wealth and splendor of eight hundred or more fine steamers, some of which may truly be called floating palaces, and the flat monotonous wilderness of uncleared land which extends for hundreds of miles on both sides of the great navigable stream. Mr. Lyell visited, in March, 1846, the region shaken for three months in 1811, 12, by the earthquake of New Madrid. One portion of it, situated in the States of Missouri and Arkansas, is now called "the sunk country." It extends about seventy miles north and south, and thirty east and west, and is for the most part submerged. Many dead trees are still standing erect in the swamps, a far greater number lie prostrate. Even on the dry ground in the vicinity, all the forest trees which are of prior date to 1811, are leafless: they are supposed to have been killed by the loosening of their roots by the repeated shocks of 1811, 12. Numerous rents are also observable in the ground where it opened in 1811; and many "sink holes," or cavities, from ten to thirty yards wide and twenty feet or more in depth, now interrupt the general level of the plain, which were formed by the spouting out of large quantities of sand and mud during the earthquake. In attempting to compute the minimum of time required for the accumulation of the alluvial matter in the delta and valley of the Mississippi, Mr. Lyell referred to a series of experiments, made by Dr. Riddell, at New Orleans, showing that the mean annual proportion of sediment in the river was, to the water $\frac{1}{1245}$ in weight, or about $\frac{1}{3000}$ in volume. From the observations of the same gentleman, and those of Dr. Carpenter, and of Mr. Forshey, (an eminent engineer of Louisiana,) the average width, depth, and velocity of the Mississippi, and thence the mean annual discharge of water, are deduced. In assuming 528 feet (or the tenth of a mile) as the probable thickness of the deposit of mud and sand in the delta, Mr. Lyell founds his conjecture on the depth of the Gulf of Mexico, between the southern point of Florida and the Balize, which equals on an average 100 fathoms. The area of the delta being about 13,600 square statute miles, and the quantity of solid matter annually brought down by the river 3,702,758,400 cubic feet, it must have taken 67,000 years for the formation of the whole; and if the alluvial matter of the plain above be 264 feet deep, or half that of the delta, it has required 33,500 more years for its accumulation,—even if its area be estimated as only equal to that of the delta, whereas it is, in fact, larger. If some deduction be made from the time here stated, in consequence of the effect of drift wood, which must have aided in filling up more rapidly the space above alluded to, a far more important allowance must be made, on the other hand, for the loss of matter, owing to the finer particles of mud not settling at the mouth of the river, but being swept out far to sea, and even conveyed into the Atlantic by the Gulf Stream. Yet the whole period during which the Mississippi has transported its earthy burthen to the ocean, though perhaps far exceeding 100,000 years, must be insignificant, in a geological point of view, since the bluffs or cliffs bounding the great valley, (and therefore older in date,) and which are from 50 to 250 feet in perpendicular height, consist in great part of loam, containing land, fluviatile, and lacustrine shells of species still inhabiting the same country. These fossil shells, occurring in a deposit resembling the *loess* of the

Rhine, are associated with the bones of the mastodon, elephant, tapir, mylodon, and other megatherioid animals; also a species of horse, ox, and other mammalia, most of them of extinct species. The loam rests at Vicksburg and other places on eocene or lower tertiary strata, which in their turn, repose on cretaceous rocks. A section from Vicksburg to Darien, through the States of Mississippi, Alabama and Georgia, exhibits this superposition, as well

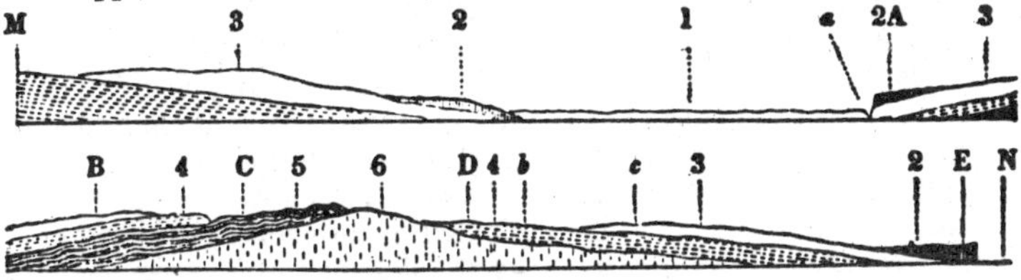

Section (M to N) about 750 miles in length from west to east, from Louisiana (on the west) through Jackson, Mississippi, to Tuscaloosa, in Alabama, and thence, by Montgomery, to the Atlantic, near Darien, in Georgia.*

as that of the cretaceous strata on carboniferous rocks at Tuscaloosa. Mr. Lyell ascertained that the huge fossil cretacean, named Zeuglodon, by Owen, is confined to the eocene deposits. In the cretaceous strata, the remains of the mosasaurus, and other reptiles, occur without any cetacea.

The coal-fields of Alabama were next alluded to; from which fossil plants have been procured, by Prof. Brumby and Mr. Lyell, of the genera Sphenopteris, Neuropteris, Calamites, Lepidodendron, Sigillaria, Stigmaria, and others, most of them identical in *species*, as determined by Mr. C. Bunbury, with fossils of Northumberland. This fact is the more worthy of notice, because the coal of Tuscaloosa—situated in lat. 33° 10′ N.—is farther south than any region in which this ancient fossil Flora had previously been studied, whether in Europe or North America; and it affords, therefore, a new proof of the wide extension of a uniform Flora in the carboniferous epoch. Mr. Lyell—adverting to the opinion recently adopted by several able botanists, that the climate of the coal period was remarkable for its moisture, equability, and freedom from cold, rather than the intensity of its tropical heat—stated that this conclusion, as well as the oscillations of temperature implied by the glacial period, are confirmatory of the theory first advanced by him, in 1830, to explain the ancient geological changes of climate, by

* 1. The alluvium of the Mississippi.—2. Post-pliocene loam and sand, with recent shells and bones of extinct mammalia; the shells in this deposit, on the borders of the valley of the Mississippi, are of land and fresh-water species, those near Darien, of marine species.—3. Eocene formation.—4. Cretaceous strata.—5. Carboniferous rocks.—6. Hypogene or granite, gneiss, mica schist, &c.

M to a. Louisiana.—A. Vicksburg, Mississippi.—B. Jackson.—C. Tuscaloosa, Alabama.—D. Montgomery.—E. Darien, Georgia.—N. Atlantic.

a. Mississippi river.—b. Chatahoochie river.—c. Flint river.

geographical revolutions in the position of land and sea. The lapse of ages, implied by the distinctness of the fossils of the eocene, cretaceous, carboniferous, and other strata, is such, that, were we to endeavor to give an idea of it, we must estimate its duration, not by years, as in the case of the delta, but by such units as would be constituted by the interval between the beginning of the delta and our own times. "It is now fifty years," said Mr. Lyell, "since Playfair, after studying the rocks in the neighborhood of Edinburgh, in company with Dr. Hutton and Sir James Hall, was so struck with the evidence they afforded of the immensity of past time, that he observed, 'How much farther reason may go, than imagination can venture to follow!' These views were common to the most illustrious of his contemporaries; and since that time have been adopted by all geologists, whether their minds have been formed by the literature of France, or of Germany, or of Italy, or Scandinavia, or England;—all have arrived at the same conclusion respecting the great antiquity of the globe, and that, too, in opposition to their earlier prepossessions and to the popular belief of their age. It must be confessed that, while this unanimity is satisfactory as a remarkable test of truth, it is somewhat melancholy to reflect, that, at the end of half a century, when so many millions have passed through our schools and colleges since Playfair wrote that eloquent passage, there is still so great a discordance between the opinions of scientific men and the great mass of the community. Had there been annual gatherings, such as this, where those who are entitled to speak with authority, address themselves to a numerous assembly, drawn from the higher classes of society, who, by their cultivation and influence, must direct the education and form the opinions of the many of humbler station, it is impossible that so undesirable and unsound a state of things should have now prevailed as that where there is one creed for the philosopher and another for the multitude. Had there been meetings like this, even for a quarter of a century, we should already have gained for geology the same victory that has been so triumphantly won by the astronomer. The earth's antiquity, together with the history of successive races of organic beings, would have been ere this as cheerfully and universally acknowleged as the earth's motion, or the number, magnitude, and relative distances of the heavenly bodies. I am sure it would be superfluous if I were to declare, in an assembly like this, my deep conviction, which you —all of you—share, that the further we extend our researches into the wonders of creation in time and space, the more do we exalt, refine, and elevate our conceptions of the Divine Artificer of the Universe."—Mr. Lyell concluded this discourse by announcing his corroboration of the discovery, recently made by Dr. King, at Greensburg, thirty miles from Pittsburg, in Pennsylvania, of the occurrence of fossil foot-prints of a large reptilian, in the middle of the ancient coal-measures. They project, in relief, from the lower surfaces of slabs of sandstone; and are also found impressed on the subjacent layers of fine unctuous clay. This is the first well-established example of a vertebrated animal, more highly organized than fishes, being met with in a stratum of such high antiquity.

ON THE ALLEGED COEXISTENCE OF MAN AND THE MEGATHERIUM

4. *On the alleged Coëxistence of Man and the Megatherium*; by C. Lyell, (addressed as a letter to the Editor of the Times, and published in the Times of Dec. 8, 1846.)—Sir: A considerable sensation appears to have been caused in the minds of the scientific, and part of the unscientific public, by the announcement in many of the newspapers of the discovery in America of a fossil human bone associated with the remains of the megatherium and other extinct quadrupeds. I have only just read in the last number of the Athenæum, (December 5, 1846,) a brief notice of the reported discovery, copied from an American paper, and which the editor has inserted with a judicious caution to his readers that no competent authority has yet vouched for all the facts. I feel sure that the story relates to part of a human pelvis (a fragment of the *os innominatum*) which was shown to me at Natchez last spring, together with several very remarkable remains of megatheroid animals. I visited Natchez in March last, on which occasion I was informed of the antiquity assigned to the human relic, and having examined carefully into the evidence, came to the conclusion that the proofs of the coëxistence of the human individual with the megatheroid and other extinct quadrupeds found in the vicinity was altogether unsatisfactory. We are indebted to the skill and indefatigable industry of Dr. Dickerson of Natchez, for having formed a splendid collection of organic remains, derived from the superficial deposits of that neighborhood. All these he had the kindness to show me, and afterwards accompanied me to the principal spots where they had been procured. One of these, distant about six miles from Natchez, has been named by him the Mammoth ravine, from the skeleton of the mastodon discovered there; and it was there also that the fragment of the human pelvis, supposed to have been contemporaneous with the megatherium, was picked up.

In order to explain its position, a few words on the geology of the region will be indispensable. The broad, flat, alluvial plain of the Mississippi is bounded on its eastern side by a table-land about two hundred feet higher than the river, and having a very gentle upward slope to the eastward or towards the interior. This elevated platform ends abruptly at Natchez, in a line of perpendicular cliffs, or bluffs, the base of which is continually undermined by the Mississippi. The vertical section here exposed to view affords the geologist a fine opportunity of studying the composition and succession of the strata for a vertical thickness of rather more than two hundred feet. The whole deposit is of comparatively modern date, the upper sixty feet consisting of clayey loam, containing shells of recent species, and the lower of sand and gravel without organic remains, except some wood and silicified corals washed out of older rocks. The yellow loam at the top bears a singularly close resemblance to the fluviatile silt, or "loess," as it is termed, of the valley of the Rhine, between Cologne and Basle, and, like it, contains abundance of freshwater and land shells, of which I myself obtained more than twenty species, now in my cabinet in London. They belong to the genera Helix, Helicina, Pupa, Succinea, accompanied or rather replaced in a few places where the loam passes into shell-marl, by Lymnea, Planorbis, Physa, and Cyclas. All the species are identical with testacea, now inhabiting the same part of the United States. With these shells are found, at different depths, some in the loam, and some in the clay at the bottom of the loam, the scattered bones and occasionally entire skeletons of the mastodon, megatherium, mylodon, casteroides, equus, bos, and other quadrupeds. If in any part of this formation, whether at the depth of ten feet or one hundred, any remains of man lie buried by natural causes in undisturbed loam or clay, the antiquity of the human race is thereby established to an extent far greater than has ever yet been made out in any instance which I have had an opportunity of investigating. It would follow that the human race had survived the extinction of one group of terrestrial mammalia and seen another succeed and replace it; and also that man had lived in North America when the geographical configuration of the country was very different,—in other words before the valley of the Mississippi, or even some of the strata forming its boundary rocks, were in existence. Every deposit entering into the composition of the bluffs of the Mississippi valley must be older than the alluvial plain and delta of that river; and we thus obtain, independently of any evidence attesting great changes in the animate world, a measure for estimating the *minimum* of time which has elapsed since the bluffs originated. For we must allow a sufficient series of years for the deposition of all the sedimentary matter forming the plain and delta of the great river. I have endeavored to calculate the period, (one of enormous duration relatively to the historical epoch,) and explain my views on the subject, in a lecture delivered to the British Association at their meeting at Southampton, and a correct, though incomplete, abstract of that communication appeared in the Atheneum, for Sept. 26, 1846.*

To come then to the point—what proof is there that the geological position of the human pelvic bone was such as to demonstrate its anteriority in date to the entire hydrographical basin of the Mississippi, and the formation of its alluvial deposits and delta? The yellow shelly loam or loess, before mentioned, extends for twelve miles inland or eastward from the river; and in consequence of its incoherent and destructible nature, every streamlet flowing over it cuts out for itself, in its way to the Mississippi, a deep gully or ravine. This denudation has of late years proceeded with accelerated speed, especially in the

---

* This volume, p. 34.

course of the last thirty or thirty-five years. Some attribute the increased erosive action to partial clearings of the native forest; others to the effects of the great earthquake of New Madrid in 1811–12, by which this region was much fissured, ponds being dried up, and some landslips caused. Colonel Wiley, a proprietor in this part of the state of Mississippi, and who well remembers the district before the year 1812, assured me that the Mammoth ravine although now seven miles long, and in some parts sixty feet deep, with its numerous ramifications, has been entirely formed since that year when the earthquake occurred. He has himself ploughed some of the land exactly over the spot where the ravine is now situated.

It is however enough for our purpose to affirm that whatever be the date of the origin of this watercourse, it has of late years been considerably enlarged and lengthened, its banks presenting everywhere precipices in which the loam, unsolidified as it is, retains its verticality, as is the case with its counterpart, the loess of the Rhine. Land shells are seen in great numbers at the depth of about thirty feet from the top, and the fossil bones of the mastodon, and other extinct quadrupeds, are usually picked up in the bed of the stream, after they have been washed out of the undermined cliffs, where, however, some few have also been observed *in situ.* Under these circumstances, as I was given to understand, the human pelvis was procured at the base of the cliff. Even if it had been dug out in the presence of a practical geologist, it would have been necessary for him to be more than usually on his guard against deception, for landslides have detached large masses from the cliffs, and these may easily cover human bones previously washed down by the stream or dislodged from the soil near the top of the cliff, where some old Indian graves, so common throughout the country, may have been undermined. It is not rare to find on shoals and on the shores of the islands in the Mississippi at low water, numerous bones of man mingled with those of extinct animals washed out of the bluffs. In these cases the human bones are as black as the quadrupedal fossils, having been apparently stained with peaty matter, in the soil where they were buried; but no geologist has ever ventured on this evidence, to infer the contemporaneousness of man and the fossil specimens thus accidentally associated.

After I had made up my mind that the remote antiquity of the human bone at Natchez was questionable, and that its occurrence in the ravine might be explained in the manner above suggested, I found that Colonel Wailes, a friend of Dr. Dickerson's, who accompanied us in part of our excursion, and who has also made a fine collection of the fossils of this neighborhood, not only shared my doubts, but had made the same conjecture respecting the probable manner in which the fossil may have been conveyed to the spot where it was found. I have the honour to be, Sir, yours, &c., CHARLES LYELL.

11, Harley street, Dec. 7.

ON THE RELATIVE AGE AND POSITION OF THE SO-CALLED NUMMULITE LIMESTONE OF ALABAMA

3. *On the relative Age and Position of the so-called Nummulite Limestone of* ALABAMA. By C. LYELL, F.R.S. and V.P.G.S.

IN a former paper published in the Quarterly Journal of the Geological Society of London (vol. ii. p. 405, May 1846), I stated that the limestone containing abundantly the *Nummulites Mantelli*, Morton, which occurs near Suggesville, Clarksville, and other places between the rivers Alabama and Tombecbee, in the State of Alabama, was a member of the Eocene tertiary group, and that so far from constituting any part of the cretaceous formation, as had formerly been imagined, it holds in reality a place high up in the Eocene series of the South. In the same memoir I gave a section extending from Claiborne through Suggesville and Macon to the west of Clarksville, Alabama, in which the position of the so-called nummulitic limestone was explained. It was stated to be newer than all the beds of the well-known Claiborne Bluff, and I mentioned that "the bones of the gigantic cetacean called Zeuglodon by Owen were everywhere found in Clarke County, in a limestone below the level of the nummulitic rock and above the beds which contain the greater number of perfectly preserved eocene shells, such as *Cardita planicosta* and others."

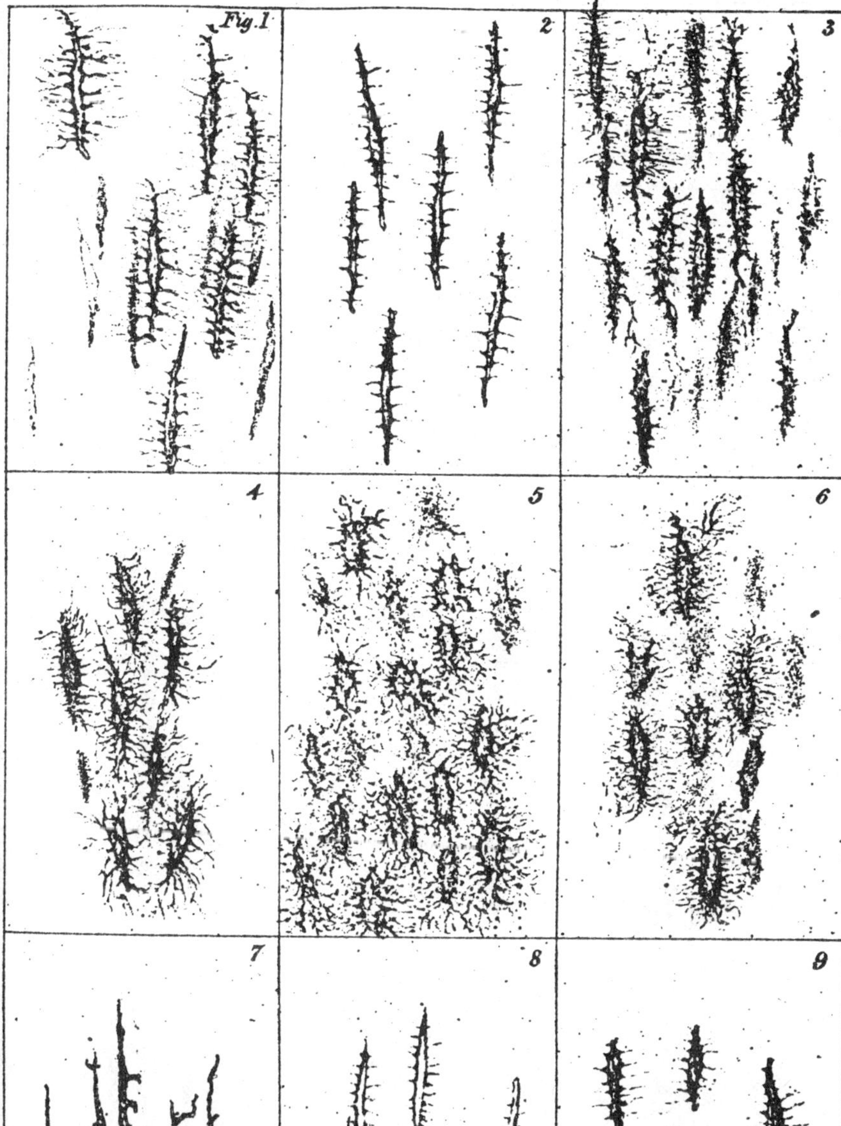

Lens Aldous del. et zinc. Day & Son lith^rs to the Queen

M^r. BOWERBANK ON BONE CELLS, P. 2. &c.

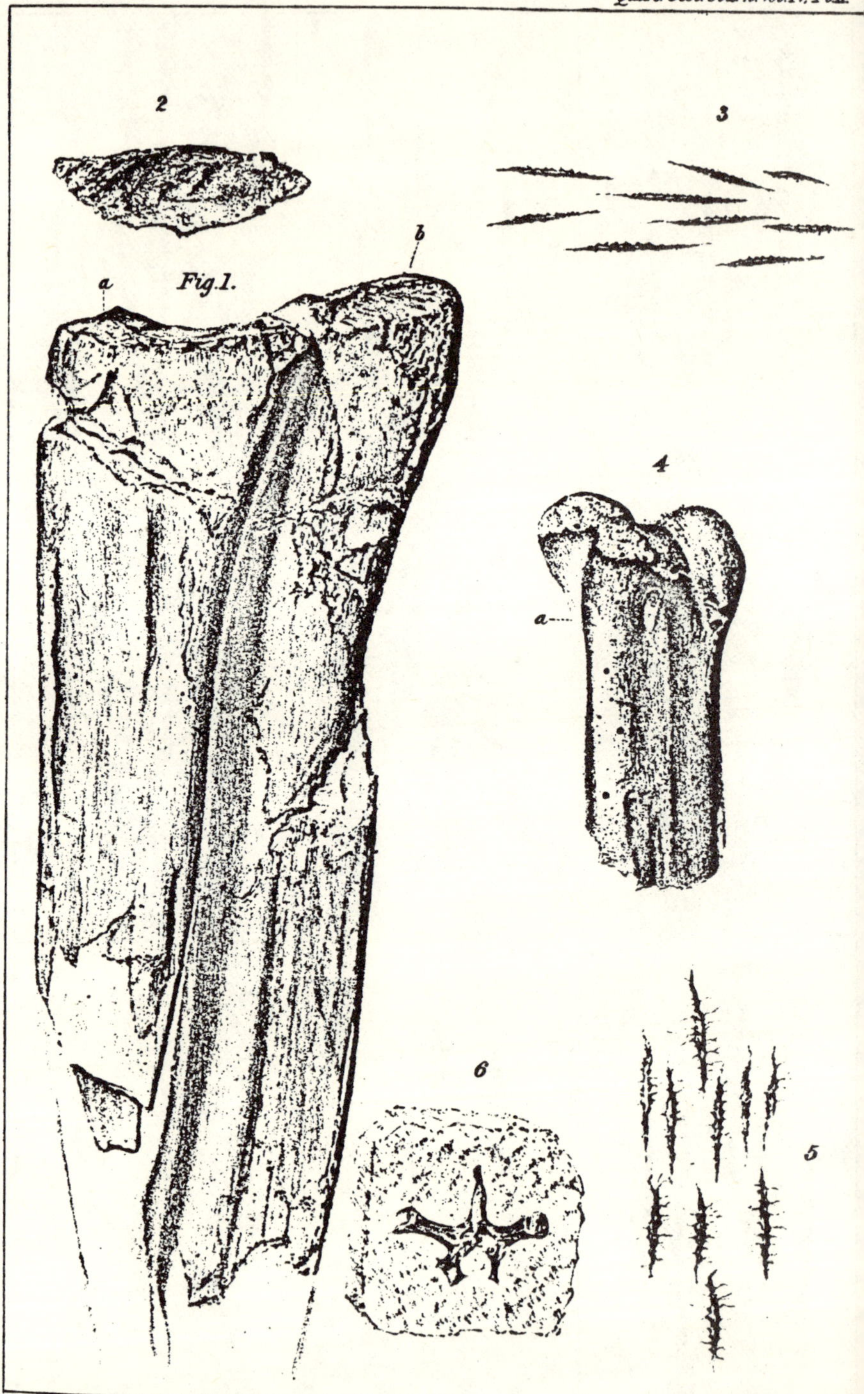

Lens Aldous del. et zinc.

Day & Son, lith^rs to the Queen

Mr BOWERBANK ON BONE CELLS, P. 2 &c.

(Quart. Journ. Geol. Soc. vol. ii. p. 409, May 1846.) At the time that my first communication was written, I had not finished my explorings in Alabama, nor visited St. Stephen's Bluff on the Tombecbee river, where I afterwards obtained additional proofs of the order of superposition above indicated; nor had I then compared the Eocene strata at Vicksburg with those of Jackson in the State of Mississippi, which throw light on the same question of relative position. Before adverting to these last-mentioned localities, I will first offer a few observations on the country between Claiborne and Clarksville, for I understand that doubts have been lately thrown on the correctness of the views which I have expressed relatively to the true age and place in the series to be assigned to the "rotten limestone of Alabama," and the associated rock in which the fossil first named *Nummulites Mantelli* by Morton abounds*.

Before restating the grounds of my former opinion and corroborating them with fresh proofs, it may be well to say something of the nature and zoological relations of the *discoid* bodies from Alabama which have passed under the name of Nummulites, and which constitute the chief part in bulk of considerable masses of limestone in certain districts. Having obtained many specimens both from Alabama and from Vicksburg in Mississippi, in which the structure of this fossil was beautifully preserved, I first showed them to Prof. E. Forbes, who at once pronounced them not to be Nummulites, but related to some living plants or zoophytes which Mr. Jukes had brought from Australia. Mr. Lonsdale, who examined them immediately afterwards, said, "They are certainly not Nummulites, but allied to some of the bodies usually termed Orbitolites, and are, I believe, *corals*, in the usual acceptation of that word." Afterwards Mr. Forbes having compared the American fossil with the living species from Australia, and satisfied himself of its near affinity, sent me the following note, dated June 14th, 1847:—

"*On the so-called* NUMMULITES MANTELLI.

"The American '*Nummulites Mantelli*,' judging from Mr. Lyell's specimen, is not a Nummulite, nor is it a foraminiferous shell. It is a species of *Orbitolites*, and consequently a Zoophyte (probably Ascidian). The genus *Orbitolites* was established by Lamarck for the reception of a fossil of the Paris basin, the *Orbitolites complanata*, which may be regarded as the type. Other tertiary species and a Maestricht fossil were associated by Lamarck in the same genus, in which he also placed the '*Orbitolites marginalis*' of the European seas. Respecting the true position of the last-named body, however, there is considerable doubt.

"The *Orbitolites complanata* is very nearly allied to the American

* Sir R. Murchison announced to the Geological Society of London at their meeting May 26th, 1847, that he had just received a letter from M. Agassiz, in which he stated "that M. Desor had clearly shown that the rotten limestone of Alabama was not cretaceous, as Morton and Conrad had supposed, nor Eocene, as Lyell had considered it, but was of the age of the *Terrain Nummulitique* of Biaritz."

fossil. The *Orbitolites elliptica* of Michelin, from near Nice, and that author's *Orbitolites Pratti*, are also closely allied species.

"In British strata, species of *Orbitolites* are recorded from the greensand of Milber Down, from the chalk of Lewes and from the coralline crag of Sutton. It is possible however that bodies belonging to distinct genera have been placed together in our lists.

"Mr. Jukes has collected at Swan River in Australia numerous disciform bodies, apparently Ascidian Zoophytes, which occur there in great numbers upon marine plants resembling Zostera, and when dead are found in great abundance in mud, procured by the dredge from various depths under seventeen fathoms. These discs are usually about half an inch in diameter and are composed of minute cells. They appear to me to belong to the same generic group with the tertiary Orbitolites, and such appears also to have been the opinion of Defrance, for we can scarcely doubt that these are the bodies alluded to by him (in the following passage) as living in the seas of New Holland: 'Cette espèce (i. e. *Orbitolites complanata* of the Paris basin) a les plus grands rapports avec celle que l'on trouve . . . . . . vivant dans les mers de la Nouvelle Hollande.' (Dict. des Sc. Nat. t. 36, Art. Orbitolite.) Marginopora of Quoy and Gaimard seems to be a similar if not identical body.

"As the subject stands at present, then, we have no right to infer from the presence of an Orbitolite, however abundant, that the stratum in which it occurs belongs to one period more than another, between the commencement of the cretaceous epoch and our own times*."

A few days after I had received this communication from Mr. Forbes, a letter reached me from M. D'Orbigny, of which I subjoin a translation:—

"Dear Sir, "Paris, 18th June, 1847.

"I have been long acquainted with the fossil body which you forwarded to me, and at this moment I am printing, in an elementary work, all the mistakes concerning it; it is, in fact, of all genera that perhaps which has been most often misunderstood, and I should call it the greatest culprit in geology. It is a genus nearly allied to *Orbitolina*, and which I have named, in consequence of this analogy, *Orbitoides*. It has always been taken for a nummulite, though it differs from it by the most marked characters. I have known many species, such as the *O. media*, *papyracea*, and that which you have forwarded to me, and which I had designated by the name of *Americana*. The *Orbitoides* are found in the cretaceous and tertiary formations, the *Nummulina* in the tertiary only. Such at least is the result of my numerous investigations on this subject. The species that you have forwarded to me had been sent me from North America, with a great number of tertiary and cretaceous shells; it came to me without any information respecting it, and I am anxious to know where you found it. "Yours, &c.,

"*To C. Lyell, Esq.*" "Alcide D'Orbigny."

* The *Plagiostoma dumosum* of Morton is decidedly a Spondylus.

The American fossil therefore now under consideration will henceforth be called *Orbitoides Mantelli*, retaining the specific name first given to it by Dr. Morton.

In my former paper I endeavoured to point out the cause of the obscurity in which the true age of the nummulitic or orbitoidal limestone of Alabama had been involved, it having been considered sometimes as an upper cretaceous group, and at others as intermediate between the cretaceous and the Eocene formations. The accompanying section from Claiborne Bluff to Bettis' Hill near Macon in Alabama may serve to explain the relations which I found to exist between the white limestone group of the south, comprising the successive formations 1. 2. 3, and the overlying group 4, which is perhaps of equal thickness, but which, from the absence of calcareous matter, rarely yields organic remains, and those consisting only of silicified casts of shells and corals. This upper formation (4) is composed of variously coloured red, pink and white sands, and yellow ochreous coloured sands, white quartzose gravel and sand with beds of chert and flint, blood-red and pink clays, and clays of white kaolin or porcelain earth, all horizontally stratified. I could find no fossils in those in Alabama, and only conjectured that they are of Eocene date from the analogy of Georgia, where a deposit of the like aspect and nature and occupying a similar position contains Eocene shells and corals. I formerly explained in 1841–42 the relative position of the upper clays and sands with flint (the Burr-stone formation of Georgia) to the underlying white limestone and marl of the State of South Carolina, in a diagram published in the Journal of the Geol. Soc. vol. i. p. 438, where the newer group is represented as resting on the eocene limestone of Jacksonborough near the Savannah river. It appeared in that case as in Alabama, that the older calcareous strata of limestone and marl had undergone great denudation, and had acquired a very uneven surface, having been shaped into hills and valleys before the incumbent clays and sands were thrown down.

At the bluff on the Alabama river at Claiborne, where so rich a harvest of fossils has been obtained, especially in the beds of No. 2, we see at one spot called "The Old Landing," that nearly the whole precipice in its lower 160 feet, exposes to view the calcareous beds 1. 2. covered with about 20 feet of red clay and sand, whereas at the distance of less than a mile from this spot, the upper formation No. 4 occupies more than 100 feet of the face of the same cliff from its summit, while at the base the lower members of the calcareous series crop out from beneath the horizontal and incumbent beds of sand and clay. This twofold composition of the mass of strata in the bluff at Claiborne is expressed at A in the annexed woodcut, and I verified a similar mode of juxtaposition of the two series of beds in several places in the interior of Clarke County, where the limestone often ends abruptly, and is succeeded sometimes in the same ridge or hill by the newer beds (No. 4), the latter having evidently filled up the inequalities of a previously denuded deposit, after which the whole was again denuded.

I have suppressed several details and repetitions of the same phænomena in the country represented in the annexed diagram, and have

Fig. 1.

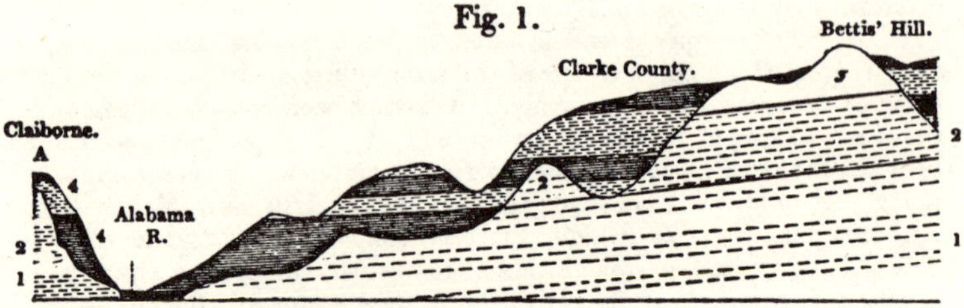

1. Sand, marl, &c. with numerous fossils.
2. White or rotten limestone with Zeuglodon. } Eocene.
3. Orbitoidal or so-called nummulitic limestone.
4. Overlying formation of sand and clay (Eocene?).

been obliged to give a considerable inclination to the strata, because in the distance of twelve or more miles between Claiborne and Bettis' Hill, although the dip is not perceptible to the eye, the same beds are at the latter place more than twice as high above the Alabama river as at Claiborne. The lowest mass, above 100 feet thick, No. 1, which constitutes the lowest visible member of the Eocene series, comprises marly beds with *Ostrea sellæformis*, seen at the base of the cliff at Claiborne, and an argillaceous stratum with impressions of leaves, and sandy beds with marine shells, among which are found *Cardita alta*, *C. planicosta*, *C. parva*, *Crassatella prætexta*, *Cytherea æquorea*, *Oliva Alabamensis*, *Pleurotoma* (several species), *Solarium canaliculatum*, *Crepidula lyrata*, *Endopachys alatum*, Lonsdale, and 200 other species. No. 2, about 50 feet thick, is the white or rotten limestone, which is sometimes soft and argillaceous, but in parts very compact and calcareous, and contains *Flabellum cuneiforme*, Lonsdale, *Scutella Lyelli*, Conrad, *Lunulites*, and several shells, some peculiar, others common, to the formation below. Mr. Conrad has already described this section at Claiborne, and I hope soon to give a fuller notice of it with the observations which I made there in 1846. Of the limestone No. 2, only the lower portion is seen here, for it is cut off at the top of the bluff by the newer series of beds No. 4; but in many parts of Clarke County, as near Bettis' Hill and near Clarksville, the same No. 2 is found more largely developed. It is characterized among other organic remains by a large Nautilus allied to *N. ziczac*, and by the gigantic Zeuglodon of Owen. Near the junction of the mass with the incumbent orbitoidal limestone we find *Spondylus dumosus* (*Plagiostoma dumosum*, Morton), *Pecten Poulsoni*, *Pecten perplanus*, and *Ostrea cretacea* in abundance.

No. 3 is a pure limestone, sometimes hard and full of *Orbitoides Mantelli*. At Bettis' Hill the formation is about 70 feet thick, and the upper beds are composed of a cream-coloured soft stone which hardens on exposure, is not divided by lines of stratification, and is for the most part made up of orbitoides of various sizes with occasionally a lunulite and other small corals, and specimens of *Pecten Poulsoni*. The origin of this limestone, like that of our white chalk, the softer varieties of which it much resembles, is I believe due to the decomposition of corals, and like our chalk downs, the surface of the country where it prevails is sometimes marked by the absence of wood, by which all the other deposits in this part of Alabama are continuously covered. The spots where few or no trees appear are called "bald Prairies," but in some places, and at Bettis' Hill among others, the orbitolite rock produces what is termed a "cedar knowl," the red cedar, *Juniperus virginiana*, having exclusive possession of the ground. I was much struck with the resemblance of such calcareous tracts, covered with the tree above-mentioned, to certain chalk regions in the south of England, where the only wood which grows on the white rock consists of yew-trees accompanied here and there by shrubs of juniper.

At St. Stephen's, on the left bank of the Tombecbee river in Alabama, a similar limestone with orbitoides forms a perpendicular bluff. The water of the river at the time of my visit was too high to enable me to collect fossils from the beds at the base of the cliff, but I was afterwards furnished with them through the kindness of Prof. Brumby of Tuscaloosa. They are imbedded in a ferruginous ochreous-coloured sand, and consist in part of shells common to Claiborne Bluff, such as *Terebra costata*, Conrad, *Cardita parva*, *Dentalium thalloides*, *Flabellum cuneiforme*, Lonsd., *Scutella Lyelli*, Con., and several more. I shall now conclude by adverting briefly to the result of a comparison which I made of the fossils contained in the eocene strata of Vicksburg on the left bank of the Mississippi river, the position of which is indicated at 4 *a* in the annexed woodcut, with those of other eocene beds forty-five miles farther inland or eastward, at Jackson in the same State (3 *b*). In the former of these at 4 *a*, the *Orbitoides Mantelli* abounds, together with *Pecten Poulsoni*, *Dentalium thalloides*, *Sigaretus arctatus*, Con., *Terebra costata*, Con., and a few others common to Claiborne, but the great bulk of the associated fossils do not agree specifically with those of Claiborne Bluff. I found these distinct species of Vicksburg obtained by me to be referable to the genera *Voluta*, *Conus*, *Terebra*, *Fusus*, *Murex*, *Cassis*, *Pleurotoma*, *Oliva*, *Solarium*, *Natica*, *Turritella*, *Corbula*, *Panopæa*, *Crassatella*, *Lucina*, *Venus*, *Cardium*, *Arca*, *Pinna*, *Pecten* and *Ostrea*, with several corals, the whole having a decidedly tertiary and eocene aspect. The genus *Pleurotoma* for example, which is represented by several species, is one of the forms most characteristic of tertiary as distinguished from secondary formations.

At Jackson, which as before stated is more than forty miles to the eastward, older eocene beds crop out near to the area occupied by cretaceous deposits, as at *b*, woodcut No. 2. Here, on the Pearl river, I found no specimens of *Orbitoides Mantelli*, although some are said to have been met with in the vicinity; but I observed that a larger proportion among the fossils were of species common to Claiborne Bluff than at Vicksburg. Among these may be mentioned *Cardita planicosta*, *Cardita rotunda*, *Cytherea æquorea*, *Natica* like one which I collected at Claiborne, *Flabellum cuneiforme*, Lonsd., and *Endopachys alatum*, Lonsd. (*Turbinolia Maclurii* of Lea); these I found in strata of yellow loam, sand and marl on the Pearl river and in the banks and beds of one of its tributaries. The other shells

Fig. 2.

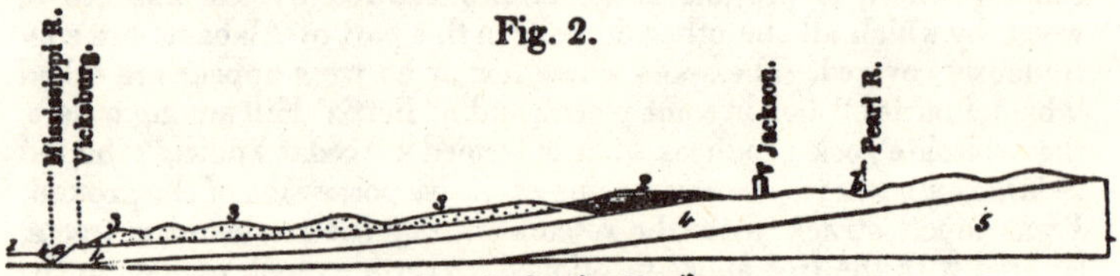

Length of section 50 miles.

1. Mud of alluvial plain of Mississippi.
2. Superficial drift.
3. Freshwater loam with land shells, &c.
4. Eocene strata.
5. Cretaceous strata.

collected by me at the same place, several of them I believe identical with Claiborne species, belong to the genera *Voluta, Oliva, Terebra, Rostellaria, Murex, Umbrella, Natica, Turritella, Crepidula, Dentalium, Corbula, Mactra, Lucina, Cytherea, Cardium, Cardita, Pectunculus, Nucula, Pinna, Pecten* and *Ostrea*. With these are many corals, teeth of fish, &c. I was shown the remains of Zeuglodon procured from the neighbourhood at a place five miles south of Jackson on the right bank of the Pearl river, but as I did not visit the locality, I cannot point out the precise place in the Eocene series in which it occurs. Some of the accompanying corals however were the same specifically as those occurring with the shells above-mentioned at Jackson, and one of my informants stated that this Zeuglodon bed was immediately under "the rotten limestone."

Nov. 5, 1847.—Since the above was read to the Geological Society, I have seen two papers on the Vicksburg deposits by Mr. Conrad, the first in Silliman's Amer. Journ. 2nd Ser. July 1846, No. 4, p. 124, the second in the same Journ. Sept. 1846, No. 5, p. 210, which I had previously overlooked. Mr. Conrad had not visited Jackson, Mississippi, but his results in regard to the fossils of Vicksburg, of which he evidently obtained a larger collection than mine, agree in the main with my own. I cannot however reconcile some of his statements in regard to the want of identity of species at Claiborne and Vicksburg, and suspect some errors of the press and of the dates of his two memoirs, no reference being made from the one to the other. In the first paper, p. 124, he affirms that not one species is common to Vicksburg and Alabama, yet he mentions *Pecten Poulsoni*, a Claiborne shell, as abundant at Vicksburg with a Nummulite. In the second paper, p. 211, he says that ten species will be found on comparison common to Vicksburg and Claiborne, Alabama.

ON THE FOSSIL FOOTMARKS OF A REPTILE IN THE COAL FORMATION OF THE ALLEGHANY MOUNTAINS

SOCIETIES

Royal Institution.—*Feb.* 4.—The Duke of Northumberland, President, in the chair.—'On the Fossil Footmarks of a Reptile in the Coal Formation of the Alleghany Mountains,' by C. Lyell, F.R.S.—Mr. Lyell began by observing that, notwithstanding the numerous remains of land plants in the carboniferous strata and the evidence they afford of the existence of large tracts of dry land (the exact position of which is often indicated by seams of coal and buried forests), no monuments of any air-breathing creatures had been detected in rocks of such high antiquity until Dr. King, in 1844, published his account of the foot-prints of a reptile occurring in sandstone in Pennsylvania (see *Silliman's Journal*, vol. 48, page 343. These fossil tracks were found in a stone quarry five miles S.E. of Greensburg and about twenty miles E. of Pittsburg, appearing on the under surfaces of slabs of argillaceous sandstone extracted for paving. They project in relief, being casts of impressions formed in a subjacent layer of fine unctuous clay, and they are accompanied by numerous casts of cracks of various sizes, evidently produced by the drying and shrinking of the clayey mud. These cracks occasionally traverse the foot-prints, showing that the shrinkage took place after the animal had walked over the soft mud, and before it had begun to dry and crack. Mr. Lyell exhibited a slab which he had brought from the quarries, having visited them with Dr. King; and then proceeded to point out the differences between these foot-prints and those of the European Cheirotherium found in Saxony and in Warwickshire and Cheshire, always in the upper part of the New Red Sandstone or Trias. In the European hand-shaped foot-marks, from the form of which the animal was called by Kaup, Cheirotherium, both the hind and fore feet have each five toes, and the size of the hind foot is about five times as large as the fore foot. In the American fossil the posterior foot-print is not twice as large as the anterior, and the number of toes is unequal, being five in the hinder and four in the anterior foot; as in the European Cheirotherium the fifth toe stands out nearly at a right angle with the foot and somewhat resembles the human thumb. On the external side of all the Pennsylvanian tracks, both the larger and smaller, there is a protuberance like the rudiment of another toe. The average length of the hind foot is 5½ inches and of the fore foot 4½. The fore and hind feet being in pairs follow each other very closely, there being an interval of about one inch only between them. Between each pair the distance is 6 to 8 inches, and between the two parallel lines of tracks there is about the same distance. In the case of the English and German Cheirotherium the hind and fore feet occur also in pairs, but they form only one row, in consequence of the animal having put its feet to the ground nearly under the middle of its body, and the thumb-like toes are seen to turn to the right and to the left in the alternate pairs; while in the American tracks, which form two parallel rows, all the thumb-like toes in one set turn to the right and in the other set to the left. Mr. Lyell infers, therefore, that the American Cheirotherium belongs to a new genus of reptilian quadrupeds wholly distinct from that which characterizes the triassic strata of Europe; and such a generic diversity, he observes, might have been expected in reptilian fossils of such different ages. The geological position of the sandstone of Greensburg is perfectly clear, being situated in the midst of the Appalachian coal-field, having the main bed of coal, called the Pittsburg seam, a hundred feet above it worked in the neighbourhood, and several other seams of coal at lower levels. The impressions of Lepidodendron, Sigillaria, Stigmaria, and other carboniferous plants, are found both above and below the level of the reptilian footsteps. Mr. Lyell then adverted to some spurious fossil foot-prints of dogs, hoofed quadrupeds, birds, and other creatures seen on the surface of ledges of a soft quartzose sandstone in the neighbourhood of Greensburg, which had been confounded with the fossil ones. He pointed out the proofs that these had been carved by the ancient inhabitants of America, whose graves are seen in the vicinity; and that the Indian hunters had sculptured similar bird-tracks, together with human foot-prints, in solid limestone of the State of Missouri,—the true origin of which was first explained by Mr. D. D. Owen of Indiana.* To illustrate the mode of interpreting fossil foot-prints in geology, Mr. Lyell gave a sketch of the discovery of three distinct species of Cheirotherium in Europe,—and explained how after it had been conjectured by Link that they might belong to gigantic Batrachians, Mr. Owen found, by examining the teeth and bones of reptiles of triassic age, that three different species of air-breathing reptiles of the Batrachian order, referable to a new genus, Labyrinthodon, had existed, both in Germany and England, at that period; their fossil bones indicating that they were air-breathers, and there being as great a disparity in size between the bones of their anterior and posterior extremities as between the fore and hind foot-prints of the several Cheirotheria. To account for the sharpness of the casts of Cheirotherium on the under surfaces of slabs of sandstone, Mr. Lyell adverted to the manner in which he had seen, on the sea-beach, near Savannah in Georgia, a cloud of fine sand drifted by the wind filling up the foot-prints of racoons and opossums, which, a few hours before, had passed along the shore after the retreat of the tide. Allusion was also made to the recent foot-prints of birds called sandpipers (*Tringa minuta*), which Mr. Lyell saw running, in 1842, over the red mud thrown down every tide along the borders of estuaries connected with the Bay of Fundy, in Nova Scotia. These consist both of impressions on the upper surfaces and of casts in relief on the under sides of successive layers of red mud (see Lyell's 'Travels in North America,' vol. ii. p. 166),—of which he has presented a specimen to the British Museum. The ancient foot-prints of more than thirty species of birds found fossil in the New Red Sandstone or Trias of the valley of the Connecticut river, in Massachusetts, were next stated to be analogous to these modern bird-tracks; and the size of the largest, although they indicate a biped more huge than the ostrich, is exceeded in magnitude by the gigantic Deinornis of New Zealand,—of which nearly the entire skeleton has just been found fossil by Mr. Walter Mantell. The absence hitherto of the bones of birds in the ancient American strata of the triassic period appears to Mr. Lyell quite intelligible; for the circumstances which combine to cause foot-prints of sand-pipers in the recent mud of the Bay of Fundy, repeated throughout many superimposed layers, have no tendency to preserve any bones of the same birds,—and none have yet been ever observed in cutting trenches through the red mud, where it has been laid dry by artificial embankments and drained. In all the cases of foot-prints, both fossil and recent, and whether made by quadrupeds or bipeds, the lecturer insisted on the necessity of assuming that the creatures were air-breathers, for their weight would not have been sufficient under water to have made impressions so deep and distinct. The same conclusion is borne out by the evidence derived from the casts of cracks produced in the same strata, by shrinkage, and so generally accompanying the impressions of feet; and it was remarked that similar effects of desiccation are observable in the recent red mud of Nova Scotia, where thousands of acres are dried by the sun in summer, between the spring and neap tides. The ripple mark also so common in strata of every age, and among others in the coal measures, and New Red Sandstone of Germany, England, and America, exemplifies the accurate preservation of superficial markings of strata, often less prominent than those caused by the tread of reptiles or large birds. As the discovery of three species of Cheirotheria was soon followed by the recognition of as many species of Labyrinthodon, so the announcement by Dr. King, in 1844, of reptilian foot-prints in the coal strata of Pennsylvania has been followed by the news lately received from Germany, that in the ancient coal measures of Saarbruck, near Treeves, the antiquity of which is vouched for by Von Dechen, Prof. Goldfuss has found the skeleton of a true saurian. Dr. Falconer, after a cursory examination of the original specimen, has stated his opinion in favour of its reptilian character, and although the evidence has not yet been rigorously tested by the most eminent comparative osteologists of Europe, Mr. Lyell believes that the opinion of Prof. Goldfuss and Dr. Falconer will be confirmed. Such facts should serve to put us on our guard against premature generalizations founded on mere negative evidence, and caution us not to assume the present limits of our knowledge of the time of the first appearance of any class of beings in a fossil state to be identical with the date of the first creation of such beings.

---

* For some account of Mr. Lyell's observations on the Pennsylvanian Cheirotherium see *Quarterly Geological Journal*, 1846, vol. 2, p. 417, and *Silliman's American Journal*, 1845, vol. 48, p. 343.

NOTES ON SOME RECENT FOOT-PRINTS ON RED MUD IN NOVA SCOTIA

3. *Notes on some Recent Foot-prints on Red Mud in* NOVA SCOTIA, *collected by* W. B. WEBSTER *of* KENTVILLE. By Sir CHARLES LYELL, P.G.S.

I BEG to present to the Society some impressions of footsteps which Mr. Webster has sent me, at my request, from the mud-flats of Kentville in Nova Scotia.

In my 'Travels in North America,' vol. i. p. 168. plate 7, I have given a plate and description of some of these foot-prints made by the sandpiper (*Tringa minuta*), which I saw daily running along the water's edge in the Bay of Fundy. The deposit there consists usually of red mud, with which the waters are charged by the undermining of cliffs of red sandstone and soft red marl. The tides rise very high, and when they are lowest, large areas recently overspread with red mud are laid dry, and are often baked in the sun for many days, so that the mud becomes consolidated and retains permanently the impressions of rain-drops, and the tracks of birds and animals which walk over it.

Mr. Webster tells me, that the divisibility of the solid mud into layers arises from the deposit of each tide being separated by a layer of sand or loam thrown down when the tide first rushes along the bank. The sandy particles being the heaviest are first deposited, and then the thin layer of mud on which the birds walk when the tide recedes. On examining these specimens, I perceive that while some of the foot-prints standing out in relief on the under-sides of the slabs are casts of impressions made in a subjacent layer, and therefore do not correspond with the imprints on the upper side, some of the projecting feet-marks, on the contrary, which are almost equally sharp, are formed by the forcing down of the soft mud, beneath which there was a sandy layer.

On some of the specimens will be seen the foot-print of a cat, which was no doubt in pursuit of the birds, and which was walking with its claws drawn in. In this case the weight was so great that the foot has displaced several of the underlying laminæ of mud and sand, and has caused them to be uneven, without effacing the bird-tracks previously formed there. On other specimens, worm-like tracks, similar to those we often meet with on older rocks, are visible.

Some of the specimens are in their natural state, and have remained entire; but most of them have been slightly baked in an oven by Mr. Webster to enable them the better to bear their weight in travelling.

LOWER SILURIAN REPTILE IN CANADA

2. *Lower Silurian Reptile in Canada*, (From Lyell's Anniv. Address before the Geol. Soc. of London, Feb., 1851, p. 59.)—I have not alluded in this Address to the recent discovery of the track of a quadruped imprinted on a lower silurian sandstone in North America. We are indebted to Mr. Logan, now at the head of the Government Survey in Canada, for having carefully determined the position of the rock containing it. The locality is the village of Beauharnois, on the south side of the St. Lawrence, twenty miles above Montreal. The rock, a fine-grained whitish sandstone, quarried for building, belongs to the group called the Potsdam sandstone by the New York surveyors, and lies at the base of the whole fossiliferous series of North America. The markings were first pointed out to Mr. Logan by Mr. Abraham, editor of the Montreal Gazette, who appreciated their geological importance. Assuming the Chelonian origin of these foot-prints, they constitute the earliest indication of reptile-life yet known, and are not only anterior to the most ancient memorials of fish hitherto detected, but agree in date with the first known signs of well-defined organic bodies, such as Lingulæ, met with in the same rock. Professor Owen, of the College of Surgeons, has examined a slab of the sandstone, on the upper surface of which the foot-prints are impressed, together with a plaster cast of the remainder of the continuous trail, in all 12½ feet long, brought to London by Mr. Logan; and the Hunterian Professor has had the kindness to communicate to me the following description.*

"The impressions are more numerous in regular succession than any that have been previously discovered; so that the evidence of their having been made by successive steps, afforded by this succession of corresponding prints at regular intervals, is the strongest we possess. They are in pairs, and the pairs extend in two parallel linear series with a groove midway between the two series. The outer impression of each pair is the largest, and it is a little behind the inner one. Both are short and broad, with feeble indications of divisions at their fore part. They succeed each other at intervals much shorter than that between the right and left pair.

"The median groove is well defined and slopes down more steeply at its sides than towards the bottom, at some parts of the track. I conclude from these characters that the animal which left the track was a quadruped, with the hind-feet larger and wider apart than the fore-feet; with both hind and fore-feet very short, or impeded by some other part of the animal's structure from making any but short steps; that the fore and hind limbs were near each other, but that the limbs of the right and those of the left side were wide apart: consequently, that the animal had a short but broad trunk, supported on limbs either short or capable only of short steps; and that its feet were rounded and stumpy, without long claws.

"As to the median impression, that may be due either to a thick heavy tail, or to the under surface of the trunk, dragged along the ground. The shape of the body and the nature of the limbs, indicated by the above-described characters of the steps, accord best with those of the land or freshwater tortoises, and the median groove might have been scooped out by the hard surface of a prominent plastron.

"The disproportion in the size of the fore and hind-feet is such as we find in some existing Terrapenes, e. g., the *Emys geographica*."

ON IMPRESSIONS OF RAIN-DROPS IN ANCIENT AND MODERN STRATA

WEEKLY EVENING MEETING,

Friday, April 4.

H. R. H. Prince Albert, Vice Patron, in the Chair.

Sir Charles Lyell,

On Impressions of Rain-drops in Ancient and Modern Strata.

Foot-prints of reptiles and birds have been observed on the surface of several ancient strata, accompanied by cracks resulting from the shrinkage of mud during desiccation, and it had been fairly inferred that the rocks bearing these marks must have been formed on a beach, between the level of high and low tide. It might therefore have been presumed that the same combination of circumstances would favour the preservation of impressions left by rain-drops, if any rain had fallen on the surface of the same strata, when in a state of mud or sand. Accordingly, memorials of rain have been met with, and Sir Charles Lyell exhibited specimens of fossil rain and hail-prints, collected by Mr. Redfield of New York, from the New Red Sandstone of triassic age in New Jersey, and others of still older date, obtained by Mr. Richard Brown, from green slabs and sandstones of the Coal Measures of Cape Breton in Nova Scotia.

Casts of rain-drops were first recognized in 1828 by Dr. Buckland on the lower surfaces of slabs of quartzose sandstone, found by Mr. Cunningham in the Storeton Hill quarries in Cheshire, where they are accompanied by shrinkage cracks, foot-prints of Cheirotherium, and ripple marks. Mr. Redfield and Sir C. Lyell observed others at Newark in New Jersey in 1841, in red sandstone and shale; and still finer examples have been since met with at Pompton, in the same state, twenty-five miles from New York, by Mr. Redfield. The Lecturer had also an opportunity of observing, in 1842, that a shower of rain had left numerous impressions on the mud-flats exposed at low water, in estuaries communicating with the Bay of Fundy; and he afterwards obtained a collection of specimens of the hardened mud from Dr. Webster of Kentville, some of which are marked by the drops of a heavy but transient shower which fell on the 21st of July, 1849. The average size of the hemispherical cavities is small, but some of them are no less than half an inch in diameter. Many of them are circular, but in some the longest diameter exceeds the shortest by $\frac{1}{5}$th, or even $\frac{1}{3}$rd. They are surrounded by a small rim of mud, consisting of the matter which has been forcibly expelled from the pit by the falling drop; and this marginal rim sometimes projects as much above the plane of the stratum, as the bottom of the pit extends below it. In those impressions which have been made when the wind was blowing, and when the rain fell obliquely, the cavities are not only of an oval shape, but all deeper at one end than at the other. Foot-prints of birds, and the winding tubular tracks of annelids are seen on the same surface with the rain-prints. On splitting open slabs formed by numerous thin layers deposited by successive tides, impressions of previous showers are seen, and casts of the same, standing out in relief on the under surface of incumbent layers.

The Lecturer next considered the nature of certain small protuberances, which might, on a cursory view, be mistaken for casts, which project from the upper surface of certain layers of mud, and are caused, some of them by dried bubbles of mud, and others by small particles of solid matter, covered with a film of mud. He also distinguished between the cavities produced by air-bubbles rising up through the mud, which give rise to cavities differing in shape from those formed by rain, as he has proved by several experiments.

In illustration of the foot-tracks of quadrupeds, such as the musk-rat, the minc, the dog and others, so common on the recent red sand of Kentville, on the borders of the Bay of Fundy in Nova Scotia, Sir C. Lyell exhibited a copy of a brick, one foot square from Babylon, now in the British Museum, on which the track of a small animal of the Ichneumon tribe, apparently the Asiatic Mongoose, is distinctly seen. This brick has been sun-dried (not baked in a kiln,) and must have been traversed by the creature, when the clay mixed with straw was still very soft. In the middle of the brick is an inscription in the Babylonian cuneiform character, which according to Colonel Rawlinson's interpretation signifies that Nabokodrossor, King of Babylon, built certain cities, &c. This king is the same as the Nebuchadnezzar of scripture, so that the brick is twenty-four centuries old.

When the tidal waters densely charged with fine sediment creep gently over a slightly inclined sand or mud-bank, they do not disturb the surface, especially when it has been baked hard in the sun, as happens in summer in Nova Scotia; and the new layer of matter which is thrown down, fills up all superficial indentations, which serving as moulds are protected from further disturbance by the casts thus taken from them.

Mr. Cunningham threw out as a conjecture, that the fine-grained quartzose substance of Storeton Hill, might have resulted from blown sand. That such was really its origin, Sir C. Lyell, who has himself examined the quarries on the Mersey, entertains little doubt; for on the sea-shore near Savannah in Georgia, he saw the foot-tracks of raccoons and opossums which had been made in sandy mud at low water, in the course of being gradually filled up with blown sand, clouds of which were swept along by the wind from adjoining cliffs. This layer of sand when the tide rose again would in its turn be overspread by a new deposit of mud.

After describing both the impressions and casts of rain occurring in the recent red mud of the Bay of Fundy, the Lecturer pointed out their close analogy with markings inscribed on triassic slabs of sandstones in New Jersey, on which also ripple-marks, shrinkage cracks and foot-prints of birds have been observed. The character of these ancient impressions may sometimes be seen to vary where the rain has fallen obliquely on rippled surfaces, the cavities being deeper on the windward and shallower on the leeward sides of the ridges formed by the ripple. Casts of rain-prints are seen on the lower surface of several sandstone strata. The direction of the rain is usually distinguishable, the longest diameters of the cavities being all parallel, and their deepest ends all on the same side. The markings attributed by Mr. Redfield to hail, are deep, irregular in form, and extremely angular in outline; and the walls are steeper, especially at the deepest extremity of the excavation where they often overhang.

The carboniferous rain-prints of Sydney, Cape Breton, observed by Mr. Richard Brown, are some of the most delicately sculptured on the laminæ of shale. In some specimens they are quite separate from each other, most of them oval and with distinct rims. Mr. Brown remarks that they only extended over a certain narrow zone, disappearing when the stratum containing them was traced further in each direction; so that they appear to have constituted a narrow belt, as might be expected if they were formed on a sea beach. For when rain falls on recent mud bordering the Bay of Fundy, impressions are only made on one portion of the exposed surface, the upper part of the bank (left dry for ten days or more after the highest spring tides,) being too hard to receive any imprints, and the lower part near the water's edge being too soft. In some shales from Cape Breton, perfect casts are seen projecting from an under surface where the drops are few in number, while in another stratum distinct casts of a heavy shower are preserved in a fine-grained sandstone which presents a warty and blistered appearance. The casts also of small cracks, which must have traversed the subjacent clay, stand out in relief. Together with these memorials of rain are seen numerous winding cylindrical cavities, open at the top and precisely resembling these now formed by annelids on the recent mud of the Bay of Fundy. These strata occur in the same series of beds, in which so many examples of buried forests occur, with the trunks of trees standing erect, and having their roots attached to them. There are also numerous rippled sandstones at different levels in the same formation.

On re-examining the slab which he brought in 1846 from the coal-strata of Greensburg, Pennsylvania, on which Dr. King first found impressions of a carboniferous reptile, Sir C. Lyell finds not only shrinkage-cracks but a multitude of small tubercles covering the surface resembling the casts of rain-prints, and which he can scarcely doubt are referable to pluvial action.

In conclusion, the Lecturer enlarged on the important inferences deducible from the discovery of rain-prints in rocks of such remote antiquity. They confirm the ideas entertained of the humid climate of the carboniferous period, the forests of which we know were continuous over areas several hundreds of miles in diameter. The average dimensions of the drops indicate showers of ordinary force; and show that the atmosphere corresponded in density as well as in the varying temperature of its different currents with that which now invests the globe. The triassic hail, moreover, implies that some regions of the atmosphere were at this epoch intensely cold; and coupled with the foot-prints, worm tracks, ripple-marks, and the casts of cracks formed by the drying of mud, these impressions of rain clearly point to the existence of sea-beaches where tides rose and fell, and therefore lead us to presume the joint influence of the moon and sun. Hence we are led on to infer that at this ancient era, the earth with its attendant planet was revolving, as now, round the sun, as the centre of our system, which probably belonged then as now to one of those countless clusters of stars with which space is filled.

C. L.

ON FOSSIL RAIN-MARKS OF THE RECENT, TRIASSIC, AND CARBONIFEROUS PERIODS

2. *On* Fossil Rain-marks *of the* Recent, Triassic, *and* Carboniferous Periods. By Sir Charles Lyell, F.R.S., G.S.

Mr. John Cunningham, F.G.S., in the year 1839, read a paper before the Geological Society on Impressions and Casts of Drops of Rain discovered in the quarries of Lower New Red Sandstone at Storeton Hill, Cheshire*. After he had inferred their pluvial origin, he pointed out the indentations on the spot to Dr. Buckland, who recognized the correctness of his interpretation.

When, in 1841, I visited the quarries of new red sandstone at Newark, in New Jersey, in company with Mr. W. C. Redfield, of New York, we observed some very distinct rain-prints on ripple-marked shales. Afterwards, in 1842, I saw similar impressions of recent date, which had been made between high- and low-water mark on the red sand and mud bordering the Basin of Mines, in the Bay of Fundy. Since that period I have been enabled to form a collection of specimens of this mud, hardened in the sun, through the kindness of Dr. Webster of Kentville, to which I shall presently allude. In 1843, Mr. Redfield, in a letter to the author which was read to this Society, stated that he had found impressions of rain-drops in another locality of the new red sandstone, called Pompton, in New Jersey, twenty-five miles from New York†; and in the same year he published in Silliman's Journal an account of the sandstone strata of that place, and of the Ichthyolites contained in them‡. In these beds, many of which are frequently ripple-marked, and which exhibit the foot-prints of

* Proceedings Geol. Soc. vol. iii. p. 99.

† Ibid. vol. iv. p. 23.

‡ Amer. Journ. of Science, vol. xliv. p. 136.

birds, shrinkage-cracks are seen, together with fossil impressions and casts of rain-drops and of hail.

Early in the present year I received from Mr. Richard Brown some fine specimens of rain-marks from the greenish shales of the coal-measures of Cape Breton, Nova Scotia, to which he has made a passing allusion in his excellent description of the Sydney coal-field in our Quarterly Journal*. A comparison of all these specimens has convinced me that the impressions of triassic and carboniferous date, above mentioned, have been correctly referred to the action of rain, and that they are distinguishable from such cavities as are sometimes made by the rising of air-bubbles through mud or sand, with which Mr. Desor, in a memoir recently published, has declared many, if not all, the supposed fossil rain-marks to have been confounded†.

Recent Rain-prints of the Bay of Fundy.

In my 'Travels in North America'‡ some notice is taken of the peculiar combination of circumstances which render the mud-flats, exposed at low-tide in the Bay of Fundy, so peculiarly fitted to receive and retain the foot-prints of animals, or any impressions which may happen to be made on their surface. The sediment with which the waters are charged is extremely fine, being derived from the destruction of cliffs of red sandstone and shale, belonging chiefly to the coal-measures. On the borders even of the smallest estuaries communicating with a bay, in which the tides rise sixty feet and upwards, large areas are laid dry for nearly a fortnight between the spring and neap tides, and the mud is then baked in summer by a hot sun, so that it solidifies and becomes traversed by cracks, caused by shrinkage. Portions of the hardened mud may then be taken up and removed without injury. On examining the edges of each slab, we observe numerous layers, formed by successive tides, usually very thin, sometimes only one-tenth of an inch thick,—of unequal thickness, however, because, according to Dr. Webster, the night-tides, rising a foot higher than the day-tides, throw down more sediment. When a shower of rain falls, the highest portion of the mud-covered flat is usually too hard to receive any impressions; while that recently uncovered by the tide near the water's edge is too soft. Between these areas a zone occurs, almost as smooth and even as a looking-glass, on which every drop forms a cavity of circular or oval form, and, if the shower be transient, these pits retain their shape permanently, being dried by the sun, and being then too firm to be effaced by the action of the succeeding tide, which deposits upon them a new layer of mud. Hence we often find, on splitting open a slab an inch or more thick, on the upper surface of which the marks of recent rain occur, that an inferior layer, deposited perhaps ten or fourteen tides previously, exhibits on its under side perfect casts of rain-prints, which stand out in relief, the moulds of the same being seen on the layer below. But in some cases, especially in the more sandy layers, the markings have

* Vol. vi. p. 119.

† Edinb. New Phil. Journ. for 1850, p. 246.

‡ Vol. ii. p. 126.

been somewhat blunted by the tide, and by several rain-prints having been joined into one by a repetition of drops falling on the same spot; in which case the casts present a very irregular and blistered appearance. Dr. Webster has also sent me some specimens, showing the result of long-continued rain, in which all signs of circular or oval cavities have disappeared, and where no one would suspect the pluvial origin of that slight unevenness which may still be traced on the otherwise level surface.

The finest examples of rain-prints sent to me from Kentville were made by a heavy shower, which fell on the 21st of July, 1849, when the rise and fall of the tides were at their maximum in that small estuary which opens into the Basin of Mines. The impressions (see fig. 1) consist of cup-shaped or hemispherical cavities, the largest being

Fig. 1.—*Recent Rain-prints, formed July* 21, 1849, *at Kentville, Bay of Fundy, Nova Scotia.*

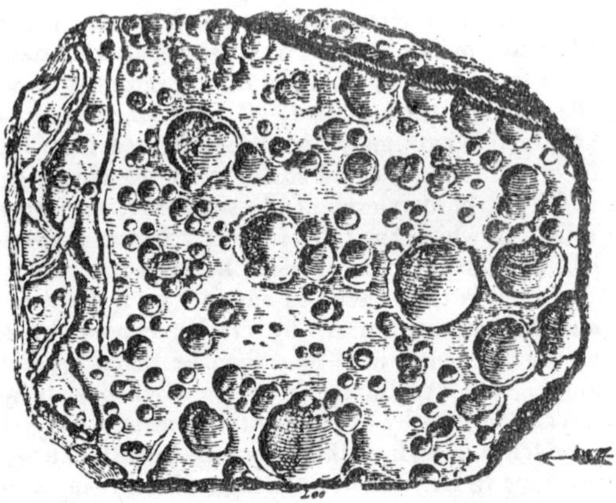

The arrow represents the direction of the shower.

fully half an inch in diameter, and from one-tenth to one-sixth of an inch deep; but there are very few of such dimensions. The depth is chiefly below the general plane of stratification, but the walls of the cavity consist partly of a prominent rim of sandy mud, formed of the matter which has been forcibly expelled from the pit, and this margin or lip sometimes projects as much above the plane of the stratum as the bottom of the pit extends below it. The rim of the largest rain-prints is sometimes no less than one-twelfth of an inch broad, but it is usually much narrower. The outer side of it is often perpendicular or almost overhanging. In the same shower in which the largest drops are half an inch in diameter, the average impressions are only from one-eighth to one-tenth of an inch across. Many of every size are circular, but others are oval, the largest diameter exceeding the shortest by one-third or one-fifth. All the cavities having an elliptical form are deeper at one end, where they have also a higher rim, and all the deep ends have the same direction, showing towards which quarter the wind was blowing. Two or more drops are sometimes seen to have interfered with each other; in which case it is usually possible to determine which drop fell last, its rim being unbroken.

On some of the specimens the winding tubular tracks of worms are seen, which have been bored just beneath the surface (see fig. 1). They occasionally pass under the middle of a rain-mark, having been formed subsequently. Sometimes the worms have dived beneath the surface, and then reappeared. Occasionally the same mud is traversed by the foot-prints of birds (*Tringa minuta*) and of musk-rats, mink, dogs, sheep, and cats. The leaves also of the elm, maple, and oak trees have been scattered by the winds over the soft mud, and, having been buried under the deposits of succeeding tides, are found on dividing the layers. When the leaves themselves are removed, very faithful impressions, not only of their outline, but of the minutest veins, are left imprinted on the clay.

On one specimen of dried mud from Kentville I observed numerous small protuberances which on a hasty view seemed not unlike the casts of rain, but which I satisfied myself could not be such, because they stood out in relief from the upper surface of the mud, on which foot-marks of birds were *indented*. On examining this slab more closely, the protuberances were seen to be irregular in form, and beneath them were found small pellets of shale and crystals of salt, which had evidently lain on the beach, and then been covered with a film of sediment. This solid matter not having shrunk when the muddy layer dried in the sun, a small projection was caused. Small cracks were usually visible round the base of each of these protuberances.

Another set of small convexities, also protruding from the upper surface of the mud, proved to be the crusts of small cavities, each cracked at the top, and were suspected by Mr. Faraday, to whom I showed them, to be bubbles of mud which had dried without bursting. He succeeded in producing similar convex protuberances experimentally, by pounding up the Kentville mud, moistening it with water, and then, by means of glass tubes, introducing air below, which rose to the top in bubbles. Some of these being dried, consolidated without breaking, until finally the crust which covered the cavity where the air had been imprisoned, cracked at the top on shrinking, like the convex protuberances from the Bay of Fundy.

Being desirous of ascertaining whether air-bubbles, rising through mud and bursting as they reached the surface, could give rise to cavities similar to those caused by the fall of rain, I poured some pounded mud from Kentville on a small quantity of water, and shook the basin containing it, upon which numerous bubbles of entangled air rose through the mud, and, on bursting at the surface, left cavities resembling in size the ordinary rain-prints from Nova Scotia, but very different in character. Nearly all of them were perfectly circular, with a very sharp edge, and without any rim projecting above the general surface. In a few cases, however, there was a slight, narrow rim, sharper and more even than that of a rain-print. In no instance was this rim connected with a greater depression at one end of an oval concave depression. Most of the pits produced by these air-bubbles were different also from rain-prints, in being deeper than they were wide. Their sides were very steep, and often overarching, the cavity below the surface being wider than the opening at the top. The axis of some few of these deeper cavities was oblique to the surface of the mud. Where two bubbles had touched, a vertical thin parting wall of mud was left between them.

I also exposed the same mud in a soft state to a shower of rain, and reproduced markings like the smaller rain-prints which had been sent me from the Bay of Fundy. By causing single large drops of water to fall from a sponge upon the same mud from a height of about ten feet, circular pits were formed, similar in size and depth to the largest impressions of rain from Kentville, and having similar rims.

The occasional discovery of a few isolated cavities of the size of rain-prints in ancient strata, seems to have led some geologists to ascribe them to air-bubbles, simply on the assumption that rain would cause the mud or sand to be pitted all over by numerous impressions. But, if the hollows have characters proper in other respects to rain-prints, we ought not to seek for a distinct cause, even although only one or two may appear on a wide area. I saw such isolated cavities, which I feel sure were due to the fall of drops of water, on dried mud on the beach at Baltimore in 1842, and brought away with me a solid specimen. A few heavy drops from a thunder-shower, which had spent its force elsewhere, may have caused them; or one of the numerous water-birds which were flying about the shore may have let fall a drop of salt-water from its wing, or from a fish which it was carrying off in its mouth.

Triassic Rain-prints.

I have alluded to my visit to the quarries of New Red Sandstone or Trias at Newark, in New Jersey, in 1841, where rain-prints occur, and to still finer examples observed by Mr. Redfield in the same formation at Pompton, in New Jersey. In that locality many of the layers of red shale or red clay are ripple-marked and traversed by shrinkage-cracks, and exhibit the foot-prints of tridactylous birds, so common in the shale and sandstone of the valley of the Connecticut. In their average size these ancient rain-prints agree with those of modern date, although none of them equal the largest of those, before mentioned, from the Bay of Fundy. As in specimens of recent mud from Kentville, the triassic strata exhibit every gradation from transient rain, where a moderate number of drops are well preserved (see fig. 2), to a pelting shower, which by its continuance has almost obliterated the circular form of the cavities. In the more perfectly preserved examples, smaller drops are often seen to have fallen into cavities previously made by larger ones, and to have modified their shape. In some cases of partial interference, the last drop has obliterated part of the annular margin of a former one; but in others it has not done so, for the two circles are seen to intersect each other. Most of the impressions are elliptical, having their more prominent rims at the deeper end. We often see on the under side of some of

Fig. 2.—*Rain-prints on red shale, from the New Red Sandstone or Trias of Pompton, New Jersey, U. S.*—W. C. Redfield.

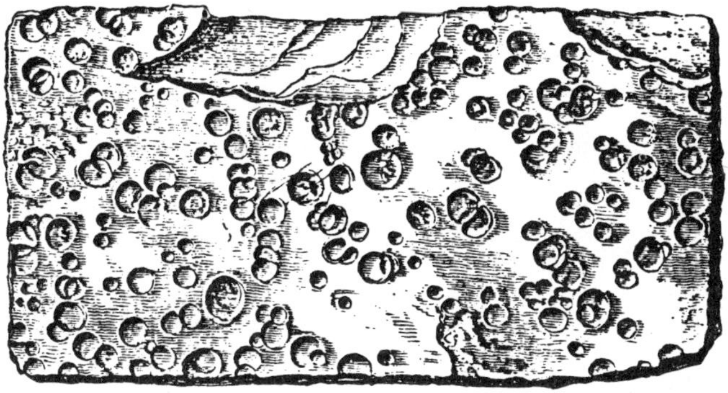

these slabs, which are about half an inch thick, casts of the rain-prints of a previous shower, which had evidently fallen when the direction of the rain was not the same. Mr. Redfield, by carefully observing the obliquity of the imprints in the Pompton quarries, ascertained that most of them implied the blowing of a strong westerly wind in the triassic period at that place. The form of the undulating surface, or the ridges and furrows of the ripple, has sometimes modified the depth of the impressions, which are fainter on the leeward side of each ridge and stronger on the windward slope of the same.

One class of superficial indentations at Pompton has, I believe, been correctly referred by Mr. Redfield to hail. These hail-marks (see figs. 3 and 4) are deeper and much more angular and jagged in

Figs. 3 & 4.—*Supposed Hail-prints on red shale, from Pompton, New Jersey.*—W. C. Redfield.

Fig. 3.

Fig. 4.

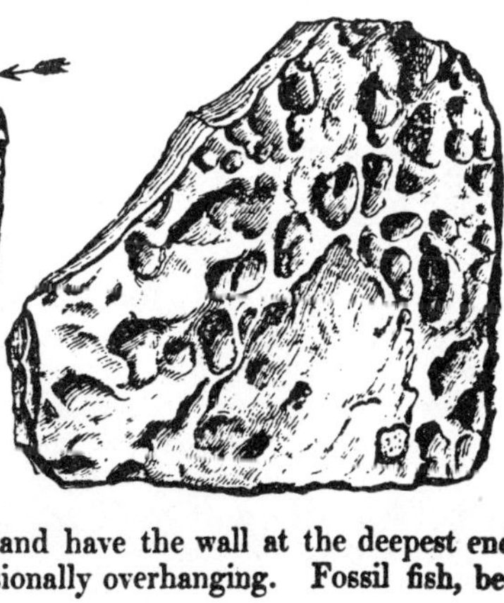

their outline than rain-prints, and have the wall at the deepest end more perpendicular, and occasionally overhanging. Fossil fish, belonging to the genus *Ischypterus*, Egerton, are found in the red sandstone at Pompton.

As some geologists imagine that a great part of the impressions, imputed to rain, may have been simply the effect of air-bubbles rising through mud, I may state that I agree with Mr. Redfield in thinking that, in such examples as those above-described, no practised observer can fall into a mistake. The oval form of so many of the cavities, and their greater depth at one end, are characters alone sufficient to prove the real origin of the imprints on the Pompton shales. It has indeed been objected that where triassic shales alternate with quartzose sandstones, as at Storeton Hill, on the Mersey, near Liverpool (a spot which I have myself visited), we might have expected the rain-marks to have been more blunted by a current of water having exerted sufficient power to spread grains of sand over a large surface of mud. Mr. Cunningham gave an answer by anticipation to this objection, by suggesting that the fine sand of the white quartzose sandstone, in which the casts of rain-prints are preserved, may have been blown by the wind over an area which we may infer to have been above water at the time, because of the depth of the Cheirotherian foot-prints traversing the rain-marked layers. I have noticed in my 'Travels in North America,' that on the beach at Beauly, in the delta of the Savannah river, in Georgia, I saw numerous foot-tracks of racoons and opossums made on the sandy mud, where the animals had come down to the sea-shore to feed on oysters. These trails had been formed during the four preceding hours or after the ebbing of the tide. The surface of the mud had, by exposure to the air and sun, already acquired in that short time a considerable degree of firmness and consistency, and, while some of the moulds remained empty, others were half-filled with fine blown sand, which had already quite covered up a portion of each trail. The quartzose sand was in this instance derived from a low cliff, formed of tertiary strata so incoherent that clouds of minute grains were swept along by a gentle wind, —admirably exemplifying a process by which perfect casts of foot-marks or of rain may be taken in a matrix capable of being afterwards converted into the hardest quartzose sandstone*.

Carboniferous Rain-prints.

In the sixth volume of the Quarterly Journal of the Geological Society, Mr. Richard Brown communicated to us an accurate and detailed account of the Sydney coal-field in Cape Breton, Nova Scotia. In that paper he called our attention to the occurrence in the coal-measures of thirty underclays with *Stigmaria*, and eighteen examples, at different levels, of the fossil trunks of trees standing erect, or perpendicular to the inclined planes of stratification, the greater part of the trees having their roots attached. Together with these proofs of the existence of forest-covered land on that area at different periods during the accumulation of the coal, he has described several intercalated strata of a marine or brackish-water character. We are pre-

* Travels, vol. i. p. 166.

pared, therefore, to hear of the discovery in such a region of the signs of all the phænomena which characterize sea-beaches or banks of sand or mud laid dry between high and low water. Accordingly Mr. Brown alludes in his memoir not only to ripple-marked surfaces and shrinkage-cracks, but to fossil rain-prints, and of these last he has had the kindness to send me some fine examples, which I shall now describe. They consist of delicate impressions on greenish slates (see fig. 5),

Fig. 5.—*Carboniferous Rain-prints, with Worm-tracks, on green shale, from Cape Breton, Nova Scotia.*—Richard Brown.

Fig. 6.—*Casts of the same Rain-prints as No. 5, seen on the under side of an incumbent layer of shale.*

Fig. 5.

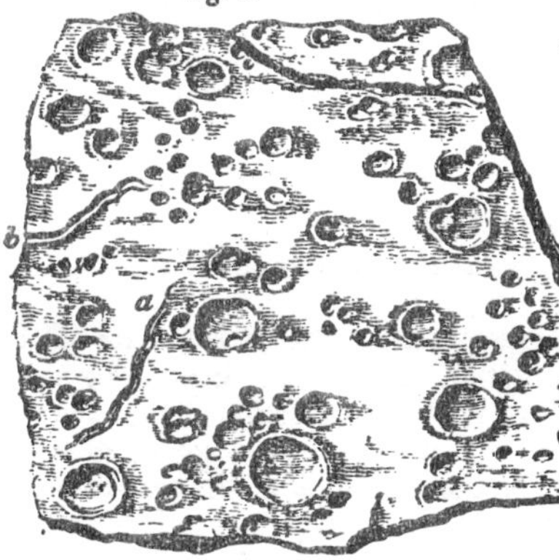

Fig. 6.

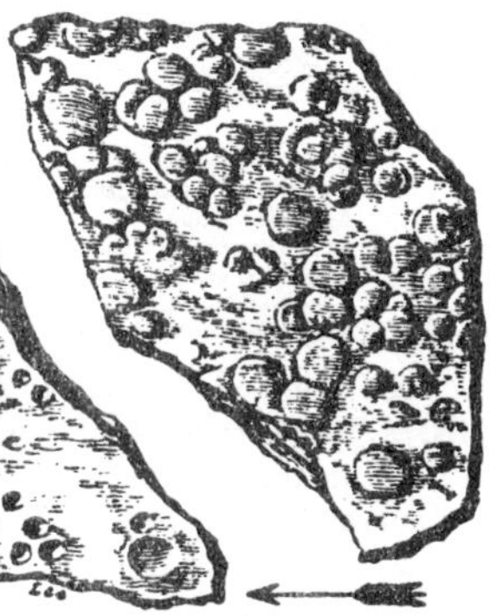

and of casts projecting from the under side of similar shale resting upon them (see fig. 6). Some of these casts present the same warty or blistered surface (see fig. 7) which Mr. Cunningham has mentioned,

Fig. 7.—*Casts of Carboniferous Rain-prints and Shrinkage-cracks on the under side of a layer of sandstone, Cape Breton, Nova Scotia.*—Richard Brown.

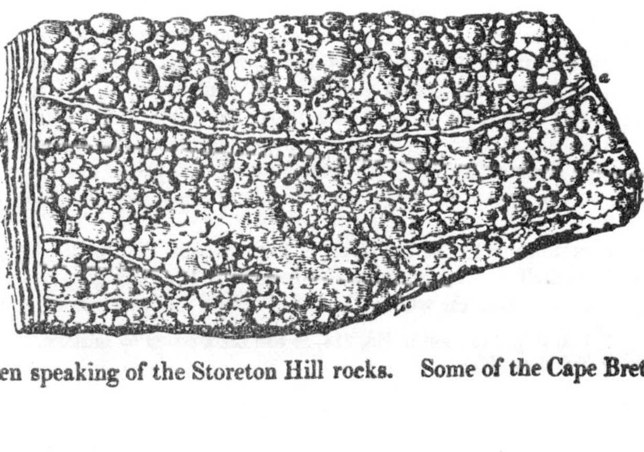

when speaking of the Storeton Hill rocks. Some of the Cape Breton specimens came from the bed No. 213 of the section published in our Quarterly Journal*. In the annexed diagram, fig. 8, Mr. Brown

Fig. 8.—*Section showing the position of Rain-prints* (*b, b*) *in the inclined Coal-measures of Cape Breton, Nova Scotia.*—Richard Brown.

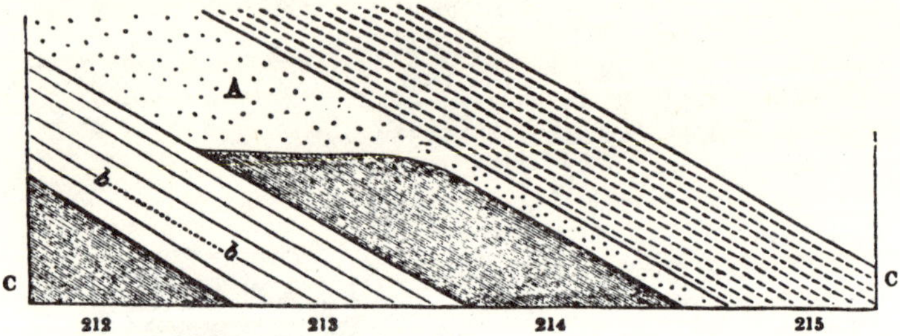

b, b. Position of the Rain-prints.
C, C. High-water mark.
212. Red shale.
213. Greenish arenaceous shales with Rain-prints.
214. Red shale.
A. Sandstone, the upper layers of which pass over the red shale 214.
215. Arenaceous shale.

has been enabled to give, in consequence of the altered state of the cliffs, a more correct representation of the beds containing rain-prints than that before published, for he has discovered that the upper layers of sandstone, A, about 1 foot in thickness, spread over the shale 214, towards the dip. The exact position of the rain-drops in the laminated and rippled arenaceous shale, No. 213, is indicated in the annexed diagram, at *b. b.* fig. 8. Some of the layers in No. 213 consist of almost pure sandstone. Here, as in another instance in the same line of coast, the rain-prints, which are very perfect in a higher part of the cliff, become first indistinct lower down and then disappear, showing, says Mr. Brown, that they constituted originally a narrow zone, as they would naturally do on a sea-beach. In the drawing given of some of these drops (fig. 5), many of them are seen to be very oval in form, and their deepening at one end shows the direction of the wind, as indicated by the arrow.

In the same manner as in the recent mud of the Bay of Fundy, the tracks of annelids are seen in the carboniferous shale on the same surface with the rain-prints (see fig. 5), and some of the ancient worm-tracks disappear and rise again to the surface as do those of modern date. A drawing is also annexed of one of the sandstones from Cape Breton (fig. 7), with casts both of rain-prints and of small cracks, which must have traversed the subjacent clay on which the rain fell.

In re-examining the slab figured in my 'Second Visit to the United States†,' which I brought in 1846 from the coal-strata of Greensburg, Pennsylvania, on which Dr. King found impressions of a car-

* Vol. vi. p. 119; not in No. 214, as was there stated by mistake.
† Vol. ii. p. 306.

boniferous reptile*, I find not only shrinkage-cracks, but a multitude of small tubercles covering a part of the surface, much resembling the casts of rain-prints, and which I strongly suspect to have been due to pluvial action.

The luxuriant vegetation of the coal-period, and especially the continuity of its forests for hundreds of miles, as well as the botanical character of its flora, had previously led botanists to infer a humid climate,—still it is satisfactory to obtain positive proofs of showers of rain, the drops of which resembled in their average size those which now fall from the clouds. From such data we may presume that the atmosphere of one of the remotest periods known in geology corresponded in density with that now investing the globe, and that different currents of air varied then, as now, in temperature, so as to give rise by their mixture to the condensation of aqueous vapour. The hail, moreover, of the triassic period, if correctly deciphered by Mr. Redfield (which I see no reason to question), affords another point of analogy between the meteoric agency of ancient and modern times, implying that certain regions of the atmosphere were occasionally intensely cold.

Since the above was in type, my attention has been called to a notice by Dr. Buckland, "On cavities caused by air-bubbles on the surface of soft clay," which, he justly observed, "must be carefully distinguished from impressions made by rain†."

ON THE DISCOVERY OF SOME FOSSIL REPTILIAN REMAINS AND A LAND-SHELL IN THE INTERIOR OF AN ERECT FOSSIL-TREE IN THE COAL MEASURES OF NOVA SCOTIA

[FOR THE USE OF MEMBERS.]

Royal Institution of Great Britain.

1853.

WEEKLY EVENING MEETING,

Friday, March 18.

William Pole, Esq., M.A., F.R.S., Treasurer, and Vice-President, in the Chair.

Sir C. Lyell,* F.R.S., V.P.G.S., &c.

On the discovery of some fossil reptilian remains, and a land-shell in the interior of an erect fossil-tree in the Coal measures of Nova Scotia, with remarks on the origin of Coal-fields, and the time required for their formation.

The entire thickness of the carboniferous strata, exhibited in one uninterrupted section on the shores of the Bay of Fundy, in Nova Scotia, at a place called the South Joggins and its neighbourhood, was ascertained by Mr. Logan, to be 14,570 feet. The middle part of this vast series of strata having a thickness of 1400 feet, abounds in fossil forests of erect trees together with root-beds, and thin seams of coal. These coal-bearing strata were examined in detail by Mr. J. W. Dawson of Pictou, and Sir C. Lyell in September last (1852), and among other results of their investigations they obtained satisfactory proof that several Sigillariæ standing in an upright position, or at right angles to the planes of stratification, were provided with stigmariæ as roots. Such a relation between Sigillaria and Stigmaria had, it is true, been already established by Mr. Binney of Manchester, and had been suspected some years before on botanical grounds by M. Adolphe Brongniart; but as the fact was still doubted by some geologists both in Europe and America, it was thought desirable to dig out of the cliffs, and expose to view, several large trunks with their roots attached. These were observed to bifurcate several times, and to send out rootlets in all directions into the clays or ancient soils in which they had grown. Such soils or underclays with Stigmaria afford more conclusive evidence of ancient terrestrial surfaces than even erect trees, as the latter might be conceived to have been drifted and fixed like snags in a river's bed. In the strata 1400 feet thick above mentioned root-bearing soils were observed at sixty-eight different levels; and like the seams of coal which usually cover them, they are at present the most destructible masses in the whole cliff, the sandstones and laminated shales being harder and more capable of resisting the action of the waves and the weather. Originally the reverse was doubtless true, for in the existing delta of the Mississippi the clays in which innumerable roots of swamp trees, such as the deciduous cypress, ramify in all directions, are seen to withstand far more effectually the excavating power of the river or of the sea at the base of the delta, than do beds of loose sand or layers of mud not supporting trees.

This fact may explain why seams of coal have so often escaped denudation, and have remained continuous over wide areas, since the roots, now turned to coal, which once traversed them would enable them to resist a current of water, whilst other members of the coal formation, when in their original and unconsolidated state consisting of sand and mud, would be readily removed.

The upright trees usually inclose in their interior pillars of sandstone, or shale, or both these substances alternating, and these do not correspond in the thickness of their layers, or in their organic remains, with the external strata, or those enveloping the trunks. It is clear therefore that the trees were reduced while yet standing to hollow cylinders of mere bark, (now changed to coal,) into which the leaves of ferns and other plants with fragments of stems and roots were drifted together with mud and sand during river inundations. The stony contents of one of these trees, nine feet high and twenty-two inches in diameter, on being examined by Messrs. Dawson and Lyell yielded, besides numerous fossil plants, some bones and teeth which they believed were referable to a reptile; but not being competent to decide that osteological question they submitted the specimens to Dr. Jeffries Wyman of Harvard University in the United States. That eminent anatomist declared them to be allied in structure to certain perennibranchiate batrachians of the genera Menobranchus and Menopoma, species of which now inhabit the lakes and rivers of North America. This determination was soon afterwards confirmed by Professor Owen of London, who pointed out the resemblance of some of the associated flat and sculptured bones, with the cranial plates, seen in the skull of the Archegosaurus and Labyrinthodon.* In the same dark-coloured rock, Dr. Wyman detected a series of nine vertebræ, which from their form and transverse processes he regards as dorsal, and believes them to have belonged to an adult individual of a much smaller species, about six inches long, whereas the jaws and bones before mentioned are those of a creature probably two and a half feet in length. The microscopic structure of these small vertebræ was found by Professor Quekett to exhibit the same marked reptilian characters as that of the larger bones.

The fossil remains in question were scattered about the interior of the trunk near its base among fragments of wood, now converted into charcoal, which may have fallen in while the tree was rotting

---

* Professors Wyman and Owen have named the reptile Dendrerpeton Acadianum, Acadia being the ancient Indian name for Nova Scotia.

away, having been afterwards cemented together by mud and sand stained black by carbonaceous matter. Whether the reptile crept into the hollow tree while its top was still open to the air, or whether it was washed in with mud during a flood, or in whatever other manner it entered, must be matter of conjecture. Foot-prints of two reptiles of different sizes have been observed by Dr. Harding and Dr. Gesner on ripple-marked flags of the lower coal measures in Nova Scotia, evidently made by quadrupeds walking on the beach, or out of the water, just as the recent Menopoma is sometimes observed to do. Other reptilian foot-prints of much larger size had been previously noticed (as early as 1844) in the coal of Pennsylvania by Dr. King; and in Europe three or four instances of skeletons of the same class of animals have been obtained, but the present is the first example of any of their bones having been met with in America in rocks of higher antiquity than the Trias. It is hoped however that other instances will soon come to light, when the contents of upright trees, so abundant in Nova Scotia, have been systematically explored; for in such situations the probability of discovering ancient air-breathing creatures seems greater than in ordinary subaqueous deposits. Nevertheless we must not indulge too sanguine expectations on this head when we recollect that no fossil vertebrata of a higher grade than fishes, nor any land-shells, have as yet been met with in the Oolitic coal-field of the James River, near Richmond, Virginia, a coal-field which has been worked extensively for three-quarters of a century. The coal alluded to is bituminous, and as a fuel resembles the best of the ancient coal of Nova Scotia and Great Britain. The associated strata of sand-stone and shale contain prostrate zamites and ferns, and erect calamites and equiseta, which last evidently remain in the position where they grew in mud and sand. Whether the age of these beds be Oolitic as Messrs. W. B. Rogers and Lyell have concluded, or Upper Triassic as some other geologists suspect, they still belong clearly to an epoch when saurians and other reptiles flourished abundantly in Europe; and they therefore prove that the preservation of ancient terrestrial surfaces even in secondary rocks does not imply, as we might have anticipated, conditions the most favourable to our finding therein creatures of a higher organization than fishes.

In breaking up the rock in which the reptilian bones were entombed, a small fossil body resembling a land shell of the genus Pupa, was detected. As such it was recognized by Dr. Gould of Boston, and afterwards by M. Deshayes of Paris, both of whom carefully examined its form and striation. When parts of the surface were subsequently magnified 250 diameters, by Professor Quekett of the College of Surgeons, they were seen to exhibit ridges and grooves undistinguishable from those belonging to the striation of living species of land-shells. The internal tissue also of the shell displayed, under the microscope, the same prismatic and tubular arrangements which characterize the shells of living mollusca.

Sections also of the same showed what may be part of the columella and spiral whorls, somewhat broken and distorted by pressure and crystallized. The genus cannot be made out, as the mouth is wanting. If referable to a pupa or any allied genus it is the first example of a pulmoniferous mollusk hitherto detected in a primary or palæozoic rock.

Sir Charles next proceeded to explain his views as to the origin of coal-fields in general, observing that the force of the evidence in favour of their identity in character with the deposits of modern deltas, has increased, in proportion as they have been more closely studied. They usually display a vast thickness of stratified mud and fine sand without pebbles, and in them are seen countless stems, leaves, and roots of terrestrial plants, free for the most part from all intermixture of marine remains, circumstances which imply the persistency in the same region of a vast body of fresh water. This water was also charged like that of a great river with an inexhaustible supply of sediment, which had usually been transported over alluvial plains to a considerable distance from the higher grounds, so that all coarser particles and gravel were left behind. On the whole the phenomena imply the drainage and denudation of a continent or large island, having within it one or more ranges of mountains. The partial intercalation of brackish water-beds at certain points is equally consistent with the theory of a delta, the lower parts of which are always exposed to be overflowed by the sea even where no oscillations of level are experienced.

The purity of the coal itself, or the absence in it of earthy particles and sand throughout areas of very great extent, is a fact which has naturally appeared very difficult to explain if we attribute each coal-seam to a vegetation growing in swamps, and not to the drifting of plants. It may be asked how during river inundations capable of sweeping away the leaves of ferns, and the stems and roots of Sigillariæ and other trees, could the waters fail to transport some fine mud into the swamps? One generation after another of tall trees grew with their roots in mud, and after they had fallen prostrate and had been turned into coal were covered with layers of mud (now turned to shale), and yet the coal itself has remained unsoiled throughout these various changes. The Lecturer thinks this enigma may be solved, by attending to what is now taking place in deltas. The dense growth of reeds and herbage which encompasses the margins of forest-covered swamps in the valley and delta of the Mississippi, is such that the fluviatile waters in passing through them, are filtered and made to clear themselves entirely before they reach the areas in which vegetable matter may accumulate for centuries, forming coal if the climate be favourable. There is no possibility of the least intermixture of earthy matter in such cases. Thus in the large submerged tract call the "Sunk Country," near New Madrid, forming part of the Western side of the valley of the Mississippi, erect trees have been standing ever since the year 1811-12, killed by the great earthquake of that date; Lacustrine and swamp plants have been growing there in the shallows, and several rivers have annually inundated the whole space, and yet have been unable to carry in any sediment within the outer boundaries of the morass.

In the ancient coal of the South Joggins in Nova Scotia, many of the underclays show a network of stigmaria roots, of which some penetrate into or quite through older roots which belonged to the trees of a preceding generation. Where trunks are seen in an erect position buried in sandstone and shale, rooted Sigillariæ or Calamites are often observed at different heights in the enveloping strata, attesting the growth of plants at several successive levels, while the process of envelopment was going on. In other cases there are proofs of the submergence of a forest under marine or brackish water, the base of the trunks of the submerged trees being covered with serpulæ or a species of spirorbis. Not unfrequently seams of coal are succeeded by beds of impure bituminous limestone, composed chiefly of compressed modiolæ with scales and teeth of fish, these being evidently deposits of brackish or salt water origin.

The Lecturer exhibited a joint of the stem of a fresh water reed (*Arundinaria macrosperma*) covered with barnacles, which he gathered at the extremity of the delta of the Mississippi or the Balize. He saw a cane-brake (as it is called in the country) of these tall reeds killed by salt water, and extending over several acres, the sea having advanced over a space where the discharge of fresh water had slackened for a season in one of the river's mouths. If such reeds when dead could still remain standing in the mud with barnacles attached to them, (these crustacea having been in their turn destroyed by a return of the river to the same spot,) still more easily may we conceive large and firmly rooted Sigillariæ to have continued erect for many years in the Carboniferous Period, when the sea happened to gain on any tract of submerged land.

Submergence under salt water may have been caused either by a local diminution in the discharge of a river in one of its many mouths, or more probably by subsidence, as in the case of the erect columns of the Temple of Serapis, near Naples, to which serpulæ and other marine bodies are still found adhering.

Sir Charles next entered into some speculations respecting the probable volume of solid matter contained in the carboniferous formation of Nova Scotia. The data he said for such an estimate are as yet imperfect, but some advantage would be gained could we but make some slight approximation to the truth. The strata at the South Joggins are nearly three miles thick, and they are known to be also of enormous thickness in the district of the Albion Mines near Pictou, more than one hundred miles to the eastward. There appears therefore little danger of erring on the side of excess, if we take half that amount or 7500 feet as the average thickness of the whole of the coal measures. The area of the coal-field, including part of New Brunswick to the West, and Prince Edward's

Island and the Magdalen Isles to the North, as well as the Cape Breton beds together with the connecting strata which must have been denuded or must still be concealed beneath the waters of the Gulf of St. Lawrence, may comprise about 36,000 square miles, which with the thickness of 7500 feet before assumed will give 7,527,168,000,000,000 cubic feet, (or 51,136.4 cubic miles) of solid matter as the volume of the rocks Such an array of figures conveys no distinct idea to the mind; but is interesting when we reflect that the Mississippi would take more than two million of years (2,033,000 years) to convey to the Gulf of Mexico, an equal quantity of solid matter in the shape of sediment, assuming the average discharge of water, in the great river, to be as calculated by Mr. Forshey, 450,000 cubic feet per second, throughout the year, and the total quantity of mud to be as estimated by Mr. Riddell, 3,702,758,400 cubic feet in the year.*

We may, however, if we desire to reduce to a minimum the possible time required for such an operation, (assuming it be one of fluviatile denudation and deposition,) select as our agent, a river flowing from a tropical country, such as the Ganges, in the basin of which the fall of rain is much heavier, and where nearly all comes down in a third part of the year, so that the river is more turbid than if it flowed in temperate latitudes. In reference to the Ganges, also, it may be well to mention, that its delta presents in one respect a striking parallel to the Nova Scotia Coal-field, since at Calcutta at the depth of eight or ten feet from the surface buried trees and roots have been found in digging tanks, indicating an ancient soil now underground; and in boring on the same site for an Artesian well to the depth of 481 feet, other signs of ancient forest-covered lands and peaty soils have been observed at several depths, even as far down as 300 feet and upwards below the level of the sea. As the strata pierced through contained fresh-water remains of recent species of plants and animals, they imply a subsidence, which has been going on contemporaneously with the accumulation of fluviatile mud.

Captain Strachey of the Bengal Engineers has estimated that the Ganges must discharge 4½ times as much water into the Bay of Bengal, as the same river carries past Ghazipore, a place 500 miles above its mouth, where experiments were made on the volume of water and proportion of mud by the Rev. Mr. Everest. It is not till after it has passed Ghazipore, that the great river is joined by most of its larger tributaries. Taking the quantity of sediment at one-third less than that assigned by Mr. Everest for the Ghazipore average, the volume of solid matter conveyed to the Bay of Bengal would still amount to 20,000 millions of cubic feet annually. The Ganges therefore might accomplish in 375,000 years the task which it would take the Mississippi, according

* See Principles of Geology, 8th Ed. p. 219.

to the data before laid down, upwards of two million years to achieve.

One inducement to call attention to such calculations is the hope of interesting engineers in making accurate measurement of the quantity of water and mud discharged by such rivers as the Ganges, Brahmapootra, Indus, and Mississippi, and to lead geologists to ascertain the number of cubic feet of solid matter, which ancient fluviatile formations, such as the coal-measures, with their associated marine strata, may contain. Sir Charles anticipates that the chronological results, derived from such sources, will be in harmony with the conclusions to which botanical and zoological considerations alone might lead us, and that the lapse of years will be found to be so vast as to have an important bearing on our reasonings in every department of geological science.

A question may be raised, how far the co-operation of the sea in the deposition of the Carboniferous Series might accelerate the process above considered. The Lecturer conceives that the intervention of the sea would not afford such favourable conditions for the speedy accumulation of a large body of sediment within a limited area, as would be obtained by the hypothesis before stated, namely, that of a great river entering a bay in which the waves, currents, and tides of the ocean should exert only a moderate degree of denuding and dispersing power.

An eminent writer, when criticizing, in 1830, Sir Charles Lyell's work on the adequacy of existing causes, was at pains to assure his readers, that while he questioned the soundness of the doctrine he by no means grudged any one the appropriation of as much as he pleased of that "least valuable of all things, past time." But Sir Charles believes, notwithstanding the admission so often made in the abstract of the indefinite extent of past time, that there is practically speaking a rooted and perhaps unconscious reluctance on the part of most geologists to follow out to their legitimate consequences the proofs daily increasing in number of this immensity of time. It would therefore be of no small moment could we obtain even an approach to some positive measure of the number of centuries which any great operation of nature such as the accumulation of a delta or fluviatile deposit of great magnitude may require, inasmuch as our conceptions of the energy of aqueous or igneous causes or of the powers of vitality in any given geological period must depend on the quantity of time assigned for their development.

Thus, for example, geologists will not deny that a vertical subsidence of three miles took place gradually at the South Joggins, during the carboniferous epoch, the lowest beds of the Coal of Nova Scotia like the middle and uppermost consisting of shallow-water beds. If then this depression was brought about in the course of 375,000 years it did not exceed the rate of four feet in a century, resembling that now experienced in certain countries where whether the movement be upward or downward it is quite insensible to the inhabitants, and only known by scientific inquiry. If, on the other hand, it was brought about in two millions of years according to the other standard before alluded to, the rate would be only six inches in a century. But the same movement taking place in an upward direction would be sufficient to uplift a portion of the earth's crust to the height of Mont Blanc or to a vertical elevation of three miles above the level of the sea. In like manner, if a large shoal be rising, or attempting to rise, in mid-ocean at the rate of six inches or even four feet in a hundred years, the waves may grind down to mud and sand and readily sweep away the rocks so upraised as fast as they come within the denuding action of the waves. A mass having a vertical thickness of three miles might thus be stripped off in the course of ages, and inferior rocks laid bare. So in regard to volcanic agency a certain quantity of lava is poured out annually upon the surface, or is injected into the earth's crust below the surface, and great metamorphic changes resulting from subterranean heat accompany the injection. Whether each of these effects be multiplied by 50,000, or by half a million or by two million of years, may entirely decide the question whether we shall or shall not be compelled to abandon the doctrine of paroxysmal violence in ancient as contrasted with modern times. Were we hastily to take for granted the paroxysmal intensity of the forces above alluded to, organic and inorganic, while the ordinary course of nature may of itself afford the requisite amount of aqueous, igneous, and vital force, (if multiplied by a sufficient number of centuries,) we might find ourselves embarrassed by the possession of twice as much mechanical force and vital energy as we require for the purposes of geological interpretation.

[C. L.]

ON THE REMAINS OF A REPTILE (DENDRERPETON ACADIANUM, WYMAN AND OWEN) AND OF A LAND SHELL DISCOVERED IN THE INTERIOR OF AN ERECT FOSSIL TREE IN THE COAL MEASURES OF NOVA SCOTIA

January 19, 1853.

John Brogden, Jun., Esq., was elected a Fellow.

The following communications were read:—

1. *On the* Remains *of a* Reptile (Dendrerpeton Acadianum, *Wyman and Owen*) *and of a* Land Shell *discovered in the* Interior *of an* Erect Fossil Tree *in the* Coal Measures *of* Nova Scotia. By Sir Charles Lyell, F.R.S., V.P.G.S. &c., and J. W. Dawson, Esq.

[Plates II., III., IV.]

Descriptions and sections of the Coal-formation of Nova Scotia, and particularly of the cliffs called the South Joggins, have been published by Messrs. Jackson and Alger*, Mr. Brown†, Dr. Gesner‡, Sir C. Lyell§, and Mr. Logan‖. In September last (1852) the authors of the present notice re-visited and re-examined the strata in a part of these cliffs, with a view of ascertaining what may have been the particular circumstances which favoured the preservation of so many fossil trees, at so many different levels, in an erect position; such a position being a rare and very exceptional fact in the coal strata of North America generally. We were also desirous of obtaining additional evidence on a point which is still a matter of controversy in the United States and elsewhere (although we ourselves were already satisfied with the evidence adduced by Mr. Binney¶ and others), namely, the relation of *Stigmaria*, as a root, to *Sigillaria*. We also directed our special attention to the difference of the deposits enveloping the upright trees, and those which fill the trunks themselves, forming casts within a cylinder of bark now turned to coal, the central wood of the trunk having decayed. We searched diligently for fossils in these stony casts, suspecting that organic bodies preserved in so peculiar a situation might differ from such as are buried in ordinary subaqueous strata. With this object we employed a labourer to dig out many vertical trees from the cliffs, so that we might break them up and carefully examine their stony contents. To clear out the trees, so as to expose to view both their trunks and roots, is a work of no small bodily labour.

We reserve for a future communication to the Society an account of the general results which we obtained from an inspection of a series of coal-measures 1400 feet in thickness, and shall confine our present remarks to a few strata, not more than 10 feet thick, which enclose certain vertical trees;—one of these trees having Stigmarian roots,

* See Jackson and Alger on the Mineralogy and Geology of Nova Scotia, Sill. Amer. Journ. xiv. p. 305, 1828, and Mem. of Amer. Acad. of Arts and Sciences, vol. i. New Ser. 1833. Cambridge, Mass.

† Reports on Nova Scotia Coal Fields, 1829.

‡ Remarks on the Geology and Mineralogy of Nova Scotia, 1836.

§ Travels in North America, 1845, vol. ii. p. 180.

‖ First Report of Survey of Canada, Appendix, 1845.

¶ See E. W. Binney, on the Dukinfield and St. Helen's *Sigillaria*. Quart. Journ. Geol. Soc. vol. ii. p. 392. 1846.

and another containing in its interior the first examples yet known in America of Reptilian bones, together with what we believe to be a Land Shell, of which last also no instance seems to have been previously observed in rocks of the Carboniferous æra. After we had already, in the course of this investigation, met with fragments of Plants, such as Ferns, *Poacites?*, *Nœggerathia?*, *Sigillaria*, *Stigmaria*, and Calamite in the inside of several trees, we at length found near the base of an upright trunk some dermal plates, together with bones, one of which we conjectured to be the femur of a Reptile, feeling sure at least that it differed from any Fish bones with which we were acquainted. This led us to return next day to examine the same trunk more thoroughly. We accordingly broke up the remaining portion, and found what we imagined might possibly be the jaws and teeth of a *Labyrinthodon*, which we supposed to belong to the same individual as the bones. These osseous remains were scattered about in the interior of the trunk among fragments of wood converted into charcoal, as if they had accumulated while the tree was rotting away. All the fragments were cemented together by a dark-coloured calcareo-argillaceous and sandy matrix. From their position in the tree, we concluded, or rather guessed, the fossil relics to be those of an air-breather, but pretending to no sufficient osteological knowledge to determine a point of such importance, we lost no time in submitting the collection to Professor Jeffries Wyman, of Harvard University, at Boston, who soon declared his opinion that the bones of the extremities resembled those of the Batrachian Reptile called *Menobranchus*, which now inhabits the Ohio river and Lake Champlain. In regard to its dimensions, he estimated this fossil ichthyoid quadruped to have been between 2 and 3 feet in length. In breaking up the fragments of rock, Professor Wyman detected in the same matrix a series of nine small vertebræ, two of them having what appeared to be short ribs connected with them (like those of a Salamander). These vertebræ he believes to be dorsal or lumbar, and to have belonged to an adult individual of a different and much smaller species of reptile, not exceeding 6 inches in length, whereas the *Dendrerpeton Acadianum* was probably 2½ feet long.

Professor Wyman, feeling considerable diffidence as to his determination, in consequence of his limited means of osteological comparison, requested one of the authors to show his notes, together with the specimens, to Prof. Owen, before they were laid before the Geological Society. To this proposal Prof. Owen has promptly and kindly acceded, fully confirming the principal conclusion to which the American anatomist had arrived, namely, that the characters of the fossils are those of Perennibranchiate Batrachians. The Hunterian Professor has also, in compliance with a suggestion of Dr. Wyman, added notes of his own on several points where he differed somewhat in opinion as to anatomical details. Prof. Quekett, of the College of Surgeons, has had the kindness to examine microscopically longitudinal and transverse sections of one of the larger bones; and he states that it exhibits very decidedly Batrachian structure, and that the bone-cells (Pl. III. fig. 8) closely resemble those seen in *Menopoma* and *Menobranchus** (fig. 10). Prof. Quekett has also found similar batrachian structure in the small vertebræ above alluded to. (See Pl. III. fig. 9.)

In breaking up the rock in which the reptilian remains were imbedded, Dr. Wyman found another fossil body, which he and Sir C. Lyell immediately suspected to be a Land-Shell allied to *Pupa* or *Clausilia*, having the central whorls larger than the anterior and posterior ones, and exhibiting the same kind of striation as characterizes the shells of these pulmoniferous mollusks. Dr. Gould, of Boston, detached the same fossil more completely from the matrix (see Pl. IV. figs. 1, 2, 3, 4); but he was unable to discover the mouth, so as to enable a conchologist to determine the genus with precision. He observed that in its form and striation it bore no resemblance to any known marine shell. M. Deshayes, when the fossil was shown to him, declared at once his opinion that it was the shell of a pulmoniferous and terrestrial mollusk, of the same family as that to which *Helix* and *Pupa* belong.

[Doubts, however, having been expressed by other palæontologists (since this paper was read) as to the correctness of this conclusion, the specimen was submitted to Prof. Quekett for microscopical examination. Part of the striated or finely grooved surface, on being magnified 50 diameters (see Pl. IV. fig. 5), presented exactly the same appearance as a portion, corresponding in size, of the surface of the common English *Pupa juniperi* (see Pl. IV. fig. 6). On making sections of such parts of the fossil shell as were not in too crystalline a condition to exhibit structure, Prof. Quekett detected the same prismatic or hexagonal tissue and the same tubular structure as are generally characteristic of the shells of mollusca, and as may be seen in the recent *Pupa*, *P. juniperi*, before alluded to. See Pl. IV. figs. 7–14. One of the sections, moreover, exhibited what may probably have been a portion of the columella and spire somewhat crushed (Pl. IV. fig. 7, upper part), for, as will be seen by the magnified representation of the entire shell (Pl. IV. figs. 2–4), it had been subjected to considerable pressure, and had been slightly flattened and fractured.—March 21, 1853.]

More than usual hesitation has been felt in acceding to the conclusion that this fossil belonged to the family *Helicidæ*, inasmuch as no land-shell had previously been observed in any palæozoic rock; but the mode of occurrence of this body appears in perfect accordance with the idea of its being referable to a pulmoniferous mollusk, since it was lying among the rotten wood in the interior of a hollow tree, the upper part of which doubtlessly projected into the air 8 or 9 feet above the roots, or into the water during river inundations.

In connexion with the reptilian remains, it may not be irrelevant to observe, that Mr. Logan and Dr. Harding have previously called attention to the foot-prints of a quadruped imprinted on sandy slabs,

* In a plate of the volume of the Histological Catalogue, Mus. Coll. Surg., on which Prof. Quekett is now engaged, sections of the bones of recent Batrachians are magnified on a uniform scale, namely 440 diameters, and the identity of size and character of the bone-cells can at once be recognized.

in a lower part of the coal-measures of Nova Scotia, at Horton Bluff, which foot-prints might well correspond in size with the *larger* animal found in the South Joggins tree, assuming that the small vertebræ indicate a distinct species. Dr. Gesner has also mentioned in a letter to one of the authors, that he has discovered in the lower coal-measures of Parsborough a smaller series of foot-prints supposed to be those of an individual about 5 inches long, or of the size of the common small lizard of Nova Scotia. The mark of the tail between the rows of tracks is, he says, very perfect. These impressions deserve notice, as they may possibly have some relation to the smaller of the two reptiles above noticed.

The only indication of reptilian life detected in America previously to this discovery, in rocks of such high antiquity, were confined to foot-prints in the coal of Pennsylvania.

The following section, extracted from our field-notes, being part of a more complete list of the South Joggins strata, is given to enable the reader to understand the exact position of the fossil tree in which the bones occur. We have also subjoined a description of the strata immediately above and below the bed where the bones were found. The locality is that part of the South Joggins coast where vertical trees are most frequent, about half a mile eastward of the Coal Pier. A thickness of 600 feet of strata intervenes between the main coal, called the King's Vein, which is higher in the series, and the reptile-bearing bed, No. 5, see fig. p. 62, there being several thousand feet of well-characterized carboniferous strata both above and below the rocks containing the fossil forests. The dip at the place where the bones were met with is to the south 25° west, at an angle of 20°. The beds in the following section are mentioned in the descending order.

	ft.	in.
1. Grey sandstone (Grindstone), with prostrate carbonized trees and fragments of vegetable matter .... (This bed is about 550 feet below the main coal.)	25	0
2. Alternations of shale and bluish argillaceous sandstone, with two erect fluted trees, one of them with Stigmarian roots spreading in shale 3 feet above next bed	14	0
3. Coal and bituminous shale, with *Sigillaria, Stigmaria, Lepidodendron, Calamites,* and *Poacites* or *Nœggerathia?*..	0	14
4. Underclay, with rootlets of *Stigmaria,* resting on grey shale, with two thin layers of coaly matter..........	8	10
5. Grey sandstone, with three erect trees, one having Stigmarian roots; erect *Calamites* and stems of unknown plants. (*One of the erect trees in this bed contained the bones, teeth, &c. and land-shell above-mentioned.*)	9	0
6. Coal	0	6
7. Underclay, with *Stigmaria* rootlets..................	2	0

The bed No. 1 forms a high cliff and projecting reef, behind which the shore recedes and takes for a short space the direction of the strike of the beds. Owing to this circumstance an unusually large portion of the beds immediately underlying is exposed and accessible. Thus exposed, the sandstone No. 5 shows, on the beach, a stump 1 foot in height and 1 foot in diameter, having its surface-markings destroyed by the sea, but retaining two distinctly marked *Stigmaria* roots, one of them bifurcating at the distance of 2 feet from the stump. In the cliff are seen, in the same bed, *Calamites*, rooted and erect, a plant-stem, somewhat inclined, but apparently having roots attached, and a large vertical tree, 2 feet in diameter at base, and fluted irregularly, but without leaf-scars. This tree springs from the coal No. 6, and extends to the top of No. 5, but the *Calamites* and stems of some unknown plant are rooted at a higher level, and there are *Stigmaria* roots, apparently *in situ*, at three distinct levels in the sandstone. One of these levels corresponds to the broken tops of the Calamite stems, another to their base, and the third apparently to the roots of the first stump seen on the beach. These appearances are shown in the following section, which represents these fossils as seen from the shore above mentioned, which follows the strike of the beds.

Section of part of the South Joggins Strata.

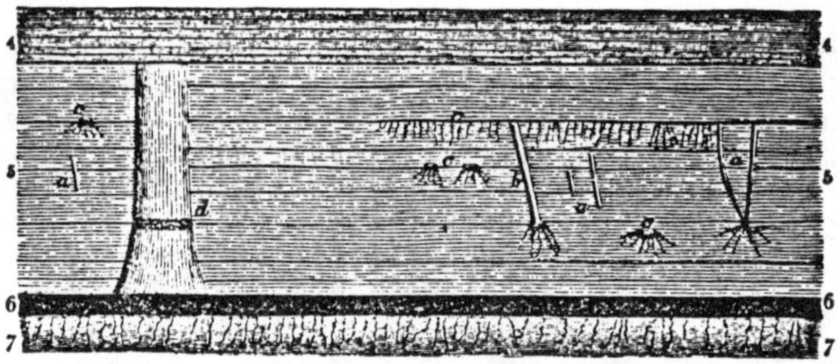

4, 5, 6, 7, beds referred to in the foregoing list of strata.
a, Calamites. c, Stigmaria roots.
b, Stem of plant, undetermined. d, Sigillaria trunk, 9 feet high.

In tracing this bed a little beyond the large tree represented in the section, see fig., we found on the beach another fossil tree, 22 inches in diameter, which had fallen from the bed No. 5, and had no doubt occupied a position similar to that of the large tree still standing. It was a cylindrical cast in sandstone, having on the surface flutings like those of *Sigillaria*, but without leaf-scars. In the lower part of the cast, the sandstone contained a large quantity of vegetable fragments, as above mentioned (p. 59), principally pieces of carbonized wood, leaves of *Nœggerathia* or *Poacites*?, and stems of *Calamites*. With these vegetable remains we found the bones, jaws, teeth, &c. before alluded to, all distinctly *within* the lower part of the cast, and scattered among the vegetable fragments contained in it, as if either washed in in separate pieces, or, more probably, mixed with the woody matter, when the animal fell to pieces through decay. A part of the vegetable matter present must have been introduced after the tree became hollow. The creature to which the bones belonged may, therefore,

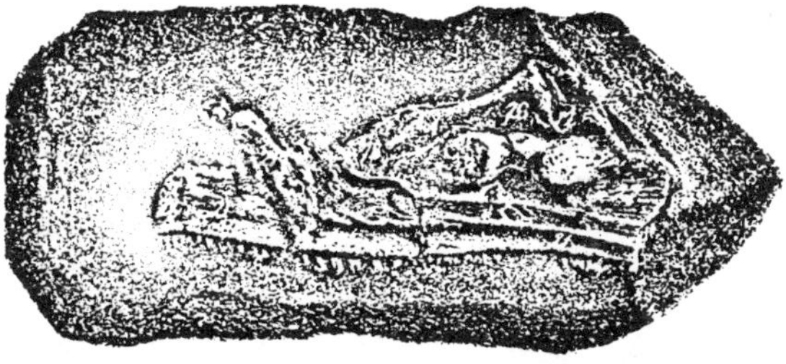

PARABATRACHUS COLEI. ½ nat size

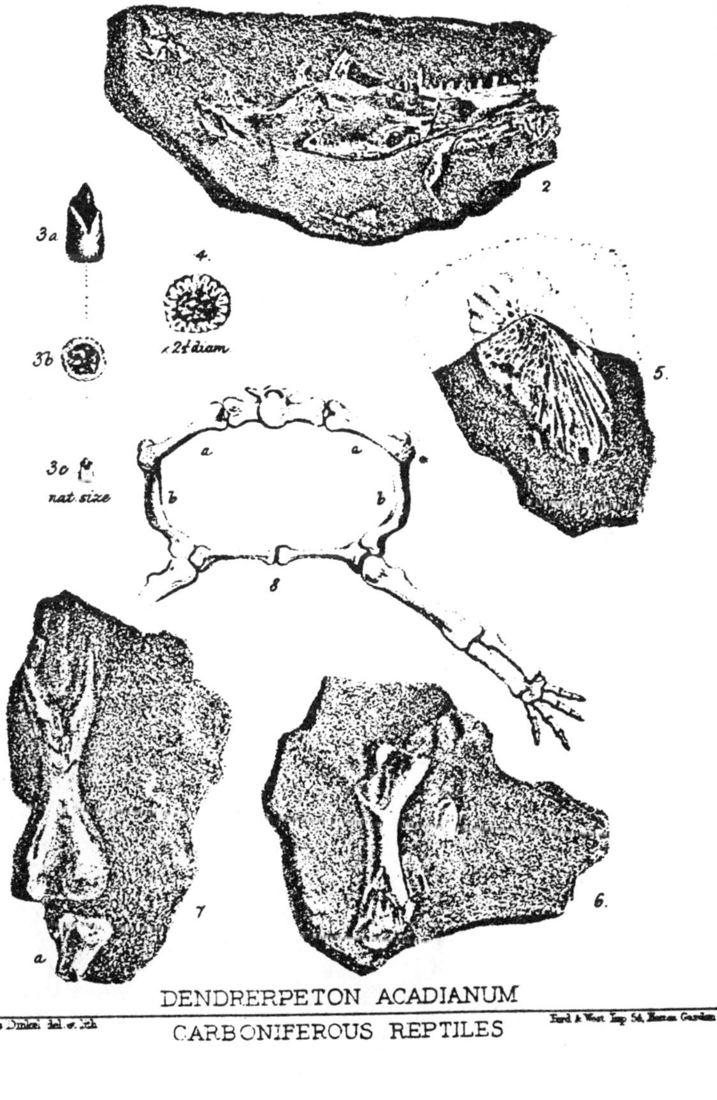

DENDRERPETON ACADIANUM

CARBONIFEROUS REPTILES

Jos Dinkel del et lith

Ford & West Imp 54, Hatton Garden

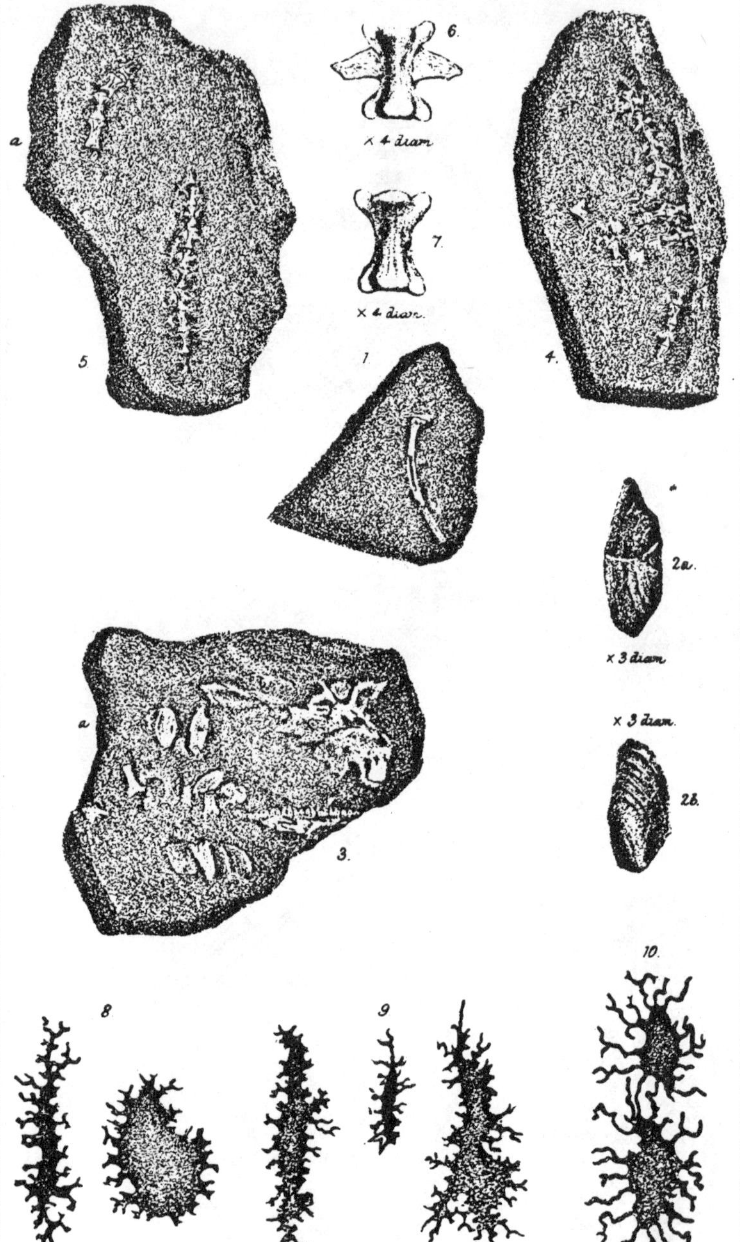

DENDRERPETON ACADIANUM

CARBONIFEROUS REPTILE

Jos. Dinkel del. et Lith.

Ford & West Imp. 54, Hatton Garden

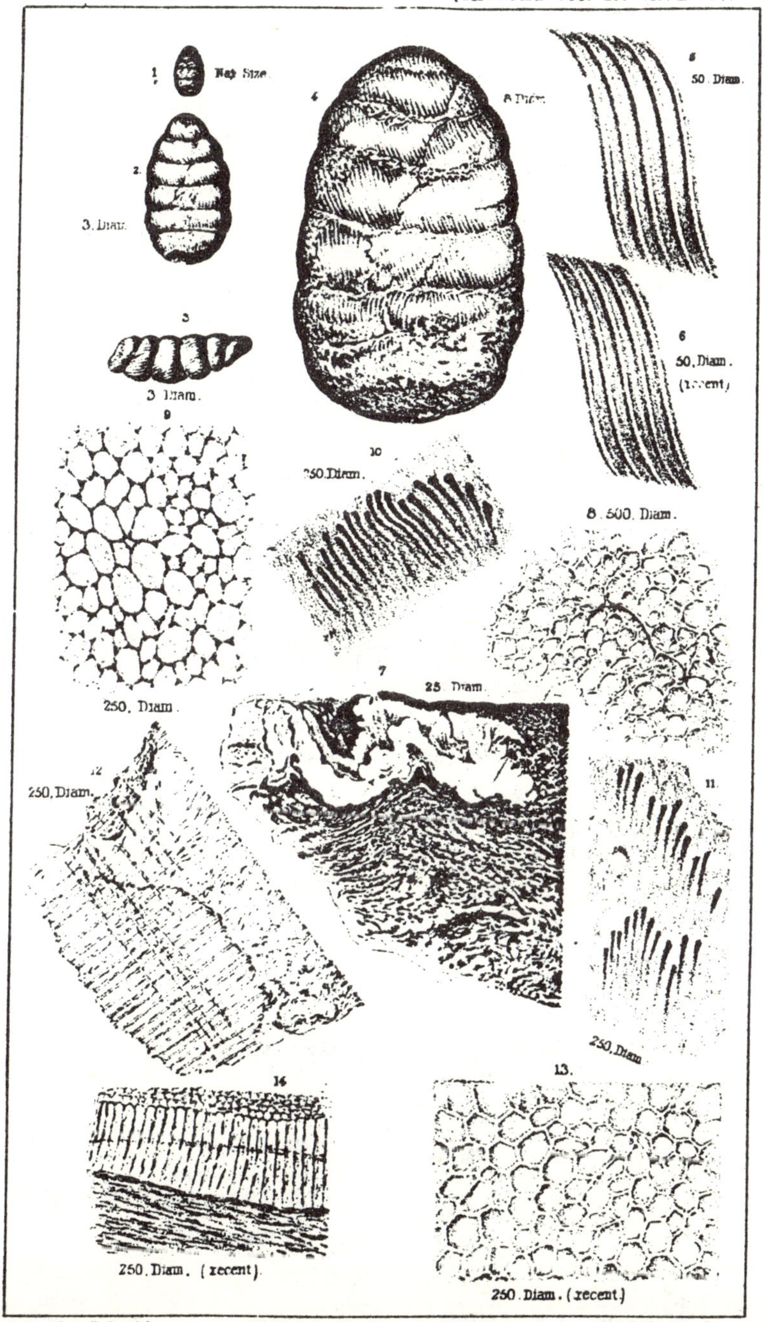

Lens Aldous Del et lith.

M & N Hanhart, Imp.

"CARBONIFEROUS" LAND SHELL.

Nova Scotia.

either have been washed in after death, or may, when creeping on the surface, have fallen into the open pit caused by the decay of the tree, or it may have crept into some crevice in the trunk before it was finally buried in the mud and sand. The fragments of wood found with the bones, when sliced, show a dense irregular cellular tissue, with scalariform ducts, like those of Ferns. They therefore probably belong to *Sigillaria,* and may be parts of the woody tissue of the tree within which they were found.

The other vegetable fossils contained in these beds, and the circumstances under which they grew and were entombed, will be fully discussed in another communication to the Geological Society.

DESCRIPTION OF PLATES II., III., IV.

Plate II.

Fig. 1. Parabatrachus Colei. Portion of skull; half nat. size. 12, Post-frontal bone; 21, maxillary; 22, premaxillary; 26, malar; 27, squamosal; 73, lacrymal.
Fig. 2. Dendrerpeton Acadianum. Portion of lower jaw, with teeth; nat. size.
Fig. 3. Small tooth; *a, b,* magnified; *c,* nat. size.
Fig. 4. Base of larger tooth, magnified 2½ diam.
Fig. 5. Cranial bone; nat. size.
Fig. 6. Iliac bone; nat. size.
Fig. 7. Humerus, with proximal extremity of the radius at *a*; nat. size.
Fig. 8. The bones of the pelvis and hind leg of the *Menobranchus*: *a a, b b,* the two elements of the iliac bone.

Plate III.

Dendrerpeton Acadianum.

Fig. 1. Rib bone; nat. size.
Figs. 2 *a*, 2 *b*. Dermal scutes, magnified 3 diam.
Fig. 3. Pharyngeal (?) teeth, and dermal scutes at *a*; nat. size.
Fig. 4, 5. Vertebræ of the smaller reptile; nat. size.
Fig. 6. Single vertebra; magnified 4 diam.
Fig. 7. Single vertebra from *a*, fig. 5; magnified 4 diam.
Fig. 8. Bone-cells from the iliac bone of the *Dendrerpeton Acadianum*, magnified 440 diameters.
Fig. 9. Bone-cells from the vertebræ of the smaller reptile.
Fig. 10. Bone-cells from the recent *Menobranchus*.

Plate IV.

Fig. 1. Land Shell from the interior of fossil tree, nat. size.
Fig. 2. ———, magnified 3 diam.
Fig. 3. ———, side view, magnified 3 diam.
Fig. 4. ———, magnified 8 diam.
Fig. 5. ———, striæ of the surface. } Magn. 50 diam.
Fig. 6. Recent *Pupa Juniperi*, ditto. }
Fig. 7. Fossil Land Shell, section of part of, magnified 25 diam.
Fig. 8. ————, microscopical structure of the same, showing hexagonal cells, magnified 500 diam.
Fig. 9. ————, section of a chipping, with hexagonal cells. }
Fig. 10. } ————, tubuliferous structure. }
Fig. 11. } } Magn. 250 diam.
Fig. 12. ————, prismatic structure. }
Fig. 13. Recent *Pupa Juniperi*, hexagonal cell-structure of. }
Fig. 14. ————, prismatic cell-structure of. }

NEW YORK INDUSTRIAL EXHIBITION.

SPECIAL REPORT

OF

SIR CHARLES LYELL

Presented to the House of Commons by Command of Her Majesty, in pursuance of their address of February 6th, 1854.

LONDON:

PRINTED BY HARRISON AND SONS.

CONTENTS.

[When "Exhibited products" are not given separately, they are noticed under the several heads of "Economic products."]

CLASS I.—MINING AND MINERAL PRODUCTS.

	Page
INTRODUCTION	3
Mining products	4
Geological maps and surveys ..	4
General geology and physical geography ..	5
Introduction	5
List of formations	7
Alluvial and drift	8
Economical products from the alluvial and drift formations	8
Gold	8
Copper	9
Iron ore	9
Sands and clays	9
Peat and shell-marl, &c. ..	9
Exhibited products	9
Tertiary formations	9
Economical products of the tertiary formations	10
Brown hæmatite and ochreous iron ore..	10
Manganese	10
Fossil fuel	10
Fire-clays and porcelain earth..	10
Shelly marls and green-earth ..	11
Building-materials	11
Cretaceous formation	11
Economic products and agricultural conditions	11
Oolitic or Liassic formation (Jurassic) ..	12
Economical products: coal, &c. ..	12
New red sandstone, shale, and conglomerate (Triassic)	12
Economic products of the new red sandstone formation	13
Building-stones	13
Copper and lead	13
Exhibited products	13
Coal-formation or coal-measures ..	14
Economical products	15
Coal	15
Iron-ores	16
Limestones	17
Sandstones	17
Fire-clays	17
Carboniferous limestone	17
Economical products	18
Lead	18
Building-stones	18
Red shale of the carboniferous limestone formation	19
Economical products: iron	19
Devonian formation	19
Economical products	21
Iron-ore, building-materials, &c. ..	21
Upper Silurian formation	21
Economical products of the Upper Silurian rocks	23
Iron-ore	23
Lead, zinc, and copper ..	23
Building-stone and millstones..	24
Salt	24
Exhibited products	24
Lower Silurian formation	24
Economical products of the Lower Silurian rocks	26
Iron ore	26
Lead ores	26
Copper ores	27
Marble, limestone, and building-stones	27
Grindstones and flagstones ..	27
Exhibited products	27
Upper metamorphic rocks	27
Economical products of the upper metamorphic rocks	29
Gold	29
Iron ores	29
Chromic iron ores and serpentines	29
Copper ores	29
Tin ore	29
Lead ores	29
Silver	29
Building stones	29
Flagging stones	30
Millstones..	30
Whetstones	30
Statuary marble	30
Mica	30
Exhibited products	30
Older metamorphic rocks	30
Economical products of the older metamorphic rocks	31
Iron ores	31
Copper ores	32
Lead ores	32
Zinc ores	32
Phosphate of lime	32
Porcelain earth	33
Marble and building stones ..	33
Firestones, hones, &c. ..	33
Plumbago and crystalline minerals	33
Exhibited products	33
The Trappean formations	33
Economical products of the Trappean formations	35
Copper	35
Silver	35
Building-stone	35
General remarks	35
INDEX to Report on mining and mineral products and general geology of the United States	36 to 38

[*Continued.*

	Page
CLASS X.—Topographical and Hydrographical surveys	39
Introduction	39
Coast survey	39
History of its establishment ..	39
System at present pursued	39
Plan of the survey	40
Constitution of the bureau ..	40
Progress and method of work ..	40
Measurement of base-lines, and apparatus	40
Measurement of angles, and apparatus	41
Observations of latitude, and apparatus	41
Observations of azimuth, and apparatus	43
Observations of local time, and apparatus	43
Observations of longitude, and apparatus	43
Observations of magnetism, and apparatus	44

CLASS X.—(*continued.*)	Page
Observations of height ..	44
Topographical survey	44
Hydrographical survey, and charts ..	45
Soundings	45
Discoveries	45
Observations on tides	45
Exploration of the Gulf Stream ..	46
Survey of the Pacific Coast of the United States	46
Reduction and publication of observations	46
Execution and publication of the charts and maps	46
Exhibited articles	47
Natural History charts	47
Lieut. Maury's Charts	48
Wind and current charts ..	48
Whale Charts	48
Maritime conference at Brussels	48
Support given to Lieut. Maury's plan	50

SIR CHARLES LYELL'S SPECIAL REPORT

ON THE GEOLOGICAL, TOPOGRAPHICAL, AND HYDROGRAPHICAL DEPARTMENTS OF THE EXHIBITION.

Introductory Remarks.

HAVING been requested, when I visited New York as Commissioner in 1853, to report on the Geological Department of the Industrial Exhibition, I have endeavoured to discharge this duty as well as was in my power; but the object has been less fully accomplished than I could wish, owing partly to the delay experienced in the arrangement and display of the collections of mineral and mining products and of geological and geographical maps sent for exhibition. Explanations have already been offered in the General Report of the steps taken by me to overcome the difficulties presented by this delay; and mention has been made of the assistance liberally afforded me by several eminent men of science in the United States. I take this opportunity of repeating my acknowledgments, in particular, to my friend, Mr. James Hall, Geological Surveyor of the State of New York, for the assistance which he gave me, both during my visit and after my return to England. We had previously made, together, in the years 1841, 1846, and 1852, several exploring excursions over different parts of the United States, with a view of studying the geographical and geological features of the country. On the present occasion, Professor Hall's more extensive acquaintance with the geology of the Western States, and his professional experience in all that relates to the economical applications of his science, have enabled me to complete a task which, without such aid, I could scarcely have undertaken.

CLASS I.—MINING AND MINERAL PRODUCTS.

The classification of objects in the New York Crystal Palace being essentially the same as that adopted in the London Exhibition of 1851, we have first to notice the objects of raw production, which belong to the department of Class I, entitled "Mining and Mineral Products." The New York Association for the Exhibition of the Industry of all Nations stated in their prospectus, that it was their wish so to arrange the objects belonging to this class in a separate space assigned to them, as to make them afford information similar to that derived from mineralogical and geological maps, with the additional advantage of having presented to the eye the specimens themselves derived from each mining and mineral region. These collections were also to be accompanied and illustrated by geological maps and sections and by mining plans. Such intention has been to a considerable extent carried into execution under the superintendence of Professor Silliman, jun. When we regard the immense territory of the United States, the incipient condition of many of the mining operations, and the difficulty of interesting parties immediately concerned therewith, we are more surprised at the extent and variety of objects presented, than disposed to dwell on deficiencies observable in the collections which have been brought together and arranged within a few months.

Mining Products.—The mining products which form the most prominent feature of the Exhibition consist of the different varieties of coal and metals, such as iron, lead, copper, zinc, and gold. The coal and iron-ores are derived chiefly from the eastern or Alleghany coal-field; the lead, from limestone and other rocks of silurian age, as well as from the metamorphic formations; the zinc, from metamorphic rocks in New Jersey; and the copper, from the metamorphic regions in the eastern part of the United States, but chiefly from the sandstone and trap regions of Lake Superior, the age of which has until recently been a subject of great difference of opinion among eminent geologists.

Geological Maps and Surveys. — We may here properly mention the geological maps which have been published from time to time, since they give a general view of the several districts which have yielded the various substances used for manufactures, building, ornamental purposes, or agriculture. These maps consist partly of the results of surveys, published and unpublished, undertaken by different States and by the General Government; and partly of those of more general character, in which the geological information obtained during such public surveys, and other knowledge acquired by private exertions, have been set forth.

These maps, and the sections and plans accompanying them, were but meagrely represented in the Exhibition, but some of the more important ones, already in the hands of the public, may be alluded to in connection with the mineral products of the country, the geographical position of which will thus be more readily comprehended.

Within the last twenty years geological surveys have been commenced, and completed in eight of the States, viz., Massachusets, Rhode Island, Connecticut, New Hampshire, New York, New Jersey, Delaware, and South Carolina. During the same period geological surveys have been ordered by legislative authority, and carried forward in a greater or less degree, in fifteen of the States, viz.: Maine, Vermont, Pennsylvania, Maryland, Virginia, North Carolina, Georgia, Alabama, Tennessee, Michigan, Ohio, Indiana, Illinois, and Missouri.

At the present time the surveys of Pennsylvania,* North Carolina,† Alabama, Illinois, Wisconsin, and Missouri are in progress. The others have been suspended after one or more reports of progress had been made. Of those States in which surveys are completed, we have geological maps; and in connection with the survey of Alabama, though still in progress, there is published in the first biennial Report a geological map of the State.

Various surveys and reconnaissances, which have been undertaken by the United States Government with other purposes as the main object, have nevertheless contributed to our knowledge of the geology of the regions over which they have passed. A geological survey of Oregon has recently, however, been instituted by the General Government; and similar surveys of Lake Superior and Chippewa Land Districts have been completed. The former, under the charge of Messrs. J. W. Foster and J. D. Whitney, published in 1851, has furnished us with a geological map of the southern shore of Lake Superior, and of all the northern peninsula of Michigan, and incidentally of some of the adjoining territory, affording a link which shows in a very satisfactory manner the connection between the eastern and western areas of the great silurian formation of the Northern States. The surveys of Dr. D. D. Owen in the Chippewa Land District, published in 1852, have furnished us with a geological map of Wisconsin, Minnesota, Iowa, and a part of Missouri.

Notwithstanding the incomplete condition of many of the surveys of the different States, we have from the reports of progress, and from other information obtained by individual enterprise, a very good general knowledge of the limits of the several geological formations throughout that part of the United

* Renewed after a suspension of several years.

† A geological survey of North Carolina was made many years since by Professor Olmstead, but we believe no map was published at that time. The present survey has been in progress two years.

States east of the Rocky Mountains, and to some extent also of the country west of this range. Of the geological maps which are the results of individual enterprise, or combining the results of public surveys and private investigations, may be mentioned a geological map of North Carolina, by Professor Mitchell, published in 1842; a geological map of the Western States,* by Dr. D. D. Owen, published in the Quarterly Journal of the Geological Society of London, 1842; and a similar map, by Byram Lawrence, in 1842; a geological map of the Middle and Western States, by James Hall, published in the Geological Report of New York in 1843; a geological map of the United States and Canada, by Sir Charles Lyell, 1845; and of part of Canada and New York State, by Mr. W. E. Logan, in the Geological Society's Journal for 1852; a geological map of the United States and British Provinces, by Jules Marcou, and another by Professor Hitchcock, in 1853.

By consulting the Reports of the various States and their accompanying maps, as well as the more general maps here noticed, where this scattered information has been connected and arranged, we are able to refer to their respective geological formations all the materials presented in the Exhibition.

It is much to be regretted that during these geological surveys, a system of triangulation had not been adopted, by which means the topography and geology could have been accurately laid down at the same time. As an unavoidable consequence, the geological maps cannot present in detail the exact limits and variations of outline of the several geological formations, nor the physical features dependent on geological agencies.

General Geology and Physical Geography.

Introduction.—The leading geographical and geological features of the United States exert, as in other countries, a powerful influence on the industrial occupations developed there, and on the resources of different portions of the Union. But it is in the vastness of this influence, and in the great extent of certain formations yielding particular products and devoted to particular industrial pursuits, that the United States present in many respects a different aspect from every other country of the globe. The immense and uninterrupted areas of pre-eminently agricultural character are, for the most part, destitute of valuable mineral products, except limestone and ordinary architectural materials. Certain portions of the country are thus placed at great distances from the resources of manufacturing pursuits, and consequently precluded from their natural advantages. There are other areas, of vast extent, which combine great mineral wealth, and other conditions adapted for manufacturing enterprise, with an excellent soil. Again, there are areas of mineral production situated upon the extreme confines of the inhabited parts of the country; and others which remain untouched from their remote and inaccessible position. These, however, are gradually being added to the available resources of the nation, and are destined one day, by the rapid advance of new settlements, to be in the midst of a populous and wealthy country.

In the existing condition of a country like the United States, it would be impossible to present, in any Exhibition or any Report, more than an approximation to a sketch of its industrial resources. Mining operations are for the most part in their infancy, and in certain regions new mineral resources are constantly coming to light. Indeed, so long as vast tracts of country of great agricultural capacity remain unoccupied, and while the population continues sparse and labour dear, it cannot be expected that mining and manufacturing enterprises can be carried on to the extent which the wants of the country require. The lands of highest agricultural value will necessarily be occupied first, and those regions offering less inducements to the agriculturist will remain for a long time comparatively unknown.

Beyond the mining of coal and iron-ore, and, we may perhaps now add, copper and lead, the mining enterprises undertaken have been more the result of chance and speculation, than of any systematic effort to develop the mineral resources of the country; consequently, much of that part of the United States

* The Western States here spoken of are those situated on the western side of the Appalachian chain.

where we may look for mineral resources of value is very little known, beyond its general geological features, and a few attractive or accidentally discovered localities of mineral wealth.

The geological and mineralogical surveys have mostly had for their object the determination of the limits of geological formations, with a general indication of the contained minerals or metals of value. It has not been the province of the surveyors to examine each district with the detail essential to a determination of the existence of certain ores or minerals in workable quantities, except where these substances formed a prominent feature in the surface-characters of the formation.

The private surveys undertaken with a view to mining operations have been generally limited to a particular locality or moderate area, and these reports do not often go beyond the immediate object of the investigation, viz., the present or prospective value of the tract under immediate examination.

In regard to the mining interests, beyond those of coal and iron, it may be said that the most valuable general information possessed is of a negative character. It is well ascertained that certain great areas, from the nature of the geological formations, are destitute of valuable mineral products. Theoretically, we are prepared to go further, and to say, that with the exception of a certain area comprising parts of Illinois, Wisconsin, and Iowa, where lead is known to exist, ores of gold, silver, copper, lead, zinc, and other metals cannot be sought for successfully beyond certain igneous or metamorphic formations, the limits and extent of which are ascertained.

We therefore know the areas within the limits of which mining operations must be carried on in future; but the present state of these by no means gives an adequate idea of the future magnitude of this department of national industry. In a word, the yearly experience of the United States, and the advances of the nation in this direction, prove very conclusively the broad foundations of her mineral resources.

The physical and geological character of the Eastern* United States may be comprehended in a few sentences. From the north-eastern extremity, stretching south-westerly, nearly to the Gulf of Mexico, lies a belt of mountainous or elevated and more or less broken country, sometimes rising to the height of more than 6000 feet above the level of the sea. Throughout the whole extent it presents evidences of uplifting agencies, which have operated in lines parallel to the direction assumed by the mountain chains. This mountainous country occupies nearly the whole of New England, its numerous indentations affording excellent harbours along the coast. It crosses New York at the Highlands, and to the southward gradually recedes from the coast, leaving a widening belt of more recent formations between it and the Atlantic. The States are limited on the north mainly by the course of the St. Lawrence and the Great Lakes; we have, nevertheless, only to look a little beyond this limit to find a chain of mountains running parallel to the St. Lawrence as far as its source in Lake Ontario, and thence stretching westward, and, penetrating in a few places the geographical limits of the States, it extends, with few interruptions, to the Rocky Mountains. The latter chain has a general southward course, and reaches the southern limits of the United States at about the 105th degree of longitude. The great basin, or rather plain, bounded by these ranges of mountains, and opening to the southward on the Gulf of Mexico, is, with very limited exceptions, occupied by sedimentary strata in their normal condition, and presenting, for the greater part, only slight deviations from their original horizontal position. West of the great eastern or Appalachian chain, we have a wide area of the older formations, from the lowest sedimentary rock to the coal, inclusive. These formations reach, in their widest extent, from the valley of the Hudson to beyond Council Bluffs on the Missouri River, or over a space of twenty-four degrees of longitude. Further south their breadth is considerably reduced, and they finally reach their most southerly termination, as occupying a continuous area, in the northern part of Alabama. From north to south, these formations extend uninterruptedly over nearly fifteen degrees of latitude, or from the south shore of Lake Superior, in latitude 47° 50′, to below 33° of north latitude. Such is the vast extent of the Palæozoic (or primary fossiliferous) formations in the United States, leaving out of view smaller areas not directly connected with this great plain. Of this immense territory, it is scarcely too

* The Eastern States are those on the eastern side of the Appalachian chain.

much to say, that nearly one-half of the entire area is occupied by the coal-measures of the carboniferous period.

Flanking the Appalachian chain on the east, and forming a broad belt or plateau along the Atlantic coast and the Gulf of Mexico, we have the cretaceous and tertiary formations. These extend into Texas, thence, trending to the northward, they form a wide belt between the western limit of the palæozoic formations and the Rocky Mountains chain.

Such simply are the great geological features of this vast country, the details of which, however, would require many volumes to describe.

In the midst of the cretaceous formation in Texas is an isolated mountain of older igneous or metamorphic rocks, surrounded by those of palæozoic age (*vide* Dr. F. Roemer's "Cretaceous Formation of Texas"). In Arkansas there is a belt of similar character and of greater extent, near the junction of the palæozoic and cretaceous rocks; and again, in Missouri, we have, in the midst of silurian and carboniferous strata, mountains of the older metamorphic rocks.

It is within the limits above described, and from the geological formations indicated, that the industrial resources of the Eastern United States, both agricultural and mineral, are to be chiefly derived.

In viewing the mountain barrier which lies to the east of the great central and western plain, it must not be supposed that it is impenetrable except by difficult roads, or that the country lying to the west of it presents an unbroken monotony of surface. Nearly the whole extent of the northern frontier is occupied by a chain of great lakes, navigable for the largest steamers and other vessels, and presenting, within the United States alone, a line of coast of nearly 3,000 miles, exclusive of smaller bays and rivers which communicate with the same. The several outlets from these to the Atlantic render communication as easy between the remote interior regions and the coast as between equally distant maritime cities.

The western and central portions of this great area are traversed by navigable rivers, whose great outlet, by the Gulf of Mexico on the south, invites trade and commerce in that direction. Towards the north and east the waters of these rivers are connected by canals with those of the great lakes; and numerous railroads traverse the entire territory on the east of the Mississippi River, and will soon extend to the Rocky Mountains.

It will thus readily be seen, that in considering the industrial resources of a country of such vast extent, and which is still in its infancy, a juster idea of its capabilities will be formed by studying its leading physical and geological features, than by examining collections of its minerals and rocks in any place of exhibition. With this view it will be necessary to preface the special notice of raw mineral products, by a short sketch of the principal geological formations; noticing the characteristics of each, its geographical extent, and its products.

The geological formations of the United States, which either afford materials of use, or give character to the portions of country which they occupy, may be enumerated in the following order, omitting the minor subdivisions described in detail in the Geological Reports of the various States:—

Alluvial and drift formations.
Tertiary.
Cretaceous.
Oolitic or liassic (jurassic).
New red sandstone (triassic ?).
Coal measures.
Carboniferous limestone, with other rocks below the coal-bearing strata.
Old red sandstone, or Devonian formation.
Upper silurian.
Lower silurian.
Crystalline, or metamorphic rocks of the Appalachian chain, chiefly recognised as metamorphic rocks of Palæozoic age.
Older metamorphic or crystalline rocks.
Trappean rocks.

Alluvial and Drift Formations.

The superficial deposits referable to the Drift-period are almost universally distributed over the larger part of the United States. The presence of large boulders is usually regarded as indicative of this deposit; but these, in many instances, are absent, or not conspicuous, over large areas where the country is deeply covered with water-worn materials. In the mountainous regions the superficial deposit, lying next to the older rocks, is usually unstratified, and fragments of all sizes, more or less worn, lie confusedly mingled together; while, throughout a large part of the great plain west of the Alleghanies, this deposit is more or less regularly stratified, and the materials sorted. Everywhere, in fact, water-worn materials cover the surface of the older subjacent rocks. In all localities these older formations bear evidence of powerful erosive action, and whether in the northern portion of the United States, where boulders are frequent, or in the south, where the materials are less massive, the results are similar, and all seem referable to the same agency.

The various modifications which this deposit has subsequently undergone need not now detain us; since it is only in reference to the general character and economical relations that we are to treat of it.

As a general rule, obtaining over the Northern and Middle States of the Union, the character of this drift is dependent on the strata lying immediately beneath it, or a short distance north of the locality. Although conspicuous masses or boulders of northern rocks are found one or two hundred miles south of their origin, yet it is not to be supposed that any considerable proportion of the superficial mass has been derived from equally distant points. It is only along a few valleys or watercourses that large quantities of water-worn materials have been carried to the south. The area occupied by them is, therefore, small, compared to that which is occupied by màterials derived from rocks in close proximity. Were it otherwise, or were it true that the causes transporting these northern materials were as powerful as sometimes supposed, the influence of underlying geological formations could not be relied on for producing their legitimate results upon the soil above. Over comparatively small areas of mountainous countries, it is true, this disturbing influence can hardly be over-estimated; but in the general extent of the United States the soil corresponds in character with the nature of the rocks below. This superficial deposit, when found undisturbed and resting on the older rocks from which it has been derived, usually contains a large amount of coarse materials; subsequent changes, produced by water in more or less rapid motion, have modified the original condition, separating the coarser from the finer materials, and depositing each in regular layers, unmixed with the others. At a still later period small bodies of water, remaining in depressions of this formation, have given origin to a growth of vegetation, which in its decay has been preserved in the numerous swamps and marshes, together with the bones of animals, shells, and other fossil remains. There have been also incursions of the sea, and deposits of marine sand and clays, with organic forms of marine origin, mingling with the water-worn materials of the drift or superficial accumulations.

Economical Products from the Alluvial and Drift Formations.

Gold.—This metal, which is the only one of value occurring in the formations of this age, is derived from the destruction of certain auriferous metamorphic rocks. In the Eastern* United States the gold-bearing drift is derived from a series or group of sub-crystalline metamorphic rocks, which may be denominated newer metamorphic in contradistinction to the older system of metamorphic rocks, which will be described under that name in the present Report This drift is co-extensive with the formation from which it is derived,

* See ante, p. 6, for explanation of Eastern States.

and is found from the northern limits of the United States in Vermont and New Hampshire, along the Green Mountain range, and passes through the southeastern part of New York, extending thence along the eastern flank of the Appalachian chain to Georgia and Alabama. This drift has proved productive of gold in considerable quantities in Georgia, South Carolina, North Carolina, and Virginia; but further north it has not been wrought for gold within the limits of the United States, although it has long attracted attention in the same direction in Canada, and a considerable quantity of gold has been obtained from washings on the Chaudière river. In California the gold-bearing drift or alluvium is derived from rocks of similar metamorphic character, and probably of the same age. These auriferous sands and gravels are very extensively distributed; the collections in the Exhibition showed samples of gold from nearly 200 different washings or localities in California alone.

Copper.—This metal, though occurring in the metallic form and in that of some of its ores in the drift, has never been found of economic value in this formation.

Iron Ore.—Magnetic iron-sand is a very general accompaniment of the drift in the vicinity of mountain ranges. It has not, however, been applied to economic uses. Bog-iron-ore is almost universal, though in quantities to be valuable only in comparatively few places. The hæmatite, so abundant along the base of the Alleghanies and in other localities is sometimes, it would appear, associated with clays of this formation. Specimens of these from various places were shown in the Exhibition.

Sands and Clays—Fit for the purposes of making bricks and coarse pottery, are widely and abundantly distributed; many millions of bricks are annually produced. Owing to the iron contained in the clay, they are usually of a reddish brown; but in Wisconsin, particularly in the vicinity of Milwaukie, the clay produces bricks of a drab or yellowish colour, which are so highly prized that they are transported many hundreds of miles.

Peat and Shell Marl, &c.—In a few districts of the New England States peat is used for fuel to some extent; but generally very little use is made of it for that purpose in the United States. The peaty deposit or "muck" of many of the swamp lands is much valued as manure, together with shell-marl derived from the decomposition of the shells of existing species of mollusks. Sometimes alternating strata of muck and shell-marl occur in swamps of large magnitude. It consists of an aggregation of small shells with granular or pulverulent carbonate of lime, which is deposited from springs having their origin in limestone formations, or which have percolated the calcareous drift or alluvium, carrying thence the calcareous matter dissolved in the water. In some places where the drift contains an abundance of pebbles, they are used with cement for building walls, and even houses and other buildings. In many places, in consequence of the great depth of the drift and alluvial formation, or from it overlying shaly deposits, which afford no good building material, these large pebbles and boulders become of much importance to the inhabitants of the district.

Exhibition.—The specimens in the Exhibition from this formation consisted principally of native gold, from about two hundred different washing or localities in California, and a few from other places; a boulder of native copper from Wisconsin; bog-iron-ore; ochre; marl; soil, &c.

Tertiary Formations.

The tertiary formations proper are of marine, estuary, and fresh-water origin; consisting of numerous argillaceous, arenaceous, and calcareous beds, and of various admixtures of the like materials. These strata, containing remains of numerous shells, corals, &c., land plants and animals and fresh-water shells, have a wide range in the United States. They commence at the northeastern extremity of Maine, if we include the deposits in the valleys of Lake Champlain and the St. Lawrence, and, stretching south-westerly, become well defined south of New Jersey, thence extending south and south-west, they occupy a broad belt between the Atlantic and the base of the mountain range. Still farther to the south and west the same formations continue to expand; and, trending westward past the southern extremity of the Highlands, spread out over the southern part of Georgia, Alabama, Mississippi, and a large part of Louisiana and Texas bordering the Gulf of Mexico. To the northward from Texas the tertiary formations extend along the base of the Rocky Mountains, forming a broad belt, flanked on both sides by the cretaceous rocks. The deposits of this age reach even to the sources of the Missouri, and spread out into the interminable wastes of that region known as the *Mauvaises Terres.*

West of the Rocky Mountains, and between that region and the Sierra Nevada, tertiary fossils have been collected by Captain Fremont and Captain Stansbury from widely separated localities, leaving no doubt of the existence of the formation on a large scale in that part of the continent. On the western coast the tertiary deposits are equally extensive and well developed, as we have learned from Dance and others (see United States Exploring Expedition: Geology by T. D. Dance); but the details have not been studied.

Economical Products of the Tertiary Formations.

Brown Hæmatite and Ochreous Iron Ore.—It has been long known that there occurs in the United States an almost continuous belt from Canada to Alabama, containing hæmatitic iron-ores, ochres, &c., and clays of a peculiar character. This ore has been regarded as an aqueous or oceanic deposit, and as having been derived from the subjacent strata through the agency of thermal springs. The discovery of a bed of brown-coal associated with the ores and clays has induced Professor Hitchcock to refer the whole to the Pliocene period; but the data for determining their precise geological age are as yet very incomplete.

The hæmatite is usually compact, fibrous, and stalactitic, and always more or less enveloped in clays of various colours. Throughout New England this formation lies at the western base of the Green Mountains, and rests upon a metamorphic limestone, which is more or less associated with talcose or micaceous slate. From these beds the iron may have been derived, as it probably has been, by the action of thermal waters at the period of the deposition of the clays.

In Pennsylvania these ores occur in depressions or fissures in the unaltered limestones of the lower silurian period, but associated with or enveloped in clays; while farther to the south, however, the hæmatites of this character are associated with metamorphic strata.

This deposit is one of the most important sources of iron-ore, and the first worked in this country, furnishing an excellent material, and readily wrought. Specimens of the ore and of the manufactured iron were presented from Vermont, Massachusetts, New York, New Jersey, Connecticut, Pennsylvania, and Maryland.

Manganese.—Black oxide of manganese occurs in numerous localities, associated with the hæmatite above-mentioned. It occurs at Brandon and Bennington, Vermont, in large quantities. The black oxide of manganese exhibited from several southern localities is probably from this formation.

Fossil Fuel.—A bed of brown-coal associated with the hæmatite and fire-clays occurs at Brandon, Vermont, and is there used as fuel. Lignites are found in many localities. Specimens of tertiary coal have also been brought from the country west of the Rocky Mountains, and from the Pacific coast.

The only specimens exhibited at the Exhibition were from Brandon, Vermont.

Fire Clays and Porcelain Earth.—Kaolin, and various coloured clays, of greater or less purity, occur in this formation along the western base of the Green Mountains. They are mainly used in the manufacture of fire-bricks and the stronger kinds of pottery. The clays are pure white, yellowish from iron, and pink-coloured from mangenese. Clays of similar character are known in several localities in New Jersey.

Another variety of clay, perhaps a little more recent in its origin, is, however, the most abundant in the State of New Jersey. It occurs in the neighbourhood of South Amboy, and is known as "potter's clay:" from this source are supplied many of the manufactories of "stone-ware" in the United States. It is associated with, or succeeded by beds of mottled clay and sand. Fire-bricks are also extensively manufactured from these beds.

Specimens were exhibited from Brandon, Vermont, and also some specimens of kaolin and fire-brick from New Garden, Pennsylvania.

Shelly Marls and Green Earth.—In New Jersey, Maryland, Virginia, and farther to the south, shelly marls, sometimes containing green-earth, but frequently without, are used by agriculturists for enriching the soil.

Building-materials.—In this formation, sand, clay, and brick-earth supply the want of good building-stones. A coarse friable sandstone occurs in some localities of the tertiary formation, but it is unfit for durable structures, as is fully shown by the Capitol at Washington, and some other buildings in which it has been used.

Cretaceous Formation.

The cretaceous formation in the eastern part of the United States makes its first appearance in New Jersey, where it was long since recognized, and from this locality were derived many of its most characteristic fossils described by Dr. Morton. It occupies a broad belt in New Jersey and Delaware; and, conforming to the extent and direction of the tertiary beds, it appears at intervals in Virginia and in North and South Carolina, wherever it is exposed by the denudation of the superincumbent tertiary beds. It thence stretches across the northern part of Georgia, central Alabama, and Mississippi, and trends northward along the outcrop of the palæozoic rocks beyond the mouth of the Ohio into the State of Illinois; thence on the west of the Mississippi, following the line of the palæozoic strata through Arkansas, it expands to a great width in Texas. Northward again, it is traced, with varying breadth, along the eastern slopes of the Rocky Mountains and the tributaries of the Missouri, beyond Fort Mandan on that river. On the whole, the cretaceous deposits extend from north to south more than twenty-four degrees, or from the tropic of cancer to the 48th degree of north latitude.

Economical Products and Agricultural Conditions.—No products have been sent to the Exhibition from this formation.

Green-earth, green-sand, or, as it is usually termed, marl, of the cretaceous series is extensively used in New Jersey and further south by agriculturists. It is sometimes nearly or entirely destitute of fossil shells; while in some localities it has large numbers of them in its composition. The fertilizing effects are mainly due to the contained potash, and perhaps also to sulphate of lime, derived from the double decomposition of sulphate of iron and the carbonate of lime from the fossil shells.

Notwithstanding the immense extent of the cretaceous formation in the United States, no true chalk has been discovered. The calcareous beds thus far known are only useful for burning into lime, and are not used for building or ornamental purposes.

Small deposits of iron, often phosphate of iron and iron pyrites, occur in this formation, but they are not of economical value.

Lignite is found in small quantities, and fossil wood of several species is known. On the Upper Missouri and on the Yellow Stone Rivers the beds of lignite are extensive, and may at some future time be of economical value.

The area occupied by the cretaceous and tertiary formations, between the Mississipi river and the Rocky Mountains, is much greater than the whole area of silurian, devonian, and carboniferous strata between the Alleghanies and the Mississippi; and probably greater than the space occupied by all these rocks together in the limits of the Eastern United States. These formations, therefore, extending so far beyond the limits of the other formations, are destined to have an important influence in the future history of the country. Much of this great area is destitute of forests, and often of shrubs; the surface is barren, and much of it of such a character that it can never be converted to agricultural uses, being often a stiff clay, both in the cretaceous and tertiary formations. The climate is so dry that in summer the soil is parched and cracks to such a degree as to destroy all vegetation. Small alluvial tracts alone are capable of cultivation, and vast areas of country deserve to be, as they are, designated the *mauvaises terres.*

Oolitic or Liassic Formation.—(Jurassic.)

The formations which have been recognised as belonging to this period occupy a belt of ten or twelve miles wide by a length of fifty miles, west of Richmond (Virginia), and may be denominated the Chesterfield basin. The Deep River basin in North Carolina is a similar formation, extending, with some interruptions, nearly across the entire width of the State. These can only be regarded as separate parts of what has been one continuous formation.* Up to the present time this is the only geological group in the United States which has from its fossil remains been positively recognized as belonging to the age of the oolite or lias.

This formation rests upon a floor of gneissoid and granite rock; and the coarser grits and sand-stones of the series consist of the cemented materials of these ancient rocks. Higher up, the rock has the character of a micaceous schist, which in the upper beds becomes highly argillaceous,

Economic Products: Coals, &c.—A distinguishing feature of the formation, and one of vast economical importance, is the existence of several valuable coal-seams. These are wrought in the vicinity of Richmond, and near the Deep River, in North Carolina. The whole thickness of the formation in Virginia is 800 feet, and the coal-seams are confined to the lower 150 feet. In some places, two or three seams are known to occur, but in the Virginia basin, to the south of James River, these various seams are united into one vast bed of from 20 to 40 feet in thickness.

The quantity of coal shipped from this basin, according to the best estimates at which we are able to arrive, is more than 150,000 tons annually. The coal-seams in the same formation in North Carolina have been less explored, and thus far no considerable quantity of coal has found its way to market, for want of means of transportation. Facilities of transport, however, are likely to be soon afforded by the completion of the slackwater navigation of Deep River.

The chief economical products of this formation are coal, mill-stones, building-stone, &c. No specimens from this formation were presented in the Exhibition.

New Red Sandstone, Shale, and Conglomerate.

The conglomerates, sandstones, and shaly deposts so well known in the valley of the Connecticut River through Massachusetts and Connecticut, and along the Hudson River valley, and thence through New Jersey, Pennsylvania, Virginia, and North Carolina, have been usually regarded as belonging to the new red sandstone, or triassic period. In the two States last mentioned, the deposits now under consideration lie in similar insulated troughs or basins to the westward of those already described, and differing from the former more in negative than in positive characters.

This formation has its greatest breadth in New Jersey and Pennsylvania, and gradually diminishes to the south-west,—so that in Maryland and the northern part of Virginia it is but a narrow belt, the continuity of which becomes interrupted in the latter State. These insulated basins lie in the same line as the continuous belt, and are doubtless parts of a once continuous series.

In the Connecticut River valley there is a belt of this formation, having a length of 100 miles, and a width of from ten to twenty miles. It is in

* The question of the age of the interrupted belt farther west, which is also a continuation of the formation in New Jersey, and the same as that of the Connecticut River valley, is not yet fully determined.

this valley that it has been studied with more care than elsewhere, and from whence we have derived more of its characteristic fossils than from any other district. Its organic remains consist of fossil plants of the genus *Voltzia* and others, numerous fossil fishes, also bipedal tracks which seem for the most part referable to birds, together with those of quadrupedal animals.

Economical Products of the New Red Sandstone Formation.

Building-stones.—The Connecticut sandstone, though occupying a very limited area, furnishes at the present time more architectural and common building-materials than perhaps all other geological formations in the country. The stone is supplied to all the maritime cities of the United States, and even to the interior.

This sandstone is likewise quarried in New Jersey, and at some points on the Hudson River, in the State of New York. The stone used in the construction of Trinity Church, in New York City, is sandstone of this age, from the Passaic River, New Jersey.

Copper and Lead.—The only metals or ores of importance known in this formation are those of copper and lead. In several localities in Connecticut copper ores occur in this rock, though they have been wrought with but partial success. The principal mine in this rock in Connecticut is the Simsbury copper mine, in the town of Granby. The ore is vitreous copper, with small quantities of variegated copper and malachite. The mine has not been worked for several years.

In New Jersey it would appear that this formation affords copper ores in many localities; but, notwithstanding numerous attempts to work them at different times, they have been finally abandoned. The blue and green carbonates are the prevailing ores. The grey and blue sulphuret, with red oxides of copper, have been found in some of the mines; and, more rarely, native copper, and phosphate and silicate of copper. The ores do not occur in regular veins, but appear to be disseminated through certain layers of the slaty part of the formation; sometimes aggregated in considerable quantities, and giving promise of success to the mining operations.

In Pennsylvania, in the counties of Montgomery and Chester, some of the productive veins of lead and copper ores, which lie near the junction of this formation with the gneiss rocks, penetrate both systems, while other veins lie wholly within one or the other set of strata.* The localities mentioned afford large quantities of lead and copper ores; and some of the mines are producing valuable returns. The veins in the shales and sandstones are observed to be more prolific in copper than in lead; and the same lode, which is lead-producing in the gneiss, becomes copper-bearing in the superior rock; or the lead-ores become greatly reduced, and the copper largely preponderates.

The principal ores are sulphuret and phosphate of lead, and sulphuret and carbonate of copper.† These ores will be again referred to in the description of the older metamorphic rocks.

Exhibition.—Besides the specimens from the Wheatley mine, the products of this formation in the Exhibition were rough specimens and small blocks of dressed sandstone from the quarries at Portland, Connecticut, on the Connecticut River; also sulphate of baryta from Cheshire, Connecticut.

* Report of Professor H. D. Rogers on the Wheatley, Brookdale, and Charlestown mines.

† The Wheatley mine, which is in one of the lodes penetrating both the gneissic and sandstone formations, has afforded the following minerals, all of which were represented at the Exhibition:—

Sulphate of Lead, crystallized.
Phosphate of Lead.
Arseniate of Lead.
Molybdate of Lead.
Chromomolybdate of Lead.
Arsenio-phosphate of Lead.
Antimonial Sulphuret of Lead and Silver.
Sulphuret of Lead.
Sulphuret of Zinc.
Carbonate of Zinc.
Silicate of Zinc.
Sulphuret of Copper.
Green Malachite.
Blue Malachite.
Black Oxide of Copper.
Native Copper.
Oxide of Manganese.
Native Sulphur.
Native Silver.
Quartz, crystallized.
Cellular Quartz.
Oxide of Iron with Silver.
Brown Hæmatite.
Brown Spar.
Sulphate of Barytes.
Iron-pyrites.

Coal-formation, or Coal-measures.

The series of sedimentary and other deposits, constituting what is usually designated as the Coal-measures, or great Coal-formation, are, in the United States, extremely similar to those of Europe. The foundation or underlying rock is usually a conglomerate, or coarse sandstone, with quartz-pebbles. In some parts this rock is not present, and shaly beds with coal rest upon the limestone below. The succeeding beds consist of sandstones more or less fine, sometimes of conglomerate, or conglomeratic sandstone, shale, limestone, coal, and fire-clays. The immense thickness of this formation in some localities is truly astonishing, and the changes of condition that must have been required to produce the alternation of beds of different materials, and of such magnitude, is no less a matter of surprise.

The great coal-formation of the United States is one of its principal and most striking geological features, as it is by far the most important of all in its economical influences. The eastern Alleghany or Appalachian coal-field extends from the extreme northern part of Pennsylvania to the central part of Alabama, its length being more than 750 miles, and its greatest breadth more than 180 miles; exclusive of the anthracite basins which lie detached upon the eastern side of the main field. The superficial area of this coal-field has been estimated, by Professor H. D. Rogers, at about 63,000 square miles.

The great western coal-field, or, as it has been usually termed, the Illinois coal-field, occupies the larger part of the State of Illinois, and parts of Indiana and Kentucky. The late surveys of Dr. D. D. Owen have shown that this coal-field has, in its full limits, a much greater extension in Iowa and Missouri than had before been supposed.

It is true that the part west of the Mississippi River is not only separated from the eastern or Illinois portion by the river-channel, but that this river runs along a gentle anticlinal axis, the elevation of which has contributed in part to the separation of these two portions of the great western coal-field.

This western portion, lying in the States of Iowa and Missouri, occupies an area two-thirds as great as that part on the east of the Mississippi. It is the opinion of Mr. Hall, and other geologists of note, that this coal-field extends to the south-west beyond the limit of any careful geological explorations as yet made, though probably not reaching beyond the parallel of the silurian and metamorphic regions of southern Missouri.

Its greatest extent from the south-east to the north-west, or from the head-waters of Green River, in Kentucky, to its northern limit on the Des Moines River in Iowa, is more than 550 miles; while its greatest breadth across Illinois and Missouri, in an eastern and western direction, is more than 400 miles; and from its northern termination in Iowa, to its present known limits on the Osage River at the south, is more than 300 miles.

This western coal-field, therefore, including the area on both sides of the Mississippi, has a much greater superficial extent than the Appalachian coal-field. Indeed, that portion alone of this western field which lies east of the Mississippi has a superficial area nearly equal to the Appalachian coal-field; while that on the west of that river adds fully half as much more to the entire extent.

Still farther to the south, we know that a coal basin of considerable extent exists in Arkansas. This was perhaps originally a part of the same field as that in Iowa and Northern Missouri, and may have been separated from that by the elevation of the belt of metamorphic rocks; or this belt may have existed in its present elevated form as an island in the ancient ocean in which the sedimentary materials of the coal formation were deposited. Other coal basins, either as small detached areas, or as parts of one great field, are known to exist still farther to the south and west; and from the fact that the great underlying carboniferous limestone has been traced into New Mexico, and at intervals to the Rocky Mountains range, still farther north, we may infer that the coal measures will be found, if not co-extensive with it, to occur at least at intervals and in isolated basins over this vast area.

Besides these, there is a smaller field in Michigan, which has thus far proved unproductive, and which from its position, in a low flat country, is not likely to become available at present. Its geographical position is such as would give it superior advantages, lying as it does upon the navigable waters of Lake Huron, and being more accessible to the commerce of the great northern lakes than either of the other coal-fields. At the present time we have too little accurate knowledge regarding it to speculate as to its future comparative importance.

There is likewise a small area of the nearly unaltered coal-measures,* in Rhode Island and Massachusetts. The relative position of these beds, together with those of Nova Scotia, when viewed in connection with the age and character of the intervening country, suggests the idea that these detached portions may once have been united with the Alleghany coal-field; and further, that this was originally continuous with the western field; and, in fact, in their original or undisturbed condition, the carboniferous formations may have been nearly co-extensive with the area of that part of the United States lying east of the Rocky Mountains.

In conclusion, besides these immense and inexhaustible coal-fields on the east of the Rocky Mountains, specimens of coal have been brought by Captain Stansbury from beyond Fort Laramie; coal has also been obtained at Bellingham Bay on Puget's Sound, Washington Territory; little is known, however, of the age or geological associations of either. The specimens of the latter are so similar to the ordinary bituminous coals of the coal-measures, that no essential difference is perceptible, and we are thus led to hope that on the western coast of the United States this important source of progress and wealth may be found.

Economical Products.

Coal of the Coal Formation.—The rapid disappearance of the forests from the northern and middle portions of the United States, together with the steady and increasing advancement in manufacturing operations, have necessitated the general use of mineral fuel. Although comparatively but few years have elapsed since the introduction of coal into use as ordinary fuel, the present enormous and steadily increasing consumption renders the consideration of supply, and the means of bringing this supply within the facilities of transportation to towns and cities, a matter of great importance.

The earlier manufacturing operations of the United States began in New England, where the abundance of water-power compensated for the absence of mineral fuel, and supplied the requirements of the manufacturer. At the present time, however, since the American coal-formations are better known, and the supply has kept pace with the wants of the community, it is beginning to be felt that manufacturing operations must depend upon the abundance and unfailing supply of fossil fuel. The ratio of increase in the consumption of anthracite coal in the United States may be understood from the fact, that in 1820, 365 tons were used; while in 1847, the increased demand reached the amount of nearly 3,000,000 tons; and of bituminous coal during the same year, more than 2,000,000 of tons were consumed.†

In 1851, it was estimated that the bituminous coal consumed in Pittsburg, and shipped to southern ports, including that from the Monongahela, amounted to more than 1,000,000 tons.

The anthracite and bituminous coal-fields of Pennsylvania and Maryland, which form outliers in advance of the great coal-field, are reached with facility from the Atlantic coast. The Ohio, Monongahela, Kanawha, and other smaller but navigable streams traverse and penetrate the Alleghany field on the west, while its northern and eastern margins approach within thirty or fifty miles of Lake Erie at Cleveland, and within less than one hundred miles of Buffalo, which is the terminus of the Upper Lake navigation and the commencement of the canal navigation of New York.

* The partially altered beds of the coal-measures here spoken of are those which are productive in coal; but I am satisfied that a much larger extent of country in New England consists of the coal measures, in a more or less highly metamorphic condition.

† R. C. Taylor's "Statistics of Coal."

On the south, the same coal-field reaches within 200 miles of the Gulf of Mexico, and is there penetrated by streams which admit the transportation of coal from the mines to tide-water; and the coal for the steam-navigation of the Gulf will eventually be supplied in that direction. At the present time, large quantities of coal are shipped down the Ohio, and thence to New Orleans, making a distance of more than 2000 miles of river navigation.

The western coal-field is traversed by the Ohio, the Wabash, the Illinois, the Mississippi, the Missouri, and the Des Moines; thus the greatest possible facilities are afforded for transportation by natural navigation. The north-eastern boundary of the Illinois coal-field reaches to within 80 or 90 miles of the southern extremity of Lake Michigan, and is connected with Chicago both by canal and railroad. Add to all these facilities for transportation the numerous railways, in progress and projected, over a nearly level or slightly undulating country, and it will be obvious that the whole of this great western country can readily be furnished with an abundant and cheap supply of mineral fuel.

The coals of the small districts included in Massachusetts and Rhode Island are anthracites of a low degree of combustibility. This character of the coal, together with the irregular condition of the beds, in the midst of a much disturbed and highly metamorphic region, have caused them to be little sought for; and of the ten or twelve workings commenced at different times, few have been continued.

The anthracite basins of Pennsylvania produce anthracite of a superior quality, and from its accessibility it has furnished a large part of the fuel used in the towns and cities of the Atlantic coast.

A specimen of anthracite coal of enormous size, from the Mammoth vein, Wilkesbarre, Pennsylvania, was presented at the Exhibition by citizens of that town. This mass showed a vertical section of the vein, being a shaft 5 feet square at the base and 30 feet high; weighing about 60 tons. Several other large masses, one 10 feet long by 4 wide, and 3½ high, were exhibited from the same locality; and specimens of the same bed from other localities.

Coals from the Carbondale and Pittstown mine, a collection of about sixty varieties of anthracite from Schuylkill county, and several other specimens of coal, were presented at the Exhibition.

Bituminous coals from Maryland were exhibited in large masses, showing the thickness of beds, 11 and 15 feet respectively. Numerous specimens of bituminous coal accompanied samples of iron-ore from several places in Pennsylvania.

A collection representing the bituminous coals of Perry, Hacking, Vinton, Jackson, and Mahoning counties, Ohio, also specimens from Kentucky and Virginia, were exhibited.

Iron-Ores.—The coal-fields of the United States contain rich deposits of the iron-ores which usually occur in the coal-measures. These are chiefly the carbonate and peroxide of iron; the latter mainly resulting from partial decomposition or change of the former, which is argillaceous, calcareous, or siliceous in its composition. In Pennsylvania and in Ohio, where these ores are wrought to a greater extent than elsewhere, the beds appear to be inexhaustible, and will supply for an indefinite period the requirements of advancing physical improvement and civilization.

In Tennessee, Alabama, and Western Virginia, the coal-formation abounds in iron-ores. In the western coal-field less is known of this connection of coal and iron, the extent to which iron is manufactured being far less than in the east. The geological survey of Illinois, now in progress, has already shown that this State is richly supplied with iron-ore in the midst of her inexhaustible coal-fields; although there are as yet but two furnaces in the State. In Missouri, iron from these ores is manufactured to some extent, yet too little is known to warrant a conclusion as to their abundance.

The iron-ores from the coal-formation presented at the Exhibition were chiefly from Pennsylvania; and from that State was furnished a very full collection, not only of the ores from the coal, but from other formations.* The collection comprised the ore and furnace-products, viz., ore, coal, slag, pig-iron, as well as manufactured iron;—ores from geological formations below the coal,

* Further mention of these will be made under the headings of "Iron Ores."

and occurring in the midst of the coal-fields, being here exhibited in company with the coals by which they are smelted.

Iron-ores from the coal-formation were exhibited from Vinton, Jackson, and Hocking counties, Ohio; and also from one locality of the coal-formation of Missouri. A good series of specimens presumed to be from the same formation were exhibited from Alabama.

In estimating the relative abundance and accessibility of this metal in the coal-formation, it will be recollected that only a small part of what is known in these immense coal-fields was represented in the Exhibition; and that what is known is comparatively but a small portion of what exists in the entire area; and when we reflect that the ores belonging to this formation, with their accompanying mineral fuel, are distributed more or less abundantly over an area of more than 160,000 square miles, we may have some faint conception of its vast amount, and of the important results of its influence on the future industrial interests of a country like the United States.

The States of Tennessee and Alabama are known to be rich in iron-ores of the coal-formation; as also Virginia and Kentucky; though specimens were exhibited from only one of these States.

Limestones.—In nearly every part of the coal-fields there occur beds of limestone of marine origin, alternating with shales, sandstones, and beds of coal. In some places the coal and limestone are in actual contact; and some of the ores occur in connection with the limestones. These beds of limestone furnish the necessary flux for the reduction of ores of iron, and also lime for ordinary building-purposes.

Sandstones.—The beds of sandstone, alternating with the shales and coal-seams, often furnish an excellent and durable building-material. Some of the more friable sandstones, which are free from iron, are used for glass-making.

Fire-Clays.—Some of the coal-seams are underlaid with beds of fire-clay, of greater or less thickness; and in some parts of the formation there are extensive beds of fire-clay not directly associated with coal-seams. This material is used for fire-bricks, stoneware or pottery, water-pipes, and drain-pipes, and is everywhere co-extensive with the great coal-fields.

Carboniferous Limestone Formation.

The great limestone formation immediately underlying the coal-measures, and which in the carboniferous districts is usually a conspicuous feature in the topography of the surrounding country, is an unimportant and inconspicuous rock upon the margin of the greater part of the Alleghany coal-field. On the eastern margin through Pennsylvania, it can scarcely be said to exist; unless the thin bands of limestone associated with the red shale of No. XI. (Pennsylvania Geological Reports) may be regarded as this rock.

In Virginia this limestone becomes greatly expanded; and, in the southern part of the State, it acquires an important thickness. In Tennessee, following the same direction, the limestone forms a narrow belt along the eastern side of the coal-field, and continues to its southern extremity in Alabama. In Northern Alabama, upon the west and north-west sides of the coal-field, this limestone area is greatly expanded, and continues into Tennessee; forming a very conspicuous feature in the topography of the Cumberland Mountains. In the northward, however, the limestone gradually diminishes, and is finally lost before reaching the Ohio River, or becomes so merged in the adjacent sandstones as to pass unobserved.

When we turn to the western or Mississippi coal-field, we find the extent and importance of this formation greatly increased. Along the eastern side of this coal-field the limestone occupies a broad belt, extending through Indiana and Kentucky, and re-appearing on the western side of this part of the basin, or along the valley of the Mississippi, where it is deeply excavated for many miles by the channel of that river, and often forms high cliffs along its banks. It occupies a belt of country of varying width, on one or both sides of the Mississippi, from a short distance above the Ohio to the mouth of Rock River, more than 300 miles in extent, and passes both east and west under the coal-measures on either side. It appears at numerous intervals along the Missouri, and, as far as known, it forms a broad belt encircling the whole of the western division of the coal-field. Farther west it appears near Fort Laramie, and is the principal stratified rock in the vicinity of the Great Salt Lake. Its occurrence in Texas and New Mexico is also known, though its limits have not been defined. In Michigan, little is known of this formation, though limestone, referred to this period, occurs in several localities along the western side of the coal-field. The great variety of character exhibited by this limestone in the west, together with other considerations, has induced Dr. D. D. Owen to separate it into a lower and upper formation. The whole is 390 feet thick,* and is divisible into twelve different beds, characterised by lithological differences.

Thus, over an extensive region in the west, this group of limestone beds, with their layers of marl, or soft argillaceous sandstone, and marly partings with earthy beds of limestone, imparts a character of fertility to the country surrounding the western coal-field, unknown in the same geological associations elsewhere in the United States. On the northern, eastern, and western margins of the Alleghany or eastern coal-field, there is a broad belt of country in which no important calcareous formation exists; and here the limestone products can only be obtained from many miles beyond the outcrop of the coal-measures, or from the thin calcareous beds which alternate with the coal-seams. The effect of this geological difference is perceptible, not only in the architectural character of the buildings, but in the agricultural condition of the country; in the one case, the belt around the coal-measures producing a grazing country; while in the other, the corresponding belt, holding similar relations to the coal-measures, and in the same latitude, possesses a highly calcareous soil and gives origin to a grain-growing region of the first order.

But not only does the belt of country surrounding the coal-fields present this contrast of character, the regions within the areas of the coal-fields are not less strikingly different. This surrounding calcareous formation, together with the many alternating beds of limestone in the coal-measures themselves, all of which are frequently cut through by the river-channels and deep valleys, have served to distribute calcareous matter very generally and abundantly throughout the region. From this cause, perhaps, together with the accumulation of soils derived from the calcareous beds of lower formations in the northern districts, the area within the coal-fields of the west, in their wide extent of prairie and level lands, present a very different aspect from those parts of the country within the eastern coal-field, although the coal-fields hold similar geological relations.

Economical Products of the Carboniferous Limestone.

Lead.—The carboniferous limestone, throughout the greater part of its extent, preserves its normal characters, and is generally unproductive of mineral substances. In the southern part of Illinois, where the strata are much disturbed, it is traversed by several veins which have been worked for lead.† The lodes contain galena, blende, fluor-spar in large quantities, calc-spar, &c.

Lead-ores likewise occur in the carboniferous limestone of Tennessee and Alabama, but not in quantities sufficient to be wrought. A vein of lead-ore in proximity to the coal-measures is worked in Arkansas; but this may perhaps be found to be in the carboniferous limestone.

Building Stones.—This limestone affords an excellent and cheap material for building, easily quarried and dressed. Some of the beds are extremely fine-grained and compact in texture, admitting a good polish, as do also some portions of the lower beds of the limestone, which are oolitic in structure.

* This thickness is much less than that estimated for the limestone in the same association in Virginia and Tennessee.

† The State Geologist of Illinois, Dr. J. S. Norwood, expresses much confidence in the prospective value of these veins.

Several blocks, and some smaller polished specimens of this limestone from the vicinity of St. Louis, were shown in the Exhibition.

Red Shale of the Carboniferous Limestone Formation.

In connection with this limestone it would seem necessary to notice a very important formation of red shale, which, in the absence of the limestone, lies directly beneath the conglomerate of the coal-measures. This shale formation (No. XI of the Pennsylvania survey) has a thickness of 2,949 feet at Pottsville, and extends in a broad belt along the eastern margin of the coal-fields throughout Pennsylvania and Virginia.

Economical Products: Iron.—It is economically important from being the repository of iron-ores which are wrought in many places in Pennsylvania, not only along the borders of the coal-measures, but at numerous points within the basin, where the superincumbent strata are cut through. The ores of the red shale are chiefly carbonates of iron, with variable proportions of silica, alumina, &c. They yield from 60 to 80 per cent. of carbonate of iron, and some of them yield peroxide of iron in about the same proportion. Specimens of these ores were shown in the Exhibition.

Devonian Formation.

The group of limestones, shales, and sandstones, forming this division of the palæozoic rocks which in the United States are generally recognised as constituting the Devonian system, are a heterogenous assemblage. In the descending order, this series begins with the formation No. X of the Pennsylvania Survey; which, though it may be a debateable ground between carboniferous and devonian, will be for the present included in the latter. To this we have to add the shales and sandstones of the Cattskill Mountains, and the Chemung and Portage Groups, together with the shales and calcareous beds of the Hamilton Group of New York, and their extension westward, known as the Black Shale, and Waverly sandstone, or fine-grained sandstones of Ohio, Indiana, Iowa, &c. The whole of this series of deposits rests upon a limestone included in the same formation, and which is at least co-extensive with them.

We shall first notice the extent and general character of the sedimentary strata of the upper portion, and afterwards describe the limestone formation.

First group.—The series of strata under consideration has a far greater thickness in the eastern part of the United States, as New York and Pennsylvania, than in the western part. In the former it is clearly a littoral deposit, and the coarser materials are accumulated in this part of the formation; while towards the west, thefiner materials only have been transported, and consequently the accumulation is less. For this reason also, its economical resources, and its influence upon the topography of the country and the fertility of the soil, vary to a great degree in different parts of the formation.

In those situations where we find the greatest development of these groups, the lower beds have been composed of a fine impalpable mud, succeeded by others formed of a similar fine mud, with a large admixture of calcareous matter; and sometimes the rock becomes arenaceous, with calcareous matter in smaller proportion. Succeeding this, are often some thin calcareous bands, which, however, are not constant. To these again succeed soft shales, or shaly sandstones, which are followed by arenaceous beds, and a successive alternation of these with bands of shale, of greater or less thickness. To these succeed red and green shales and sandstones, the structure being in some cases coarser than that of the lower deposits, but often shaly, or consisting of alternating arenaceous and shaly beds. This is the formation composing the Cattskill Mountains. To this must be added the formation No. X, which is a coarse sandstone or conglomerate, and sometimes a shaly mass, with an admixture of conglomerate.

This formation may be considered as occupying nearly all the southern half of the State of New York, from the Hudson River, extending along the northern outcrop of the Pennsylvania coal-basin, and forming the southern shore of Lake Erie from near Brussels, N. Y., to Cleveland, Ohio. From this point it curves round in a broad belt, skirting the coal-measures along the entire western margin, but gradually thinning out to the southward. On the east it follows the outline of the Alleghany coal-field from Pennsylvania to Alabama.

In like manner, the black slate and fine-grained sandstones encircle the Michigan coal-field, and form a broad belt, of several hundred feet in thickness, along the entire margin of the Illinois coal-field, cropping out beneath the well-marked carboniferous limestone, into which they sometimes appear to graduate, or they alternate with the lower beds of the latter rock. Along the Mississippi valley these sedimentary rocks crop out at numerous points beneath the carboniferous limestone; and the fine sandstones, so extensive in the west of this great coal-field, are probably of the same age.

Second group.—At the base of all these sedimentary deposits lies an immense limestone formation, important less from its thickness, than from the area which it occupies, and the positive evidence which we have of its once having been a continuous and uninterrupted coral-reef, over an area of not less than 500,000 square miles.

Within the limits of the United States, it is first recognized in its unaltered condition in New York, New Jersey, and Pennsylvania, and extending thence westward. Its most conspicuous outline, however, is from the Helderberg Mountains, on the Hudson River, near Albany, westward throughout the entire extent of New York. Crossing the Niagara River at Black Rock, extending through Upper Canada in a line nearly parallel to the coast of Lake Erie, and crossing into the northern part of Michigan, it forms the western boundary of Lake Huron in its northern part, quite to the Straits of Mackinac, and appears in an enormous outlier, capping Mackinac Island and Gross Cap. Its position to the south-west of these points is in the bed of Lake Michigan, having been denuded, with the softer beds beneath it, and excavated to form the bed of this lake.

The anticlinal axis extending from the western part of Lake Erie, by Cincinnati, to the southern part of Tennessee, brings up this limestone, which occupies a broad belt on either side of the lower strata that have been here elevated. In the central part of Ohio, it extends from Sandusky, south, to the Ohio river, and thence into Kentucky. It has an equal extent in Eastern and South-eastern Indiana, reaching the Ohio River at the Falls at Louisville, and passing thence through Kentucky, where it occupies a considerable area, it extends into Tennessee, skirting the irregular outline of the silurian formation of that State.

In a north-west direction from where we leave this rock in Northern Indiana, it follows a course parallel to the northern outline of the Illinois coal-field; and extending by the head of Lake Michigan across the Mississippi River, it constitutes in part the formations of the valley of the Red Cedar River in Iowa.

Notwithstanding the remarkable persistence of this group, it has nowhere a very great thickness; and indeed it is for the most part a very thin deposit, reaching, perhaps, at its maximum, no more than 350 feet, and it is usually less than 100 feet. In its north-western extension in Iowa, Dr. Owen estimates it at no more than 60 or 70 feet, and its furthest south-western limit is scarcely more than half this thickness.

Third group.—At the present time geologists are inclined to extend the limits of the Devonian system, in America, still lower than the limestone just described, and to include within the signification of this term a series of grits and sandstones which intervene between the last-named limestone and another limestone formation below. In New York two distinct members are recognised, the Cauda-galli grit, and the Oriskany sandstone. The former is a close, compact, and fine-grained mass, of a dark colour, and nearly destitute of fossil remains; while the lower member is a coarse, porous, or loosely aggregated sandstone, containing an abundance of fossils.

These grits and sandstones have their northern termination in the Helderberg Mountain, near the Hudson River; and in their extreme western prolongation they do not reach beyond 200 miles, and present several interruptions within this distance. The entire thickness in the State of New York does not reach 200 feet. The lower member of the group is much more persistent than the upper, and extends southward through Pennsylvania, and into Virginia, and in this direction acquires in Pennsylvania, according to Professor H. D. Rogers, a thickness of 700 feet.

Economical Products of the Devonian Rocks.

Iron-ore, Building Materials, &c.—A system so varied in its composition, and occupying so great an extent of country, must necessarily present a variety of useful products. The principal materials of economic value are iron-ores, limestone for lime, hydraulic limestone, building-stones, sandstone for building and for fire-stones, flagging-stones, &c.

At the junction of the Chemung and Cattskill Groups, Nos. VIII and IX of the Pennsylvania Survey, there occur bands of fossiliferous iron-ore, not unlike one to be mentioned in the Upper Silurian formation, or Clinton Group of New York. These ores yield from 30 to 50 per cent. of peroxide of iron. Iron-ores also occur associated with the limestone at the base of No. VIII of the Pennsylvania Survey.

These ores are described as of good quality and abundant in quantity, in the State of Pennsylvania, and they occur still further to the south. In the western extension of the limestone described, no ores of importance are known to occur, though the line of junction with the rock next below it is often marked with an ochreous deposit.

The only products of this formation shown in the Exhibition were a few specimens of iron-ore and limestone from Pennsylvania.

Upper Silurian Formation.

The formations which in the United States may be included in the upper division of the Silurian system consist of limestones known in New York as the Lower Helderberg Limestone Group, which includes the following members in the descending order: Encrinal limestone, Delthyris shale limestone, Pentamerus limestone, and Water-lime. To these again succeed the Onondaga Salt Group of New York and the great limestone of Niagara, altogether constituting No. VI of the Pennsylvania survey. Below these occur a series of shales, sandstones, conglomerates, and beds of limestone, constituting the Clinton Group of New York; and the whole is terminated below by the Medina sandstone. The two latter-named formations constitute No. V of the Pennsylvania Survey.

These different members or groups of the series have a very different geographical distribution, and require to be separately noticed.

Group A.—The lower Helderberg limestones have their greatest development in the Helderberg Mountains, near the Hudson River, and in New Jersey, Pennsylvania, and Virginia. In Pennsylvania the thickness of the group is estimated at 900 feet, which is four times greater than in New York. In a south-west direction, parallel to the Appalachian chain, this group appears to be co-extensive with the other members of the palæozoic formations. Pursuing it in a westerly direction, however, we find that it thins out; and, with the exception of the lower members, it is not known beyond 100 miles west from the Hudson River. The lower member, or Water-lime, extends to the western boundary of the State.

Group B.—The second division, or Onondaga Salt Group, has a wider range, and may be traced throughout the entire length from east to west of the State of New York, and thence through Canada West, where it is often deeply covered by superficial accumulations. Its course north-westerly is in the direction of Lake Huron, where it has been entirely removed, and the basin of the lake now occupies its place. It re-appears forming the base of Mackinac Island and Gross Cap of Lake Michigan.

From these points its course is again in the direction of the basin of Lake Michigan,* from which, together with the superior limestone, it has been removed.

In Northern Ohio, west of Sandusky city, and along the Sandusky River, it has some importance; but in the southern part of that State, and in the south-eastern part of Indiana, although easily discoverable in numerous localities, its thickness is unimportant.

The most persistent member of this part of the silurian system, is the Niagara group, consisting of shale and limestone, or of limestone alone, and well known as forming the barrier at the Falls of Niagara. Although this group is often extremely thin, and with difficulty recognised, along the eastern outcrop of the palæozoic series, it becomes very important in the western part of the State of New York. In the eastern part of the State it is a thin bed of little note, but augments in thickness towards the west, producing a conspicuous feature in the topography of the country, in the great terrace, or mountain ridge, as it is termed, which extends from Rochester, by Lewiston, New York, Queenstown and St. David's, Canada West, to the head of Lake Ontario. Thence westward, this formation produces the prominent headland, Cabot's Head, between Lake Huron and Georgian Bay; and, in its continuation, it forms, in great part, the Manitoulin Islands. From these west by the Straits of Mackinac, it forms the northern shore of Lake Michigan and the peninsula between that lake and Green Bay; and continuing thence, it forms the western shore of Lake Michigan nearly to its southern extremity. Farther west, it occupies the elevated portions of the country in Southern Wisconsin and Northern Illinois, and, extending into Iowa, spreads out into a wide area between the Mississippi and the Red Cedar River.

The anticlinal axis before mentioned, which extends from Nashville to Lake Erie, brings up this limestone in the islands near the west end of that lake; and, extending thence south-westerly, it is broken through by the Lower Silurian about fifty miles north of Cincinnati, and thence extends on either side of this axis into the States of Kentucky and Tennessee, constituting the lower beds of limestone at the Falls of the Ohio, and holding the same position in the well-known fossiliferous localities of Perry county, Tennessee.

Group C.—We may for the present include in one the two succeeding formations, the Clinton group and the Medina sandstone of the New York Survey;—a series of beds consisting of shale, shaly sandstone, conglomerate, beds of limestone, iron-ore, and a massive formation of soft red sandstone at the base. This group, either entire, or represented in some of its members, is nearly co-extensive with the preceding. It follows the same course, cropping out from beneath it, and often aids in giving prominence to the bold outline of the superior limestone. This group is well developed in all its members, in the central part of New York, its most northerly limit in the United States, if we except some parts of Maine, where, however, its extent is little known, and the Green Mountains, in which it may be represented by some metamorphic rocks. From its point of greatest development in Oneida county, this formation becomes more calcareous and of less volume to the westward; and at the Niagara River, the whole formation consists of a bed of shale and a bed of limestone. North-westerly through Canada, though greatly reduced in thickness, it follows the line of the superior limestones, and is seen in the islands of Lake Huron. It is distinctly recognised on the eastern shore of Green Bay, and again in Wisconsin, which, as far as now known, is the most westerly limit of the rocks of this formation. Some of the limestone beds of this group, as defined in New York, are to the westward intimately blended with the Niagara limestone. The same is true of these rocks along the great central axis, though here the sedimentary beds of the group are scarcely to be recognised.

From the south-west end of the Shawangunk Mountain in New York, this group extends through New Jersey, and into Pennsylvania, where it attains the great thickness of 2,000 feet. It is persistent in some of its members

* For the origin and relative position of the valleys of Lakes Huron and Michigan, see a Paper by James Hall, in Foster and Whitney's Report on the Lake Superior Land District.

through Virginia and Tennessee, and reaches the southern termination of the palæozoic rocks in Alabama.

The base of the upper division of the Silurian system is marked by an immense accumulation of thick-bedded sandstone, both of coarse and fine texture, and massive strata of conglomerate, with thin beds of shale or shaly sandstone. The conglomerate often contains pebbles of large size, and frequently the materials composing these beds are but partially worn, the fragments preserving to a great degree their angular form.

Notwithstanding the coarseness of the materials of which this rock is often composed, they are so cemented as to suggest the idea that some portions of the siliceous matter may have been in partial solution at the time of the aggregation of the mass. The less rounded character of the materials, and the compact nature of the rock, together with the peculiarities of its bedding, and other features, serve to distinguish it from the conglomerates of the coal-measures.

This rock is first recognised in the valley of the Mohawk, south of Utica, where it altogether constitutes a mass of less than 20 feet in thickness, and is soon lost both towards the east and the west by thinning out, while towards the south it passes beneath the overlying formations. Seventy miles to the southward, however, it abruptly appears in great force in the Shawangunk Mountain, which commences a few miles from the Hudson River, near Kingston, and thence it ranges through the counties of Ulster, Sullivan, and Arago, to the Delaware River. The formation extends thence south-westerly, through New Jersey, Pennsylvania, Maryland, and Virginia, constituting a prominent feature in the topography of the country throughout its entire extent. Including the Shawangunk Mountain in New York, and its continuation, the Kittatinny Mountain in Pennsylvania, it forms an almost continuous mountain-range from the Hudson River in New York to the northern part of Virginia. We should not omit to observe, that although the Shawangunk Mountain terminates abruptly on the north-east a few miles from the Hudson, yet the conglomerates and coarse sandstones are infolded with the underlying slates of a preceding period, in the contorted and metamorphic rocks of Eastern New York.

This formation usually imparts a conspicuous character to the scenery and topography of the country which it traverses, and of itself it constitutes the mountain range described. The surface is usually barren, and often quite destitute of vegetation over considerable areas; and the white sandstone, often conspicuous in large naked patches, gives a peculiar aspect to some parts of the mountains.

And it is worthy of remark, that these rocks present the very significant feature of an ancient coast-line formed during a period of disturbance, and marking in the most decisive manner the period of greatest change in the fauna which occurred during the deposition of the entire series of the palæozoic rocks.

Economical Products of the Upper Silurian Rocks.

Iron-Ore.—The substance of greatest economical importance in this part of the system is iron-ore. The strata known in New York as the "Clinton Group" contain one, and sometimes two distinct beds of iron-ore, alternating with beds of limestone and shale. This iron-ore appears to be nearly co-extensive with the formation. In the western part, however, where the formation has become interrupted and thin, iron-ore is known to exist in only a single locality in Wisconsin. The ore is fossiliferous, as are the adjoining bands of limestone, and often highly calcareous in its composition. It is much prized and extensively used in the manufacture of iron in Pennsylvania and New York.

Small quantities of iron-ore occur in the superior limestones (No. VI. of Pennsylvania Survey) in Pennsylvania, but throughout the large part of the area occupied by these limestones, no ore of importance is known.

Lead, Zinc, and Copper.—Galena, blende, and copper-pyrites occur in the Shawangunk conglomerate. At the present time a lead-mine is wrought in this rock, in the Shawangunk Mountain in Ulster county; the products of the mine are galena, copper-pyrites, and a small quantity of blende. This mine has afforded some remarkably large masses of galena, one of which weighed 16,000 pounds.

Further to the southwest, within the State of New York, a vein or bed of sulphuret of zinc in the same formation has been somewhat extensively wrought.

Building-Stone and Millstones.—Limestone everywhere abounds in this part of the system, and some of the finest and most durable structures in the country are made from limestone of the Niagara group; this is evidenced in the great works along the line of the Erie Canal; and the facility with which this rock is quarried, and the proximity of navigation, have caused its extensive use; it may be recognised even in the city of Chicago, at a distance of 1,500 miles, by way of the lakes, from its source in Western New York.

The Medina sandstone and Clinton groups afford an abundance of fine flagging-stones, as well as, in some parts, excellent building-stones.

Millstones of good quality have long been manufactured from the compact conglomerate of the Shawangunk grit; this rock, in earlier treatises on American geology, was supposed to be identical with the millstone-grit of England.

Salt.—The Onondaga Salt-group, in the State of New York, yields brines of great strength, from which large quantities of salt are manufactured, affording an extensive business as well as a large revenue to the State. The salt manufactured at these *salines* in 1797, amounted to 25,474 bushels; since that time the quantity has been steadily increasing; in 1852 it amounted to 4,922,533 bushels, and in the present year the quantity will extend beyond 5,000,000 of bushels.

Although this formation yields brines for a distance of 200 miles or more along its northern outcrop, yet those which are of sufficient strength to be used in the manufacture of salt are confined within a small area in the neighbourhood of Syracuse, New York.

It should be observed, that the wells from which these brines are obtained are bored in drift or alluvium, and in only a few cases reach to the rock *in situ.* The brine occurs in the drift, having percolated from the superior beds of the group which rise above the level of this valley on the south; and its quantity or quality is not improved by penetrating the strata which lie below the gravel.

Exhibition.—The products of the Upper Silurian in the Exhibition, were specimens of iron-ore from Pennsylvania, and gypsum from New York.

Lower Silurian Formation.

The geological formations which at the present time are included in the lower division of the Silurian consist, in the descending order, of a series of argillaceous sandstones, shales, and slates, which are known as Formation III of the Pennsylvania Survey, or the Hudson River Group, including the Utica slate, of the New York Geological Reports. To these succeed limestones of Formation II of the Pennsylvania Survey, or the Trenton, Black River, Birds-eye, and Chazy limestones of the New York Report. To these succeed, as the base of the system, the Formation I, or Calciferous and Potsdam sandstones. These three divisions of the Lower Silurian period require a few words of description separately, though in geographical extent they are, with some modifications, nearly co-extensive.

Group A.—The first-mentioned, or highest division, has a great development towards the north-east; and, entering the United States from Canada, we find it continued along the eastern shore of Lake Champlain, mostly in a partially metamorphic condition. Below Whitehall it expands, occupying a belt on the west as well as on the east side of the Hudson valley, and gradually assuming its normal character. It extends through the Hudson valley as far as Newburg, and thence following a south-westerly direction, we find it continued through New Jersey, Pennsylvania, Maryland, and Virginia, sometimes occupying a single belt of country, sometimes two or three belts, and again uniting into one. In its south-western extension it follows the general trend of the Alleghany range, and corresponds in outline to the preceding and following formations. Westward and north-westward from the Hudson River, these strata pass beneath the lower limestones of the Helderberg from Kingston on the Hudson to the Mohawk; and along this valley they continue westward with numerous undulations, finally rising to the summits of the hills on either side, as the lower rocks are elevated from beneath. Finally they spread out in a broad area upon the eastern shore of Lake Ontario. In Canada they reappear upon the northern shore of the lake, and are traced uninterruptedly to the head of Georgian Bay, Lake Huron. The basin of this bay, and its continuation in the space between the Manitoulin Islands and the main shore, are caused in part by the excavation and removal of the Hudson River Group, the higher beds of which appear on the northern margins of some of these islands. To the westward, this Group reaches the northern shore of Lake Michigan, forming Pointe aux Baies, between the Little and Big Bay de Noquet. Its place is occupied by the waters of Green Bay, near the head of which, on the eastern side, the higher portions of the group are seen passing beneath the superior limestones. It is seen along the eastern shore of Winnebago Lake, and beyond this it is lost apparently from gradually thinning out. From the Georgian Bay, westward, the character of this formation is argillo-calcareous; the higher beds consist of impure limestone, and arenaceous beds are almost entirely wanting.

This formation has a great development in the region fifty miles around Cincinnati, Ohio, and south-westerly along the axis extending through Kentucky and Tennessee. In this part of the United States it consists of alternating beds of soft shale and thin layers of limestone, with occasional traces of the shaly sandstones which are so constant and characteristic a feature of this formation in its eastern extension.

On the Mississippi River, and in numerous places in Wisconsin and Iowa, there occur some thin-bedded argillaceous limestones with seams of shale, overlying the lead-bearing rock, and holding apparently the position of the Hudson River Group. Admitting this as proved, we are able to trace this group from the north-eastern boundary of the United States, in a south-west direction to Alabama; and from the eastern line of New York, through Canada, beyond the Mississippi river; and we have good reason to suppose, from its appearance in the region of Cincinnati and Nashville, that it underlies the superior formations throughout the intervening area.

Group B.—The second division of the Lower Silurian formation, or the limestone group, consists, in its eastern extension, of four distinct members: the Trenton limestone, the Black River limestone, the Bird's-eye limestone, and the Chazy limestone. Of these the Trenton limestone is the most persistent, and often gives name to the entire group.

This series of limestones is recognized as entering the United States from the north by the St. Lawrence and Champlain valleys, and is known along the St. Lawrence quite to its mouth. We find it continuing with slight interruptions through New York, New Jersey, Pennsylvania, Virginia, and Alabama. Westward from the central part of the State of New York it extends along the valley of the Mohawk by Trenton Falls and Black River to the eastern extremity of Lake Ontario. Thence, crossing the St. Lawrence, it passes through Canada, and appears on the shores of the Georgian Bay, and along the north side of Lake Huron; it forms the greater part of the Island of St. Joseph; and, extending westward between Lake Superior and Lake Michigan, it forms the western shore of Green Bay. Thence it enters the State of Wisconsin, continuing in a course nearly parallel to the western outline of Lake Michigan to within fifty miles of the southern boundary of the State, when it bends westward and occupies a broad belt in the southern part of Wisconsin and Northern Illinois. It extends into Iowa, and, following the course of the Mississippi River, occurs, at intervals, as high as the Falls of St. Anthony. In the axis of elevation which extends from below Nashville to Lake Erie, these lower limestones are seen along the Stones River in Tennessee, and at Frankfort, Kentucky; but they do not appear north of the Ohio river.

The limestones of the Lower Silurian period are co-extensive with the superior shales over the eastern part of the United States; while in the western part, or in the Mississippi valley, they extend beyond the positively known limits of the shales.

This group of limestones, whether recognized in the several sub-divisions, or only as a single mass, is everywhere important. In all the more eastern and south-eastern localities, the group consists of limestone, with occasionally interposed layers of chert and thin slaty seams. Westward the portion of shaly matter increases, and in some places, as at the Falls of St. Anthony, it prevails to such an extent as to render a large proportion of the rock nearly unfit for all economical purposes.

There is a general diminution in thickness, and change in character of the limestones of this group as we progress westward; and in this direction commences the intercalation of another member, which is of little magnitude before reaching Wisconsin, but there becomes developed into a mass of much importance, being the lead-bearing rock (Galena limestone) of Wisconsin, Illinois, and Iowa, and formerly described in the Surveys of this region as a part of the Cliff limestone.

Group C.—The lowest division of rocks included in the Lower Silurian consists of the Calciferous and Potsdam sandstones. The former known as the Lower Magnesian limestone in the States bordering the Upper Mississippi. These two rocks often alternate with each other at the line of junction, and in numerous localities we find a friable arenaceous deposit succeeding the calciferous sandstone, being obviously due to a repetition of the same operations which produced the lower arenaceous beds.

The lower division is for the greater part nearly devoid of calcareous matter; nevertheless, in its western extension it contains partially calcareous bands. The upper or calciferous sandstone was originally so named from its combining the arenaceous and calcareous composition which everywhere marks it in varying degrees. This formation enters the United States from the St. Lawrence valley on the north-east, and stretches across the northern part of New York as far as the outlet of Lake Ontario. It may be followed down the valley of Lake Champlain, and thence with interruptions through New Jersey, Pennsylvania, Maryland, Virginia, and Tennessee. Westward through Canada, it skirts the northern shores of Lake Huron, and becomes more extensively developed on Lake Superior; from the outlet of this lake westward, it spreads out to a great width in Northern Wisconsin, and extends along the shores of the Mississippi River, forming the high cliffs which are so conspicuous a feature in the scenery between Prairie du Chien and the Falls of St. Anthony.

Economical Products of the Lower Silurian Rocks.

Iron-ore.—The large quantity of iron-ore described by Professor H. D. Rogers, in his "Geological Report of Pennsylvania," as connected with the limestones of this period, does not, we conceive, compose any part of the formation itself; but the association of the ore with various clays and other deposits clearly indicate for it the same origin as for those other deposits of hæmatite ores which have been described in this report among the products of the tertiary formation.

Lead-ores.—The most important economical product of the Lower Silurian formation is galena, or sulphuret of lead. The lead-bearing rock, or "galena limestone"* of Wisconsin, Northern Illinois, Iowa, and part of Missouri, is a Lower Silurian limestone which has yielded immense quantities of lead-ore for many years past.†

* So named by Professor Hall, who first recognised it as a Lower Silurian limestone (see Geological Report of Lake Superior Land District), it having been described originally as a part of the cliff limestone of the Ohio Reports, or the upper magnesian limestone of Dr. D. Owen.

† The mode of occurrence of this lead-ore is somewhat remarkable, and may be noticed in this place. The metalliferous limestone is from 50 to 80 feet thick, and is a thick-bedded, porous, and friable limestone. This rock is traversed by fissures having a generally north-east and south-west direction, with others crossing these nearly at right angles. The galena occurs in the fissures, those having the most nearly north and south directions containing the largest quantity, or being the best lodes. The ore occurs with calcareous spar, and mixed with or surrounded by clay in the fissures, rarely, if ever, attached to the wall-rock or penetrating it. The limestone below the proper lead-bearing rock is a thin-bedded mass, some of the layers being slaty, and the whole with continuous open fissures. In following the lead-bearing fissures downward, they are found not to penetrate the rock below; and the lode of ore either thins out abruptly or spreads horizontally between the

The lead-ores of Missouri occur in the same rock as those of northern Illinois, Wisconsin, and Iowa, and under similar conditions. Veins of galena and blende are known in several localities of the Trenton limestone in the northern part of New York, but these have never been wrought with profit.

Copper Ores—Veins or lodes of sulphuret and carbonate of copper occur in several localities within the limits of the lead-region and in the same association, particularly at Mineral Point, Wisconsin, where these ores were formerly wrought, but ultimately proved unprofitable; they occur likewise in Missouri. They are confined to a rock of moderate thickness, and are productive in their linear extent rather than in depth. Copper ore has also been found in a lower position in the lower magnesian limestone of Wisconsin, but only in a single locality of small extent. On the south shore of Lake Superior, the lower or Potsdam sandstone is associated with extensive beds of conglomerate and an unbroken range of trap rocks which are copper-bearing. The sandstone and conglomerate contain some ores of copper, but the metallic copper occurs mainly in the trap rocks; none of the veins in sandstone or conglomerate are worked at this time.

Marble, Limestone, and Building-stones.—Some of the beds of limestone afford a black marble of excellent quality, which is extensively used, with marble of other colours, for flooring in halls and public buildings, and also for other purposes. For structures requiring heavy blocks, the lower limestones afford an abundant supply of excellent quality throughout their northern and eastern range, but are of far less value in this respect in their western extension.

The Potsdam sandstone furnishes excellent and durable building-stone and fire-stone.

The slates of the Hudson River group afford roofing-slates in numerous localities in Vermont, New York, and Pennsylvania, and probably farther south. These usually occur on the margin of the metamorphic belt; and, when either entirely unaltered, or too much changed by metamorphic action, they are unfit for this purpose.

Grindstones and Flagstones.—Grindstones of good quality are made fromthe sandstones of the Hudson River group. Flagging-stones are also abundant in many parts along the eastern exposures, and as far west as Lake Ontario. The thin-bedded Potsdam sandstone in many localities affords good flagging-stones.

Exhibition.—The products of the lower silurian rocks in the Exhibition were encrinal marble from Lake Champlain, and also roofing-slates from the semi-metamorphic slates of this formation, which may with equal propriety be referred to the formation next to be described.

Upper Metamorphic Rocks.

To draw the line between lower fossiliferous and upper metamorphic rocks is extremely difficult, inasmuch as the formations between the Hudson River and the Atlantic consist of the various palæozoic rocks already described, sometimes in a transitional state, and sometimes perfectly crystalline and devoid of organic remains. All these rocks, however, in whatever degree they may have

laminæ or strata of the subjacent rock. It is quite impossible that the ore could have come from any source beneath the lead-bearing beds; and the only explanation that can be offered is, that they were filled from above by sulphuret of lead from solution, which, on coming to the subjacent rock with close joints, has spread out horizontally as far as it could penetrate between the laminæ, in some instances replacing the fossils of the rock. The only objection that can be offered to this hypothesis is, that whenever the lead-bearing rock is traced beneath superincumbent beds, these are are of thin-bedded argillaceous sandstone, with shaly partings, and nearly free from joints. This rock is designated the "Cap Rock" by the miners, and the lead-ore is said always to terminate abruptly against the lower side of the "Cap Rock." The mines which have been worked are mainly in the exposed surface of the lead-bearing rock, and have never been carried very far beneath where it is covered with the "Cap Rock;" and therefore the character and condition of the lodes in the lead-bearing rock beneath the overlying formation are not so well known. If the mines in this position should continue to yield as abundantly, there are still large areas of lead-bearing rocks. which remain untouched; but, on the contrary, should they prove less prolific beneath the superincumbent rock, then we may regard the lead region of the West as destined to be soon exhausted. —*From Manuscript Notes of Prof. James Hall.*

undergone metamorphic action, may be classed as upper metamorphic rocks, and distinguished from an older series of crystalline rocks to which they are unconformable in stratification, as may be seen at Little Falls, New York, along the valleys of Lake Champlain and the St. Lawrence, and in numerous other localities. The upper metamorphic rocks comprise the quartz-rocks, or altered sandstones, the finely granular, often micaceous, white and coloured limestones, the schistose rocks, varying from soft and fissile black, green, and reddish slaty rocks, to the micaceous and talcose schists into which they graduate, and various gneissoid, syenitic, and hornblendic rocks—all of which can be recognized as clearly derived from stratified fossiliferous rocks of palæozoic age.*

These formations have been proved to occupy large portions of the New England States, and it is only in small areas, if at all, that the older metamorphic rocks exist within their limits. The extension of the same formation south-west, in a direction parallel to the older fossiliferous rocks, shows a wide area occupied by the rocks of this age, co-extensive indeed with the palæozoic formations from north to south.

The direction and dip of strata composing these formations are usually well preserved, though sometimes, within limited areas, they are obscured, and numerous dislocations and inverted axial folds render their study difficult.

The larger portion of this formation consists of rocks older than the coal-period; while to these we find superadded, in the eastern part of New England, the metamorphic carboniferous rocks. In numerous localities, these metamorphic rocks can be traced as positive continuations of the fossiliferous strata, the fossils gradually disappearing as the beds assume a more crystalline condition, from the gradually increased metamorphism; all doubt of the cause of the obliteration of organic remains being thus removed.

The rocks of this age produce very marked physical features in the topography of the countries which they occupy. Entering the United States on the north by the St. Lawrence valley, we find the rocks of palæozoic age, which thus far have been observed in their undisturbed and unaltered condition, assuming slight lithological changes, which increase until the strata become more or less crystalline; the fossils are gradually obliterated; the beds become elevated, often vertical, and even appear in overturned axial folds; and, rising into numerous hills and ridges, they finally assume the character of mountain-ranges, as in the Green Mountains of Vermont and the White Mountains of New Hampshire.

Devonian metamorphic.—The altered rocks of the age of the Chemung and Portage groups of New York, forming part of the Devonian, are recognized by Messrs. Logan and Hunt as appearing in great force at Gaspé, and are known to be highly developed in the State of Maine. These rocks, in their metamorphic condition, skirt the eastern part of New England, forming a part of the White Mountains; and, passing beneath the anthracite coal-formation of Massachusetts and Rhode Island, constitute with the latter the greater part of the metamorphic and crystalline formations on the east of the Connecticut River.

Silurian metamorphic.—In that portion of the country lying between the Hudson River on the west and the Connecticut River on the east, the more or less metamorphosed rocks consist of those formations already described as silurian. In the western part of this belt the quartz-rocks or hard sandstones, the crystalline limestones, and various slaty rocks terminate above by coarse gneissic and hornblendic beds, which are recognized as the altered forms of the Potsdam sandstone, the Trenton limestone group, and the Hudson River group, with occasional and interrupted beds of the Shawangunk conglomerate, infolded with the slates, and forming the summits of the most conspicuous ridges.

The upper silurian rocks have been traced by Messrs. Logan and Hunt from their unaltered condition in the St. Lawrence valley to within the boundary of the State of Vermont, where they assume the condition of micaceous limestones, interstratified with micaceous slates.

The lower part of this metamorphic formation, or chiefly that composed of the lower silurian slates and associated arenaceous beds, constitutes the true auriferous belt, as has been abundantly proved in the northern extension of these rocks, and we are well warranted, from numerous observations in the

* Granite, which is usually referred to true igneous origin, occupies comparatively little of the area described.

Southern States, in inferring that the auriferous belt of Virginia, North and South Carolina and Georgia is of the same age. Whether the auriferous rocks of California are of the same age as those of the upper metamorphic rocks of the Eastern United States, is a question which cannot now be decided.

Economical Products of the Upper Metamorphic Rocks.

Gold.—The most prominent and important product of the metamorphic rocks of this age is gold, which occurs more or less abundantly throughout the whole extent of the formation from Canada to Georgia. The gold is obtained either directly from the rocks themselves, or from the auriferous debris derived from their destruction by natural causes. At the present time the gravel and sand, resulting from the disintegration of these rocks, are searched for gold on the Chaudière River in Canada; and gold has been found in the rocks and in the gravel in several places in Vermont and Massachusetts; though they have never proved productive of any considerable amount. From the facts here adduced, it appears to be demonstrated that the gold is not indiscriminately distributed through all the rocks of this age, but is in reality confined to a comparatively narrow belt.

Iron Ores.—The iron ores directly connected with these rocks are not often productive; though, from the decomposition of iron pyrites and other causes, beds of iron-ore of secondary origin frequently occur. Some iron-ores of this formation occur in workable quantities in Vermont and New Hampshire; and one or more beds of little economical value are known farther to the south in Massachusetts.

Chromic Iron Ores and Serpentine.—Associated with the rocks of this age occurs chromate of iron in numerous localities, from the northern limits of the United States, as far south as Virginia, and probably still farther. These ores have been wrought in Maryland and Pennsylvania.

In connection with these ores occur chlorite beds and serpentine rocks, the one apparently co-extensive with the other. The part of the formation bearing these products, however, is not continuous, but occurs at intervals where certain beds existed in the original sedimentary deposits. The serpentines, or *Verd antique*, are of beautiful varieties; and among their localities may be mentioned Millford, near New Haven, Connecticut, and Cavendish and other towns in Vermont.

Copper Ores.—Sulphurets and carbonates of copper frequently occur in the gold region of Virginia and North Carolina, and in the same formation in Maryland. The copper ores of Bristol, Connecticut, and some other localities in that State, must also be referred to rocks of this period.

Tin Ore.—The tin ore (oxide of tin) described by Dr. Jackson as found in the town of Jackson, New Hampshire, occurs in rocks of this age. Tin-ore in small quantities often accompanies the gold in Virginia and North Carolina.

Lead Ores.—Galena, carbonate, sulphate, and phosphate of lead are known in rocks of this age, but their economical value has never been fully tested. Sulphuret of lead occurs in considerable quantities in some of the copper mines recently opened in Maryland.

Silver.—Native silver occurs in Davidson county and other localities in North Carolina.

Building Stones.—Whether we regard the syenitic rocks of New England as belonging to this period, or as igneous and intrusive rock, they are nevertheless so intimately connected with these upper metamorphic rocks, that it is necessary to speak of them under the same head. The firm and durable syenite of the neighbourhood of Boston, known as Quincy granite, so much prized as a building-material, and so extensively used, belongs to this formation. Still further eastward, in the State of Maine, building-materials of this kind are wrought to considerable extent. This syenite is usually of lighter colour than that of Quincy, but becomes darker by the action of the weather.

Flagging Stones.—In nearly every part of this formation flagging-stones and step-stones of good quality are obtained.

Millstones.—The syenite in some localities is sufficiently quartzose to produce good millstones, which are manufactured in the neighbourhood of Salem in Massachusetts, and other places.

Whetstones.—The talcose and mica-slates of this age are often sufficiently siliceous to be used as whetstones or hones.

Statuary Marble.—White and variegated marble of various qualities are found in these formations. They result mainly from the metamorphism of several beds of lower silurian limestone, which vary in character in their normal condition. In the State of Vermont some beds of statuary marble of excellent quality have been found.

Fine marble occurs in numerous localities between the northern line of Vermont and the city of New York, and large quantities are obtained from Vermont, Massachusetts, Connecticut, New York, and Pennsylvania.

The marble from these localities is extensively used for building and decoration, &c., the coarser varieties for floor-tiles and ordinary building purposes.

Mica.—This mineral, which occurs in the metamorphic rocks, or associated with belts of crystalline intrusive rocks, is a substance of economical value. Large quantities are annually used for the doors and sides of stoves, which are so extensively manufactured in the United States.

The entire supply at the present time is obtained from two localities in New Hampshire, Grafton and Acworth.

Exhibition.—The products of this formation presented in the Exhibition are specimens of gold from the rocks in place in Maine, Vermont (a crystallized specimen in quartz), Virginia, North and South Carolina, and Georgia, and some from California. The principal exhibited specimens of gold were from the gravels derived from these rocks, and which have been already enumerated under the head of Drift or Alluvium. Silver was exhibited from North Carolina; copper ores from Connecticut, Maryland, and North Carolina; lead ore from Massachusetts, Maryland, and Virginia; tin ore from New Hampshire; chrome and iron ores from Pennsylvania and Maryland; plumbago from New Hampshire; varieties of marbles from Vermont, Connecticut, New York; mica from New Hampshire.

Older or Lower Metamorphic Rocks.

In many parts of the United States there are broad areas occupied by crystalline rocks, which in age and lithological characters are quite distinct from the preceding. For these it is proposed to adopt the term Older or Lower Metamorphic, to distinguish them from those which are clearly altered rocks of the Palæozoic period. They are known to be of prior date, and in many cases, if not always, to have acquired their crystalline structure and metamorphic condition previously to the deposition of the surrounding sedimentary formations of the age of the oldest Palæozoic rocks. In numerous localities the Potsdam sandstone, which is the oldest known fossiliferous rock in the United States, lies in a horizontal, or nearly horizontal position against the inclined masses, or rests upon the vertical edges of the rocks of this age; proving the latter to be of long anterior date.

The formation consists in a great degree of syenite, gneiss, and various hornblendic, micaceous, and talcose schists, &c., with highly crystalline limestone: the whole still preserving evidence of stratification to a degree which warrants the opinion that they were originally sedimentary rocks which have been metamorphosed by igneous agencies. Among the evidences of the causes of this change, we find large intrusive masses of granite and other rocks generally regarded as of igneous origin, penetrating the stratified crystalline rocks, of which the system is in a great measure composed. The whole system is likewise penetrated by numerous dykes of trap, which are sometimes of great width and extent. The prevailing and characteristic rock of this formation is a syenitic gneiss, generally of firm and compact texture. The divisional planes are often obscure, and are frequently lost in the granitic or syenitic structure. One of the most constant and characteristic accompaniments of this formation is magnetic iron-ore.

The chain of mountains on the north of the St. Lawrence, and extending parallel with that river as far as the Thousand Islands, is mainly composed of this formation. We find the same rocks extending through Canada to the shores of Lake Superior, occurring on both sides of this lake, and stretching westerly to the sources of the Mississippi.

Some portions of the mountain-ranges of Maine and New Hampshire may, perhaps, belong to this formation; but the extent and limits of such portions have not been defined, nor have these beds been distinguished from surrounding rocks of more recent origin.*

In Northern New York this formation rises into mountains of 5000 feet in height, but generally its elevation is much less. In this region it occupies an area of a somewhat triangular shape, having a length of nearly 150 miles from north to south, and about 125 miles from east to west, or from Lake Champlain westward to the St. Lawrence river. The more elevated ridges are towards the centre of this tract. To the north and south it gradually dies out, not however without producing some disturbance among the superincumbent silurian rocks. In the southern part of New York this formation reappears, rising into elevated ridges and mountains, which, however, scarcely attain one-half the elevation of those in the northern part of the State. From the southern limit of New York it enters New Jersey, forming an almost continuous range of mountainous country from the New York boundary to the Delaware River. From the Delaware River to the Maryland State boundary, the older metamorphic rocks extend through the south-eastern part of Pennsylvania in two narrow belts, one of which is continuous, and the other occurs only in insulated patches in the midst of formations of more recent date.

The rocks of this age continue in the same general direction through Maryland, Virginia, and North Carolina, and are known as the Blue Ridge or South Mountains, the highest parts of which in Virginia reach the elevation of 4000 feet. Continuing south-westerly, this formation probably extends more or less interruptedly to the southern limits of the metamorphic formations in Alabama.

The rocks of this formation have been shown to occupy a broad belt on the southern shore of Lake Superior,† occupying, with some slight interruptions, the coast of the lake from the neighbourhood of Camp River to Point Abbaye; they extend westerly with gradually expanding width into Northern Wisconsin; and in the central part of that State reach as far southward as the parallel of the outlet of Winnebago Lake. Thence the course of the formation is to the north-west, reaching the Mississippi River north of the Falls of St. Anthony.

There is an isolated belt of this formation in Missouri, one in Arkansas, and another in Texas.

Economical Products of the Older Metamorphic Rocks.

Iron-Ores.—In point of economical resources this formation, both in Canada and the United States, is perhaps second only to the coal-formation. The magnetic iron-ores from Canada, forming immense veins or masses, so fully illustrated in the London Exhibition of 1851, were from the rocks of this older metamorphic period. This mineral, either as magnetic protoxide of iron, or mingled with anhydrous peroxide in various proportions, is charac-

* In Canada Mr. Logan recognised another series of metamorphic rocks, resting upon those just described, and of anterior date to the newer metamorphic; thus making three divisions. This middle formation is made up of sandstones, conglomerates with red jaspery pebbles, beds of slate, and limestone. "The slates are greenish, and highly chloritic, and often containing epidote; sometimes they assume the character of conglomerates, from the presence of pebbles of syenite."

† Report of Foster and Whitney on the Lake Superior Land District, 1851; and of D. D. Owen on the Chippeway Land District, 1852.

teristic of the formation everywhere. In Northern New York the mountains are traversed by veins or beds of the same ore, which have the appearance of intercalated beds running parallel to the schists, and having a width or thickness of from 2 or 3 feet to more than 100 feet, and can often be traced for a mile or more in length.* Numerous beds are opened in the vicinity of Lake Champlain, where the ore is largely used in manufacturing iron, and whence also it is shipped southward to the valley of the Hudson, where it is used for smelting with inferior ores, to improve the quality of the iron.

The ores of the Adirondack region, in Northern New York, have been successfully used in the manufacture of very fine iron and steel, which were likewise shown in the London Exhibition in 1851.† In the western part of the metamorphic region here described, the iron-ores are chiefly the peroxide or specular iron-ore. In the southern part of the same State, extensive beds of magnetic oxide of iron occur in the same rocks, and are wrought to considerable extent. In New Jersey, Pennsylvania, and Maryland, these magnetic ores are wrought; but farther south they seem to have attracted little attention.

On the southern shore of Lake Superior the rocks of this class contain extensive beds or veins of mixed magnetic ore and peroxide of iron.‡ In the western extension of the same formation, and in the vicinity of Montreal and Bad River, in Wisconsin, beds of magnetic iron-ore are known to exist.§ The iron-ores, a mixture of magnetic iron and peroxide, of the Iron Mountain and Pilot Knob in Missouri are derived from rocks of the same age. Similar ores occur in the metamorphic region of Arkansas.

Copper-Ores.—The rocks of this age in Northern New York have yielded small quantities of copper-ores in several places, but no mines have been profitably wrought. The Bruce Mine on Lake Huron, and others on the northern shore of Lake Superior, which yield copper, chiefly from the sulphurets, are in this formation, or in the intermediate metamorphic division recognized in Canada by Mr. Logan.

Lead Ores.—The lead ores of Rassie and its vicinity, in Northern New York, occur in rocks of this age. The lead and copper mines of Montgomery and Chester counties, Pennsylvania, already mentioned in connection with the new red sandstone, occur in the gneiss rocks of this system. These veins promise to yield largely of these two metals. The line of demarcation between these rocks and the newer metamorphic is not so clearly defined in the region southward as to enable us to designate with certainty the mines which may be in this or in the preceding formation. In this direction both series appear to be more prolific in ores, excepting iron, than the same formations in New England and New York.

Zinc Ores.—Zinc-blende often occurs with the lead-ores in this formation; and veins of blende, sometimes with a small admixture of galena, are found. Thus far this ore has proved of no economical value. The red oxide of zinc and franklinite of New Jersey must be regarded as of this period. The intimate connection of these products with the older gneiss rocks, and their associated minerals, point to this age; though the crystalline limestone found in the same geological association is regarded by Professor Rogers as of silurian age. The red oxide is largely used in the manufacture of the white oxide of zinc, and the mixture of this ore with the franklinite, ground in oil in its natural state, forms a brown paint which is much used. Although producing metallic zinc of excellent quality, it is not manufactured at the present time. The franklinite of the same mine has, it is said, of late been successfully used in a process by which the oxide of zinc is obtained and the iron reduced, both processes being accomplished by the same furnace operation.

Phosphate of Lime—In fine crystals, and in a massive form, occurs in rocks of this age. In one locality in Northern New York,‖ a vein or bed of this mineral was discovered, which at one time promised to prove of great value for

* The details of many of these have been given in the Geological Reports of New York, and others have subsequently been discovered.

† Jury Reports, pp. 17 and 18.

‡ Foster and Whitney's Report on the Lake Superior Land District. Part 2. 1851.

§ Dr. D. D. Owen's Report on Wisconsin, Iowa, and Minnesota.

‖ See also Mr. Logan's notice of phosphate of lime in the metamorphic rocks of Johnstown, East Canada. Geol. Journ., vol. viii., p. 210.

agricultural purposes. It has, likewise, been found in considerable quantities in New Jersey..

Porcelain Earth.—Granite proper forms but a small part of the extent of country occupied by the metamorphic rocks of the United States. It occurs in both the formations as intrusive bands, dykes, or more extended masses. Many of these granite rocks are very coarsely crystalline, and consist mainly of feldspar, with small proportions of mica and quartz. The feldspar, whether in a partially decomposed condition or in its natural state, is largely used in the manufacture of porcelain and the finer kinds of earthenware.

Marble and Building-stones.—The lower metamorphic rocks usually afford an abundance of rather coarsely crystalline limestone, with occasional beds of finer quality. None of it, apparently, can come under the designation of statuary marble; it is, however, adapted to many purposes for which the coarser varieties of white or clouded marbles are used; and in many localities it is a good building-material. In Northern New York it is sometimes mixed with serpentine, which gives it a variegated appearance. On the southern shore of Lake Superior these marbles are described as of a beautiful pink colour, traversed by veins of red; others are blue and dove-coloured, and beautifully veined. The gneiss and granite from this formation afford excellent building-materials of a great variety and colour; and in many places these may be quarried in massive blocks. The situations where the best examples occur are not easily accessible for the means of transportation to the larger cities.

Fire-stones, Hones, &c.—The talcose slates of this period often furnish good fire-stones, which, when the divisional planes of the rock are not of too frequent occurrence, and when blocks of sufficient dimensions can be obtained, may be used for the construction of iron-furnaces. Sometimes beds of a fine material, as potstone or compact steatite, are found associated with the limestone.

Some of the slates contain sufficient silica to adapt them for hones or whetstones. This material, of good quality, has been found on the south shore of Lake Superior.

Plumbago and Crystalline Minerals.—Plumbago occurs in considerable quantities in rocks of this age in Northern New York.

In some localities the rocks of this period are extremely rich in beautiful minerals. Amongst these may be mentioned, fine crystals of apatite, zircon, spinelle, sphene, augite, tourmaline, &c.

Labrador feldspar, in large crystalline masses, is not of unfrequent occurrence in Northern New York. This mineral, when polished, is very beautiful; and some of the finer specimens have been set in gold and worn as ornaments.

Exhibition.—The products of the older metamorphic rocks presented in the Exhibition were iron-ores from Northern New York, New Jersey, Pennsylvania, and Maryland; some large masses from Lake Superior and Missouri, with manufactured iron from most of the localities, and fine specimens of steel from Adirondack mines; also,—copper-ores, from Pennsylvania; lead-ores, from Northern New York and from Pennsylvania; red oxide of zinc and franklinite, with products of manufacture, from New Jersey; hone-slate, from Lake Superior.

Besides these, numerous beautiful crystalline minerals from the older metamorphic formations of Northern New York and New Jersey were exhibited.

The Trappean Formations.

New Red Trap Rocks.—The most conspicuous formation of this kind in the Eastern United States is that which is associated with the new red sandstone, the extent of which has already been described. Accompanying the sandstone of the Connecticut River valley, trappean or basaltic rocks occur in such force as to present marked features in the topography of the country. Mount Holyoke and Mount Tom, in Massachusetts, are conspicuous examples of this kind. The East and West Rock, near New Haven, are also prominent points in the landscape, which are due to this trappean formation. Less conspicuously the trap occurs in numerous, often parallel, ridges and dykes, and in long continuous belts throughout the extent of the river valley, for 100 miles above its mouth. The Palisades of the Hudson River, and numerous prominences of trap in New Jersey, are all of the same age, and have the same geological associations. Finally, this trap-formation may be traced through Pennsylvania, Maryland, Virginia, and North Carolina, everywhere accompanying the sandstone rocks.

One of the constant accompaniments of this trap-formation is native copper; but, throughout its entire extent, it has never been found in available quantities. Other ores of copper often occur in the sandstones adjacent to the trap-rocks; and there appears to have been some connexion between the intrusion of these trap-dykes and the segregation of the copper ores. The trap-rocks of the period of this new red sandstone are often columnar in character, and their mode of weathering is by vertical joints. This feature is less prominent in the trappean rocks next to be described.

Lake Superior Trap-rocks.—The second great trap-formation is that of Lake Superior, which occurs in connexion with the sandstone of the Lower Silurian period. This formation is most conspicuously developed upon Keweenaw Point, and extends thence in a broad belt west and south-west, conforming generally to the shore of the lake. It is flanked on the northern side by conglomerates, and on the southern by sandstones; and a narrow belt of trap runs through the conglomerate parallel with the main central formation. Towards the western extremity of Lake Superior, this trap-formation becomes interrupted, and occurs only in insulated dykes or ridges piercing the sandstone. Isle Royale presents many features in common with Keweenaw Point; the range of the trap is similar and parallel, though there is but a small development of the conglomerate. On the north shore of the lake, within the territory of the United States, the trap extends in numerous narrow belts or dykes entirely to the western shore of the lake. Thence, at several points between this and the Mississippi River, above the Falls of St. Anthony, belts of trap are of frequent occurrence.

Beyond this river less is known of its extent or character; but, from the little information we have, it appears probable that the same formation occurs at intervals between the Mississippi and the Rocky Mountains.

Throughout the whole, the associations are so similar as to leave no doubt that this trap is all of the same age. On the St. Croix River it is thrust through the lingula-beds of the Potsdam sandstone in great bosses or knobs; and in Northern Wisconsin it occurs in nearly similar situations, more or less connected with the lower sandstone, but frequently associated with syenite or syenitic gneiss.

The trap-formation of Lake Superior, where it appears in its greatest development, is the repository of immense quantities of native copper, which in that region everywhere characterizes the presence of trap of a certain variety. In a map of Keweenaw Point, recently published by Mr. Whitney, United States Geologist, we find the trap of Keweenaw Point thus described. Amygdaloidal trap, bearing native copper; porphyritic trap, bearing sulphuret of copper; and crystalline trap, in which the veins are not productive. The principal part of the formation of Keweenaw Point is composed of the former kind of rock; a narrow strip passing longitudinally through the centre, represents the crystalline portions; and a broader belt on the south side, adjoining the sandstone, is of the latter variety.

Trap-rocks of the Western Regions.—From the borders of Texas to the country of the Blackfoot Indians, and still further north, trappean and basaltic rocks are found at frequent intervals. Westward, in the great plain between the Rocky Mountains and the Sierra Nevada, rocks of this character are of frequent occurrence, though we do not know their geological relations. Again, to the north-west, in the Rocky Mountains and along the Columbia as far as the Pacific coast, there are numerous and extensive basaltic and volcanic rocks. But it would require too much detail for a Report like the present, to give a connected view of all the observations that have been made upon these igneous rocks in the great area here indicated.

Trap-rocks of the Metamorphic Series.—A system of the trappean rocks of a different age, consisting of trap, greenstone, porphyry, &c., traverse the metamorphic rocks of the periods above described. These, however, rarely produce any distinct features in the topography of the country, nor do they appear to have had any connection with the upheaval of the mountain ranges in which they occur. They traverse these rocks in various directions, apparently filling fissures or joints, of which several are sometimes parallel to each other. Although these dykes occur everywhere in the United States, they are more numerous and extensive on the northern shore of Lake Huron and Lake Superior.

In the eastern part of New England, and particularly near the coast, the strata are traversed by these dykes; and more extended belts of similar rocks occupy portions of the metamorphic area. The porphyritic belts appear sometimes to have resulted frcm extreme metamorphism, or igneous fluidity, of previously existing formations.

Economical Products of the Trappean Formations.

Copper.—Native copper in large quantities is obtained from the trap-rocks of the Lake Superior region, where numerous mines are opened on the south shore of the lake. The copper occurs in sheets of greater or less thickness, in veins cutting the trap-range nearly at right angles, and associated with various vein-stones. These sheets of copper are of variable extent, weighing from a few pounds to 80 tons. The produce of copper from the mines of Lake Superior, during the year 1853, will reach 4000 or 5000 tons.

A mass of native copper in the Exhibition from one of these mines weighs 6,300 pounds; it is cut upon four sides, and the thickness between the two natural surfaces is more than two feet: this was cut from a mass weighing 40 tons.

The sulphurets of copper in the porphyritic trap, on the south shore of Lake Superior, have not yet been wrought to any great extent.

Silver.—The native copper contains a small amount of native silver, and this metal also occurs in small nodules or irregular masses in the copper, or attached to its exterior surface, and also in the accompanying trap.

In these cases the quantity of silver is usually small, but some masses of large size have been discovered in the *debris*, associated with lumps of native copper. More recently, a vein of silver has been wrought upon the north shore of Lake Superior. Specimens from this place were exhibited in the Exhibition.

Building-stone.—The trap-rocks, which are traversed by numerous joints, separate into irregular angular blocks used for rough walls, and to some extent for buildings. The rapid breaking down of these rocks along the Palisades of the Hudson River, by the action of water and frost, renders them accessible; and large quantities are annually transported to the city of New York and its vicinity.

GENERAL REMARKS.

From the foregoing sketch of the economic geology of so large a portion of North America, it will be seen that the great metalliferous belts of the United States are the two nearly parallel and contiguous ranges of metamorphic rocks, which extend almost uninterruptedly from the north-eastern extremity of the United States to Alabama. To these two principal zones must be added an area of considerable breadth in Northern New York, one of similar character on the south shore of Lake Superior, a small area in Missouri, one in Arkansas, and another in Texas. On the western side of the continent we have the metalliferous belt of the Pacific coast, extending through California and Oregon. Too little is known of the Rocky Mountains to enable us to speak with any degree of confidence with regard to their future mineral produce.

Permanent and productive mines of the metals will not be found beyond the areas which have been thus hastily sketched out. The general depositories of mineral wealth, therefore, are within moderate proximity to the coasts. The copper and iron region of Lake Superior lies upon navigable waters, communicating by artificial channels with the Atlantic. The coal formation, the source and foundation of every enterprise, stretches almost the whole length of the country, from the great lakes on the north to the Gulf of Mexico on the south.

The natural distribution, therefore, of these sources of wealth and power, combined with the physical features of the entire country, leave nothing to be desired, with respect to the materials and incentives for its physical progress and development.

INDEX

TO THE REPORT ON THE MINING PRODUCTS AND THE GEOLOGY OF THE UNITED STATES.

Acworth, N.H., 30
Adirondack iron-mines, N.Y., 32, 33
Agricultural characters of the carboniferous limestone districts, 18
—— conditions of the cretaceous districts, 11
—— use, mineral products for, 9, 11
Alabama, auriferous drift, 9
—— carboniferous limestone, 17
—— cretaceous, 11
—— iron-ores, 16, 17
—— lead-ores, 18
—— lower metamorphic, 31
—— lower silurian, 25
—— palæozoic rocks in, 6
—— survey, 4
—— tertiary, 10
—— upper silurian, 23
Albany, 20
Alleghany coal-field, 4, 15, 17, 20
—— mountains 6, 7, 9, 11, 14, 21, 25
Alluvial formation, 8
Alluvium, economic products of the, 8
Anthracite basins, 14, 16
Apatite, 33
Appalachian chain. *See* Alleghany Mountains.
Arago county, upper silurian, 23
Arkansas, cretaceous, 11
—— coal-field, 14
—— iron-ore, 32
—— lead-ores, 18
—— lower metamorphic, 13
—— metalliferous area of, 35
—— metamorphic rocks, 28
Augite, 33
Auriferous drift, 8
Bad River, 32
Barytes, new red, 13
Basaltic rocks, 34
Bay de Noquet, 25
Bellingham, Bay coal, 15
Bennington, Vermont, 10
Bird's-eye limestone, 24, 25
Black oxide of manganese, 10
—— River, 25
—— limestone, 24, 25
—— rock, Niagara, 20
—— shales, 19
Blende, Illinois, 18; New York, 27
—— upper silurian, 23
Blue ridge, 31
Bog iron-ore, 9
Boulders of the drift, 8
Boston, 29
Brandon, Vermont, 10
Bricks, 9, 10, 11, 17
Brines, New York State, 24
Bristol, Conn., 29
Brookdale lead and copper mine, 13
Brown coal, tertiary, 10
—— hœmatite, tertiary, 10
Bruce copper-mine, 32
Brussels, N.Y., 20
Buffalo, 15
Building-materials, coal-formation, 17
—— drift, 9
—— tertiary, 10, 11
Building-stone, carboniferous limestone, 18
—— devonian, 21
—— jurassic, 12
—— lower metamorphic, 33
—— lower silurian, 27
—— trap-rocks, 35
—— triassic, 13
—— upper metamorphic, 29
—— upper silurian, 24
Cabot's Head, 22
Calciferous sandstone, 24, 26
California, auriferous drift, 9
—— gold, 30
—— metalliferous area of, 35
Camp River, 31
Canada, auriferous drift, 9
—— devonian, 20
—— geological maps of, 5
—— gold, 29
—— iron-ores, 10, 31
—— lower metamorphic, 31
Canada, lower silurian, 24, 25, 26
—— phosphate of lime, 32
—— upper silurian, 21, 22
Capitol, Washington, 11
Carbondale coal-mine, 16
Carboniferous limestone, 17
—— rocks, metamorphic, 28
Carolina. *See* North Carolina and South Carolina.
Cattskill group, 21
—— Mountains, 19
Cauda-galli grit, 20
Cavendish, Vermont, 29
Chalk, absence of true, 11
Charlestown lead and copper mine, 13
Chaudière River, 9, 29
Chazy limestone, 24, 25
Chemung group, 19, 21, 28
Cheshire, Conn., new red, 13
Chester county, Penns., lead and copper, 13, 32
Chesterfield Basin, jurassic, 12
Chicago, 16, 24
Chippewa, Land district, survey, 4
Chromic iron-ore, upper metamorphic, 29
Cincinnati, 20, 22, 25
Clays, alluvium and drift, 9
—— coal-formation, 17
—— tertiary, 10, 11
Cleveland, Ohio, 20
Cleveland, 15
Clinton group, 21, 22, 24
Coal, extent, supply, and consumption of, 4, 12, 15, 35
—— fields, 12, 14
—— formation, 14
—— jurassic, 12
—— measures, 14
Columbia River, 34
Connecticut, copper-ore, 39, 30
—— iron-ore, 10
—— marble, 30
—— new red, 12
—— River, 28
—— sandstone, 13, 33
—— serpentine, 29
—— survey, 4
—— trap rocks, 33
Copper and copper-mines, 4. *See* Copper-ores.
Copper-ores, drift, 9
—— lower metamorphic, 32
—— lower silurian, 27
—— new red, 13
—— trap rocks, 34, 35
—— upper silurian, 23
—— upper metamorphic, 29
Council Bluff, on the Missouri, 6
Cretaceous formation, 11
Crystalline minerals and gems, 33
Cumberland Mountains, 17
Dance, Mr. F. D., 10
Davidson county, silver, 29
Deep River Basin, jurassic, 12
Delaware, cretaceous, 11
—— River, 23, 31
—— survey, 4
Delthyris shaly limestone, 21
Des Moines River, Iowa, 14, 16
Devonian formation, 19
—— rocks, metamorphic, 28
Drain-pipes, 17
Drift-formation and its economic products, 8
Dykes, granitic, 33
—— trappean, 34
Eastern coal-field, 4, 14
—— States, auriferous drift, 8
—— cretaceous area, 11
—— general geology, 6
Economic products of the alluvium and drift, 8
—— carboniferous limestone, 18, 19
—— coal-formation, 15
—— cretaceous series, 11
—— devonian rocks, 21
—— jurassic rocks, 12
Economic products of the lower metamorphic rocks, 31
—— lower silurian rocks, 26
—— new red series, 13
—— tertiary series, 10
—— trap-rocks, 35
—— upper metamorphic rocks, 29
—— upper silurian rocks, 23
Encrinal limestone, 21
—— marble, 27
Erie Canal, 24
Exhibited, products, 9, 10, 11, 13, 15, 18, 19, 21, 24, 27, 30, 33, 35. *See also* Economic products.
Falls of Niagara, 22
—— St. Anthony, 25, 26, 31, 34
—— the Ohio, 22
Fire-bricks, 10, 17
—— clays, coal formation, 17
—— tertiary, 10
—— stones, devonian, 21
—— lower metamorphic, 33
—— lower silurian, 27
Flagging stones, devonian, 21
—— lower silurian, 27
—— upper metamorphic, 30
—— upper silurian, 24
Formations, list of the, 7
Fort Laramie, coal, 15
—— carboniferous limestone, 18
Fort Mandan, cretaceous, 11
Fossil fuel, 10, 12, 15
—— wood, cretaceous, 11
—— tertiary, 10
Foster, Mr. J. W., 4, 22, 31
Frankfort, Kentucky, 25
Franklinite, lower metamorphic, 32
Fremont, Captain, 10
Galena limestone, lower silurian, 26
—— upper silurian, 23
Gaspé, 28
Gems and crystals, 33
Geological maps and surveys, 4
Geology of the United States, 5
Georgia, auriferous drift, 9
—— rocks, 29
—— cretaceous, 11
—— gold, 30
—— tertiary, 10
Georgian Bay, 22, 25
Glass-making, sand for, 17
Gold-bearing drift, 8, 29
—— metamorphic rocks, 28
Gold, superficial deposits, 8
—— upper metamorphic, 29
Grafton, N.H., 30
Granby, 13
Granite, 28, 30, 33
Great Salt Lake, carboniferous limestone, 18
Green Bay, 22, 25
Green-earth, tertiary, 11
Green Mountains, 9, 10, 22, 28
Green River, Kentucky, 14
Green-sand, cretaceous, 11
Grindstones, lower silurian, 27
Gross Cap, 20, 22
Gulf of Mexico, 6, 7, 10, 15
Gypsum, upper silurian, 24
Hacking county, coal, 16
Hæmatite, drift, 9
—— tertiary, 10
Hall, Prof. James, 3, 5, 14, 22, 26
Hamilton group, 19
Helderberg limestone group, 21, 25
—— Mountains, 20, 21
Highlands of the Hudson, 6, 10
Hitchcock, Prof., 5, 10
Hocking, iron-ores, 17
Hones, lower metamorphic, 33
Hudson, Highlands of the, 6, 10
Hudson River, 12, 20, 21, 23, 24, 25, 27, 28, 32, 35
—— group, 24, 25, 27, 28
Hunt, Mr. S., 28
Hydraulic lime, 21
Illinois coal-field, 14, 16
—— cretaceous, 11

Illinois, devonian, 20
——— geological survey, 16
——— lead and zinc ores, 18, 26
——— lower silurian, 25, 26
——— metalliferous area of, 6
——— River, 16
——— survey, 4
——— upper silurian, 22
Indiana, carboniferous limestone, 17
——— coal-field, 14
——— devonian, 20
——— sandstone, 19
——— upper silurian, 22
Iowa, coal-field, 14
—— devonian, 20
—— geological map of, 4
—— lead-ore, 26
—— lower silurian, 25, 26
—— metalliferous area of, 6
—— sandstone, 19
—— upper silurian, 22
Iron and iron-mines, 4. *See* Iron-ores.
—— manufactured, 10, 33
—— Mountain, 32
Iron-ores, alluvium and drift, 9
——— carboniferous red shale, 19
——— coal-formation, 16
——— cretaceous, 11
——— devonian, 21
——— lower metamorphic, 31
——— lower silurian, 26
——— magnetic, 31
——— tertiary, 10, 26
——— upper metamorphic, 29
——— upper silurian, 23
Island of St. Joseph, 25
Isle Royal, 34
Jackson County, Ohio, coal, 16
——— iron ores, 17
Jackson, Dr. 29
Jackson, N.H., 29
James River, 12
Johnstown, Canada, 32
Jurassic formation, 12
Kanawha River, 15
Kaolin, 10
Kentucky, carboniferous limestone, 17
——— coal-field, 14
——— coal, 16
——— devonian, 20
——— iron-ores, 17
——— lower silurian, 25
——— upper silurian, 22
Keweenaw Point, copper, 34
——— geological map, 34
——— trap-rocks, 34
Kingston, 23, 25
Kittatinny Mountain, 23
Labrador-felspar, lower metamorphic, 33
Lake Champlain, iron-ore, 32
——— lower metamorphic 31
——— lower silurian, 24, 25, 26
——— marble, 27
——— tertiary, 9
——— upper metamorphic, 28
Lake Erie, devonian, 20
——— lower silurian, 25
——— upper silurian, 22
Lake Huron, coal, 15
——— copper, 32
——— devonian, 20
——— lower silurian, 25, 26
——— trap-rocks, 35
——— upper silurian, 22
Lake Michigan, coal, 16
——— devonian, 20
——— lower silurian, 25
——— upper silurian, 22
Lake Ontario, flagging-stones, 27
——— lower silurian, 25, 26
——— upper silurian, 22
Lake Superior, copper, 27, 32, 34, 35
——— copper-mines, produce of the, 35
——— iron-ores, 32, 33
——— lower metamorphic, 31
——— lower silurian, 26
——— marble, 33
——— palæozoic rocks at, 6
——— silver, 35
——— trap-rocks, 34, 35
——— whetstones, 33
——— Land District, survey, 4
Lawrence, Mr. Byam, 5
Lead and lead-mines, 4. *See* Lead-ores.
Lead-ores, carboniferous limestone, 18
——— lower metamorphic, 32
——— lower silurian, 26
——— new red, 13
——— upper metamorphic, 29
——— upper silurian, 23
Lead-veins in Galena limestone, 26
Lewiston, N. Y., 22
Liassic formation, 12
Lignite, cretaceous, 11
——— tertiary, 10
Lime, phosphate of, 32
Limestone, coal formation, 17
——— devonian, 21
——— lower silurian, 27
Lingula beds, trap, 34
List of the formations, 7
Little Falls, N. Y., 28
Logan, Mr. W. E., 5, 28, 31, 32
Louisiana, tertiary, 10
Louisville Falls, 20
Lower magnesian limestone, 26
—— metamorphic rocks, 30
—— silurian formation, 24
——— trap rocks, 34
Lyell, Sir C., his geological map of U. S., 5
Mackinac Island and Straits, 20, 22
Magnesian limestone, 26, 27
Magnetic iron-ore, 31
Mahoning County, coal, 16
Maine, gold, 30
——— lower metamorphic, 31
——— syenite, 29
——— tertiary, 9
——— upper metamorphic, 28
——— upper silurian, 22
Malachite, 13
Manganese, new red, 13
——— tertiary, 10
Manitoulin Islands, 22, 25
Maps and surveys, geological, 4
Marble, lower metamorphic, 33
——— lower silurian, 27
——— upper metamorphic, 30
Marcou, M. Jules, 5
Marls, alluvium and drift, 9
—— cretaceous, 11
—— tertiary, 11
Marshes, peat and marl, 9
Maryland, chromic iron-one, 29, 30
——— coal, 15, 16
——— copper-ore, 29, 30
——— iron-ore, 10, 32, 33
——— lead-ore, 29, 30
——— lower metamorphic, 31
——— lower silurian, 24, 26
——— marls, 11
——— new red, 12
——— trap-rocks, 34
——— upper silurian, 23
Massachusetts, coal, 15, 16, 28
——— gold, 29
——— iron-ore, 10, 29
——— lead-ore, 30
——— marble, 30
——— millstones, 30
——— new red, 12
——— survey, 4
——— upper metamorphic, 29
Mauvaises terres, 10, 12
Medina sandstone, 21, 22, 24
Metalliferous rocks, range of the, 35
Metamorphic rocks, auriferous, 28
——— lower, 30
——— upper, 27
——— series, trap-rocks of the, 34
Mica, upper metamorphic, 30
Michigan, carboniferous limestone, 18
——— coal-field, 14, 20
——— devonian, 20
——— geological map of, 4
——— upper silurian, 22
Millford, Conn., 29
Millstones, jurassic, 12
——— upper metamorphic, 30
——— upper silurian, 24
Milwaukie, 9
Mineral Point, 27
Minerals, precious, 33
Mines, coal, 12, 15
——— copper, 13, 24, 27, 29, 32, 35
——— iron, 10, 16, 21, 23, 29, 32
——— lead, 13, 18, 23, 26, 29, 32
——— zinc, 23, 32
Mining products, 4. *See also* Mines.
Minnesota, geological map of, 4
Mississippi, coal-field, 17
——— cretaceous, 11
——— devonian, 20
——— River, 14, 16, 22, 25, 26, 31, 34
——— tertiary, 10
Missouri, coal-field, 14
——— geological map of part of, 4
——— iron-ores, 16, 32, 33
——— lead ore, 26
——— lower metamorphic, 31
——— metalliferous area of, 35
——— River, 6, 11, 16, 18
——— survey, 4
Mitchell, Prof., 4
Mohawk River, 25
——— Valley, upper silurian, 23
Monongahela River, 15
Montgomery County, Penns., lead and copper, 13, 32
Montreal, Wisc., 32
Mount Holyoke, 33
——— Tom, Mass., 33
Muck or peat, 9
Nashville, 22, 25
Native copper, 9, 13, 35
—— gold, 29
—— silver, 29, 35
Newburg, 24
New England, coal, 15
——— mountains, 6
——— peat, 9
——— syenite, 29
——— tertiary, 10
——— trap-rocks, 35
——— upper metamorphic, 28
New Garden, Penns., 11
New Hampshire, auriferous drift, 9
——— iron-ore, 29
——— lower metamorphic, 31
——— mica, 30
——— plumbago, 30
——— survey, 4
——— tin-ore, 29, 30
——— upper metamorphic, 28
New Haven, Conn., serpentine, 29
——— trap-rocks, 33
New Jersey, clays, 10
——— copper-ores, 13
——— cretaceous, 11
——— devonian, 20
——— iron-ore, 10, 32, 33
——— lower metamorphic, 31
——— lower silurian, 24, 25, 26
——— new red, 12
——— phosphate of lime, 33
——— sandstone, 13
——— shelly marls and green earth, 11
——— survey, 4
——— tertiary, 9
——— trap-rocks, 33
——— upper silurian, 21, 22, 23
——— zinc-ores, 32, 33
New Mexico, carboniferous limestone, 14, 18
New Orleans, 16
New Red sandstone, 12
——— trap-rocks, 33
New York, auriferous drift, 9
——— copper-ore, 13, 32
——— devonian, 19, 20 21
——— iron-ore, 10, 23, 32, 33
——— labrador-felspar, 33
——— lead-ore, 13, 27, 32, 33
——— lower metamorphic, 31
——— lower silurian, 25
——— marble, 30, 33
——— metalliferous area of, 35
——— mountains of, 6
——— phosphate of lime, 32
——— plumbago, 33
——— salines, 24
——— slates, 27
——— upper silurian, 21, 22
——— zinc-ore, 24, 27
Niagara group, 22, 24
——— limestone, 21
——— River, 20, 22
North Carolina, auriferous drift, 9
——— rocks, 29
——— copper ore, 29, 30
——— cretaceous, 21
——— geological map of, 4
——— gold, 9, 30

North Carolina, jurassic coal, 12
——— lower metamorphic, 31
——— silver, 29, 30
——— survey, 4
——— tin-ore, 29
——— trap-rocks, 34
Norwood, Dr. J. S., 18
Nova Scotia, coal, 15
Ochre, drift, 9
Ochreous iron-ore, tertiary, 10
Ohio, cretaceous, 11
—— devonian, 20
—— iron-ore, 16, 17
—— lower silurian, 25
—— sand-stone, 19
—— upper silurian, 22
—— River, 15, 16, 17, 20
Older metamorphic rocks, 30
Oneida, upper silurian, 22
Onondaga salt group, 21
Oolitic carboniferous limestone, 18
——— formation, 12
Oregon, metalliferous district, 35
——— survey, 4
Oriskany sandstone, 20
Osage River, 14
Owen, Dr., D. D., 4, 5, 14, 18, 20, 26, 31
Oxide of manganese, 10
Pacific coast, metalliferous belt of the, 35
Paint from zinc-ore, 32
Palisades, Hudson, 33, 35
Passaic River, 13
Peat and shell marl, 9
Pennsylvania coal, 15, 16, 17
——— chromic iron-ore, 29, 30
——— copper-ores, 32, 33
——— devonian, 19, 20, 21
——— geological survey, 4
——— iron-ore, 20, 16, 19, 21, 23, 29, 30, 32, 33
——— lead and copper, 13, 32
——— lower metamorphic, 31
——— lower silurian, 24, 25, 26
——— marble, 30
——— new red, 12
——— red shales, 19
——— roofing-slates, 27
——— survey, No. I., 24
——— No. II., 24
——— No. III., 24
——— No. V., 21
——— No. VI., 21, 23
——— No. VIII., 21
——— No. IX., 21
——— No. X., 19
——— No. XI., 19
——— trap-rocks, 34
——— upper silurian, 21, 22, 23
Pentamerus limestone, 21
Perry County, coal, 16
——— upper silurian, 22
Phosphate of lime, lower metamorphic, 32
Physical Geography of the United States, 5
Pilot Knob, Miss, 32
Pittsdown coal-mine, 16
Plumbago, lower metamorphic, 33
——— upper metamorphic, 30
Point Abbaye, 31
Pointe aux Baies, 25
Porcelain-earth, lower metamorphic, 33
——— tertiary, 10
Porphyritic trap-rocks, 34
Portage Group, 19, 28
Portland, Conn., new red, 13
Potsdam sandstone, 24, 26, 27, 28, 30, 34
Pot-stone, lower metamorphic, 33
Pottery clay, coal formation, 17
——— drift, 9
——— tertiary, 10
Pottsville, Penns., 19
Prairie du Chien, 26
Precious minerals, 33
Products, economic. *See* Economic products.
——— mining, 4. *See* Mines.
Puget's Sound, coal, 15
Queenstown, 22
Quincy granite, 29
Rassie, 32
Red Cedar River, 20, 22
Red shale of the carboniferous limestone formation, 19
Rhode Island, coal, 15, 16, 28
——— survey, 4
Richmond, Virginia, 12
Rochester, 22
Rocky Mountains, 7, 10, 11, 14, 15, 17, 34
Roemer, Dr. F., 7
Rogers, Prof. H. D., 13, 14, 26, 32
Roofing-slates, lower silurian, 27
St. David's, Canada, 22
St. Lawrence River, 6, 9, 18, 25, 26, 28, 31
St. Louis, 18
Salem, Mass., 30
Salines, New York State, 24
Salt, upper silurian, 24
Sand for glass-making, carboniferous, 17
Sands and clays of the alluvium and drift, 9
Sandstone, coal formation, 17
——— devonian, 21
——— jurassic, 12
——— lower silurian, 27
——— new red, 13
——— tertiary, 11
——— upper silurian, 24
Sandusky, 20, 22
Schuylkill, coal, 16
Serpentine, upper metamorphic, 29
Shawangunk conglomerate and grit, 24, 28
——— lead-mine, 24
——— millstones, 24
——— mountain, 22, 23, 24
Shell marl and peat, 9
Shelly marls, tertiary, 11
Sierra Nevada, 10, 34
Silliman, Jun., Prof., 3
Silurian formation, lower, 24
——— upper, 21
——— rocks, metamorphic, 28
Silver, new red, 13
—— trap-rocks, 35
—— upper metamorphic, 29
Simsbury copper-mine, 13
Slates, lower silurian, 27
South Amboy, 10
—— Carolina, auriferous drift, 9
——— rocks, 29
——— cretaceous, 11
——— geological survey, 4
——— gold, 9, 30
South Mountains, 31
Specular iron-ore, 32
Sphene, 33
Spinelle, 33
Statuary marble, upper metamorphic, 30
Stansbury, Captain, 10, 15
Steel, 33
Stones River, Tenn., 25
Stone-ware, 17
Sullivan County, upper silurian, 23
Sulphur, new red, 13
Superficial deposits, 8
Surveys and maps, Geological, 4
Swamps, peat and marl, 9
Syracuse, N. Y., 24
Syenite, upper metamorphic, 29
Taylor, Mr. R. C., 15
Tennessee, carboniferous limestone, 17
——— devonian, 20
——— iron-ores, 16
——— lead-ores, 18
——— lower silurian, 25, 26
——— upper silurian, 22, 23
Tertiary coal, 10
——— formations, 9
Texas, carboniferous limestone, 18
——— cretaceous, 11
——— lower metamorphic, 31
——— metalliferous area of, 35
——— tertiary, 10
Texas, tertiary and secondary rocks in, 6
——— trap-rocks, 34
Thousand Islands, 31
Tin-ore, upper metamorphic, 29
Tourmaline, 33
Trap-dykes, 34
Trappean formation, 33
Trap-rocks, distribution of the, 33
——— of Lake Superior, 34
——— of the metamorphic and silurian series, 34
——— of the new red series, 33
——— of the western regions, 34
——— silver, 35
Trenton Falls, 25
——— limestone, 24, 25, 27, 28
Triassic formation, 12
Ulster County, upper silurian, 23
United States, geological maps of, 5
——— geology and physical geography of the, 5
Upper metamorphic rocks, 27
—— silurian formation, 21
——— iron, lead, copper, and zinc ores, 23
Utica, 23
——— slate, 24
Verd antique, 29
Vermont, auriferous drift, 9
——— gold, 9, 29, 30
——— iron-ore, 10, 29
——— serpentine, 29
——— slates, 27
——— statuary marble, 30
——— upper metamorphic, 28
Vinton County, coal, 16
——— iron-ores, 17
Virginia, auriferous drift, 9
——— rocks, 29
——— carboniferous limestone, 17
——— chromic iron-ore, 29
——— coal-field, 16, 17
——— copper-ore, 29
——— devonian, 21
——— gold, 9, 30
——— iron-ores, 16, 17, 29
——— jurassic coal, 12
——— lead and tin ores, 29, 30
——— lower metamorphic, 31
——— lower silurian, 24, 25, 26
——— marls, 11
——— new red, 12
——— red shales, 19
——— trap-rocks, 34
——— upper silurian, 21, 23
Wabash River, 16
Washington Territory, coal, 15
——— the Capitol at, 11
Water-lime, 21
———pipes, 17
Waverley sandstone, 19
Western coal-field, 14, 16, 17
——— States, general geology of, 5
——— geological maps of, 5
Wheatley lead and copper mine, 13
Whetstones, lower metamorphic, 33
——— upper metamorphic, 30
Whitney, Mr. J. D., 4, 22, 31, 34
Whitehall, 24
White Mountains, 28
Wilkesbarre, Penns., coal, 16
Winnebago Lake, lower silurian, 25
Wisconsin, brick-clays, 9
——— copper-ore, 27
——— geological map of, 4
——— iron-ore, 23, 32
——— lead-ore, 26
——— lower metamorphic, 31
——— lower silurian, 25, 26
——— metalliferous area of, 6
——— survey, 4
——— trap-rocks, 34
——— upper silurian, 22
Zinc-ores, carboniferous limestone, 18
——— lower metamorphic, 32
——— upper silurian, 23
——— new red, 13
Zircon, 33

Class X.—Topographical and Hydrographical Surveys; Charts and Maps.

Introduction.—Massachusetts is, I believe, the only State which has undertaken a trigonometrical survey of its territory; and this was independent of, or subsequent to, her geographical survey. It is certainly somewhat surprising that a people, who are straining every energy for physical advancement, should yet neglect a subject of so much economical importance.

In some of the older States such a survey is more essential than in the newer ones, and in the former there exist the means of accomplishing the object. When this is effected, and maps have been constructed combining topography and geology, these must become the foundation of every new enterprise in internal improvements, which would be planned and carried forward in accordance with the knowledge there expressed.

Under Class X we have to notice the maps and charts which exhibit progress in their several departments, or which present new facts of economical or scientific interest. The only works of this kind which particularly claim our attention, are the maps of the United States Coast Survey, under the charge of Professor A. D. Bache, and the Wind and Current Charts of Lieutenant Maury.

Coast Survey.—The coast survey of the United States, under the able direction of Professor Bache, may be regarded as by far the most important scientific work now in progress in the United States. Both in its scientific results, and in its economical and utilitarian influences, its importance cannot be over-estimated; and from these considerations we may be permitted to introduce a short sketch of its history and progress, with a general summary of the objects proposed to be accomplished, and a notice of the mode of procedure, with the results thus far obtained.*

History of the Establishment of the Coast Survey.—The importance of an accurate survey of the coast was felt by the people and authorities of the United States at an early period of their history as an independent nation; and a plan for the execution of a complete geodetic and hydrographical survey was adopted by President Jefferson, who appointed the late Mr. F. R. Hassler to the office of Superintendent.

Mr. Hassler went to Europe in 1811 to procure the necessary instruments, all of which had to be constructed; but, owing to the disturbed condition of Europe and the subsequent war of the United States with England, he was unable to obtain them, or to undertake the prosecution of the work, until the year 1817, when a commencement was made on the Bay of New York. Shortly afterwards, however, the work was suspended by the failure of Congress to appropriate funds for its prosecution; and it was not until 1832 that it was revived, Mr. Hassler being again appointed Superintendent.

At the time of Mr. Hassler's death, which occurred in 1843, the survey had been extended from the base near New York, eastward to Rhode Island, and southward to the capes of the Delaware and the upper part of Chesapeake Bay.

Professor Alexander Dallas Bache was appointed Mr. Hassler's successor; and under his direction the survey has been expanded in all its branches, and vigorously prosecuted up to the present time.

System pursued by the present Superintendent.—According to the plan developed by Professor Bache, the operations are carried on simultaneously on different parts of the coast, which, for the purposes of the survey, is divided into eleven sections, in each of which the survey is executed independently on its own base line and geographical determinations. When completed the

* I take this opportunity of returning my thanks to the officers of the Hydrographical Survey at Washington, for the courteous and valuable assistance which they gave me in drawing up this part of the Report.

whole will form a continuous survey of the coast, presenting an unbroken chain of triangulation; with several base lines verifying each other, and numerous determinations of latitude, longitude, and azimuth, fixing geographical positions, and contributing largely to our knowledge of the figure of the earth.

Plan of the Survey.—The general plan of the survey may be briefly stated as follows:—

After a reconnaissance to determine the best plan of work, a base line is measured, and a primary triangulation, having sides as long as the limits of vision or the nature of the country will allow, is carried on along the coast.

Astronomical observations to determine the latitude and direction of the meridian are made at a number of the stations, and the longitude of some cardinal point in each section is determined. Magnetic observations are also made, furnishing the variation of the compass for the charts, and incidentally the elements of the earth's magnetism.

On the sides of this primary triangulation is based the secondary triangulation, determining prominent points a few miles apart along the coast and bays, furnishing points of departure and checks for the topographical survey, which is next executed. The hydrographical survey lastly furnishes soundings and observations on tides and currents, and completes the material required for the charts, which are drawn, engraved, and printed in the office of the Coast Survey.

Constitution of the Bureau of the Survey.—The *personnel* of the survey consists of civilians and military and naval officers; the first class forming a permanent nucleus, and being composed of persons of scientific and practical ability, a great part of whom have been trained to the work under the present Superintendent; while the officers of the army and navy are put upon the survey or withdrawn from it, according as their professional duties may permit. The geodetic and topographical work is performed by the civilians and army officers, while the hydrographical survey is made by officers of the navy, all acting under the immediate orders of the superintendent, and reporting their results to him. A unity of plan is thus secured highly conducive to accuracy and despatch, and which has not been attained on similar works in other countries, where the geodetic and hydrographic surveys are carried on separately; in consequence of which the latter fails to derive from the former those data which are essential to its perfection.

Progress and method of the work of the Survey.—After this brief summary, a more detailed notice of some of the methods used in the work and its results will not be uninteresting.

Measurement of Base-lines.—Up to the year 1853, six principal base-lines had been measured, two of which are now connected by a primary triangulation. These, as well as a third, were measured with the apparatus designed by Mr. Hassler, and fully described by him in the "Transactions of the American Philosophical Society," Vol. IV, the measurement being made with iron bars and by optical contact. The subsequent bases have been measured with an apparatus designed by Professor Bache, and constructed under his direction in the office of the Coast Survey. In this apparatus the measuring bars are arranged upon the compensating system, first used by Colonel Colby in Great Britain, and by Mr. Borden in the survey of the State of Massachusetts, but the mode of obtaining the compensation differs entirely. A principle is also introduced, which, though obvious enough, had not been applied before. A bar of brass and a bar of iron, of the same dimensions, exposed to the same source of heat, will not become heated equally in equal times, the difference depending on the different conducting powers of the two metals, their different specific heats, and the different powers of their surfaces to absorb heat. The bars then, if of equal sections, will not be compensating when the temperature is changing, however perfectly the compensation may be effected at any stationary temperature. The sections of the bars must, therefore, be so proportioned to their conducting power and specific heat, as to cause both to have the same temperature during variable temperatures of the atmosphere, their surfaces being easily made to absorb equally by giving them the same coating.

The lever of contact and level, first used in the adjustment of standard measures by Bessel, has been adapted to this apparatus; the contact between two sets of bars being made by a blunt knife-edge and a plane of agate. The rapidity and accuracy with which the measurement of a base-line can be made with this apparatus can leave but little to be desired. From remeasurements made of portions of the base-line on Dauphine Island, it appeared that the probable error in measuring a mile is about two-hundreths of an inch.

The Fire Island base, on the south side of Long Island, and the Kent Island base in Chesapeake Bay are connected by a primary triangulation. The former is 8·7 miles, the latter 5·4 miles long, the shortest distance between them is 208 miles; the distance through the triangulation is 320 miles: the length of the Kent Island base, as computed from the Fire Island base, through a series of 32 triangles, differs but 4 inches from the measured length.

The principal base-lines measured are in the States of Massachusetts, New York, Maryland, North Carolina, South Carolina, and Alabama. In other sections of the survey preliminary base-lines have been measured by means of wooden or iron bars and some simple apparatus: all of these will be superseded by measurements with the apparatus described.

Apparatus for measurement of the Angles.—The instruments used for measuring the angles of the primary triangulation are a 30-inch and a 24-inch theodolite, by Troughton and Simms, both fitted with reacting micrometer microscopes, and a number of 12 and 10-inch repeating theodolites, by Gambey. The latter instruments are great favourites on account of their admirable graduation, and the principle of repetition causes them to be preferred in cases where the instrument has to be elevated on wooden structures, the perfect stability of which cannot be depended on. The Atlantic coast of the United States south of New York, and the shores of the Gulf of Mexico, being without any considerable eminences, and in general densely wooded, artificial elevations are resorted to in order to carry on a triangulation having sides of from 10 to 16 miles in length. Wooden structures, varying from 25 to 40 feet in height, have been erected on various plans by the assistants of Professor Bache, and have proved amply steady for measurements of angles with the repeating instruments.

The large (30-inch) theodolite was designed by Mr. Hassler, and is used by Professor Bache in the execution of the triangulation in the north-eastern part of the United States. Mr. Hassler followed the usual rule of observing at such times only when the conditions of vision appeared to be entirely unobjectionable. This rule is attended with serious loss of time. Mr. Bache has increased the rapidity of the work, and improved rather than diminished its accuracy, by making observations under conditions ordinarily favourable and of frequent occurrence, and applying to them the mathematical test of the least probable error, the number of observations being made dependent on their agreement among themselves. For each observation, the appearance of the signal observed has been noted, and the means have thus been obtained for a discussion of the results with reference to the condition of vision, from which it has appeared that the mean of observations, taken under various conditions of of the atmosphere, is probably to be preferred to the mean of such as have been obtained under apparently the most favourable circumstances; the corrections required to fulfil the geometrical conditions of the work being less in the former case, while a given limit of probable error can generally be reached in much less time. This result may be readily accounted for by supposing that under apparently the best condition of vision along a line, a slight lateral deflection may regularly take place, which is eliminated in the mean of observations taken under different conditions of the atmosphere.

The average correction to an observed angle required in the eastern section, where the stations are on natural elevations, and the sides average over 40 miles, is about 0″·4. In the southern sections it is rather larger. The limit of error allowed in a primary triangle is one second to an angle; this, however, is seldom reached in fact.

Observations of Latitude, and the Apparatus.—Observations of *latitude* are made at numerous stations of the primary triangulation. Different methods and instruments have been used, and the results have been carefully discussed, to determine which are the best and most available. A 2-feet vertical circle and some 12-inch repeating reflecting circles, by Troughton and Simms, 6-inch and 10-inch repeating theodolites, by Gambey, were tried successfully, but the results were found to be liable to considerable instrumental errors, and their accuracy did not repay the labour bestowed on them. Finding that instruments of a larger class were requisite, Mr. Bache procured from Mr. Simms a transit instrument, a zenith telescope, made on Captain Talcott's plan, and a zenith sector, according to Mr. Airy's design, each of the instruments having a telescope of 45 inches focal length. They were compared by observations made with each at the same stations.

The *transit* mounted in the *prime vertical* gave very satisfactory results, but its use was abandoned on account of their paucity, corresponding sets of eastern and western transits being rarely obtained on account of the cloudiness of the sky.

The *zenith telescope*, or equal altitude instrument, is a great favourite as a field instrument among American observers, although but little known in Europe. Its use consists in measuring with a micrometer the differences of zenith distance of stars culminating at nearly the same altitude, on different sides of the zenith, within a short time of each other. The British Association's Catalogue furnishes any desirable number of pairs of stars having the required relative positions. The observations with the zenith telescope being considerably more accurate than the north polar distances of the stars in the catalogue, the number of pairs of stars used has been increased, rather than the number of observations on the single pairs. In practice, from 25 to 40 pairs of stars are observed at a station, with three or four observations on each pair on different nights. The average probable error of a north polar distance in the British Association's Catalogue is 1''·7, while the probable error of an observation with one of the zenith telescopes, as constructed by Simms, is about 0''·5. The recent Greenwich Twelve Years' Catalogue frequently supplies better places of the stars, and this contributes to improve the result.

A new zenith telescope, constructed recently by Mr. Würdemann, of Washington City, on an improved plan, has been found to give results with a much less probable error than the former instruments.

For further details concerning this valuable instrument, see "Notes on the Zenith Telescope," by Professor Bache.*

The *zenith sector* of the Coast Survey is constructed on the plan communicated by Mr. Airy in the Notices of the Astronomical Society.

It is regarded by the American surveyors as an admirable instrument, and its results are considered to have about the same accuracy as those of the zenith telescope: its ponderousness, however, and the greater laboriousness of the observations, make it less desirable as a field instrument than the latter. Its capabilities, however, are far greater than those of the zenith telescope, as it measures absolute zenith distances, and it is used for determining corrections to the catalogue of north polar distances of stars observed at different stations with the zenith telescope.

When the latitudes observed at different stations are referred to one another by means of the geodetic differences of latitudes, there appear discrepancies much larger than could result from the residual probable error of observation, which can only be ascribed to local irregularities in the figure and density of the earth, and which are designated as *station errors*. They are similar to those caused by the proximity of mountains, but occur where the external configuration of the earth's crust would not indicate their existence. Mr. Bache recognized this source of error in 1844, and as a consequence, determined to carry on a continuous series of astronomical stations through the primary triangulation, in order to arrive as nearly as possible at the true figure of the portion of the earth's surface he was engaged upon.

The elements of the figure of the earth, used in these comparisons, are those obtained by *Bessel* in his last discussion of the results of ten measured arcs. It is interesting to know, that the observed difference of latitude of the extremities of the arc at present passed over by a connected triangulation agrees very closely with the geodetic difference computed on Bessel's elements. This arc extends from Maine to Maryland, over about 5 degrees of latitude and 6½ degrees of longitude. In intermediate portions of this arc, the deviation from the mean elipsoid reaches three seconds of arc and over, but without any regularity, intervisible stations differing frequently with opposite signs. The average amount of station error is between one and two seconds of arc in the eastern portion of the survey, and about half a second in the sections south of the Delaware.

* American Journal of Science and Arts. Second Series. Vol. XIV. 1852.

Azimuth Observations.—Observations of *azimuth* are made at all the latitude stations. The instruments employed are the same that are used for the angles of the primary triangulation. The meridian is determined by observations on the pole-star, principally near its greatest elongations, where the result will be nearly independent of the knowledge of time, the star having very little motion in azimuth, and when observations taken within 45 minutes on either side of the elongation can be reduced to that position of the star by a most simple formula. The pole-star, however, moves so slowly in azimuth, even at its culminations, that the liability to error in pointing is found to be no greater at those periods than at elongations; and Mr. Bache has occasionally pursued the plan of observing on the star at equidistant times on both sides of the meridian near the culmination. In these cases the reduction is simplified into a mere taking of means; and it may be noticed, that here, as in every other process, it has been the aim of the Superintendent so to arrange the method of observing as to diminish, as far as practicable, the labour and liability to error of the computation.

The azimuths are liable to the same irregularities which affect the latitudes from local variations of level; and a comparison of azimuth observed at intervisible stations has shown the station errors to exist in the same measure as exhibited by the latitudes.

Instruments used in observing local time.—Observations of local time are made in connection with those of latitude and azimuth, the instruments employed being generally portable transits of from 26 to 48 inches focal length, constructed by Simms of London, and Würdemann of Washington. Chronometers by the best makers are used in preference to clocks (except for telegraphic determinations of longitude), on account of their greater portability and the half-seconds beat.

Observations of Longitude.—In order to fix American *longitude* in reference to European observations, observations of eclipses, occultations, and moon culminations, made at different American observatories previous to 1844, as well as the corresponding observations at European observatories, have been collected and reduced; and similar observations have since been continued at Cambridge in Massachusetts, Philadelphia, Washington, Charleston, and Cincinnati. All these stations being connected by the electric telegraph, their differences of longitude have been determined with great accuracy, and all the results are referred to Cambridge as a common station of reference.

The deference of longitude between Cambridge and Liverpool has also been determined by means of a large number of chronometers carried repeatedly between the two stations on the Cunard's steam-ships. This work is still in progress, and the intention is to accumulate so large a number of chronometric results, as to be entitled to the same confidence as that obtained from observations of the moon. It is remarkable, that while the results from occultations differ sensibly from those obtained by moon culminations, the chronometric results differ again from both, and by an amount which a larger number of results is not likely materially to diminish. It remains to be seen whether the recent corrections of the lunar theory will so modify the results depending on the moon as to produce a better agreement. For an interesting numerical comparison of these results, see Appendix, No. 26, of the Annual Report for 1851.

The differences of longitude between Cambridge and principal stations of the survey in other sections are determined by the aid of the electric telegraph, where the latter has extended. The manner in which the difference of longitude is obtained may be briefly stated as follows:—Transit instruments are mounted at both stations; at one of them an astronomical clock makes a record of its second's beats on a sheet of paper, by breaking a galvanic circuit which passes through both stations each time the pendulum passes the vertical; the times of a star's crossing the wires of the transit at each station are recorded on the same sheet by the observer's breaking the circuit by pressing a key the instant the star appears bisected on the wire. The passages of the same star over two different meridians are thus recorded by the same timepiece, and the difference of longitude results immediately after applying the necessary instrumental corrections.

The method of recording observed instants of time by means of an electromagnet on a time-scale graduated by a clock, designated by Mr. Airy as the American method, and its application to the problem of telegraphic differences of longitude is due to the late Mr. S. C. Walker, one of the assistants on the coast survey, to whom Mr. Bache had entrusted the charge of the longitude department. The mechanical detail of the apparatus employed was furnished by others; and, indeed, several different modes have been devised; but it was Mr. Walker who designed the new plan, and pointed out what mechanical conditions were required to be fulfilled to attain the end in view.

In the course of these telegraphic operations, it became apparent to Mr. Walker that a far more suitable time was required for the transmission of the galvanic current through the telegraphic wires than had been generally supposed. A number of experiments made on different lines to investigate this subject showed that the velocity is between 15,000 and 16,000 miles in a second. For an interesting discussion of these experiments, giving a detailed description of the apparatus, see a paper by Dr. Gould, on the Velocity of the Galvanic Current in Telegraph Wires, in Appendix No. 25 to the Annual Report for 1851.

By means of the electro-magnetic register and telegraph, differences of longitude are determined with an accuracy nearly approaching that of the latitudes, and arcs of parallels may be measured with the same facility as arcs of the meridian.

In sections where the telegraph has not reached, as Florida, the shores of the Gulf of Mexico, and the Pacific coast, the longitudes of cardinal points are determined by observations of moon culminations and by chronometers. Corresponding observations of moon culminations are made at several American observatories, at the expense of the Coast Survey, and valuable aid is derived from the series of meridian observations of the moon made at Greenwich. The promptitude with which these have been always supplied through the kindness of the Astronomer Royal, Mr. Airy, is gratefully acknowledged by Professor Bache, and both he and his colleagues express their admiration of the manner in which the reductions are made to keep pace with the observations at the British Observatory.

The late Mr. Walker bestowed great attention on the method of longitudes by moon culminations, and arranged the details of the reduction in an admirable manner. The recent improvements in the lunar theory lead to the hope that the moon's right ascension for any given time may be represented more accurately by theory than by any single observation, in which case corresponding observations would be dispensed with, and the reduction of the longitude effected in a much simpler plan, which has been developed by Professor Pierce, of Cambridge. Extensive computations are actually in progress under the direction of the Superintendent, having for their object to test theory by observation, and to produce corrected lunar tables.

Observations of *magnetic declination, dip,* and *intensity* are made at all astronomical stations, and at many other points where it is desirable to know the declination for purposes of navigation. The instruments employed are similar to those used in the British magnetic surveys, being declinometers and horizontal force magnetometers on the plans of Gauss and Weber, Lloyd, and Lamont, by Jones of London, and dip-circles by Barrow and Gambey, provided with Lloyd needles for total intensity. In this manner not only the variation of the compass required for the mariner is obtained, but also contributions to our knowledge of the earth's magnetism, the value of which can hardly be appreciated as yet.

Observations on Heights.—The *heights* of the trigonometrical stations above the level of the sea are determined generally by observations of reciprocal zenith distances, frequently checked by direct levellings from the ocean Interesting observations are in progress to compare results obtained by these methods with barometric and thermometric measurements of heights.

The Topographical Survey.—The *topographical survey* is executed with the plane-table on scales of $\frac{1}{5000}$, $\frac{1}{10000}$, and $\frac{1}{20000}$ of nature, according to the importance of the locality and the amount of detail to be represented. On the sheets drawn in the field the irregularities of the surface, hills, and depressions are indicated by horizontal curves, drawn at certain fixed vertical distances from each other. In the reduced drawing for the engraved charts, the spaces between these horizontals are filled up with *hashures*, the shade being darker in proportion to the steepness of the slope, according to a system somewhat modified from that of Lehman. Scales of shades showing the distance between the hashures, and the strength adapted to different scales of the maps, have been printed. The absence of very steep slopes and predominance of small ones rendered necessary the modification of the Lehman system.

The topographical survey is carried inland as far as required for purposes of navigation and the defence of the coast.

The Hydrographical Survey and Charts.—The *hydrography* is next in order. Soundings are taken with such frequency, as to exhibit the configuration of the bottom with sufficient minuteness for the purposes of navigation, the number of casts depending on the greater or less irregularity and slope of the bottom. The positions of the casts are determined in reference to the trigonometrical stations, generally by means of angles measured with the sextant at the place of sounding, and sometimes by the use of theodolites at two shore-stations; the latter especially in off-shore work, and in cases where great accuracy is desired. Great aid is also derived from the outlines of the topographical survey in fixing positions, as well as in determining the *lines of equal depth*, which are drawn on the charts for every fathom down to three or four fathoms.

The spaces contained within the several *fathom curves* are shaded by dotting, lighter shades corresponding to greater depths. The configuration of the bottom is thus graphically represented to the eye, and the direction and limits of channels are conveyed to the mind in a more rapid and comprehensive manner by mere inspection, than could be done by a close examination of the figures indicating the depths, which of course are also given.

Soundings.—The soundings extend from the shore and the head of tide-water generally, out to sea as far as soundings can be of use to navigators in determining their position. In this point of view, not only the depth, but also the character of the bottom, is of importance: the sounding-leads are provided with an apparatus to bring up a portion of the bottom, specimens of which are deposited in the archives of the Coast Survey. Microscopic examinations of some of the specimens from considerable depths have shown them to be crowded with organisms, the number and genera varying with the depth; and it is not impossible that science may here supply the mariner with another mode of determining his position at sea.

Discoveries by the Hydrographic Survey.—It is the hydrographical department of the survey that develops the results to which public attention are most directed, and which most immediately benefit the community. By the thorough system of hydrographic survey, new channels are discovered and hidden dangers are pointed out; the positions of many that are imperfectly known are determined with accuracy, and the proper positions for buoys, beacons, and light-houses are indicated. Among the benefits thus conferred upon commerce, we may enumerate the discovery of a new channel, straighter and deeper than those before known, over the outer bars of New York harbour; of several channels into Delaware Bay, and of one inside the bay, often free from ice when the main channel is obstructed; of channels across the Cape Fear Shoals; of a new channel into Key West Harbour, and one of an increased depth, into Mobile Bay. The discovery and accurate delineation of Davis' Shoal and of Davis' Bank off Nantucket, and of other shoals in the vicinity, in the track of the commerce between New York and Europe, have been of the greatest value. Numerous harbours of refuge along the coast have been sounded out, and charts with sailing directions have been, for the first time, furnished to navigators.

Observations on Tides.—The subject of *tides* is receiving great attention, not only with reference to local tides and currents, but the general laws of the phenomena, both on the Atlantic and Pacific coasts, are under discussion. Extensive half-hourly observations of the tides have been made at a number of stations, extending over a time of from one to three years; at intermediate points, half-hourly observations during two lunations are found sufficient to exhibit the law and constants. Within a year or two past, self-registering tide-guages have been used with excellent success. Professor Bache personally engages in the discussion of this most interesting subject, and has made several communications to the American Association for the Advancement of Science, on the nature of these researches and their results, which will be found in their Transactions.

This subject is the more important, since the tides of the Atlantic, Pacific, and Mexican Gulf coasts of the United States present three distinct types; the diurnal tide, which is moderately large on the Atlantic coast, producing a perceptible diurnal inequality in height and time,—is very great on the Pacific, —and almost obliterates the the semi-diurnal tide in the Gulf of Mexico.

The direction and velocity of tidal currents, as important to navigators as the rise and fall of the water, are also observed with care. Tide and current tables on the charts embody the information thus obtained, and convey to the mariner all the knowledge he requires of the times, heights, and durations of tides, and the set and drift of currents.

Exploration of the Gulf Stream.—In connection with the hydrography, the present Superintendent has undertaken the *exploration of the Gulf Stream.* The plan of exploration is to determine the temperature and other physical circumstances, at different depths from the surface to the bottom, at positions on lines perpendicular to the axis of the stream, drawn from capes and headlands of the coast at suitable distances from each other. In this way various sections of the stream have been made, beginning north at Cape Cod in Massachusetts, and extending south to Cape Canaveral in Florida. The cold current between the coast and the Gulf Stream, the two branches into which the stream divides south of Hatteras, and the cold water between, and the ocean beyond to a sufficient distance from the warm water of the gulf, have been thus examined. On the section across the stream from Charleston, bottom was had at less than 500 fathoms in every cast, and a bank has been discovered and traced on the east side of the stream from the Charleston to the Canaveral section. This work is still in active progress, and it is much to be desired that the exploration eastward, from the American side of the northern part of the stream, should be met by a corresponding exploration from the shores of Great Britain westward, which would complete our knowledge of the Gulf Stream throughout its whole extent from the shores of Florida to those of Ireland.

Survey of the Pacific Coast of the United States.—Within a few years past the Pacific coast of the United States suddenly assumed a great importance in a commercial point of view. Its geography was very imperfect, and great inconvenience arose to navigators from the want of reliable charts. Professor Bache with great energy immediately extended the survey to that new region. The wants of commerce required a rapid survey of the whole extent of the coast, the geographical positions of the principal headlands, lines of sounding to show how near the land might be approached, and examinations of the entrances of harbours of commerce and refuge. The more elaborate and accurate, but necessarily slow, system which we have detailed must for a time be set aside, and reconnaissance work must take the place of the refined survey. The most necessary results were furnished to the world in a surprisingly short time. This is a field where discoveries are made at every step, and to enumerate them would be to rehearse the whole of the work done.

Reduction and Publication of the Observations by the Survey.—The results of the field-work are prepared for publication at the office of the Coast Survey in Washington city. The reductions of geodetic and astronomical observations are in the first instance made by the observers themselves, and a second computation is made independently by persons having no connection with the field-work. A strict scrutiny of the observations and perfect security against errors of computation is thus ensured, the two computations being examined and compared by an experienced person, who refers discrepancies which may occur to the Superintendent. The records of observations and computations are also put in form for publication.

The Annual Report of the Superintendent for 1851 contains a table of over 3000 geographical positions of trigonometrical points determined by the coast survey.

Execution and Publication of the Charts and Maps.—The original sheets, containing the topographical and hydrographical work, are combined and reduced to the scale of publication by regular draughtsmen in the office. The beauty and accuracy of these drawings can scarcely be surpassed.

The maps and charts are of three classes,—charts of harbours, maps of portions of the coast, and sketches of discoveries, &c. The scales of publication vary from $\frac{1}{10000}$ to $\frac{1}{400000}$, according to the detail to be represented.

The charts are engraved on copper by a corps of engravers in the office, among whom are some first-rate artists. When ready for publication, the plates are multiplied by the electrotype process; the originals are preserved in the archives, and the electrotype copies are used in printing. For an interesting description of the electrotyping process, as carried on at the office of the Coast Survey, see Appendix No. 55 of the Annual Report of the Superintendent for 1851. Large charts are sometimes engraved in two or four separate pieces, and then joined on a common electrotype plate; by this means great despatch may be effected, as several engravers can be engaged on the chart at the same time.

The prices at which the charts are sold are extremely low, ranging from fifteen cents to one dollar. They are freely distributed to all educational institutions, libraries, &c.

In conclusion it may be observed that the funds for carrying on the survey are voted annually by Congress, upon estimates submitted by the Superintendent, and the work above sketched out is executed in a manner as economical as it is accurate and comprehensive.

Exhibited Articles.—The following list embraces the contributions by the Bureau of the Coast Survey of the Treasury Department of the United States to the Exhibition in New York:—

Nos. 1. Base-line apparatus for laying down the base-line in geodetic trigonometry.
2. Astronomical transit instrument.
3. Astronomical zenith telescope.
4. Half-inch theodolite, for primary triangulations.
5. Heliostat.
6. Horizontal sectors for secondary triangulation.
7. Plane-table and metre-chain.
8. Sextants used in hydrography.
9. Steelwagen's sounding-lead, and specimens of ocean bottom.
10. Saxton's deep-sea thermometer.
11. Sphere crushed by the pressure of deep sea.
12. Sexton's reflecting pyrometer.
13. Electrotypes, comprising model plates, and their electrotype moulds and duplicates.
14. Electrotype copper, exhibiting great hardness, elasticity, and soundness.
15. Hydrographic and plane-table sheets.
16. Maps and charts of the United States' Coast Survey, and progress sketches.
17. Reductions for engravings.
18. Self-registering tide-guage.

Natural History Charts.—Besides the "Whale Charts," by Lieutenant Maury, to be noticed presently, a coloured map or chart, executed by James D. Dana, A.M., was exhibited, in which the geographical distribution of marine animals, especially that of the different species of crustacea, is explained. This chart, a work of great originality, gives in one view the results of multiplied observations. It accompanies a report by Mr. Dana, made to the Government of the United States, on the Crustacea collected during the Exploring Expedition under the command of Captain Wilkes, U. S. N., during the years 1838-1842.

The author has laid down on his map the isocrymal lines on the zones of greatest cold of the surface-waters of the ocean, in preference to the isothermals usually traced on maps descriptive of climate, since he finds that they are of higher importance in limiting the range of marine species. These isocrymal lines are not precisely those of any chart previously constructed, for they mark the positions where a given temperature constitutes the mean of the coldest month of the year,—for example a zone where the coldest thirty consecutive days may be 68° or 50° or 35°, Fahrenheit, or some other given temperature. The geographical distribution of marine life is stated to be a subject of far greater simplicity than that of continental life and the lines which bound the coral-reef seas and the range of certain mollusca and radiata coincide with those which govern the distribution of the crustacea. Hence the subdivision of the frigid, temperate, and torrid zones adopted in this chart have a direct bearing on the general advancement of our knowledge of the geography of marine life.

Maury's Charts.—In this department the "Wind and Current Charts," published by Lieutenant M. F. Maury, U. S. N., Superintendent of the National Observatory at Washington, occupy a prominent place, both as regards their usefulness and the novelty of the information which they convey. It was in 1842, that Lieut. Maury conceived the idea of requiring all the masters of American vessels to keep their log-books in an improved form, so as to exhibit, in addition to the ordinary information, records of all phenomena that could be rendered available for the improvement of hydrography and navigation, and generally for the promotion of science; and of employing the materials so recorded for constructing improved charts, to be engraved at the expense of Government, and distributed, free of cost, to those captains who supplied the data for their construction, or who sent to him what he terms "abstract logs." Lieutenant Maury applied to the United States Hydrographer, who entered warmly into the scheme; and circulars, addressed by the latter to all masters of vessels, were placed in the hands of the collectors of customs, who had instructions to give a copy to every captain at the time of clearance.

For some time this measure produced no fruit, but Lieutenant Maury persevered, and having obtained several old log-books from the Navy, and diligently studied and collated them, he discovered the existence of a shorter route to Rio; which discovery he publicly announced in 1848. Vessels which tried this new route reached the line, on an average, in nineteen days, whereas the same distance had previously taken them about forty days. The success of this experiment immediately caused the whole scheme to be regarded with favour,—abstract logs came in in abundance, and in 1853 more than 1000 masters of vessels were engaged night and day in making and recording the observations required; their logs are carried to the Observatory, where the information they contain is collated and entered on charts, and the results to the latest period are published in successive editions of Maury's "Sailing Directions." In some of the Wind and Current Charts the course of the trade-wind is laid down, as also the isothermal lines traversing the ocean.

Besides the charts by Lieutenant Maury already mentioned, others by the same author were exhibited, called "Whale Charts," in which the principal statistics of the whale fishery are given in the form of maps; especially the number of different species of large cetaceans found to occur in the ocean north and south of the equator.

Lord Wrottesley, in an able speech delivered in the British House of Peers (April 26, 1853), gave a full and clear explanation of Lieutenant Maury's plan for improving navigation and for promoting meteorological science. At the same time he earnestly advocated a proposal, previously made by the Royal Society of London, that the British Government should co-operate with the United States "by furnishing every ship under the control of the Admiralty with instruments properly constructed and compared, and with proper instructions for using them."

It is satisfactory to record that in the month of July, while the British Commissioners were yet in the United States, a meeting of the leading merchants of New York, representing more especially the shipping interest, took place, in which a public acknowledgment was made of the services rendered by Lieutenant Maury to his country. At the same time a valuable testimonial was voted to him, in proof of the high appreciation entertained of the results of his patient and skilfully directed labours.

At the Brussels Congress also, presently to be noticed, a vote of thanks was passed to Lieutenant Maury; and similar acknowledgments were subsequently made to him at public meetings of merchants and shipowners in London and Liverpool.

In August and September, 1853, a scientific congress was held at Brussels, to which the chief maritime nations of Europe sent delegates for the express purpose of consulting with the American hydrographer, and concerting with him a plan of combined operations, whereby the physical features and meteorology of the ocean might best be explored, and a record kept of scientific observations, calculated to promote the progress of navigation. The nations represented at the "Maritime Conference" were Belgium, Denmark, France, Great Britain, the Netherlands, Norway, Portugal, Russia, Sweden, and the United States. This meeting was convened at the suggestion of the American Government.

consequent upon a proposition which it had previously made to the British Government, (in reply to a desire which had been expressed on the part of Great Britain that America should join in a uniform system of meteorological observation *on land*,) for extending Lieutenant Maury's plan of research. Although difficulties were at present found to exist with regard to the desired *uniformity* of the methods of observation *on land*, it was proposed that meteorological observations should be made by the military and mercantile marine, British and American, as well as by the mariners of all civilized nations, as far as possible, in a uniform manner over all navigated seas.

Meteorological observations* to a certain extent had long been made at sea, and Lieutenant Maury had turned to useful account such as had from time to time fallen into his hands; but these observations, although many of them good in themselves, were but isolated facts, deprived of much of their value from the absence of observations with which they could be compared, and above all from the want of a constant and uniform system of record, and from the rudeness of the instruments with which they had been made. The moment then appeared to him to have arrived when nations might be induced to combine in a general system of meteorological research. To use his own words, he was of opinion that "the navies of all maritime nations should co-operate, and make these observations in such a manner and with such means and implements, that the system might be uniform, and the observations made on board one public ship be readily referred to and compared with those made on board all other public ships, in whatever part of the world. And, moreover, as it is desirable to enlist the voluntary co-operation of the commercial marine, as well as that of the military of all nations, in this system of research, it becomes not only proper, but politic, that the forms of the abstract-log to be used, the description of the instruments to be employed, the things to be observed, with the manipulations of the instruments, and the methods and modes of observation, should be the joint work of the principal parties concerned."

The Conference opened its proceedings † at Brussels on the 23rd August, 1853, and was attended by twelve eminent meteorologists, hydrographers, and engineer officers, naval and military, selected by the respective Governments before mentioned. The United States were represented by Lieutenant Maury, the talented and indefatigable originator of this admirable scheme for improving navigation and advancing the progress of science; and M. Quetelet, the renowned Belgian meteorologist, presided at the meetings.

In concerting a plan of uniform observations for all nations, the Conference took into consideration the inconvenience arising from the variety of scales in use in different countries. After mature deliberation, no modification was recommended, except with regard to thermometers; it being agreed that the centigrade scale should be used in addition to the one employed, whether Fahrenheit or Réaumur, with a view to its ultimate general adoption.

The *accuracy* of the barometers and thermometers used, their comparison with standards, and the correct ascertainment of the errors of the instruments, were insisted upon. The mercurial barometer was preferred; and the attention of the public was directed to the *desideratum* of an economical marine barometer, that, at all times and in all weathers at sea, would afford the means of absolute and accurate determinations.

An anemometer also was considered necessary for the contemplated plan of research.

For the more complete observations, thermometers with dry, with wet, and with coloured bulbs are required. In addition to the implements above mentioned, and those generally used at sea, the only other instrument with which the Conference recommended the ships be provided was one of trifling cost, for obtaining the specific gravity of sea-waters.

The object of the Conference being to secure, as far as possible, uniformity of record, and such a disposition of the observations, that they would admit of ready comparison, a certain form of register was concerted and agreed upon. The first columns of this form will receive the data which the Government of the United States require merchant vessels to supply, in order to entitle them to the privileges of co-operators in this system of research, and may therefore be

* Maury's Sailing Directions, p. 55, 6th ed., January, 1854.
† Ibid., p. 54, &c.

considered as the *minimum* of what is expected of them. This condition requires that at least the position of the vessel, and the set of the current, the height of the barometer, the temperature of the air and water, should each be determined once a-day; the force and direction of the wind three times a-day; and the observed variation of the needle occasionally. The register contains other columns that are intended for men-of-war to fill up, and for such officers in the mercantile navy as may be willing to distinguish themselves in this joint action for the mutual benefit of the services.

A concise body of instructions for making certain observations and filling-up the columns of the register (or of the abstract-log, in the case of merchantmen) were drawn up by the Conference. The directions were limited to such only as seemed necessary to insure uniformity of observation.

The Conference also expressed a hope that whatever observations may be made will be turned to useful account without delay; and that, with this view, the several Governments will individually and mutually give their best attention to the means for perfecting the discussion and reduction of the observations obtained, and the interchange of materials and results.

General support of the plan of Universal Observation.—Lieutenant Maury has notified, that, with regard to the proposals that have emanated from the United States for a universal system of meteorological observation by sea and land, he has received assurances of approbation, support, and co-operation from very many sources, including scientific individuals and societies at Bombay, Calcutta, Mauritius, &c.; and in England and on the Continent; also from the Holy See, Sardinia, Holland, Prussia, Spain, Brazil, Peru, New Grenada, Chili, &c. With respect to the definite object of the Brussels Conference, namely, the adoption of a uniform system of observations by the marine of all maritime nation, the Report refers to the measures already taken by the Governments of Sweden and Norway, the Netherlands, Belgium, and Portugal, towards the establishment of a department for the object in view, and notifies that the Admiralty of Great Britain has expressed its intention of giving instructions for meteorological observations to be made throughout the Royal Navy.

Thus much we learn from Lieutenant Maury's published Report of the progress of the system so happily suggested by himself, and so warmly supported by various public and private coadjutors. With reference to the active part which the British Government is about to take in the furtherance of Lieutenant Maury's comprehensive and useful plan, it is arranged that a Parliamentary grant be made this year to the Board of Trade, *Department of Science and Art*, for carrying out the meteorological observations at sea, and for establishing an office for their discussion. This office will be located at Marlborough House, the officer in charge of it being a well-qualified captain in the Navy. By the co-operation of the departments of the Board of Trade, (that of Science and Art, and that of Mercantile Marine), ships will be provided with instruments at the public cost; these instruments being carefully compared with standards at the Kew Observatory, under the direction of the meteorological committee of the British Association, which has recommended the particular instruments that are to be used. Under these favourable auspices, and with the support of a widely spread and well-organized body of observers, actuated in their work both by public and private interests and by the love of science, Lieutenant Maury's noble plan for the study of oceanic meteorology, the benefit of navigation, and the general advancement of science, bids fair to fulfil the most sanguine hopes of its originator.

RAPPORT SUR LA PARTIE GÉOLOGIQUE DE L'EXPOSITION DE NEW-YORK

ANNALES

DES MINES.

RAPPORT

SUR LA PARTIE GÉOLOGIQUE DE L'EXPOSITION DE NEW-YORK.

Par MM. Charles LYELL et J. HALL (1).

(Traduit de l'anglais par M. A. LAUGEL, ingénieur des mines.)

INTRODUCTION.

J'ai été chargé, quand je visitai en qualité de commissaire l'exposition de New-York en 1853, de faire un rapport sur la partie géologique de cette exposition. Je me suis efforcé d'accomplir cette tâche avec tout le soin possible, mais le retard qu'on a mis à exposer et mettre en ordre les produits minéraux et les produits d'art, les cartes géologiques et géographiques, a été cause, au moins en partie, que mon but n'a pas été atteint aussi entièrement que je le désirais. Le rapport général vous a déjà expliqué de quelle manière je suis parvenu à surmonter les difficultés que m'opposait un pareil délai, et fait connaître le concours que

(1) Sir Charles Lyell a demandé que le nom de M. J. Hall fût ajouté au sien dans le titre, comme ayant été son collaborateur pour les *parties géologiques* de ce rapport. (*Note de M. Élie de Beaumont.*)

m'ont libéralement prêté quelques savants éminents des États-Unis. Je saisis cette occasion pour témoigner ici combien je suis particulièrement reconnaissant à mon ami M. James Hall, géologue général de l'État de New-York, pour son assistance pendant ma visite et après mon retour en Angleterre. Nous avions fait ensemble, en 1841, en 1846 et en 1852, plusieurs voyages d'exploration dans plusieurs parties des États-Unis, dans le but d'étudier les traits géographiques et géologiques de la contrée. Dans l'occasion actuelle, la connaissance plus intime que possède le professeur Hall de la géologie des États de l'Ouest et l'expérience qu'il a acquise de tout ce qui concerne les applications économiques de la science qu'il cultive m'ont aidé à compléter une tâche que, sans son aide, j'eusse à peine pris sur moi d'entreprendre.

CLASSE I. — MÉTAUX ET PRODUITS MINÉRAUX.

La classification des divers produits dans le palais de cristal de New-York a été presque toujours la même que celle qu'on avait adoptée à Londres en 1851 : nous avons donc à rendre compte tout d'abord de ceux qu'on a rangés sous la rubrique : « Métaux et produits minéraux. » L'association de New-York pour l'exposition universelle de l'industrie déclarait dans son prospectus que son intention était de ranger les objets appartenant à cette classe dans un espace réservé, une telle exposition étant propre à fournir des informations pareilles à celles qu'on tire de l'inspection de cartes géologiques et minéralogiques, et ayant en même temps l'avantage de mettre sous les yeux les échantillons mêmes provenant de chaque mine et de chaque région minérale. Cette intention a été remplie, en grande partie du moins, sous la haute direction de M. le professeur Silliman junior. Quand on considère l'immense territoire occupé par les États-Unis, l'état naissant de la plupart des exploitations de mines, la difficulté de faire concourir à une telle œuvre les personnes qui pourtant y ont l'intérêt le plus direct, on est plus surpris de la quantité et de la variété des objets exposés que disposé à relever des lacunes dans une collection organisée et classée en quelques mois seulement.

Produits de mines. — Les produits de mines qui forment le trait le plus frappant de l'exposition consistent en différentes variétés de charbon de terre et en métaux, tels que fer, plomb, cuivre, zinc et or. La houille et les minerais de fer proviennent principalement du district oriental ou des Alleghanys; le plomb, du calcaire et des autres roches siluriennes, et des formations métamorphiques; le zinc, des roches métamorphiques de New-Jersey; et le cuivre, des régions métamorphiques de l'est des États-Unis, mais spécialement de la région des grès et trapps du lac Supérieur, dont l'âge a été récemment l'objet d'une grande dissidence entre des géologues éminents.

Cartes et plans géologiques. — C'est ici l'occasion de mentionner les cartes géologiques qui ont été publiées à différentes époques, parce qu'elles donnent comme une vue générale des divers districts qui ont fourni les produits de toute espèce utilisés par l'industrie, l'architecture, les constructions, l'agriculture. Parmi ces cartes, les unes sont le résultat de travaux, publiés ou non, entrepris par différents États et par le gouvernement central; d'autres ont un caractère plus général et résument, outre les informations géologiques fournies par les études dont nous venons de parler, les résultats nouveaux obtenus à l'aide de recherches personnelles.

Ces cartes, avec les sections et plans qui les accompagnent, n'ont été que médiocrement représentées à l'exposition; mais je ne puis me dispenser de parler des plus importantes d'entre elles : on comprendra ainsi plus aisément la position géographique des produits minéraux de la contrée.

Pendant les vingt dernières années on a commencé vingt cartes géologiques qui ont été terminées dans huit États seulement : ceux de Massachusets, de Rhode Island, du Connecticut, de New-Hampshire, de New-York, de New-Jersey, du Delaware et de la Caroline du Sud.

Pendant la même période, des études géologiques préparatoires ont été exécutées par ordre législatif et poussées plus ou moins loin dans quinze États : Maine, Vermont, Pensylvanie, Maryland, Virginie, Caroline du Nord, Géorgie, Alabama, Tennessee, Michigan, Ohio, Indiana, Illinois, Missouri.

Ces études sont actuellement poursuivies dans la Pensylvanie, la Caroline du Nord, l'Alabama, l'Illinois, le Wisconsin et le Missouri. Elles ont été suspendues dans les autres États après la publication d'un ou plusieurs rapports sur l'avancement des travaux. Dans tous les États où les études sont achevées, nous avons des cartes géologiques; et dans l'Alabama, bien qu'elles ne le soient point, on a publié une carte géologique en même temps que le premier rapport bi-annuel.

Différents travaux et reconnaissances, entrepris par le gouvernement des États-Unis dans un but différent de celui qui nous occupe, ont néanmoins contribué à jeter du jour sur la constitution géologique des régions qui en ont été l'objet. Le gouvernement central a aussi récemment ordonné l'exploration géologique de l'état d'Orégon : et l'on a terminé celle du lac Supérieur et du district de Chippewa. MM. Foster et J. D. Whitney ont publié en 1851 une carte géologique du bord méridional du lac Supérieur et de la péninsule septentrionale du Michigan, et incidemment d'une partie des territoires adjacents : cette carte forme comme un lien naturel qui fait voir de la manière la plus satisfaisante la liaison qui existe entre les deux grandes formations orientale et occidentale des États du Nord. Les travaux du docteur D. D. Owen, dans le district de Chippewa, publiés en 1852, nous ont fourni une carte géologique de Wisconsin, Minnesota, Iowa et d'une partie du Missouri.

Quoique ces études préparatoires soient loin d'être complétées dans les divers États, les rapports qu'on publie sur leur avancement et les informations fournies par des recherches individuelles, permettent de tracer avec une suffisante certitude les limites des diverses formations géologiques dans la partie des États-Unis située à l'est des montagnes rocheuses, et même sur une certaine étendue de leur lisière occidentale. Parmi les cartes géologiques faites par des particuliers, qu'ils aient ou non profité des études ordonnées par le gouvernement, mentionnons une carte géologique de la Caroline du Nord par le professeur Mitchell, publiée en 1842 ; une carte géologique des États de l'Ouest, par le docteur D. D. Owen, publiée dans le *Quarterly*, journal de la Société géologique de Londres, en 1842 ; une carte du même pays, par Byram Lawrence, en 1842 ; une carte géologique des États du Centre et de l'Ouest, par James Hall, publiée dans les Rapports géologiques de New-York en 1843 ; la carte géologique des États-Unis et du Canada, par M. Charles Lyell, 1845 ; une partie du Canada et de l'État de New-York, par

M. W. S. Logau dans le journal de la Société géologique, en 1852; une carte géologique des États-Unis et des provinces anglaises, par Jules Marcou, et une autre par le professeur Hitchcock, en 1853.

En consultant les Rapports des divers États et les cartes qui les accompagnent, aussi bien que les cartes plus générales où ces observations éparses ont été réunies et mises en ordre, nous n'aurons pas de difficulté à rapporter à leur véritable position géologique tous les matériaux envoyés à l'exposition.

Il est extrêmement regrettable que pendant qu'on était occupé de ces études géologiques, l'on n'ait point adopté en même temps un système de triangulation qui eût permis de poursuivre d'un commun accord les explorations topographique et géologique. Il résulte évidemment de là que les cartes géologiques ne peuvent pas présenter dans le détail les limites exactes et les contours précis des diverses fonctions, non plus que les traits physiques qui ont été imprimés au pays par les agents géologiques.

GÉOLOGIE GÉNÉRALE ET GÉOGRAPHIE PHYSIQUE.

Introduction. — Les caractères généraux géographiques et géologiques des États-Unis exercent, là comme ailleurs, une influence puissante sur les industries qui s'y développent, et sur les ressources des différentes parties de l'Union. Mais l'immense rayon où elle s'exerce, et la grande étendue de certaines formations liées à des produits particuliers et vouées à des industries spéciales donne aux États-Unis une physionomie différente sous beaucoup de rapports de celle de toute autre contrée du globe. Les plaines immenses et non interrompues qui se prêtent par-dessus tout à l'agriculture, sont presque entièrement dépourvues de productions minérales importantes et ne fournissent que du calcaire et des matériaux ordinaires de construction. Certaines portions de la contrée se trouvent ainsi placées très-loin des produits qui fournissent à l'industrie ses éléments, et restent privées pour cette raison des avantages que leur offre la nature. Sur d'autres espaces, d'une grande étendue, une grande richesse minérale, jointe aux autres conditions favorables à l'industrie manufacturière, se combine avec un sol excellent. Ailleurs, des districts riches en productions minérales occupent les confins extrêmes de la partie habitée du pays : d'autres enfin, trop éloignés et inaccessibles, sont demeurés encore vierges; mais peu à peu leur richesse est ajoutée à la richesse nationale, et les nouveaux établissements pénètrent rapidement à l'intérieur de ces régions destinées à devenir un jour riches et populeuses.

Dans les conditions actuelles d'une contrée telle que les États-Unis, il serait impossible de présenter à présent, dans une exposition ou un rapport, plus qu'une esquisse approchée de ses ressources industrielles. Les exploitations de mines sont, pour la plupart, dans l'enfance, et il y a des régions où l'on découvre chaque jour de nouvelles richesses minérales. Néanmoins, aussi longtemps que de vastes territoires d'une grande fertilité demeureront inoccupés, que la population restera éparse et la main-d'œuvre coûteuse, on ne peut pas espérer que les mines et les manufactures perdront le développement que les besoins du pays peuvent exiger. Les contrées les plus riches pour l'agriculture seront occupées les premières, et celles qui ne l'attireront pas aussi puissamment resteront pendant longtemps relativement inconnues.

En exceptant les mines de houille et de fer, et nous

pourrions peut-être ajouter maintenant, de cuivre et de plomb, les exploitations ont été plutôt le fruit du hazard ou de l'esprit de spéculation, que le résultat d'efforts systématiques destinés à mettre en œuvre les ressources minérales du pays; aussi, dans la grande portion de la partie des États-Unis où l'on peut rechercher des richesses minérales et dont on a simplement reconnu les traits géologiques principaux, on ne connaît véritablement que quelques localités privilégiées ou que le hasard a fait découvrir.

Les travaux d'études géologiques et minéralogiques ont eu le plus souvent pour but de déterminer les limites des formations, et d'indiquer d'une manière générale les minerais et les métaux précieux qu'on y rencontre. Les personnes chargées de ces travaux n'ont pas eu à examiner chaque district assez en particulier pour noter tous les gîtes exploitables, à moins qu'ils ne constituent un trait important de la stratigraphie de la contrée.

Les études privées entreprises pour exploiter des mines ont été généralement limitées à une localité particulière et un espace restreint, et les rapports auxquels elles ont donné lieu vont rarement au delà de l'objet immédiat des investigations, c'est-à-dire de la valeur actuelle ou future de la région examinée.

En ce qui concerne les mines, en exceptant la houille et le fer, nous pouvons dire néanmoins que les données générales les plus valables que l'on possède revêtent un caractère négatif. Il est bien établi que certains espaces très-étendus, sont, par la nature même de leur nature géologique, dépourvus de tous produits minéraux importants. Théoriquement, nous sommes prêts à aller plus loin et à déclarer, qu'à l'exception d'un certain territoire comprenant une partie de l'Illinois, du Wisconsin et de l'Iowa, où l'on sait qu'il existe du plomb, on ne peut rechercher avec succès des gisements d'or, d'argent, de cuivre, de plomb, de zinc et d'autres métaux que dans certaines roches ignées et métamorphiques, dont les limites et l'étendue sont parfaitement établies.

Nous connaissons donc l'espace réservé aux exploitations de mines dans l'avenir : mais leur état actuel ne donne aucune idée exacte de l'importance que doit prendre un jour cette part de l'industrie nationale. L'expérience qu'apportent, chaque année, les progrès opérés dans cette direction, prouve d'une manière concluante sur quelle large base repose la richesse minérale des États-Unis.

Quelques mots suffiront à faire comprendre les caractères physiques et géologiques des États-Unis de l'Est (1). De leur extrémité nord-est, et dans la direction de sud-ouest jusque vers le golfe du Mexique, s'étend comme une ceinture une contrée haute ou montagneuse, plus ou moins fracturée, qui s'élève parfois à plus de 6000 pieds au-dessus du niveau de la mer. Sur toute sa longueur elle présente des traces évidentes d'un soulèvement qui s'est opéré parallèlement à la direction que suit la chaîne de montagnes. Cette région occupe à peu près toute la surface de la Nouvelle-Angleterre, et ses découpures présentent d'excellents abris aux vaisseaux tout le long des côtes. La chaîne traverse l'État de New-York sous le nom de *High Lands*, et s'écarte graduellement de la côte en avançant vers le sud, et en laissant ainsi une ceinture de formations plus récentes, de plus en plus large entre son versant et l'Atlantique. Les États-Unis sont limités

---

(1) Nous appelons ainsi ceux qui sont situés à l'est des monts Appalachiens.

au nord par le cours du St-Laurent et les grands lacs; mais nous n'avons qu'à dépasser un peu cette limite pour rencontrer une chaîne de montagnes qui court parallèlement au Saint-Laurent jusqu'au lieu où il prend sa source dans le lac Ontario; elle se dirige ensuite vers l'ouest et, pénétrant en quelques endroits au delà des limites géographiques de l'Union, s'étend, sauf de rares interruptions, jusqu'aux montagnes Rocheuses. Cette dernière chaîne a une direction générale vers le sud et atteint les limites méridionales des États-Unis vers le 10° degré de latitude. Le grand bassin, ou plutôt la plaine qui est entourée par ces rangées de montagnes, et qui s'ouvre au sud sur le golfe du Mexique, est, sauf de rares exceptions, occupée par des couches stratifiées qui n'ont subi aucun dérangement et qui ne présentent, dans leur longue étendue, que de légères déviations de leur position primitivement horizontale. A l'ouest de la grande chaîne de l'est, ou chaîne Appalachienne, nous trouvons une vaste étendue de formations anciennes, depuis les terrains primitifs jusqu'au terrain carbonifère. Ces formations s'étendent, dans leur plus grande largeur, de la vallée de l'Hudson jusqu'au delà des *Council Bluffs* sur le Missouri, ce qui répond à 22 degrés de longitude. Plus au midi, leur largeur est considérablement réduite, et elles se terminent par une ligne continue au nord de l'état d'Alabama. Du nord au sud, elles s'étendent sans interruption sur près de 15 degrés de latitude, soit, depuis le bord méridional du lac Supérieur, latitude 47° 50', à la latitude 33°. En ne tenant pas compte de petites surfaces qui ne se rattachent pas directement à cette grande plaine, on voit quelle immense étendue est remplie aux États-Unis par les formations palœozoïques. Dans ce vaste territoire, il ne sera pas superflu de remarquer que près de la moitié de l'espace total est occupé par les couches de houille de la période carbonifère.

Le terrain crétacé et les terrains tertiaires bordent le versant oriental de la chaîne Appalachienne, et forment une large ceinture ou plutôt un plateau le long de la côte de l'Atlantique et du golfe du Mexique. Ces formations pénètrent dans le Texas, puis, se dirigeant vers le nord, forment une vaste zone entre la limite occidentale des formations paléozoïques et les montagnes Rocheuses.

On voit quelle simplicité présentent les traits généraux de cette vaste contrée, dont la description détaillée exigerait néanmoins des volumes.

Au milieu de la formation crétacée du Texas s'élève une montagne isolée formée par des roches primitives ou métamorphiques, entourée par des roches de l'âge peléozoïque (*Voyez* Rœmer, *Terrain crétacé du Texas*). On trouve dans l'Arkansas une bande étendue qui présente les mêmes caractères, près de la jonction des terrains paléozoïque et crétacé; et le Missouri nous offre de même, au milieu des couches siluriennes et carbonifères, des montagnes de roches anciennes métamorphiques.

Les limites que nous venons de tracer et la nature des formations géologiques qu'elles embrassent doivent nous servir à déterminer les ressources que présentent les États-Unis, soit à l'agriculture, soit à l'industrie minérale.

Quand on considère la barrière montagneuse qui borde vers l'est la grande plaine centrale et occidentale, il ne faut point s'imaginer qu'on ne puisse y parvenir qu'à l'aide de routes difficiles, ni que la contrée qui s'étend à l'ouest présente une surface dont la monotonie n'est jamais interrompue. Presque toute la frontière septentrionale est occupée par une suite de grands lacs, où les plus grands bateaux à vapeur et vaisseaux peuvent naviguer, et qui présentent, seulement dans les États-Unis, une ligne de côtes de près de 3,000 miles de longueur (sans parler de plus petites baies et de petites rivières qui communiquent avec eux). La jonction établie entre ces lacs et avec l'Atlantique rend les relations entre la côte et les contrées éloignées de l'intérieur aussi faciles qu'entre des villes maritimes qui seraient situées à la même distance.

Les portions occidentales et centrales de cette grande surface sont traversées par des rivières navigables dont les larges embouchures au sud, sur le golfe du Mexique, semblent appeler à elles le commerce et la navigation. Au nord-est, les bassins de ces rivières sont reliés par des canaux à ceux des grands lacs, et de nombreux chemins de fer traversent tout le territoire à l'est du Mississipi, et s'étendent bientôt jusqu'aux montagnes Rocheuses.

On concevra aisément que si l'on veut se rendre compte des ressources industrielles d'un pays si étendu et si nouveau, on acquerra une idée plus juste de sa richesse en étudiant les principaux caractères physiques et géologiques qu'en examinant dans une exposition des collections de roches et de minerais. Nous croyons donc nécessaire de faire précéder la liste spéciale des produits minéraux par une rapide esquisse des principales formations géologiques, et d'indiquer successivement leurs traits caractéristiques, leur étendue géographique et leurs produits.

On peut énumérer dans l'ordre suivant les formations géologiques des États-Unis, soit qu'elles fournissent des matériaux exploitables, soit qu'elles impriment un caractère spécial à la portion du pays qu'elles recouvrent : nous omettrons, bien entendu, les subdivisions dont on trouvera le détail dans les rapports géologiques des divers États :

Alluvions et drift;
Terrain tertiaire;
Terrain crétacé;
Terrain jurassique;
New red sandstone (trias?);
Couches de houille;
Calcaire carbonifère, et autres roches inférieures aux couches de houille;
Old red sandstone ou terrain dévonien;
Silurien supérieur;
Silurien inférieur;
Roches cristallines ou métamorphiques de la chaîne Appalachienne, dont il est facile d'établir le rapport avec la période paléozoïque;
Roches anciennes métamorphiques ou cristallines;
Roches trappéennes.

ALLUVIONS ET DRIFT.

Les dépôts superficiels qu'on rapporte au drift sont distribués sur presque toute l'étendue des États-Unis. Ils sont ordinairement caractérisés par la présence de grands blocs erratiques; mais dans beaucoup de cas ils manquent ou restent invisibles sur de larges étendues, où la contrée est recouverte par une puissante couche de matériaux remaniés par les eaux. Dans les régions montagneuses, le dépôt superficiel reposant près des roches anciennes ne présente aucune stratification, et des fragments de toute dimension, plus ou moins usés, sont confusément assemblés; tandis que dans une grande partie de la vaste plaine qui s'étend à l'ouest des Alleghanys, le dépôt est stratifié plus ou moins régulièrement, et il s'y est opéré une sorte de séparation entre les éléments. Partout, en fait, des matériaux de transport couvrent la surface des roches anciennes sous-jacentes. Dans toutes les localités les formations anciennes présentent les traces évidentes d'une puissante action érosive, et dans le nord des États-Unis où les blocs erratiques sont abondants, comme dans le sud, où les matériaux sont moins volumineux, les résultats sont d'une nature semblable et semblent devoir être rapportés à l'action d'un agent commun.

Nous n'avons pas à nous appesantir ici sur les diverses modifications que ces dépôts ont subies ultérieurement; car nous n'avons qu'à examiner les caractères généraux de la contrée en relation avec les circonstances économiques qu'ils ont créées.

En règle générale, dans les États du nord et du centre de l'Union, le caractère de ce drift dépend de la nature des couches immédiatement sous-jacentes ou situées plus au nord à une petite distance. Bien que des blocs erratiques, provenant des montagnes du nord, puissent être observés à un ou deux cents miles de leur origine, on ne peut néanmoins supposer qu'une proportion considérable de la masse superficielle ait subi un transport aussi considérable. Ce n'est que le long de quelques vallées ou thalwegs que de grandes quantités de matériaux, usés par les eaux, ont été transportés vers le sud. La surface qu'ils occupent est petite quand on la compare à celle qui est recouverte par des matériaux qui sont en relation avec des roches très-voisines. S'il en était autrement, ou s'il était vrai que les agents qui ont transporté ces matériaux vers le sud eussent été aussi puissants qu'on le suppose quelquefois, on ne saurait légitimement chercher à rendre compte de ce fait par l'influence du sous-sol géologique. Ce n'est que dans des districts montagneux très-restreints qu'on peut attribuer à cette influence une valeur qu'alors même on peut à peine exagérer : mais c'est sur toute la surface des États-Unis que les caractères du sol sont en rapport avec la nature de la roche qui lui sert d'appui. Ce dépôt superficiel, quand il n'a pas été dérangé et repose sur les roches anciennes qui en ont fourni les éléments, contient ordinairement une grande quantité de matériaux grossiers : des changements ultérieurs, produits par des courants plus ou moins violents, ont modifié leurs conditions primitives, ont séparé les éléments et opéré parmi eux une stratification régulière. Durant une période encore plus récente, des eaux ont séjourné dans les petites dépressions de la formation; une riche végétation y prit naissance, et les débris de sa décomposition ont été conservés dans des marécages avec des os d'animaux, des coquilles et autres débris fossiles. Les eaux marines ont aussi envahi la formation et déposé du sable, de l'argile et des débris organiques qui se sont mélangés avec les matériaux accumulés du drift.

Produits exploitables du drift et des alluvions.

Or. — Ce métal, le seul précieux qui se rencontre dans cette formation, provient de la destruction de certaines roches métamorphiques aurifères. Dans les États-Unis de l'Est, le drift aurifère est en relation avec une série ou un groupe de roches métamorphiques subcristallines, qu'on pourrait appeler roches nouvelles métamorphiques, que nous décrirons sous ce nom dans notre Rapport. Ce drift s'étend sur la formation même d'où il dérive; on le trouve sur la limite nord des États-Unis, dans le Vermont et le New Hampshire, le long des montagnes vertes; il traverse la partie sud de l'État de New-York, et s'étend sur le versant oriental de la chaîne Appalachienne jusqu'en Géorgie et dans l'Alabama.

On y trouve des quantités d'or considérables en Géorgie, dans la Caroline du Sud, la Caroline du Nord et la Virginie : il n'a pas été exploité plus au Nord, dans les limites des États-Unis, mais il a longtemps attiré l'attention dans le Canada, et une quantité d'or très-importante a été obtenue à l'aide de lavages sur les bords de la rivière de la *Chaudière*. En Californie, le drift ou l'alluvion aurifère dérive de roches du même caractère métamorphique, et probablement du même âge. Ces sables et cailloux aurifères sont étendus sur une très-grande surface : on voyait, à la collection de l'exposition, des échantillons d'or provenant de près de 200 localités pour la Californie seulement.

Cuivre. — Ce métal, qui se rencontre à l'état natif ou dans quelques-uns de ses minerais ordinaires dans le drift, n'a jamais été trouvé en quantité suffisante pour être exploité.

Minerai de fer. — Des sables avec fer magnétique accompagnent généralement le drift dans le voisinage des chaînes de montagnes ; mais on n'a point cherché à utiliser ce minerai. Le fer limoneux est plus répandu, mais ce n'est que dans peu d'endroits qu'il est en assez grande quantité. L'hématite, si abondante à la base des Alleghanys et en d'autres localités, est parfois, à ce qu'il paraît, associée à des argiles de cette formation. Plusieurs échantillons de diverses provenances se trouvaient à l'exposition.

Sables argileux. — Propres à la briqueterie et à la poterie grossière, sont extrêmement répandus ; on en fait tous les ans des millions de briques. Le fer contenu dans l'argile leur donne ordinairement une couleur rougeâtre ; mais dans le Wisconsin, surtout aux environs de Milwaukie, on fait avec ces argiles des briques d'une couleur grise ou jaunâtre qui sont si estimées qu'on les transporte à plusieurs centaines de milles.

Tourbe et marne coquillière, etc. — Dans un petit nombre de districts de la Nouvelle-Angleterre, la tourbe est utilisée comme combustible; mais, en général, on s'en sert très-peu dans ce but aux États-Unis. Le dépôt tourbeux de beaucoup de contrées marécageuses est très-estimé comme engrais, et employé en même temps qu'une marne coquillière formée aux dépens de mollusques, dont les espèces sont encore vivantes. Quelquefois on rencontre des couches alternatives de tourbe et de cette marne sur des terrains marécageux d'une grande étendue. Cette marne est un mélange de petites coquilles avec du carbonate de chaux en grains ou pulvérulent, déposé par des sources qui prennent naissance dans des formations calcaires, ou des eaux qui ont traversé le drift ou l'alluvion, et dissous les matières solubles qu'elle renferme. Dans les endroits où le drift contient des cailloux en abondance, on s'en sert pour construire avec du ciment des murs, et même des maisons et d'autres constructions. En beaucoup de points, le drift ou les dépôts coquilliers qui le recouvrent sont très-puissants, et ne peuvent fournir de bons matériaux de construction; ces larges cailloux et les blocs erratiques deviennent alors d'une grande importance pour les habitants du district.

Exposition. — Les spécimens qui représentaient cette formation à l'exposition consistaient principalement en or natif provenant de deux cents localités de la Californie et de quelques autres endroits, un bloc de cuivre natif du Wisconsin, du fer limoneux, de l'ocre, de la marne, des échantillons de sols, etc.

FORMATIONS TERTIAIRES.

Les formations tertiaires proprement dites consistent en couches marines, littorales et lacustres, argiles de toute espèce, lits calcaires et arénacés et mélanges divers des matériaux précédents. Ces couches, qui contiennent des débris nombreux de coquilles, de polypiers, de plantes, d'animaux terrestres et de coquilles d'eau douce, sont très-développées aux États-Unis. Elles prennent naissance à l'extrémité nord-est du Maine, en y comprenant les dépôts de la vallée du lac Champlain et du Saint-Laurent; elles s'étendent ensuite au sud-ouest et prennent leurs caractères les mieux définis dans le New-York; puis, se prolongeant au sud et au sud-ouest, occupent la zone qui s'étend entre l'Atlantique et le pied des montagnes. Les mêmes formations continuent à se développer au sud et à l'ouest, et, s'infléchissant au pied méridional des Highlands, s'étendent au loin sur la partie sud de la Géorgie, d'Alabama, du Mississipi et sur une grande partie de la Louisiane et du Texas, tout le long du golfe du Mexique. Au nord du Texas, les formations tertiaires suivent la base des montagnes Rocheuses, et forment une large ceinture bordée des deux côtés par le terrain crétacé. Les dépôts tertiaires atteignent même les sources du Missouri et se perdent dans les interminables prairies de cette région connues sous le nom des *Mauvaises Terres.*

A l'ouest des montagnes Rocheuses, entre cette région et la sierra Névada, le capitaine Fremont et le capitaine Stransbury ont trouvé des fossiles tertiaires dans des localités très-éloignées, ce qui permet de croire que cette formation est largement développée dans cette partie du continent. Les dépôts tertiaires sont également très-étendus sur les côtes occidentales, ainsi que Dance et d'autres nous l'ont appris (voyez *Exploration des États-Unis*, géologie, par T.-D. Dance); mais on n'y a pas encore étudié la formation avec détail.

Produits exploitables dans le terrain tertiaire.

Hématite brune et minerai de fer ocreux. — On sait depuis longtemps qu'on trouve aux États-Unis une bande continue depuis le Canada jusqu'à l'Alabama, contenant de l'hématite, de l'ocre, etc., et des argiles d'un caractère particulier. Ces minerais ont été rapportés à un dépôt aqueux ou marin, et considérés comme ayant été formés aux dépens des couches sous-jacentes par l'action de sources thermales. La découverte d'une couche de lignite, associée à ces argiles et minerais, a conduit le professeur Hicthcock à les rapporter à la période pliocène; mais les données qui doivent servir à établir leur âge précis sont encore incomplètes.

L'hématite est ordinairement compacte, fibreuse, stalactiforme, et toujours plus ou moins enveloppée par des argiles de couleur variée. Dans la Nouvelle-Angleterre, cette formation repose sur le flanc occidental des montagnes Vertes et surmonte un calcaire métamorphique, associé plus ou moins avec des schistes talqueux ou micacés. Le fer provient sans doute de ces couches, et en a été retiré par l'action de sources thermales lors du dépôt des argiles.

En Pensylvanie, ces minerais remplissent les dépressions ou fissures de calcaires siluriens qui n'ont subi aucune altération, et sont associés à des argiles ou enveloppés par elles; tandis que, plus au sud, les hématites sont encore associées à des roches métamorphiques.

Ce dépôt est un des gîtes les plus importants et a été

exploité le premier dans le pays; il fournit un excellent minerai facile à traiter.

Des échantillons de minerai et de fer ont été présentés par les États de Vermont, Massachussets, New-York, New-Jersey, Connecticut, Pensylvanie et Maryland.

Manganèse.— De l'oxyde blanc de manganèse se rencontre dans beaucoup de localités avec le minerai de fer ci-dessus mentionné. On le trouve à Brandon et Bennington, en Vermont, en abondante quantité. L'oxyde blanc de manganèse, provenant de quelques localités du sud, appartient probablement à cette formation.

Combustibles fossiles.— Un banc de lignite est associé avec l'hématite et les argiles réfractaires à Brandon, en Vermont, et est utilisé comme combustible. On a trouvé des lignites en différentes autres localités, spécialement à l'ouest des montagnes Rocheuses et sur les côtes de l'océan Pacifique.

Les seuls échantillons de l'exposition étaient ceux de Brandon, en Vermont.

Argile réfractaire et terre à porcelaine. — On trouve du kaolin et des argiles de couleur variée, plus ou moins pures, tout le long du versant occidental des montagnes Vertes. On les utilise dans les briqueteries et pour les poteries grossières. Ces argiles sont blanches ou colorées en jaune par du fer, en rose par du manganèse. On connaît des argiles de même caractère dans plusieurs localités de New-Jersey.

Une autre variété, peut-être d'un âge un peu plus ancien, existe en plus grande abondance dans cet état. On la trouve aux environs de South-Amboy; elle est connue sous le nom d'*argile à poterie*, et alimente la plupart des fabriques des États-Unis. Elle est associée à des lits d'argile bigarrée et de sable. On se sert aussi de ces argiles pour faire des briques réfractaires.

Les échantillons provenaient de Brandon, Vermont; New-Garden, en Pensylvanie, avait envoyé des kaolins et des briques réfractaires.

Marne coquillière et glauconie. — Dans les États de New-York, de Maryland, de Virginie et plus loin vers le sud, des marnes coquillières contenant parfois de la glauconie sont employées comme engrais.

Matériaux de construction. — Dans cette formation, le sable, l'argile, la terre à briques remplacent les bonnes pierres de construction. On rencontre bien en quelques points un grès grossier et friable, mais on ne peut s'en servir pour les constructions de longue durée, comme l'ont trop appris le Capitole de Washington et quelques autres constructions.

TERRAIN CRÉTACÉ.

Le terrain crétacé dans la partie orientale des États-Unis apparaît d'abord dans l'état de New-Jersey, où on l'a reconnu depuis longtemps et où les fossiles les plus caractéristiques de la craie ont été découverts par M. Morton. Il occupe une longue étendue dans le New-Jersey et dans le Delaware; en suivant les sinuosités du terrain tertiaire, il apparaît à divers intervalles dans la Virginie, les Carolines du nord et du sud, où les dénudations du terrain tertiaire l'ont mis au jour. Il traverse ensuite le nord de la Géorgie, le centre d'Alabama, le Mississipi, et se dirige vers le nord le long des roches paléozoïques jusqu'au point où l'Ohio entre dans l'état d'Illinois; de là il s'étend à l'ouest du Mississipi en suivant la direction des couches paléozoïques à travers l'Arkansas, et prend une grande largeur dans le Texas. Dans leur étendue totale, les dépôts crétacés occupent du nord au sud plus de 22 degrés, depuis le tropique du Cancer jusqu'au 48e degré de latitude nord.

Produits exploitables et caractères agricoles.

On n'avait envoyé à l'exposition aucun produit appartenant à cette formation.

La glauconie, le sable vert ou, comme on l'appelle ordinairement, la marne de la série crétacée, sont employés sur une vaste échelle par les agriculteurs de l'état de New-York et plus loin vers le sud. Parfois elle ne contient que peu ou point de coquilles fossiles; tandis qu'en d'autres localités, elles sont très-abondantes. Les effets fertilisants de cet amendement sont dus principalement à la potasse qu'il contient et peut-être aussi au sulfate de chaux provenant de la double décomposition du sulfate de fer et du carbonate de chaux opérée par les coquilles fossiles.

Malgré l'immense étendue de la formation crétacée aux États-Unis, on n'y a point découvert de craie véritable. Les lits calcaires, partout où on les a reconnus, ne peuvent servir qu'à fabriquer de la chaux et ne peuvent être employés dans les constructions et l'architecture.

On rencontre dans cette formation de petits dépôts de fer, souvent de pyrite et de phosphate de fer, mais ils n'ont pas assez de valeur pour être exploités.

On trouve aussi de petites quantités de lignite et du bois fossile de différentes espèces. Sur le Missouri supérieur et sur les bords du Yellow-Stone, les lits de lignite sont étendus et seront sans doute un jour exploités.

L'espace occupé par les formations crétacées et tertiaires, entre le Mississipi et les montagnes Rocheuses, est beaucoup plus vaste que toute la zone silurieuse, dévonienne et carbonifère comprise entre les Alleghanys et le Mississipi. Ces formations, qui s'étendent si loin au delà des dernières, sont certainement destinées à jouer un rôle important dans l'avenir de la contrée. La plus grande partie de cette immense surface est dépourvue de forêts et souvent de buissons; elle est stérile, et les argiles qui se rencontrent dans le terrain crétacé et tertiaire la rendent souvent impropre à la culture. Le climat est si sec qu'en été le sol se brûle et se fend de manière à détruire toute végétation. Quelques points, recouverts par des alluvions, se présentent seuls à l'agriculture, et la plus grande étendue de cette région mérite bien son nom de *Mauvaises Terres*.

TERRAIN JURASSIQUE.

Les formations qui appartiennent à ce terrain, et qui ont été reconnues jusqu'ici, forment une ceinture de 10 à 12 milles de largeur et de 50 mètres de longueur, à l'ouest de Richmond (Virginie), et peuvent être comprises sous le nom de bassin de Chesterfield. Le bassin de la Rivière-Profonde dans la Caroline du Nord est d'une nature semblable, et s'étend, sauf quelques interruptions, presque à travers tout cet État. Ces deux bassins doivent être considérés comme les parties séparées d'une seule et même formation (1). C'est là jusqu'aujourd'hui le seul groupe géologique que l'étude des fossiles ait autorisé à ranger avec certitude dans le terrain jurassique aux États-Unis.

Cette formation repose sur le gneiss ou le granite, et les poudingues et grès de la série ont été formés par les matériaux cimentés de ces roches anciennes. Plus haut,

(1) La question de l'âge de la ceinture non interrompue qui s'étend plus loin à l'ouest et suit la formation de New-Jersey, et celle des terrains de la vallée du Connecticut, ne sont pas encore résolues.

la roche prend les caractères d'un schiste micacé qui, dans ses portions supérieures, devient très-argileux.

Produits exploitables.

Charbon, etc. — Un trait remarquable de cette formation, et qui n'est pas moins intéressant au point de vue économique, est l'existence de plusieurs veines charbonneuses. Elles sont exploitées aux environs de Richmond et près de la Rivière-Profonde, dans la Caroline du Nord. L'épaisseur totale de la formation, en Virginie, est de 800 pieds, et les veines charbonneuses sont concentrées dans les 150 pieds inférieurs. Dans plusieurs endroits, on en trouve deux ou trois; mais dans le bassin de Virginie, au sud de la rivière James, ces différentes veines se réunissent en un vaste lit de 20 à 40 pieds d'épaisseur.

La quantité de charbon extraite de ce bassin, d'après les meilleures évaluations auxquelles nous avons pu arriver, est de plus de 150.000 tonnes par an. Toutes les parties de la formation n'ont pas été encore étudiées dans la Caroline du Nord, et jusqu'à présent d'ailleurs une faible partie seulement du charbon a pu être amenée sur les marchés, par suite du manque de voies de communication. Le transport, néanmoins, va sans doute devenir bientôt plus facile quand on aura terminé la canalisation de la Rivière-Profonde.

Les principaux produits fournis par cette formation consistent en charbon, grès, pierres de construction, etc. Elle n'était pas représentée à l'exposition.

NOUVEAU GRÈS ROUGE, SCHISTES ET CONGLOMÉRATS.

Les conglomérats, grès et dépôts schisteux si bien connus de la vallée du Connecticut, dans le Massachusets et le Connecticut, et le long de la vallée de l'Hudson et qui traversent le New-Jersey, la Pensylvanie, la Virginie et la Caroline du Nord, ont été ordinairement rapportés au nouveau grès rouge ou à la période du trias. Dans les deux États que nous venons de mentionner, les dépôts que nous considérons reposent dans des dépressions ou bassins isolés à l'ouest de ceux que nous avons déjà décrits, et en diffèrent par des caractères plutôt négatifs que positifs.

Cette formation a une grande largeur dans le New-Jersey et la Pensylvanie, et diminue graduellement vers le sud-ouest; elle ne forme plus dans le Maryland et le nord de la Virginie qu'une zone étroite, dont la continuité est interrompue dans ce dernier État. Ces bassins isolés sont rangés sur la même ligne que cette ceinture et n'en sont sans aucun doute que des prolongements.

Dans la vallée du Connecticut, on trouve cette formation sur une longueur de 100 milles et une largeur de 10 à 20 milles. C'est dans cette vallée qu'elle a été étudiée avec le plus de soin, et qu'on a trouvé la plus grande partie des fossiles les plus caractéristiques. Les restes organiques consistent en plantes fossiles du genre *voltzia* et autres, en nombreux poissons fossiles; on y trouve aussi des empreintes de pas, dont la plupart paraissent devoir être rapportés à des oiseaux et quelques-uns à des quadrupèdes.

Produits exploitables de la formation du nouveau grès rouge.

Pierres de construction. — Le grès du Connecticut, bien qu'occupant une faible étendue, fournit peut-être actuellement plus de matériaux à l'architecture et aux constructions communes qu'aucune autre formation géologique de la contrée. Cette pierre est envoyée à toutes les cités maritimes des États-Unis et même à l'intérieur des terres.

Ce grès est aussi exploité dans le New-Jersey et sur quelques points, le long de l'Hudson, dans l'État de New-York. La pierre employée pour construire l'église de la Trinité, dans la ville de New-York, est un grès de cet âge de la rivière Passaie, New-Jersey.

Cuivre et plomb. — Les seuls minerais de quelque importance trouvés dans cette formation sont des minerais de cuivre et de plomb. On a trouvé des mines de cuivre en plusieurs endroits dans le Connecticut, mais on n'a réussi à les exploiter qu'avec peu de succès. La plus importante est la mine Simsbury, près de la ville de Grauby (Connecticut). Le minerai est un cuivre silicaté avec de petites quantités de cuivre panaché et de malachite. La mine est abandonnée depuis plusieurs années.

Il paraît que dans l'État du New-Jersey on a trouvé du cuivre en plusieurs points de cette formation; mais, après des essais nombreux et répétés, on s'est décidé à abandonner les exploitaiions. Les carbonates bleu et vert sont les minerais dominants. Les sulfures gris et bleu, associés à des oxydes rouges, ont été trouvés dans quelques mines, et plus rarement du cuivre natif, du phosphate et silicate de cuivre. Les minerais ne se rencontrent pas en filons réguliers, mais semblent disséminés dans certaines couches de la partie schisteuse de la formation; quelquefois ils forment des amas puissants qui semblent offrir aux exploitations un gage de succès.

En Pensylvanie, dans les couches de Montgomery et Chester, quelques-unes des veines productives de plomb et de cuivre, qui se trouvent à la séparation de cette formation et du gneiss, pénètrent les deux systèmes, tandis que d'autres veines restent exclusivement dans l'un ou dans l'autre. Les localités que nous venons de mentionner offrent de grandes quantités de minerai de plomb et de cuivre, et, dans quelques-unes des mines, l'exploitation est très-fructueuse. Les veines qui traversent les schistes et les grès paraissent plus riches en cuivre qu'en plomb, et la même veine qui produit surtout du plomb dans le gneiss devient cuprifère dans les roches supérieures.

Les minerais principaux sont le plomb sulfuré, le plomb phosphaté, le sulfure et le carbonate de cuivre(1).

Nous reparlerons de ces minerais quand nous décrirons les roches métamorphiques.

Exposition. — Outre les échantillons de la mine Wheatley, on voyait à l'exposition des fragments et des petits blocs de grès provenant des carrières de Portland, Connecticut, sur la rivière Connecticut, ainsi que de la baryte sulfatée de Cheshire (Connecticut).

TERRAIN HOUILLER OU COUCHE DE HOUILLE.

La série de dépôts sédimentaires ou autres, qui constitue ce qu'on nomme généralement le terrain houiller, présente aux États-Unis les mêmes caractères qu'en Europe. A la base, on trouve ordinairement un grès

---

(1) La mine de Wheatley, où l'on exploite un filon qui pénètre à la fois le gneiss et le grès, a envoyé les matériaux suivants à l'exposition :

Sulfate de plomb cristallisé.	Malachite verte.
Plomb phosphaté.	Malachite bleue.
Plomb arséniaté.	Cuivre oxydé blanc.
Plomb molybdaté.	Cuivre natif.
Plomb chromomolybdaté.	Manganèse oxydé.
Plomb arsénio-phosphaté.	Soufre natif.
Sulfure de plomb et argent antimonial.	Argent natif.
Plomb sulfuré.	Quartz cristallisé.
Zinc sulfuré.	Quartz cellulaire.
Zinc carbonaté.	Oxyde de fer avec argent.
Zinc silicaté.	Hématite brune.
Cuivre sulfuré.	Baryte sulfatée.
	Pyrite de fer.

grossier ou conglomérat, à cailloux de quartz. En quelques parties cette roche manque, et ce sont des couches schisteuses avec charbon qui reposent sur le calcaire inférieur. Les couches suivantes consistent en grès à grain plus ou moins fin, parfois en conglomérats ou grès d'une nature analogue, calcaire, houille, argiles réfractaires. L'épaisseur énorme de cette formation, en quelques points, est réellement étonnante, et l'on demeure confondu en songeant aux changements qui ont été nécessaires pour produire la succession de couches d'une nature si différente et d'une pareille puissance.

La grande formation carbonifère des États-Unis constitue l'un de ses traits géologiques les plus frappants, en même temps qu'elle se rattache aux conditions économiques de l'ordre le plus important. La région carbonifère de l'est des Alleghanys se prolonge depuis la limite extrême nord de la Pensylvanie jusqu'au centre de l'Alabama; sa longueur est de plus de 750 milles et sa longueur maximum dépasse 180 milles, sans compter les bassins anthraciteux qui forment des îlots, le long du bord oriental de cette vaste surface, évaluée par le professeur H.-D. Rogers à près de 63.000 milles carrés.

Le grand district occidental, ou, comme on l'appelle généralement, le district de l'Illinois, occupe la plus grande partie de l'état de l'Illinois, ainsi qu'une portion de l'Indiana et du Kentucky. Les derniers travaux du docteur D. D. Owen ont fait voir qu'il s'étend plus loin dans l'Iowa et le Missouri qu'on ne l'avait supposé auparavant.

Il est vrai que sa partie située à l'ouest du Mississipi n'est pas seulement séparée de la partie orientale ou district de l'Illinois par le cours du fleuve; celui-ci, en effet, coule le long d'une douce ligne antéclinale, dont l'élévation a contribué à séparer les deux portions de la région carbonifère de l'ouest.

Cette portion, située dans l'Iowa et le Missouri, occupe un espace dont la partie qui est à l'est du Mississipi n'égale que les deux tiers. Suivant l'opinion de M. Hall et d'autres géologues marquants, ce district houiller s'étend beaucoup plus loin, vers le sud-ouest, que les limites, aujourd'hui encore restreintes, des explorations géologiques faites avec soin, bien qu'il ne dépasse point sans doute le parallèle des régions siluriennes et métamorphiques qu'on rencontre au sud du Missouri.

Sa plus grande étendue du sud-est au nord-ouest, ou des sources de la Rivière-Verte, en Kentucky, à sa limite nord, sur la rivière des Moines, dans l'Iowa, est de plus de 550 milles; sa largeur maximum, à travers l'Illinois et le Missouri, dans la direction est-ouest, est de plus de 400 milles; et on peut compter 300 milles depuis sa limite septentrionale, dans l'Iowa, à ses limites actuelles méridionales sur la rivière Osage.

Cette région de l'ouest, en y comprenant les deux côtés du Mississipi, a donc une plus grande superficie que celle qui est située à l'est de la chaîne Appalachienne. La portion seule qui est à l'est du Mississipi est presque aussi grande que cette dernière, et celle qui est à l'ouest la dépasse de plus de moitié.

Plus loin, vers le Sud, on sait qu'il existe un bassin houiller d'une étendue considérable dans l'Arkansas. Il est probable que dans l'origine il était dans le prolongement de celui de l'Iowa et du Missouri, et qu'il en a été séparé depuis par le soulèvement d'une ceinture de roches métamorphiques; peut-être aussi ces roches déjà soulevées formaient-elles une île allongée dans la mer, où se sont déposées les couches de la formation carbonifère. On connaît plus loin encore, au Sud et à l'Ouest, d'autres petits bassins isolés ou en relation avec le bassin principal; on a retrouvé d'ailleurs la grande couche inférieure du calcaire carbonifère dans le Nouveau-Mexique, et plus au nord, en certains points, le long des montagnes Rocheuses : on peut donc en inférer qu'on retrouvera les couches de houille, sinon sur toute l'étendue, au moins par intervalles et dans des bassins isolés de cette vaste surface.

En outre, il existe dans le Michigan un petit district jusqu'ici improductif, et qui ne semble pas tirer aujourd'hui profit de sa position dans une plaine basse. Sa position géographique lui donne cependant de grands avantages; il est situé au-dessus du niveau des eaux navigables du lac Huron, et est plus accessible au mouvement commercial des grands lacs qu'aucun autre bassin houiller.

On n'a encore aujourd'hui que trop peu de renseignements précis pour être à même de juger de l'importance que lui prépare l'avenir.

On trouve aussi dans le Rhode Island et le Massachusetts, sur une petite étendue, des couches de houille qui n'ont presque subi aucune altération (1). En étudiant la position relative de ces couches et de celles de la Nouvelle-Écosse, en connexion avec l'âge et les caractères de la contrée où elles apparaissent, on est conduit à penser que ces portions détachées ont pu autrefois être en liaison avec le district des Alleghanys; que celui-ci même ne faisait qu'un avec celui de l'ouest, et

(1) Les couches de houille un peu altérée auxquelles il est fait ici allusion sont celles que l'on exploite aujourd'hui; mais je suis convaincu qu'il existe sur une beaucoup plus grande étendue dans la Nouvelle-Angleterre des couches qui ont été plus ou moins métamorphisées.

qu'ainsi, en fait, dans leur condition primitive, et avant qu'aucune action perturbatrice ne masquât leur unité, toutes ces formations carbonifères pouvaient avoir recouvert toute la partie des États-Unis qui est située à l'est des montagnes Rocheuses.

En résumé, outre l'immense et inépuisable région qui s'étend à l'est de ces montagnes, des échantillons de houille ont été rencontrés par le capitaine Strausburg au delà du fort Laramie : on en a trouvé à Bellingham Bay, à Puget's Sound, dans le district de Washington; mais on sait peu de chose de son âge et des roches qui lui sont associées. Ces derniers échantillons ressemblent tellement aux charbons bitumineux des couches de houille ordinaires, qu'on ne saurait établir entre eux aucune différence, et que nous sommes ainsi fondés à espérer qu'on trouvera sur les côtes occidentales des États-Unis cette source importante de progrès et de richesse nationale.

Produits exploitables.

Houille. — La rapide disparition des forêts dans le nord et le centre des États, ainsi que le développement constant et de plus en plus marqué des opérations industrielles, ont rendu nécessaire l'emploi général du combustible minéral. Bien qu'il y ait comparativement un petit nombre d'années que le charbon soit devenu d'un usage commun, la quantité absorbée par une constante consommation est aujourd'hui si énorme, que l'extraction et le transport à bon marché de ce combustible aux villes et aux cités devient une matière de la plus grande importance.

Les premiers essais industriels aux États-Unis eurent lieu dans la Nouvelle-Angleterre, où l'abondance des eaux compensait le manque de combustible minéral.

Actuellement, depuis que les couches de charbon sont mieux connues, et que l'extraction s'est mise au niveau des besoins de la communauté, on commence à sentir que le succès des opérations industrielles est surtout lié à l'abondance et à la consommation toujours assurée du combustible minéral. Pour faire saisir jusqu'à quel point a augmenté l'emploi du charbon anthraciteux, il suffira de dire qu'en 1820 le chiffre en était borné à 365 tonnes, tandis qu'en 1847, la demande toujours croissante s'élevait à 3,000,000 tonnes; et que dans la même année on consommait plus de 2,000,000 tonnes de charbon bitumineux.

En 1851, on a estimé que le charbon bitumineux consommé à Pittsburg et expédié aux ports du Sud, en y comprenant le charbon de Monongahela, s'élevait à plus de 1,000,000 tonnes. On arrive facilement des côtes de l'Atlantique aux bassins d'anthracite et de charbon bitumineux de Pensylvanie et du Maryland, qui forment comme des points avancés (outliers) du grand district carbonifère. L'Ohio, le Monongahela, le Kanawna, et d'autres rivières plus petites, quoique navigables, pénètrent et le traversent à l'ouest, tandis que ses bords septentrionaux et occidentaux ne sont éloignés que de 30 ou 50 milles du lac Érié à Cleveland, et de moins de 100 milles de Buffalo, où s'arrête la ligne de navigation du lac Supérieur, et commence le canal de New-York.

Au Sud, la même région arrive jusqu'à 200 milles du golfe de Mexico, et est traversée par des fleuves qui permettent de transporter le charbon depuis la mine jusqu'à la mer : le charbon qui sert à la navigation maritime dans le golfe peut donc éventuellement suivre cette direction. Actuellement, de grandes quantités de charbon descendent l'Ohio et sont dirigées vers la Nouvelle-Orléans, et parcourent ainsi plus de 2.000 milles en rivière.

La région carbonifère de l'Ouest est traversée par l'Ohio, le Wabash, l'Illinois, le Mississipi, le Missouri et la rivière des Moines : les plus grandes facilités sont donc apportées aux transports sur les cours d'eau naturels. La limite nord est du district de l'Illinois arrive à 80 ou 90 milles de l'extrémité sud du lac Michigan, et est reliée avec Chicago par un canal et un chemin de fer. Ajoutez à toutes ces voies de transport les nombreux chemins de fer, en construction ou en projet, qui traversent cette contrée presque unie ou doucement ondulée, et l'on acquerra la certitude que cette vaste région de l'Ouest peut être fournie abondamment et d'une manière permanente de combustible minéral.

Le charbon des petits districts du Massachussets et de Rhode Island est un anthracite médiocrement combustible. Ce caractère, ainsi que l'irrégularité de ses couches, au milieu d'une contrée extrêmement bouleversée, et qui a subi le métamorphisme le plus puissant, n'ont pas permis des exploitations importantes; les travaux commencés sur dix ou douze points ont été abandonnés.

Les bassins d'anthracite de la Pensylvanie produisent de l'anthracite de qualité supérieure, et leur position leur permet d'alimenter pour une forte part les villes et cités de la côte de l'Atlantique.

Un échantillon d'anthracite de dimensions énormes, de la couche Mammoth, à Wilkesbarre (Pensylvanie), a été envoyé à l'exposition par des habitants de cette ville. Cette masse présente une section verticale de cette couche, et forme une colonne de 5 pieds carrés à la base et de trente pieds de hauteur, et pèse environ 60 tonnes. On voyait quelques autres grandes masses de la même localité, l'une d'elles de 10 pieds de long sur 4 de large et 3 1/2 de haut, et des échantillons de la même couche provenant d'autres localités.

On avait envoyé des charbons de la mine Carbondale et Pittstown, une collection d'environ soixante variétés d'anthracite du comté de Schuylkill et quelques autres échantillons de charbon.

Les charbons bitumineux du Maryland étaient représentés par de larges échantillons de l'épaisseur de la couche (11 et 15 pieds). De nombreux échantillons de charbon bitumineux accompagnaient les minerais de fer de diverses localités de la Pensylvanie.

Ceux des comtés de Perry, Hacking, Vinton, Jackson et Mahoning, de l'Ohio, du Kentucky et de la Virginie, étaient aussi représentés.

Minerais de fer. — Les districts carbonifères des États-Unis contiennent de riches dépôts de minerai de fer, ordinairement associé aux couches de charbon. Ces minerais consistent principalement en fer carbonaté et en peroxyde de fer, ce dernier résultant sans doute d'une décomposition partielle du premier, qui contient une certaine quantité d'argile, de calcaire et de silice. En Pensylvanie et dans l'Ohio, où ces minerais sont exploités plus que partout ailleurs, les gisements paraissent inépuisables, et suffiront pendant une période illimitée aux exigences d'une civilisation dont le progrès ne se ralentit pas.

Dans le Tennessee, l'Alabama et l'ouest de la Virginie, la formation carbonifère abonde en minerai de fer. Dans la région de l'Ouest, cette relation entre le charbon et le fer a été moins reconnue, parce que la fabrication du fer y est beaucoup moins développée. Les travaux géologiques poursuivis actuellement dans l'Illinois ont déjà fait voir que cet État est richement doté de minerai de fer qui se rencontre avec ses inépuisables couches de houille; et cependant il ne possède encore que deux hauts fourneaux. Dans le Missouri, on fabrique une certaine quantité de fer avec ces minerais, mais on n'est pas encore assuré de leur abondance. Les échantillons envoyés à l'exposition provenaient principalement de la Pensylvanie; et cet État a envoyé une collection très-complète de minerais, non-seulement du terrain houiller, mais encore des autres formations. La collection comprenait le minerai et les divers matériaux et produits : charbon, laitiers, fonte, fer; les minerais des formations inférieures à la houille, et qui se rencontrent dans les couches de houille, accompagnés du charbon qui a servi à les traiter.

On avait envoyé des minerais de fer de Vinton, Jackson et Hacking, Ohio, ainsi que d'une localité du Missouri. Une grande collection d'échantillons qui appartiennent probablement à la même formation avait été envoyée de l'Alabama.

Pour se rendre compte de l'abondance de ce métal dans ce terrain et du parti qu'on peut en tirer, il faut se rappeler qu'une faible partie seulement de ce que l'on connaît de cette immense région carbonifère était représentée à l'exposition, et que ce qu'on en connaît n'est qu'une faible partie de la surface entière; en songeant que les minerais qui en font partie, avec le charbon qui les accompagne, sont distribués plus ou moins abondamment sur un espace de plus de 160,000 milles carrés, nous pourrons concevoir une idée éloignée de cette immense quantité et des résultats importants qui pourront en résulter pour les intérêts industriels et l'avenir d'une contrée telle que les États-Unis.

On sait que les États de Tennessee et d'Alabama sont riches en minerai de fer appartenant au terrain houiller, aussi bien que la Virginie et le Kentucky : mais un seul de ces États a envoyé des échantillons.

Calcaires. — Dans presque toutes les parties du district carbonifère, on trouve des calcaires d'origine marine, alternant avec des schistes, des grès et des lits de houille. En quelques endroits, le charbon et le calcaire se touchent, et parfois le minerai est en relation avec ces calcaires. Ces lits calcaires fournissent le fondant nécessaire pour traiter les minerais : on les exploite aussi pour fabriquer de la chaux pour les constructions.

Grès. — Les lits de grès alternant avec les schistes et les couches de houille, fournissent souvent une pierre excellente et très-durable. Quelques-uns des grès les plus friables, qui souvent ne contiennent point de fer, sont employés dans les verreries.

Argiles réfractaires. — Quelques-uns des lits de houille reposent sur des couches d'argile réfractaire plus ou moins épaisses ; et dans quelques parties de la formation, il existe des couches étendues de cette argile qui ne leur sont point directement associées. On l'emploie pour fabriquer des briques réfractaires, pour la poterie, pour faire des tuyaux de conduite, des drains. Partout ces couches ont la même étendue que le bassin houiller.

CALCAIRE CARBONIFÈRE.

La grande formation du calcaire carbonifère inférieure aux couches de houille, et qui, dans les districts houillers, forme ordinairement le trait le plus apparent de la topographie de la contrée, perd toute son importance et son relief sur les bords de la plus grande partie de la région houillère des Alleghanys. Sur la limite orientale, en Pensylvanie, c'est à peine si l'on peut dire qu'elle existe encore ; et l'on ne peut y rapporter que de minces lits calcaires associés au schiste rouge dont nous parlerons plus loin.

Dans la Virginie, cette formation prend un grand développement, et devient très-puissante dans le sud de cet État. Dans le Tennessée, et toujours dans la même direction, le calcaire forme une étroite ceinture le long du bord oriental du bassin houiller, et se prolonge jusqu'à sa limite méridionale dans l'Alabama. Vers le nord, cependant, il disparaît graduellement et se perd complétement vers la rivière de l'Ohio, ou n'occupe plus au milieu des grès qu'une place insignifiante.

Dans le district de l'Ouest ou du Mississipi, nous retrouvons la formation beaucoup plus puissante et plus étendue. Le calcaire occupe, sur le bord oriental, une zone étroite qui traverse l'Indiana et le Kentucky, et reparaît sur le côté occidental du bassin ou le long de la vallée du Mississipi ; le lit du fleuve y est creusé à une grande profondeur, sur la longueur de plusieurs milles, et ses bords sont souvent élevés comme des falaises. Le calcaire s'étend avec une largeur variable sur un des côtés du Mississipi, à une petite distance au-dessus de l'Ohio jusqu'au confluent du Rock River, sur une longueur de plus de 300 milles par conséquent, et vient affleurer, à l'est et à l'ouest, sous les couches de houille. Il paraît, à de nombreux intervalles, le long du Missouri, et, partout où on l'a observé jusqu'ici, forme une large ceinture qui entoure toute la division occidentale du terrain houiller. Plus à l'ouest, il apparaît au Fort-Laramie, et forme le dépôt sédimentaire le plus important aux environs du grand lac Salé. On sait aussi qu'il se rencontre dans le Nouveau-Mexique et le Texas, mais ses limites n'y ont pas encore été tracées. Dans le Michigan, on sait peu de chose à l'égard de cette formation : on a seulement trouvé en plusieurs localités, sur le bord occidental du bassin houiller, un calcaire qu'on lui rapporte. La grande variété des caractères que présente ce calcaire dans l'Ouest, jointe à d'autres considérations encore, a engagé le docteur D. D. Owen à le séparer en deux étages, l'un inférieur, l'autre supérieur. Sa puissance totale est de 390 pieds, et on peut y observer douze couches de caractères lithologiques différents.

Sur une vaste région dans l'ouest, ce groupe de lits calcaires, avec leurs couches de marne ou de grès tendre argileux, et des intercalations marneuses avec lits de calcaire terreux, imprime à la contrée qui s'étend autour du district houiller un caractère de fertilité qu'on ne retrouve plus au même niveau géologique dans le reste des États-Unis. Sur les bords septentrionaux, occidentaux et orientaux du district de l'est ou des Alleghanys, il existe une zone étroite où l'on ne trouve aucune formation calcaire importante ; et il faut chercher les calcaires bien loin des affleurements houillers ou dans les lits qui sont intercalés entre les couches de charbon. L'effet de cette différence géologique se lit, non-seulement dans les caractères différents des constructions, mais encore dans les conditions agricoles des deux contrées ; dans l'une, on ne trouve que des prairies aux environs de la houille, tandis que dans l'autre, à la même latitude et dans les mêmes conditions stratigraphiques, le sol très-calcaire fournit des céréales de première qualité.

Ce ne sont pas seulement les ceintures extrêmes qui entourent les deux bassins qui présentent un tel contraste : celui qui se manifeste entre leurs parties centrales n'est pas moins frappant. Cette formation calcaire, avec les nombreux lits de nature analogue qui alternent avec les couches de houille elle-même, fréquemment coupée sur toute la profondeur par le lit des rivières et de profondes vallées, a servi à distribuer les matériaux calcaires avec abondance sur toute la surface de la contrée. C'est par ce motif peut-être et aussi par suite de l'accumulation de matériaux provenant de la destruction de couches calcaires des formations inférieures dans les districts du nord, que le bassin de l'ouest, avec ses grandes plaines et ses prairies, présente un aspect bien différent de la partie de la contrée qui appartient au bassin de l'est, malgré l'identité de leur position géologique.

Produits exploitables du calcaire carbonifère.

Plomb. — Le calcaire carbonifère dans la plus grande partie de son étendue conserve ses caractères normaux et est généralement dépourvu de substances minérales. Dans le sud de l'Illinois, où les couches ont été violemment bouleversées, il est traversé par des filons qui ont été exploités pour plomb (1). Ils contiennent de la galène, de la blende, du spath fluor en grande quantité, du spath calcaire, etc.

On trouve aussi des filons de plomb dans le calcaire carbonifère du Tennessee et de l'Alabama, mais ils ne peuvent donner lieu à une exploitation. On exploite dans l'Arkansas une veine très-rapprochée des couches de houille et qui peut-être doit être rapportée aussi au calcaire carbonifère.

Pierres de construction. — Ce calcaire fournit des pierres excellentes et à bon marché, faciles à tailler et à dresser.

---

(1) Le géologue de cet État, docteur J. S. Nerwood, professe une grande confiance dans l'avenir industriel de ces filons.

Quelques-uns des lits sont à grain très-fin et ont une structure extrêmement compacte, ce qui permet de leur donner un bon poli; quelques portions des couches inférieures, d'une structure oolithique, admettent aussi le poli.

On avait envoyé à l'exposition plusieurs blocs et quelques échantillons polis des environs de Saint-Louis.

SCHISTE ROUGE DE LA FORMATION DU CALCAIRE CARBONIFÈRE.

Il paraît nécessaire de signaler ici en même temps que ce calcaire une formation très-importante de schiste rouge qui est en relation avec lui et qui, lorsque le calcaire manque, se trouve directement sous les conglomérats qui accompagnent les couches de houille. Cette formation (n° XI de la carte de Pensylvanie) a une épaisseur de 2.949 pieds à Pottsville et s'étend sur une grande largeur sur la limite orientale du bassin houiller dans la Pensylvanie et la Virginie.

Produits exploitables.

Fer. — Cette formation est importante au point de vue économique, et contient du minerai de fer qui est exploité en un grand nombre d'endroits en Pensylvanie, non-seulement le long des affleurements de la houille, mais encore en beaucoup de points intérieurs du bassin où l'on perce, pour arriver à lui, les couches supérieures. Le minerai consiste essentiellement en fer carbonaté avec des proportions variables de silice, d'alumine, etc. Il contient de 60 à 80 p. 100 de carbonate de fer et quelquefois du peroxyde de fer dans la même proportion. On en voyait des échantillons à l'exposition.

TERRAIN DÉVONIEN.

Le groupe de calcaires, schistes et grès constituant la division du système paléozoïque qu'aux États-Unis on range généralement dans le terrain dévonien, forme un assemblage hétérogène. Dans l'ordre descendant, la série commence par la formation n° X de la carte de Pensylvanie; on le compte actuellement dans le terrain dévonien, bien que sa position dans ce terrain ou le terrain carbonifère soit encore un sujet de débats (1). Nous y ajouterons les schistes et grès des monts Cattskill, et les groupes de Chemung et de Portage, ainsi que les grès et lits calcaires du groupe d'Hamilton dans l'État de New-York, ainsi que leur prolongement à l'ouest, connu sous le nom de schiste noir et grès de Waverly, ou grès à grain fin de l'Ohio, l'Indiana, l'Iowa, etc. La série entière de ces dépôts repose sur un calcaire qui appartient à la même formation et qui occupe la même étendue qu'eux.

Nous allons tracer d'abord les limites et les caractères des couches de la portion supérieure, puis nous décrirons la formation calcaire.

Premier groupe. — La série des couches que nous considérons est beaucoup plus puissante dans la partie orientale des États-Unis, dans l'État de New-York et la Pensylvanie que dans l'ouest. Dans la première région, elle forme un dépôt littéral où les matériaux les plus grossiers sont accumulés : tandis que vers l'ouest, où les plus ténus ont été transportés, l'accumulation est moins puissante. Pour ce motif, les ressources économiques que la formation peut fournir et son influence

---

(1) Cette formation de psammites micacés est rangée par M. de Verneuil dans le terrain carbonifère. A. L.

sur la topographie et la facilité du sol varient d'une manière sensible dans ses diverses parties.

Aux points où le développement du groupe est le plus considérable, les lits inférieurs ont été formés par un limon fin impalpable et sont suivis par d'autres couches formées par un limon semblable, mais mélangé d'une grande quantité de matière calcaire; et parfois, quand celle-ci diminue, la roche prend un caractère arénacé. On trouve, en montant plus haut, de minces bandes calcaires, mais elles n'apparaissent pas constamment. Elles sont suivies de schistes tendres ou de grès schistoïdes, puis de lits arénacés qui dans le haut alternent avec des bancs de schiste plus ou moins épais. Viennent après des schistes rouges et verts et des grès, dont la structure est parfois plus grossière que celle des dépôts inférieurs, mais qui sont souvent schistoïdes ou alternent avec le schiste.

C'est cette formation qui constitue le massif des monts Cattskill. Il faut y ajouter encore le n° X, qui est un grès grossier ou conglomérat, et parfois une masse schisteuse avec mélange de conglomérat.

Cette formation peut être considérée comme occupant près de la moitié méridionale de l'État de New-York, depuis la rivière d'Hudson : elle s'étend le long de la limite septentrionale du bassin houiller de Pensylvanie et forme le bord méridional du lac Érié depuis les environs de Brussels, New-York, à Cleveland, Ohio. En ce point elle se recourbe pour former une large ceinture qui borde le bassin sur toute sa limite occidentale, mais qui s'amincit graduellement vers le sud. A l'est, elle suit le bord du bassin houiller des Alleghanys, depuis la Pensylvanie jusqu'à l'Alabama.

Le schiste noir et le grès fin entourent de la même

manière le bassin houiller du Michigan et forment autour de lui une vaste zone de plusieurs centaines de pieds d'épaisseur, tout le long du bord du bassin houiller de l'Illinois; ils affleurent partout sous le calcaire carbonifère nettement dessiné : parfois les deux roches semblent néanmoins passer de l'une à l'autre, ou leurs couches extrêmes présentent quelques alternances. Le long de la vallée du Mississipi, ces dépôts affleurent en beaucoup de points sous le calcaire carbonifère, et les grès fins, si répandus à l'ouest de ce grand bassin houiller, sont probablement du même âge.

Second groupe. — A la base de toutes ces couches sédimentaires repose une immense formation calcaire, moins importante par sa puissance que par la surface qu'elle occupe et les traces évidentes qui permettent de reconnaître qu'elle formait autrefois un banc corallien continu et non interrompu sur un espace qui n'a pas moins de 500.000 miles quarrés.

Dans les limites des États-Unis, on l'a retrouvée dans sa position primitive dans les États de New-York, de New-Jersey et de Pensylvanie, d'où elle s'étend ensuite vers l'est. Le trait le plus apparent qu'elle forme, cependant, s'étend depuis les montagnes de Helderberg jusqu'à l'Hudson, près Albany, vers l'ouest en traversant ainsi tout l'État de New-York. Elle franchit le Niagara aux Roches-Noires, s'étend à travers le Canada supérieur le long d'une ligne à peu près parallèle à la côte du lac Érié, et traversant la partie septentrionale du Michigan, forme la limite occidentale du lac Huron vers le nord, au passage de Mackinac, et forme encore un énorme dépôt isolé qui recouvre l'île Mackinac et Gross-Cap. Au sud-ouest de ces points, elle occupe le lit du lac Michigan : elle a été dénudée avec les couches plus tendres qui lui sont inférieures et excavée pour recevoir les eaux du lac.

L'axe anticlinal qui s'étend depuis l'ouest du lac Érié à Cincinnati, à la partie méridionale du Tennessee, permet de voir ce calcaire qui remplit une large zone des deux côtés des couches inférieures qui ont été soulevées. Dans la partie centrale de l'Ohio, il s'étend de Sandusky, au sud, à la rivière d'Ohio, et de là au Kentucky. Il est également étendu dans l'est et le sud-est de l'Indiana, atteint l'Ohio aux chutes de Louisville, et passant ensuite dans le Kentucky, où il occupe une surface considérable, il s'étend dans le Tennesee, en circonscrivant le bord irrégulier de la formation silurienne de cet État.

Après avoir suivi la direction nord-est dans le nord de l'Indiana, il court parallèlement à la limite septentrionale du bassin houiller de l'Illinois; et s'étendant au haut du lac Michigan et au delà du Mississipi, il forme en partie le fond de la vallée de la rivière du Cèdre-Rouge dans l'Iowa.

Malgré la remarquable persistance de ce groupe, il n'atteint nulle part une grande épaisseur : sa puissance, dont le maximum est peut-être de 350 pieds, est généralement plus petite que 100 pieds. Dans son prolongement nord-ouest dans l'Iowa, le docteur Owen estime qu'il n'a pas plus de 60 ou 70 pieds, et à sa limite sud-ouest la plus éloignée, il possède à peine la moitié de cette épaisseur.

Troisième groupe. — Les géologues sont aujourd'hui très-disposés, en Amérique, à étendre les limites de la formation dévonienne encore plus bas que le calcaire que nous venons de décrire, et à y comprendre une série de grès et de roches arénacées comprise entre ce calcaire et un autre calcaire appartenant à la formation inférieure (1). Dans l'état de New-York, on y distingue deux membres différents, le cauda-galli grit et le grès d'Oriskany. Le premier forme une masse compacte, à grain fin et serré, d'une couleur sombre, et ne contient presque pas de fossiles ; le second est un grès grossier, poreux, friable très-riche en roches organiques.

Ces roches atteignent leur limite septentrionale à la montagne d'Helderberg, près de l'Hudson ; et dans leur plus grand développement vers l'ouest, ne vont pas au delà de 200 miles, et présentent de fréquentes interruptions jusqu'à cette distance. Leur puissance totale dans l'état de New-York n'atteint pas 500 pieds. Le membre inférieur de ce groupe est beaucoup plus persistant que le supérieur : il s'étend au sud à travers la Pensylvanie et dans la Virginie, et acquiert dans cette direction, dans le premier de ces états, suivant le professeur H. D. Rogers, une épaisseur de 700 pieds.

Produits exploitables du terrain dévonien.

Minerai de fer, matériaux de construction, etc. — Un système si varié dans sa composition et qui occupe une telle superficie, doit nécessairement fournir une grande quantité de produits exploitables. Les principaux consistent en minerais de fer, calcaire destiné à fabriquer de la chaux, chaux hydrauliques, pierres à bâtir, grès pour les constructions, grès réfractaires, dalles, etc.

A la jonction des deux groupes de Chemung et de Cattskill, n°s VIII et IX de la carte de Pensylvanie, on trouve des bancs de minerai de fer fossilifère, qui n'est pas sans ressemblance avec un minerai que nous mentionnerons dans le groupe de Clinton, New-York, appartenant à la formation silurienne supérieure. Ces mi-

(1) M. de Verneuil rapporte de même au terrain dévonien le cauda-galli grit et le grès d'Oriskany. A. L.

nerais contiennent de 30 à 50 p. 100 de peroxyde de fer. On trouve aussi des minerais de fer associés avec calcaire à la base du n° VIII de la carte de Pensylvanie.

Ces minerais passent pour être de bonne qualité et abondants dans la Pensylvanie, et se rencontrent aussi plus loin vers le sud. Dans le prolongement occidental du calcaire dont nous avons parlé, on ne trouve point de minerai important, bien que la jonction du calcaire avec la roche voisine soit souvent marquée par un dépôt ocreux.

On n'avait envoyé à l'exposition, eomme représentant de cette formation, que quelques échantillons de minerai de fer et de calcaire de la Pensylvanie.

TERRAIN SILURIEN SUPÉRIEUR.

Les formations qui, aux États-Unis, peuvent être comprises dans la partie supérieure du système silurien consistent en calcaire connus dans l'État de New-York sous le nom de groupe calcaire inférieur du Helderberg qui renferme en allant de haut en bas les membres suivants : calcaire à encrines, schiste calcaire à Delthyris, calcaire à Pentamères, calcaire d'eau douce. Vient ensuite le groupe salin d'Onóndaga, New-York, et le groupe calcaire du Niagara, qui est compris dans le n° VI de la carte de Pensylvanie. On trouve encore plus bas une série de schistes, de grès, de conglomérats et de lits calcaires, constituant le groupe de Clinton, New-York, et le tout se termine par le grès de Médina. Ces deux dernières formations répondent au n° V de la carte de Pensylvanie.

Les différents membres ou groupes de la série ont une distribution géographique différente et méritent d'être traités à part.

Groupe A. — Les calcaires inférieurs d'Helderberg ont leur plus grand développement dans les montagnes d'Helderberg, près de l'Hudson, et dans le New-Jersey, la Pensylvanie et la Virginie. Leur puissance en Pensylvanie est évaluée à 900 pieds, et est quatre fois plus grande que dans l'État de New-York. Dans la direction sud-ouest, et parallèlement à la chaîne Appalachienne, ce groupe semble avoir la même étendue que les autres membres des formations paléozoïques. Cependant, nous le voyons s'amincir dans la direction de l'ouest; et à l'exception des membres inférieurs, on ne le connaît plus à 100 milles à l'ouest de l'Hudson. Le membre inférieur, ou calcaire d'eau douce, s'étend jusqu'aux limites occidentales de l'état.

Groupe B. — Le second groupe ou groupe salifère d'Onondaga a une plus vaste étendue; il traverse entièrement de l'est à l'ouest l'État de New-York et entre ensuite dans le Canada occidental, où il est souvent masqué par des dépôts superficiels. Il court au nord-ouest vers le lac Huron, dont le lit est pratiqué dans une de ses dénudations et apparaît de nouveau à la base de l'île Mackinac et de Gross Cap dans le lac Michigan.

Il se dirige ensuite de nouveau dans la direction du lac Michigan, où il a été raviné ainsi que le calcaire qui lui est supérieur (1).

Dans le nord de l'Ohio, à l'ouest de la ville de Sandusky, et le long de la rivière du même nom, il prend une certaine importance : mais dans le sud de cet État et dans le sud-est de celui d'Indiana, où on le retrouve

(1) Voyez, sur l'origine et la position relative des vallées des lacs Huron et Michigan, un travail de James Hall dans le Rapport de Foster et Whitney sur le distrit du lac Supérieur.

encore en beaucoup de points, son épaisseur devient insignifiante.

Le membre le plus persistant de cette partie du système silurien, est le groupe du Niagara, qui consiste en schiste et calcaire, ou en calcaire seulement, et est bien connu pour former la cataracte du Niagara. Bien que ce groupe soit extrêmement mince, et qu'on le reconnaisse avec difficulté, sur la limite de la série paléozoïque, il devient très-important dans l'ouest de l'État de New-York. Il ne forme dans l'est de cet État qu'un lit mince de peu de conséquence, mais il augmente graduellement d'épaisseur vers l'ouest, et constitue un des traits les plus apparents de la topographie de la contrée sous forme d'une grande terrasse, ou d'une crête montagneuse, comme on l'appelle, qui s'étend de Rochester, près Lewiston, New-York, Queenstown et Saint-David, Canada oriental, à l'entrée du lac Ontario.

Se dirigeant ensuite vers l'ouest, cette formation constitue le pays élevé, nommé Cabot's Head, entre le lac Huron et la baie de Géorgie; et dans son prolongement forme, en grande partie, les îles Manitoulin. De là, vers l'ouest, en passant par le détroit de Mackinac, elle constitue le bord septentrional du lac Michigan et la péninsule qui le sépare de la baie Verte; et plus loin, la rive occidentale du lac Michigan, près de son extrémité méridionale. Elle occupe, encore plus à l'ouest, les régions élevées du sud du Wisconsin et du nord de l'Illinois, et s'étendant dans l'Iowa, remplit une vaste surface entre le Mississipi et la rivière du Cèdre-Rouge.

L'axe anticlinal déjà mentionné, qui s'étend de Nashville au lac Érié, amène au jour le calcaire dans les îles qui sont près du bord occidental de ce lac; et il s'étend ensuite au sud-ouest, est interrompu par le terrain siliceux inférieur environ à 50 milles au nord de Cincinnati, et s'étend de là sur les deux côtés de l'axe dans les États de Kentucky et de Tennessee; les couches calcaires inférieures qu'on observe aux chutes de l'Ohio se rapportent à cette formation, ainsi que les calcaires fossilifères bien connus du comté de Perry, Tennessee.

Groupe C. — Nous pouvons pour le moment réunir les deux formations suivantes, le groupe de Clinton et le grès de Médina de la carte de New-York; la série comprendra ainsi des schistes, des grès schisteux, des conglomérats, des couches calcaires, des minerais de fer et une formation massive de grès rouge tendre à la base. Ce groupe, dans sa totalité ou représenté par quelques-uns de ses membres, a presque la même étendue que le précédent. Il suit les mêmes contours, affleure sous lui, et sert souvent à supporter la saillie du calcaire supérieur qui forme ainsi une suite de proéminences hardies. Tous les membres du groupe sont bien développés dans la partie centrale de l'état de New-York, qui forme sa limite la plus septentrionale, si nous exceptons quelques parties du Maine où, d'ailleurs, son étendue est mal connue, et les montagnes Vertes où il est représenté par quelques roches métamorphiques. Depuis le point de son plus grand développement aux environs d'Oneida, la formation diminue et devient plus calcaire en avançant vers l'ouest, et près du Niagara elle ne consiste plus qu'en un lit de schiste et un lit calcaire. Au nord-ouest et en traversant le Canada, sa puissance décroît rapidement; elle suit la ligne du calcaire supérieur, et on la trouve dans les îles du lac Huron. On la reconnaît distinctement sur la côte orientale de la baie Verte et aussi dans le Wisconsin, qui est la limite la plus occidentale qu'on puisse lui assigner aujourd'hui. Quelques-uns des lits calcaires de ce groupe, si bien définis dans l'État de New-York, sont, à l'ouest, entièrement confondus avec le calcaire du Niagara. On peut dire la même chose des roches qui longent le grand axe central, bien que l'on puisse à peine reconnaître les couches qui appartiennent au groupe.

Du sud-ouest du mont Shawangunk, dans l'État de New-York, ce groupe s'étend à travers le New-Jersey et dans la Pensylvanie, où il atteint l'épaisseur énorme de 2.000 pieds. Quelques-uns de ses membres se trouvent encore dans la Virginie et le Tennessee, et atteignent la limite méridionale de la région paléozoïque dans l'Alabama.

La base du terrain silurien supérieur est marquée par un immense dépôt de grès en couches épaisses, d'une structure tantôt fine, tantôt grossière, et des conglomérats puissants avec lits minces de schiste ou de grès schistoïde. Ces conglomérats contiennent fréquemment des cailloux d'une grande dimension, et souvent les matériaux qui les composent ne sont que peu usés, et les fragments conservent à un degré frappant leur forme angulaire.

Malgré la nature grossière des matériaux dont la roche est ordinairement composée, ils sont cimentés au point de faire naître l'idée qu'une certaine partie de la matière siliceuse pouvait se trouver en dissolution au moment de l'agrégation de la masse. La forme moins arrondie des fragments et la nature compacte de la roche, jointes aux particularités de sa stratification et de ses autres traits, servent à distinguer ce conglomérat de ceux du terrain houiller.

La roche a été d'abord reconnue dans la vallée du Mohawk, au sud d'Utica, où elle forme une masse qui a moins de 20 pieds d'épaisseur, et qui se perd en s'amincissant à la fois vers l'est et l'ouest, tandis que vers le sud elle plonge sous les formations supérieures. A 70 milles plus au sud, cependant, elle apparaît brusquement avec une grande puissance dans le mont Shawangunk, qui commence à quelques milles de l'Hudson, près Kingston, et de là se dirige par les comtés Ulster, Sullivan et Arago jusqu'au Delaware. La formation s'étend ensuite au sud-ouest, à travers New-Jersey, la Pensylvanie, le Maryland et la Virginie, et imprime son caractère principal à la topographie de la contrée sur toute sa longueur. En y comprenant le mont Shawangunk dans l'État de New-York, et son prolongement, le mont Kittatinny en Pensylvanie, elle forme une chaîne de montagnes continue depuis la rivière d'Hudson, dans l'État de New-York, à la partie septentrionale de la Virginie. Nous ne devons pas oublier d'observer que, bien que le mont Shawangunk se termine d'une manière abrupte au nord-est, à quelques milles de l'Hudson, cependant les conglomérats et grès grossiers sont mélangés avec les schistes inférieurs de la période précédente, dans les roches contournées et métamorphiques de l'est de l'État de New-York.

Cette formation forme ordinairement le trait le plus apparent dans les paysages et la topographie de la contrée qu'elle traverse, et constitue la chaîne de montagnes que nous avons décrite. Sa surface est ordinairement stérile et souvent complétement dépouillée de végétation sur une vaste étendue; et les grès blancs, qu'on voit souvent sous forme de grandes taches blanches, donnent un aspect particulier à quelques parties de ces montagnes.

Il est aussi digne de remarque que ces roches présentent tous les caractères d'une ancienne ligne de côtes pendant la période des soulèvements, et marque de la manière la plus tranchée, dans le dépôt de la série entière des roches paléozoïques, la période à laquelle la faune de cette époque a subi les changements les plus considérables.

Produits exploitables du terrain silurien supérieur.

Minerai de fer. — La substance la plus importante au point de vue économique, dans cette portion du système, est le minerai de fer. Les couches connues dans l'État de New-York sous le nom de groupe de Clinton contiennent un et quelquefois deux lits distincts de minerai de fer attenant avec des lits de calcaire et de schiste. Ce minerai paraît avoir la même étendue que la formation entière. Dans l'ouest, cependant, où elle est mince et interrompue, on n'a reconnu son existence que dans une localité de Wisconsin. Le minerai est fossilifère, ainsi que les bancs contigus de calcaire et souvent très-calcaire dans sa composition. Il est très-estimé et employé pour la fabrication du fer en Pensylvanie et dans l'État de New-York.

On trouve de petites quantités de minerai de fer dans les calcaires supérieurs (n° 6 de la carte de Pensylvanie); mais aucun gisement important n'a été reconnu sur la vaste surface occupée par ces calcaires.

Plomb, zinc et cuivre. — On trouve de la galène, de la blende et des pyrites de cuivre dans le conglomérat de Shawangunk. On y exploite actuellement une mine de plomb dans le comté d'Ulster: la même produit de la galène, des pyrites de cuivre et une petite quantité de blende. On y a rencontré quelques masses de galène de dimension remarquable: l'une d'elles pesait 16.000 livres.

Plus au sud, dans l'État de New-York, on a exploité

sur une échelle assez large une veine ou lit de zinc sulfuré.

Pierres à bâtir et grès. — Le calcaire abonde dans cette partie du système, et quelques-unes des plus belles et des plus durables constructions de la contrée ont été faites avec le calcaire du groupe du Niagara, comme on peut l'observer dans les grands travaux qu'on voit tout le long de la ligne du canal Érié; la facilité avec laquelle cette roche est extraite et la proximité des voies navigables ont rendu son usage très-général: on peut le trouver même à Chicago, à une distance de 1.500 milles, par la voie des lacs, de son gisement dans l'ouest de l'État de New-York.

Le grès de Médina et le groupe de Clinton donnent une grande quantité de dalles, et, en certains points, de pierres de construction.

Des pierres de bonne qualité ont été extraites du conglomérat compacte de Shawangunk; la roche qui les forme, dans les anciens traités publiés en Amérique, avait été identifiée à tort avec le millstone-grit d'Angleterre.

Sel. — Le groupe salifère d'Onondaga, dans l'état de New-York, renferme des eaux salées sur une vaste étendue : les grandes quantités de sel qu'on en retire forment une des industries importantes et un des plus larges revenus de l'état. La quantité de sel obtenu dans ces salines, en 1797, était de 25.474 bushels (1) : cette quantité a toujours été depuis en augmentant; elle s'élevait, en 1852, à 4.922.533 bushels, et dans la présente année dépassera 5.000.000 bushels.

Bien que cette formation contienne des eaux salées jusqu'à une distance de 200 milles ou plus de son

(1) Le bushel vaut 36,347664 litres.

affleurement septentrional, néanmoins celles qui sont assez salines pour donner lieu à une exploitation sont confinées dans un petit espace aux environs de Syracuse, New-York.

Il faut observer que les cavités où l'on puise ces eaux sont creusées dans le drift ou l'alluvion, et ne s'étendent que dans peu de cas jusqu'à la roche en place. Les eaux arrivent dans le drift après avoir traversé les couches supérieures du groupe qui s'élèvent au-dessus du niveau de la vallée vers le sud, et ni leur quantité ni leur teneur ne sont modifiées en pénétrant dans les couches qui reposent sous le gravier.

Exposition. — Le système silurien supérieur était représenté à l'exposition par des échantillons de minerai de fer de Pensylvanie et de gypse de l'État de New-York.

TERRAIN SILURIEN INFÉRIEUR.

Les formations géologiques qui, aujourd'hui, sont comprises dans l'étage inférieur du terrain silurien consistent, en allant de haut en bas, en une série de grès argileux et de schistes connus sous le nom de formation III dans la carte de Pensylvanie, ou groupe de la rivière d'Hudson, y compris le schiste d'Utica dans l'État de New-York. Après eux viennent les calcaires de la formation II de Pensylvanie, ou calcaires de Trenton, Black-River, Bird's-eye et Chazy de l'État de New-York. A la base du système vient la formation Ire ou les grès calcifères et grès de Potsdam. Ces trois divisions dans le terrain silurien inférieur nécessitent une courte description séparée, bien que, sauf peu de modifications, elles occupent à peu près la même étendue géographique.

Groupe A. — La division mentionnée en premier lieu, ou division supérieure, acquiert un grand développement vers le nord-est; elle pénètre aux États-Unis en venant du Canada et suit continuellement le bord oriental du lac Champlain, ordinairement dans un état de métamorphisme partiel. Elle s'étend au-dessous de Whithall et forme une lisière sur les deux côtés de la vallée de l'Hudson où elle prend peu à peu son caractère normal. Elle s'étend dans cette vallée jusqu'à Newbury, et suivant ensuite la direction du sud-ouest, continue à travers les États de New-Jersey, de Pensylvanie, de Maryland et de Virginie, tantôt sur une seule ligne, tantôt sur deux ou trois qui reviennent plus loin se confondre. A son extrémité sud-ouest, elle suit la direction générale des Alleghanys, et ses contours correspondent aux sinuosités des précédentes formations. A l'ouest et au nord-ouest de la rivière d'Hudson, les couches passent sous les calcaires inférieurs de l'Helderberg, depuis Kingston sur l'Hudson au Mohawk; elles continuent à suivre cette vallée avec de nombreuses ondulations, et on les voit s'élever sur ses deux flancs, par suite du soulèvement des roches inférieures. Enfin elles finissent par s'élever sur une vaste surface sur la côte orientale du lac Ontario. Dans le Canada elles reparaissent sur la rive septentrionale du lac et peuvent être suivies sans interruption jusqu'au fond de la baie de Géorgie, sur le lac Huron. Le bassin de cette baie, et sa continuation dans l'espace compris entre les îles Manloulin et les terres, a été creusé en partie dans le groupe de la rivière d'Hudson, dont les lits supérieurs apparaissent au bord septentrional de quelques-unes de ces îles. A l'ouest, ce groupe atteint le bord septentrional du lac Michigan, qui forme la Pointe aux baies entre la grande et la petite baie de Noquet. Il est recouvert par les eaux de la baie Verte, près du fond de laquelle, sur la côte orientale, on voit passer les portions supérieures du groupe sous les calcaires supérieurs. On le revoit sur la côte orientale du lac Winnebago, et au delà il se perd peu à peu en s'amincissant. Depuis la baie de Géorgie, à l'ouest, la formation prend un caractère argilo-calcareux : les lits supérieurs consistent en calcaire impur, et les lits arénacés manquent presque absolument.

Cette formation est extrêmement développée dans la région qui entoure Cincinnati, dans l'Ohio, à une distance de 50 milles et au sud-ouest le long de l'axe qui s'étend à travers le Canada et le Tennessee. Dans cette partie des États-Unis, elle consiste en lits alternatifs de schiste tendre et de lits minces calcaires avec traces accidentelles des grès schistoïdes qui forment un trait si constant et si caractéristique de cette formation dans son extension orientale.

Sur le Mississipi et en beaucoup de points du Wisconsin et de l'Iowa, on trouve quelques lits minces de calcaire argileux avec couches de schiste, surmontant la roche riche en plomb et occupant probablement la position du groupe de la rivière d'Hudson. En admettant cette dernière correspondance, nous pouvons suivre le groupe depuis la frontière nord-est des États-Unis, au sud-ouest dans l'Alabama, et nous avons de bonnes raisons de supposer, par suite de son apparition dans la région de Cincinnati et de Nashville, qu'il existe sous les formations supérieures sur toute l'étendue intermédiaire.

Groupe B. — La seconde division de la formation du terrain silurien inférieur, ou le groupe calcaire, consiste dans sa partie orientale en quatre membres distincts : le calcaire de Trenton, le calcaire de Black-River, le calcaire de Bird's-eye et le calcaire de Chazy. Celui de Trenton est le plus persistant et donne souvent son nom au groupe entier.

La série de ce calcaire pénètre aux États-Unis au nord par les vallées du Saint-Laurent et du lac Champlain, et on les reconnaît le long du Saint-Laurent jusqu'à son embouchure. Nous les retrouvons avec quelques légères interruptions à travers les États de New-York, de New-Jersey, de Pensylvanie, de Virginie et d'Alabama. Le groupe s'étend à l'ouest de la partie centrale de l'État de New-York le long de la vallée du Mohawk, par les chutes de Trenton et la rivière Noire, à l'extrémité orientale du lac Ontario. De là, en franchissant le Saint-Laurent, il traverse le Canada et apparaît sur les bords de la baie de Géorgie et le long du bord septentrional du lac Huron; il forme la plus grande partie de l'île de Saint-Joseph, et s'étendant à l'ouest entre le lac Supérieur et le lac Michigan, forme la côte occidentale de la baie Verte. De là il entre dans l'état de Wisconsin et continue à peu près parallèlement au bord occidental du lac Michigan jusqu'à 50 milles de la frontière méridionale de l'État : il s'infléchit à l'ouest et occupe une large surface dans la partie méridionale du Wisconsin et le nord de l'Illinois. Il s'étend dans l'Iowa et, suivant le cours du Mississipi, se montre par intervalle au niveau des chutes de Saint-Antoine. Le long de l'axe de soulèvement qui s'étend de Nashville au lac Érié, on retrouve ces calcaires sur les bords de la rivière des Pierres dans le Tennesse et à Francfort dans le Kentucky; mais ils n'apparaissent plus au nord de l'Ohio.

Les calcaires de la période du terrain silurien inférieur occupent la même étendue que les schistes qui leur sont supérieurs dans l'est des États-Unis, tandis que dans l'ouest, ou dans la vallée du Mississipi, ils s'étendent au delà de leurs limites positivement reconnues.

Ce groupe calcaire, soit qu'on y retrouve les diverses sous-divisions, soit qu'il ne forme qu'une seule masse, est partout très-important. Dans toutes les localités de l'est et du sud-est, il consiste en calcaires parfois avec intercalation de couches de silex et de lits minces schisteux. A l'ouest, les schistes augmentent d'épaisseur, et en certains endroits, comme aux chutes de Saint-Antoine, prennent une telle importance qu'une grande partie de la roche n'est plus propre à aucun emploi.

On observe en avançant vers l'ouest une diminution générale dans la puissance et un changement dans les caractères des calcaires de ce groupe, et dans cette direction commence à s'intercaler un autre membre, qui n'est encore que peu important avant d'arriver dans le Wisconsin, mais qui se développe alors beaucoup dans cet État et devient très-intéressant à cause du gisement qu'il renferme (calcaire à galène) dans le Wisconsin, l'Illinois et l'Iowa, et qui a été autrefois représenté dans les cartes de cette contrée sous le nom de Cliff limestone.

Groupe C. — La dernière division du terrain silurien inférieur est formée par le grès calcifère et le grès de Potsdam. Le premier étage est représenté par le calcaire magnésien inférieur dans les États qui bordent le haut Mississipi. Ces deux roches alternent fréquemment à leur point de jonction, et en beaucoup de localités on trouve un dépôt friable arénacé succédant au grès calcifère, qui est sans aucune doute dû à la répétition des mêmes actions auxquelles il faut rapporter les lits arénacés inférieurs.

La division inférieure est pour la plus grande partie dépourvue de matière calcaire; cependant dans sa partie occidentale, elle contient quelques bancs calcaires. Le grès supérieur ou calcifère a reçu originairement ce nom, parce que partout on y observe à différents degrés la combinaison des éléments calcaires et arénacés. Cette formation pénètre dans les États-Unis par la vallée du Saint-Laurent au nord-est, et traverse le nord de l'État de New-York jusqu'au bord du lac Ontario. On peut la suivre en descendant la vallée du lac Champlain et de là sans interruptions à travers les Etats de New-Jersey, de Pensylvanie, de Maryland, de Virginie et du Tennessee. A l'ouest, elle traverse le Canada, circonscrit le bord septentrional et se développe davantage sur le lac Supérieur; du bord occidental de ce lac, elle prend une grande extension dans le nord du Wisconsin et s'étend le long des bords du Mississipi, en formant les roches élevées et pittoresques qu'on voit entre la prairie du Chien et les chutes de Saint-Antoine.

Produits exploitables du terrain silurien inférieur.

Minerai de fer. — La grande région ferrugineuse décrite par le professeur H.-D. Rogers, dans son rapport géologique sur le Canada, en connexion avec les calcaires de cette période, n'est point, à proprement parler, comme on le conçoit, une partie de la formation; mais l'association du minerai avec différentes argiles et d'autres dépôts indiquent que leur origine est la même que celle des autres dépôts d'hématite que nous avons décrite dans ce rapport avec les autres produits de la formation tertiaire.

Minerais de plomb. — Le produit le plus important de la formation du terrain silurien inférieur est la galène ou plomb sulfuré. La roche qui la renferme ou «calcaire à galène (1)» du Wisconsin, de l'Iowa, du

(1) Ainsi nommé par le professeur Hall, qui l'a rangé dans

nord de l'Illinois et d'une partie du Missouri, est un calcaire silurien qui en a déjà fourni d'immenses quantités depuis un grand nombre d'années.

Les minerais de plomb du Missouri se trouvent dans la même roche que ceux du nord de l'Illinois, du Wisconsin, de l'Iowa et dans de semblables conditions. On sait qu'il existe des veines de galène et de blende en plusieurs endroits dans le calcaire de Trenton, dans la partie septentrionale de l'état du New-York, mais elles n'ont jamais été exploitées avec profit.

Minerais de cuivre. — Des veines ou des filons de cuivre sulfuré ou carbonaté se rencontrent dans plusieurs localités, dans les limites de la région plombifère et dans la même association de roches, particulièrement à Mineral Point, dans le Wisconsin, où on a commencé à les exploiter, mais où elles ont ensuite été abandonnées ; on en trouve aussi dans le Missouri. Ces filons sont confinés dans une roche d'épaisseur médiocre, et sont plus étendus dans le sens horizontal que dans le sens vertical. On trouve aussi du cuivre à un niveau plus bas dans le calcaire magnésien inférieur du Wisconsin, mais seulement dans une localité de peu d'étendue. Sur la limite méridionale du lac Supérieur, le grès inférieur, ou grès de Potsdam, est associé à des lits étendus de conglomérats et à une rangée non interrompue de roches trappéennes cuprifères. Le grès et le conglomérat contiennent certains minerais de cuivre, mais le cuivre métallique se trouve principalement dans les trapps : on n'exploite aujourd'hui aucun de ces minerais.

---

le terrain silurien inférieur. (Voyez le Rapport géologique sur district du lac Supérieur.) Il avait été décrit auparavant comme faisant partie du clifflimestone de l'Ohio, ou sous le nom de calcaire magnésien supérieur par le docteur D. Owen.

Marbre, calcaire et pierres de construction. — Quelques-uns des lits de calcaire fournissent un marbre noir d'excellente qualité, qui est extrêmement répandu, ainsi qu'un marbre bariolé, dont on pave les salles et les bâtiments publics, et qui sert encore à d'autres emplois. Les calcaires inférieurs donnent en quantité abondante des pierres de très-bonne qualité qu'on peut obtenir en blocs très-épais : ces pierres, excellentes sous ce rapport dans le nord et l'est de la formation, le sont beaucoup moins dans la partie ouest. Le grès de Potsdam fournit une pierre excellente et très-durable et des pierres réfractaires.

On extrait des schistes de la rivière d'Hudson, une grande quantité d'ardoises dans le Vermont, l'État de New-York, la Pensylvanie, et probablement encore plus au sud. On la trouve ordinairement sur les bords de la ceinture métamorphique; et quand elles n'ont subi aucune altération, ou quand l'action métamorphique a été trop puissante, elles ne peuvent servir.

Pierres meulières et dalles. — On fait des meules de bonne qualité avec des grès du groupe de la rivière d'Hudson. Les dalles sont aussi abondantes en beaucoup d'endroits, le long des parties orientales et à l'ouest jusqu'au lac Ontario. Le grès de Potsdam, en couches minces, en fournit souvent d'excellentes.

Exposition. — Le terrain silurien inférieur était représenté par le marbre à encrines du lac Champlain, par des ardoises, des schistes demi-métamorphiques de la formation, qu'on pourrait avec autant de propriété rapporter à celle que nous allons décrire.

ROCHES MÉTAMORPHIQUES SUPÉRIEURES.

Il est extrêmement difficile de tracer une ligne de séparation entre les dernières roches fossilifères et les roches métamorphiques supérieures, parce que les formations qui séparent l'Hudson de l'Atlantique sont formées par les diverses roches paléozoïques déjà décrites, tantôt présentant les caractères des terrains de transition, tantôt parfaitement cristallines et dépourvues de restes organiques. Toutes ces roches, cependant, quel que soit le degré de l'action métamorphique qu'elles ont subie, peuvent être classées sous le nom de roches métamorphiques supérieures, et distinguées d'une série plus ancienne de roches cristallines dont la stratification est discordante, comme on peut l'observer aux Petites-Chutes, New-York, le long de la vallée du lac Champlain et du Saint-Laurent et dans beaucoup d'autres localités. Les roches métamorphiques supérieures comprennent les roches quartzeuses, ou grès modifiés; les calcaires à grain fin, souvent micacés, blancs ou colorés; les roches schisteuses, variant des schistes tendres et fissiles, verts ou rougeâtres, aux schistes micacés et talqueux auxquels les premiers passent successivement, ainsi que diverses roches gneissiques, syénitiques et hornblendiques, qui toutes peuvent être clairement reconnues comme dérivant de roches stratifiées fossilifères de la période paléozoïque (1).

On a reconnu ces formations dans une grande partie de la Nouvelle-Angleterre, et ce n'est que sur de petites surfaces que les roches métamorphiques anciennes, si encore elles ont bien ce caractère, se rencontrent entre leurs limites. Le prolongement de la même formation au sud-ouest, dans une direction parallèle aux anciennes roches fossilifères, fait voir une large surface

---

(1) Le granite rapporté à une origine ignée ordinairement occupe comparativement un petit espace.

couverte par des roches de cet âge, occupant la même étendue que les formations paléozoïques du nord au sud.

La direction et l'inclinaison des couches qui composent cette formation sont ordinairement bien constantes, bien qu'en certains points et sur de petites étendues leur étude soit rendue difficile et obscure par des dislocations, des plissements et des contournements nombreux.

La plus grande partie de cette formation est formée de roches plus anciennes que celles de la période carbonifère, et nous leur trouvons superposées, dans l'est de la Nouvelle-Angleterre, ces roches carbonifères métamorphiques. Dans beaucoup de localités on peut suivre d'une manière nette le passage des roches métamorphiques aux couches fossilifères, et les restes organiques disparaissent graduellement à mesure que les lits prennent une structure plus cristalline, par suite d'une action métamorphique de plus en plus puissante : toute espèce de doute sur la cause de l'oblitération des fossiles se trouve ainsi levé.

Les roches de cet âge forment le trait physique le plus marqué dans la topographie de la contrée qu'elles occupent. Quand on entre dans les États-Unis au nord, par la vallée du Saint-Laurent, on trouve les roches paléozoïques d'abord dans leur condition normale et sans aucune altération, ne présentant que de légers changements lithologiques ; ces changements deviennent ensuite plus marqués, et les couches deviennent plus ou moins cristallines ; les fossiles s'oblitèrent graduellement, les lits s'inclinent parfois jusqu'à la verticalité, et souvent présentent une série de plissements ; enfin les roches s'élèvent en nombreuses collines et en chaînons, qui finissent par prendre le caractère de chaînes de montagnes, telles que les montagnes Vertes de Vermont et les montagnes Blanches du New-Hampshire.

Roches dévoniennes métamorphiques. — Les roches altérées de la période des groupes de Chemung et de Portage, qui forment une partie du terrain dévonien, ont été reconnues en grandes masses à Gaspi par MM. Logan et Hunt, et l'on sait qu'elles sont très-développées dans l'état de Maine. Ces roches circonscrivent la partie orientale de la Nouvelle-Angleterre, en formant une partie des montagnes Blanches et en plongeant sous la formation carbonifère du Massachussets et de Rhode Island, constituent avec cette dernière la plus grande partie des formations métamorphiques et cristallines à l'est de la rivière du Connecticut.

Roches siluriennes métamorphiques. — Dans la portion de la contrée située entre la rivière d'Hudson, à l'ouest, et la rivière du Connecticut, à l'est, les roches plus ou moins métamorphiques appartiennent aux formations que nous avons décrites déjà dans le terrain silurien. Dans la partie occidentale de cette ceinture, les quartzites ou grès durs, les calcaires cristallins, et diverses roches schisteuses, reposent sur des lits de gneiss grossier et de hornblende, reconnus pour être de simples modifications et altérations du grès de Potsdam, du groupe du calcaire de Trenton et du groupe de la rivière d'Hudson, avec des lits accidentels et interrompus du conglomérat de Shawangunk, mélangé avec les schistes, et formant les sommets des éminences les plus relevées.

Les roches siluriennes supérieures ont été suivies par MM. Logan et Hunt, depuis les parties non altérées dans la vallée du Saint-Laurent jusqu'aux limites de l'État de Vermont, où elles prennent le caractère de calcaires micacés, alternant avec des schistes micacés.

La partie inférieure de cette formation métamorphique, ou celle qui se compose principalement de schistes siluriens inférieurs associés à des lits arénacés, constitue la zone aurifère proprement dite, comme il a été abondamment prouvé par l'étude de leur prolongement septentrional, et les nombreuses observations qui ont été faites dans les États du Sud, autorisent à attribuer le même âge à la zone aurifère de Virginie, des deux Carolines et de la Géorgie. Il est impossible aujourd'hui de décider si le gisement aurifère de Californie est du même âge que ces roches métamorphiques supérieures des États-Unis de l'Est.

Produits exploitables des roches métamorphiques supérieures.

Or. — Le produit principal et le plus important des roches métamorphiques de cette période est l'or, qu'on rencontre en quantité plus ou moins abondante à travers toute l'étendue de la formation, depuis le Canada jusqu'à la Géorgie. L'or est extrait en partie de la roche même qui lui sert de gangue, ou des débris formés par suite de leur destruction par les agents naturels. Actuellement, les cailloux et le sable provenant de cette désagrégation sont exploités sur la rivière de la Chaudière dans le Canada; on a trouvé de l'or dans les roches et dans les galets de certaines localités de Vermont et du Massachusets ; mais on n'en a jamais retiré une considérable quantité. Ces faits semblent démontrer que l'or n'est point distribué indistinctement dans toutes les roches de cet âge, mais est en réalité confiné dans une ceinture comparativement étroite.

Minerais de fer.—Les gisements de fer directement en relation avec ces roches ne sont pas souvent productifs;

mais on trouve fréquemment des lits de minerai de fer d'une origine secondaire, provenant de la décomposition des pyrites de fer et d'autres motifs accidentels. On rencontre quelques minerais de fer en quantité suffisante pour permettre leur exploitation dans le Vermont et le New Hampshire; et plus au loin au Sud, dans le Massachusets, un ou plusieurs lits de peu de valeur.

Fer chromé et serpentine. — On trouve en beaucoup d'endroits, associé aux roches de cet âge, du fer chromé, des limites septentrionales des États-Unis jusqu'au Sud, en Virginie et probablement plus loin encore. Ces minerais ont été exploités dans le Maryland et la Pensylvanie. En connexion avec ces minerais, on rencontre des lits de chlorite et des roches de serpentine, probablement occupant la même étendue. La part de la formation qui les renferme néanmoins n'est pas continue, mais se montre à différents intervalles, et certains lits s'y trouvent dans la position où ils ont été originairement disposés. La serpentine, ou *Vert antique*, offre de magnifiques variétés, et parmi les localités qui le fournissent, on peut mentionner Millford, près New Haven, Connecticut et Cavendish, et d'autres villes du Vermont.

Minerais de cuivre. — On trouve fréquemment du cuivre sulfuré et carbonaté dans la région aurifère de Virginie et de la Caroline du Nord, et dans la même formation dans le Maryland. Les minerais de cuivre de Bristol, Connecticut et d'autres localités de cet état, doivent aussi leur être rapportés.

Minerai d'Étain. — Le docteur Jackson décrit de l'étain oxydé qu'il a trouvé dans la ville de Jackson, New Hampshire, et qui se trouve dans les roches de cette période. Du minerai d'étain en petite quantité accompagne souvent l'or en Virginie et dans la Caroline du Nord.

Minerai de plomb. — On sait que la formation renferme de la galène, du plomb carbonaté, sulfaté et phosphaté, mais on n'a jamais essayé sérieusement de les exploiter. On trouve du plomb sulfuré en considérable quantité dans quelques-unes des mines de cuivre récemment ouvertes dans le Maryland.

Argent. — On trouve de l'argent natif dans le comté de Davidson et d'autres localités de la Caroline du Nord.

Pierres de construction. — Soit qu'on regarde les roches syénitiques de la Nouvelle-Angleterre comme appartenant à cette période, ou comme une roche éruptive ignée, elles sont néanmoins si intimement en liaison avec les roches métamorphiques dont nous nous occupons, qu'il est nécessaire de les envelopper dans un même examen. La syénite solide et durable des environs de Boston, connue sous le nom de calcaire de Quincy, si appréciée comme pierre de construction et d'un usage si général, appartient à cette formation. Plus à l'est, dans l'état du Maine, on exploite des matériaux de construction de la même espèce sur une vaste échelle. Cette syénite est d'une couleur plus claire que celle de Quincy, mais devient plus foncée par suite des actions atmosphériques.

Dalles. — On extrait des dalles et des pierres pour escalier dans presque toutes les parties de la formation.

Meules. — La syénite en certaines localités est suffisamment quartzeuse pour donner de bonnes meules qu'on dresse dans les environs de Salem, dans le Massachusetts, et en d'autres endroits.

Meules à repasser. — Les schistes talqueux et micacés de cette époque sont parfois assez siliceux pour être employés pour meules à repasser et pierres à aiguiser.

Marbre statuaire. — On trouve dans ce terrain des marbres blancs et bigarrés de diverses qualités. Ils proviennent principalement du métamorphisme de quelques lits de calcaire silurien inférieur qui présentent des caractères différents dans leurs conditions normales. Dans l'État de Vermont, on a trouvé des couches d'excellent marbre statuaire.

On en trouve aussi d'excellent dans beaucoup de localités entre la frontière septentrionale de Vermont et la ville de New-York, et on en extrait de grandes quantités dans les États de Vermont, Massachusetts, Connecticut, New-York et de Pensylvanie.

Le marbre de ces localités est très-employé pour les constructions et les décorations, etc.; les variétés les plus grossières le sont pour paver et pour d'autres emplois analogues.

Mica. — Ce minéral, qui se rencontre dans les roches métamorphiques ou bien est associé à des roches éruptives cristallines, est exploité et l'on en retire tous les ans de grandes quantités pour les portes et les côtés des poêles qu'on fabrique en si grande quantité aux États-Unis.

Toute la quantité qu'on emploie est extraite à présent de deux localités seulement dans le New-Hampshire, Grafton et Acworth.

Exposition. — Les produits de cette formation envoyés à l'exposition étaient des échantillons d'or provenant de la roche en place dans le Maine, le Vermont (échantillon cristallisé dans du quartz), la Virginie, les deux Carolines et la Géorgie, et quelques-uns de Californie; d'autres, plus importants et plus nombreux, étaient des cailloux provenant de la destruction des roches et que nous avons déjà décrits dans le chapitre du drift ou des alluvions. La Caroline du Nord avait envoyé de l'argent; le Connecticut, le Maryland et la Caroline du Nord, du minerai de cuivre; le Massachusetts, le Maryland et la Virginie, du minerai de plomb; le New-Hampshire, du minerai d'étain; la Pensylvanie et le Maryland, du fer chromé et des minerais de fer; le New-Hampshire, du plomb sulfuré; Vermont, Connecticut, New-York, différentes variétés de marbre; le New-Hampshire, du mica.

ROCHES MÉTAMORPHIQUES ANCIENNES OU INFÉRIEURES.

Dans beaucoup de parties des États-Unis, il existe de larges espaces recouverts de roches cristallines que leur âge et leurs caractères lithologiques distinguent complétement des précédentes. On a proposé d'adopter pour les désigner le nom de roches métamorphiques anciennes ou inférieures, pour les distinguer des roches altérées qui appartiennent évidemment à la période paléozoïque. On sait qu'elles sont d'une date antérieure et dans beaucoup de cas, sinon toujours, elles ont acquis leur structure cristalline et pris leur caractère métamorphique avant le dépôt des formations sédimentaires environnantes qu'il faut ranger parmi les roches anciennes paléozoïques. Dans beaucoup de localités, le grès de Potsdam, qui est la roche fossilifère la plus ancienne que l'on connaisse aux États-Unis, s'appuie dans une position horizontale, ou peu s'en faut, contre les masses inclinées, ou repose sur les hautes cimes formées par les roches de cet âge, dont la date antérieure est ainsi établie d'une manière évidente.

La formation consiste en une longue série de syénites, gneiss, schistes divers hornblendiques, micacés et talqueux, etc., avec calcaire très-cristallisé; toute la série conserve encore des traces évidentes de stratification qui permettent de croire que les roches primitives sédimentaires ont été métamorphosées par des agents ignés. Comme témoins et preuves de ce changement, nous trouvons d'énormes masses éruptives de granite et d'autres roches généralement regardées comme d'origine ignée, qui ont pénétré les roches cristallines stratifiées, dont se compose en grande partie le système. Il est de même traversé par de nombreux dykes trappéens, qui sont quelquefois très-puissants et très-étendus. La roche principale et caractéristique de cette formation est un gneiss syénitique d'une texture généralement fermé et compacte. Les plans de stratification sont souvent obscurs et voilés par la structure granitique ou syénitique. L'une des substances qui accompagnent de la manière la plus constante et qui caractérisent cette formation est le minerai de fer magnétique.

Elle compose principalement la chaîne de montagnes qui s'étend au nord du Saint-Laurent, parallèlement au fleuve jusqu'aux Mille-Iles (*Thousand Islands*). On la retrouve à travers le Canada jusqu'aux bords du lac Supérieur; on l'observe sur les deux bords du lac, et de là elle se dirige à l'ouest jusqu'aux sources du Mississipi.

Quelques portions de la chaîne de montagnes du Maine et du New-Hampshire appartiennent peut-être à cette formation; mais leur étendue et leurs limites n'ont pas été bien définies, et les roches qui les composent n'ont pas été nettement distinguées des roches avoisinantes d'origine plus moderne (1).

(1) Au Canada, M. Logan a reconnu une autre série de roches métamorphiques, qui reposent sur celle que nous venons de décrire, et d'une date postérieure à nos roches métamorphiques supérieures, ce qui constitue trois divisions. Cette formation moyenne se compose de grès, de conglomérats avec

Dans le nord de l'État de New-York, cette formation forme des montagnes qui atteignent 5 000 pieds d'élévation, mais dont la hauteur est généralement moindre. Dans cette région, elle couvre un espace d'une forme à peu près triangulaire, qui a environ 150 milles de longueur du nord au sud et 125 milles de l'est à l'ouest ou du lac Champlain au fleuve du Saint-Laurent. Les points les plus élevés sont au centre de cet espace. La formation expire graduellement au nord et au sud, non sans produire néanmoins quelque dérangement parmi les roches siluriennes qui la recouvrent. Dans la partie méridionale de l'État de New-York, elle reparaît et forme des crêtes et des montagnes élevées qui cependant atteignent à peine la moitié de la hauteur de celle du nord. De la limite sud de cet État, elle pénètre dans celui de New-Jersey, en formant une rangée presque continue de montagnes depuis la frontière jusqu'au Delaware. De là à la frontière du Maryland, les roches métamorphiques anciennes s'étendent à travers le sud-est de la Pensylvanie, en formant deux ceintures étroites: l'une continue, l'autre formée seulement de chaînons isolés par les formations de date plus récente.

Les roches de cette période continuent à conserver la même direction dans le Maryland, la Virginie et la Caroline du Nord, et sont connues sous le nom de montagnes Bleues ou montagne du Sud. Les points les plus élevés, en Virginie, atteignent la hauteur de 4 000 pieds. La formation, qui continue à s'étendre au sud-ouest, se prolonge probablement, avec des interruptions plus

---

fragments de jaspe rouge, lits de schiste et calcaire : « Les schistes sont verdâtres, très-chlorités, et contiennent souvent de l'épidote; souvent ils prennent le caractère de conglomérats, par suite de la présence abondante des fragments de syénite. »

ou moins prononcées, jusqu'aux limites méridionales des formations métamorphiques de l'Alabama.

Nous l'avons déjà montrée sur le bord du lac Supérieur; elle y forme une large zone qui s'étend, sauf de légères interruptions, sur toute la côte, depuis le voisinage de la rivière Camp à Point-Abbaye; elle s'étend ensuite à l'ouest et sur une largeur de plus en plus croissante dans le nord du Wisconsin, et dans la partie centrale de cet État descend vers le sud jusqu'à la hauteur du bord du lac Winnebago. La formation se dirige ensuite vers le nord-ouest et atteint le Mississipi au nord, aux chutes du Saint-Antoine.

On en trouve un rameau isolé dans le Missouri, un autre dans l'Arkansas et un troisième dans le Texas.

Produits exploitables des anciennes roches métamorphiques.

Minerais de fer. — Cette formation, considérée au point de vue économique, dans le Canada aussi bien qu'aux États-Unis, vient peut-être en seconde ligne après le terrain carbonifère. Les minerais de fer oxydé magnétique du Canada, formant des veines immenses ou des amas, ont occupé une place importante à l'exposition de Londres de 1851, et provenaient des roches anciennes métamorphiques. Ce minerai, soit formé d'oxyde magnétique pur, soit mélangé avec du peroxyde anhydre en proportions diverses, est partout caractéristique de la formation. Dans le nord de l'État de New-York, les montagnes sont traversées par des veines ou des lits de ce même minerai, qui présentent l'apparence de couches intercalées entre les schistes, et dont la puissance varie de 2 à 3 pieds à plus de 100 pieds, et peuvent souvent être poursuivies sur une longueur de plus d'un mille. On en exploite plusieurs aux environs du lac Champlain, où le minerai est converti directement en fer pour une part et pour une autre embarqué dans la vallée de l'Hudson pour le sud, où il est fondu avec des minerais de qualité inférieure pour améliorer la qualité du fer.

Les minerais de la région Aridondack, au nord de l'État de New-York, ont été traités avec succès dans les fabriques de fer de qualité supérieure et d'acier, qui ont envoyé aussi leurs produits à l'exposition de Londres en 1851. Dans l'ouest de la contrée métamorphique que nous avons décrite, les minerais de fer consistent principalement en fer peroxydé ou fer spéculaire. Dans le sud du même État, on trouve dans les mêmes roches des lits étendus de fer oxydé magnétique qui sont largement exploités. Il en est de même dans les États de New-Jersey, de Pensylvanie et de Maryland; mais plus loin au sud, ils ne semblent point avoir attiré l'attention.

Sur le bord méridional du lac Supérieur, les roches de cette catégorie contiennent des lits étendus et des veines de fer magnétique mélangé de fer peroxydé. Dans le prolongement occidental de la même formation et dans le voisinage de Montréal et de Bad-River, dans le Wisconsin, on sait qu'il existe des couches de fer oxydé magnétique. Les minerais de fer, mélange de fer magnétique et peroxydé, de la Montagne-de-Fer et de Pilot-Knob dans le Missouri, appartiennent à des roches du même âge. On trouve des minerais semblables dans la région métamorphique de l'Arkansas.

Minerais de cuivre. — Les roches de cette période, dans le nord de l'État de New-York, contiennent de petites quantités de minerai de cuivre en différents endroits, mais aucune mine n'a donné lieu à une exploitation fructueuse. La mine Bruce sur le lac Huron et d'autres sur le bord septentrional du lac Supérieur, où l'on trouve du cuivre, surtout à l'état de sulfure, appartiennent à cette formation ou à la division métamorphique intermédiaire reconnue par M. Logan au Canada.

Minerais de plomb. — On trouve dans le même terrain des minerais de plomb à Rassie et dans son voisinage, dans le nord de l'État de New-York. Il faut rapporter aux gneiss du système les mines de cuivre et plomb des comtés de Montgomery et Chester, Pensylvanie, déjà mentionnées en connexion avec le nouveau grès rouge. Les veines exploitées permettent de fournir en abondance ces deux métaux. La ligne de démarcation entre ces roches et les nouvelles roches métamorphiques n'est point assez clairement définie dans la région méridionale pour nous permettre de désigner avec certitude les mines qui se trouvent dans l'une ou l'autre des deux formations. Dans cette direction, les deux séries paraissent être plus riches en minerais, sauf le minerai de fer, que dans les mêmes formations dans la Nouvelle-Angleterre et l'État de New-York.

Minerais de zinc. — La blende se trouve fréquemment associée aux minerais de plomb; mais jusqu'ici on n'a pu l'exploiter utilement. L'oxyde rouge de zinc et le franklinite de New-Jersey doivent être regardés comme appartenant à cette période. L'intime relation de ces produits avec les roches anciennes de gneiss, et les minéraux qui leur sont associés, semblent fixer leur âge, bien que les calcaires cristallins qu'on trouve dans la même association géologique soient regardés comme siluriens par le professeur Rogers.

Le zinc oxydé rouge est très-employé pour fabriquer de l'oxyde blanc (blanc de zinc), et le mélange de ce minerai avec le franklinite en poudre fine sert à faire avec de l'huile une couleur brune très-estimée. Bien que donnant du zinc d'excellente qualité, il ne reçoit pas aujourd'hui d'emploi métallurgique. Récemment, dit-on, on a traité le franklinite de la même mine avec succès, de façon à obtenir l'oxyde de zinc, et à réduire le cuivre : les deux opérations étant d'ailleurs conduites dans le même fourneau.

Chaux phosphatée. — Se trouve en cristaux fins ou en masse dans les roches de cet âge. On a trouvé dans une localité du nord de l'État de New-York une veine ou un lit de ce minerai, qu'on a un instant espéré utiliser comme amendement agricole. On en a trouvé une grande quantité dans le New-Jersey.

Terre à porcelaine. — Le granite proprement dit ne forme qu'une faible partie de la contrée occupée par les roches métamorphiques des États-Unis. On le trouve dans les deux formations à l'état de filons éruptifs, de dykes ou de masses plus étendues. Beaucoup de ces granites ont une structure grossièrement cristalline, et sont formés principalement de feldspath avec une petite proportion de mica et de quartz. Le feldspath, soit dans son état de décomposition partielle ou dans son état naturel, est largement employé dans les fabriques de porcelaine et de fine poterie.

Marbre et pierres de construction. — La formation métamorphique inférieure présente ordinairement une quantité abondante de calcaire grossièrement cristallin avec quelques lits de qualité supérieure. Aucun de ces calcaires ne mérite réellement le nom de marbre statuaire; on les emploie cependant pour beaucoup d'usages auxquels on destine ordinairement les variétés communes de marbre blanc ou taché, et dans beaucoup de localités elles donnent de bons matériaux de construction. Dans le nord de l'État de New-York, le calcaire est souvent mélangé à de la serpentine, ce qui lui donne une apparence bigarrée. Sur la côte méridionale du lac Supérieur, ces marbres ont une belle couleur rose et sont traversés par des veines rouges; d'autres sont bleus et gorge de pigeon et admirablement veinés. Le gneiss et le granite de cette formation donnent d'excellents matériaux de construction d'une grande variété de couleur, et, dans beaucoup d'endroits, on peut les tailler en blocs très-massifs. Les endroits où les plus belles variétés se rencontrent ne sont pas facilement accessibles et sont éloignés des voies de communication importantes.

Pierres réfractaires, pierres à aiguiser, etc. — Les schistes talqueux de cette période fournissent souvent de bonnes pierres réfractaires qui peuvent être utilisées dans la construction des hauts fourneaux, quand les plans de division ne sont point trop nombreux et qu'on peut obtenir des blocs suffisamment épais. Quelquefois des lits plus fins de stéatite compacte sont associés au calcaire.

Les schistes contiennent parfois assez de silice pour qu'on puisse en faire des pierres à aiguiser ou repasser. On en a trouvé de bonne qualité, pour servir à cet emploi, sur le bord méridional du lac Supérieur.

Galène et minerais cristallisés. — On trouve de la galène cristallisée en quantité considérable dans les roches de cette période au nord de l'état de New-York.

Dans quelques localités elles sont extrêmement riches en beaux minéraux. Mentionnons seulement l'apatite, le zircon, le spinelle, le sphène, l'augite, la tourmaline, etc.

Le labrador, en grandes masses cristallines, n'est point rare dans le nord du New-York. Ce minéral, quand il est poli, a une très-belle apparence, et quelques-uns des plus beaux échantillons sont montés en or et employés dans la bijouterie.

Exposition. — Les roches anciennes métamorphiques étaient représentées par des minerais de fer du nord du New-York, du New-Jersey, de Pensylvanie et de Maryland; quelques larges masses de minerai venant du lac Supérieur et du Missouri, avec du fer de la plupart des mêmes localités, et des échantillons d'acier des mines d'Adirendack; des minerais de cuivre de Pensylvanie; des minerais de plomb du nord de New-York et de Pensylvanie; de l'oxyde rouge de zinc et de la franklinite, avec les produits qui en dérivent, du New-Jersey; des schistes durs du lac Supérieur.

On avait envoyé en outre de nombreux et magnifiques échantillons de minerais cristallisés provenant des anciennes roches métamorphiques du nord de l'État de New-York.

FORMATION TRAPPÉENNE.

Trapps rouges les plus modernes. — La formation de cette nature, la plus apparente dans les États-Unis de l'Est, est celle qui est associée au nouveau grès rouge dont nous avons déjà décrit l'étendue. Les roches basaltiques ou trappéennes accompagnent le grès dans la vallée du Connecticut et sont assez puissantes pour marquer les traits les plus apparents de la topographie de la contrée. Le mont Holgoke et le mont Tom, dans le Massachusetts, en sont des exemples frappants. Les roches de l'est et de l'ouest, près New-Haven, forment aussi des points élevés dans le pays et appartiennent à la formation trappéenne. Les trapps forment de nombreux dykes et des crêtes moins apparentes, souvent parallèles, et de longues ceintures continues qui suivent toute la longueur de la vallée jusqu'à 100 milles au-dessus de son embouchure. Les palissades de la rivière d'Hudson et de nombreuses éminences trappéennes dans le New-Jersey appartiennent toutes à la même époque et présentent les mêmes relations géologiques. Enfin cette formation peut être encore suivie dans la Pensylvanie, le Maryland, la Virginie et la Caroline du Nord, et partout accompagne les grès.

Une des matières qui se trouve toujours associée à ces trapps est le cuivre natif; mais, sur toute leur étendue, on ne l'a jamais trouvé en quantité exploitable.

On trouve fréquemment des minerais de cuivre dans les grès qui sont au contact des trapps, et il semble y avoir quelque rapport entre la présence de ces amas minéraux et l'éruption des dykes. Les roches trappéennes qui appartiennent à cette période du nouveau grès rouge présentent souvent la structure en colonnes et sont divisées naturellement par des plans verticaux. Ce caractère est moins apparent dans les roches que nous allons décrire.

Trapps du lac Supérieur. — La deuxième grande formation trappéenne est celle du lac Supérieur, qui est en relation avec le grès du terrain silurien inférieur. Elle est développée de la manière la plus frappante à Keweenaw-Point et s'étend de là, en occupant une grande largeur, à l'est et à l'ouest, en suivant généralement les contours du lac. La formation est flanquée au nord par des conglomérats et au sud par des grès, et une ceinture étroite de trapp traverse le conglomérat sur toute sa longueur parallèlement à la formation centrale principale. Vers l'extrémité occidentale du lac Supérieur, la formation trappéenne est interrompue et ne se présente plus qu'en dykes isolés ou en crêtes qui percent le grès. L'île Royale présente à peu près les mêmes traits que Keweenaw-Point; le trapp y forme une rangée semblable et parallèle, mais le conglomérat y est peu développé. Sur le bord septentrional du lac, dans les limites des États-Unis, le trapp s'étend en nombreux rameaux étroits et en dykes qui tous sont dirigés vers l'ouest. Plus loin, en différents points situés entre le lac et le Mississipi, au-dessus des chutes de Saint-Antoine, on trouve encore fréquemment des lits de trapp.

Au delà de ce fleuve, on sait peu de chose de leur étendue et de leurs caractères; mais, d'après les informations que l'on possède, il paraît probable que la même formation se retrouve par intervalles entre le Mississipi et les montagnes Rocheuses.

Dans toute cette étendue, les associations sont si semblables qu'il est permis de rapporter avec certitude tous ces trapps à la même formation. Sur les bords de la rivière Sainte-Croix, ils forment de grandes protubérances et des nœuds à travers les lits à lingules du grès de Potsdam; et dans le nord du Wisconsin, on les trouve à peu près dans une position semblable, plus ou moins en relation avec le grès inférieur, mais fréquemment associés à de la syénite ou à du gneiss syénitique.

La formation trappéenne du lac Supérieur, là où elle atteint son principal développement, contient des quantités immenses de cuivre natif qui dans cette région caractérise toujours une certaine variété de trapp. Dans une carte de Keweenaw-Point, publiée récemment par M. Whitney, géologue des États-Unis, nous trouvons la description suivante: Trapp amygdaloïde, contenant du cuivre natif; trapp porphyrique, contenant du cuivre sulfuré, et trapp cristallin, où les filons ne sont point productifs. La principale partie de la formation en ce point est formée par la variété précédente; une ligne mince qui passe par le centre représente la portion cristalline, et une étroite bordure au sud, qui s'unit au grès, appartient à la première variété.

Trapps de l'ouest. — On trouve des trapps et des basaltes à des intervalles fréquents depuis la frontière du Texas jusqu'au pays des Indiens Blackfoot et plus loin encore vers le nord. A l'ouest, dans la grande plaine qui s'étend entre les montagnes Rocheuses et la Sierra Nevada, des roches du même caractère se rencontrent fréquemment, mais leur position géologique n'est point connue. On trouve de même souvent et sur de vastes étendues des roches basaltiques et volcaniques au nord-ouest, dans les montagnes Rocheuses et le long de la Colombie jusqu'à la côte de l'océan Pacifique. Mais il serait impossible, sans dépasser les bornes de ce rapport, de rendre un compte général de toutes les observations qu'on a faites sur les roches ignées de cette vaste surface.

Trapps de la série métamorphique. — Un système de roches trappéennes d'un âge différent, consistant en trapps, diorites, porphyres, etc., traverse les roches métamorphiques de la période que nous avons précédemment indiquée. Elles n'affectent pas généralement d'une manière bien sensible la topographie de la contrée et ne paraissent pas avoir de relation avec le soulèvement de la chaîne montagneuse où on les rencontre. Elles le traversent en différentes directions et remplissent des fissures ou joints qui sont souvent parallèles entre eux. On rencontre partout ces dykes dans les États-Unis, mais ils sont plus nombreux sur le bord septentrional du lac Huron et du lac Supérieur.

Dans la partie orientale de la Nouvelle-Angleterre, et surtout près de la côte, ces dykes traversent les couches; et des ceintures plus étendues de roches semblables occupent des portions de la région métamorphique. Les bordures porphyriques semblent parfois le résultat d'un métamorphisme extrême qui aurait réduit en fusion des roches préexistantes.

Produits exploitables de la formation trappéenne.

Cuivre. — On a ouvert des mines nombreuses sur le bord méridional du lac Supérieur et l'on en extrait des quantités considérables de cuivre natif. Celui-ci est en feuilles plus ou moins épaisses ; il remplit des veines qui coupent la rangée trappéenne presque à angle droit et il est associé avec divers silicates de filons. Ces feuilles de cuivre ont une étendue variable et pèsent depuis quelques tonnes jusqu'à 80 tonnes. Pendant l'année 1853, les mines du lac Supérieur ont produit de 4.000 à 5.000 tonnes de cuivre.

L'une de ces mines avait envoyé à l'exposition une masse de cuivre natif pesant 6.300 livres : elle était coupée sur les quatre côtés et l'épaisseur entre les deux faces intactes était de plus de 2 pieds ; cet échantillon provenait d'une masse pesant 40 tonnes.

On n'a pas encore exploité sur une grande échelle le cuivre sulfuré du trapp porphyrique de la côte méridionale du lac Supérieur.

Argent. — Le cuivre natif contient une petite quantité d'argent natif, et ce métal se présente aussi en nodules ou en masses irrégulières dans le cuivre, à la surface extérieure, soit du cuivre, soit du trapp qui l'environne.

Dans ce cas l'argent n'est qu'en petite proportion, mais on en a trouvé quelques masses importantes dans les débris, associées avec des fragments de cuivre natif. Plus récemment, on a exploité une veine d'argent sur le bord septentrional du lac Supérieur. Des échantillons avaient été envoyés à l'exposition.

Pierres de construction. — Les roches trappéennes, séparées par des points nombreux et divisés ainsi en blocs irréguliers, fournissent des matériaux grossiers qui peuvent être employés pour des murs ou d'autres parties des constructions. La rapide désagrégation de ces roches le long des palissades de la rivière d'Hudson, par l'action des eaux et de la gelée, permet d'en enlever une grande quantité qu'on transporte tous les ans à la ville de New-York et dans tous ses environs.

REMARQUES GÉNÉRALES.

L'esquisse que nous venons de tracer de la géologie d'une grande portion des État-Unis, envisagée au point de vue économique, peut faire voir que les grandes régions métallifères sont les deux rangées parallèles et presque contiguës de roches métamorphiques, qui s'étendent presque sans interruption depuis l'extrémité nord-est des États-Unis jusqu'à l'Alabama. A ces zones principales, il faut ajouter une surface d'une étendue considérable dans le nord de l'état de New-York, une autre de même caractère sur le bord méridional du lac Supérieur, une petite surface dans le Missouri, une dans l'Ackamas et une autre dans le Texas.

Sur la côte occidentale du continent, nous trouvons la ceinture métallifère qui borde le Pacifique et s'étend à travers la Californie et l'Oregon. On sait trop peu de chose des montagnes Rocheuses pour être autorisé à prédire avec quelque confiance l'avenir qu'elles promettent à l'industrie minérale.

On ne pourra ouvrir de mines permanentes et productives que dans les régions que nous avons décrites avec tant de rapidité. Les vastes dépôts où repose la richesse minérale sont donc situés à une distance modérée des côtes. La région si riche en cuivre et en fer du lac Supérieur est traversée par des voies navigables qui communiquent par des canaux avec l'Atlantique. La formation carbonifère, source et fondement de toute activité industrielle, s'étend sur presque toute la longueur de la contrée, depuis les grands lacs au nord jusqu'au golfe du Mexique au sud.

La distribution naturelle de ces éléments de la puissance et de la prospérité, jointe aux caractères physiques de la contrée entière, répond à toutes les exigences, et partout l'on retrouve les matériaux et les stimulants nécessaires au progrès et au développement matériel d'une nation.

EXTRAIT D'UN RAPPORT
DE M. CHARLES LYELL
SUR LA PARTIE GÉOLOGIQUE
DE L'EXPOSITION DE NEW-YORK
EN 1853

Extrait d'un Rapport de M. Charles Lyell sur la partie géologique de l'Exposition de New-York, en 1853, par M. A. Laugel.

Géologie générale des États-Unis.

De l'extrémité nord-est des États-Unis et dans la direction du sud-ouest, jusque vers le golfe du Mexique, s'étend comme une ceinture une contrée haute et montagneuse, plus ou moins fracturée, qui s'élève parfois à plus de 6000 pieds au-dessus du niveau de la mer. Sur toute sa longueur elle présente les traces évidentes d'un soulèvement qui s'est opéré parallèlement à la direction que suit la chaîne. Cette région occupe à peu près toute la surface de la Nouvelle-Angleterre, et ses découpures présentent d'excellents abris aux vaisseaux tout le long des côtes. La chaîne traverse l'État de New-York sous le nom de *high lands*, et s'écarte graduellement de la côte en avançant vers le sud, et en laissant ainsi une ceinture de formations plus récentes de plus en plus large entre son versant et l'Atlantique. Les États-Unis sont limités au nord par le cours du Saint-Laurent et les grands lacs ; mais nous n'avons qu'à dépasser un peu cette limite pour rencontrer une chaîne de montagnes qui court parallèlement au Saint-Laurent, jusqu'au lieu où il prend sa source dans le lac Ontario; elle se dirige ensuite vers l'ouest, et, pénétrant en quelques endroits au delà des limites géographiques de l'Union, s'étend, sauf de rares interruptions, jusqu'aux Montagnes-Rocheuses. Cette dernière chaîne a une direction générale vers le sud, et atteint les limites méridionales des États-Unis vers le 105ᵉ degré de longitude. Le grand bassin ou plutôt la plaine qui est entourée par ces rangées de montagnes, et qui s'ouvre au sud sur le golfe du Mexique, est, sauf de rares exceptions, occupé par des couches stratifiées qui n'ont subi aucun dérangement et qui ne présentent, dans leur longue étendue, que de légères déviations de leur position primitivement horizontale. A l'ouest de la grande chaîne de l'est ou chaîne appalachienne, nous trouvons une vaste étendue de formations anciennes, depuis les terrains primitifs jusqu'au terrain carbonifère. Ces formations s'étendent, dans leur plus grande largeur, de la vallée de l'Hudson jusqu'au delà des *Council Bluffs* sur le Missouri, ce qui répond à 22° de longitude. Plus au midi, leur largeur est considérablement réduite, et elles se terminent par une ligne continue au nord de l'État d'Alabama. Du nord au sud, elles s'étendent sans interruption sur près de 15° de latitude, soit, depuis le bord méridional du lac Supérieur, latitude 47° 50′ à la latitude 33°. En ne tenant pas compte de petits lambeaux qui ne se rattachent pas directement à cette grande plaine, on voit quelle immense étendue est remplie, aux États-Unis, par les formations paléozoïques. Dans ce vaste territoire, il ne sera pas superflu de remarquer que près de la moitié de l'espace total est occupée par les couches de houille de la période carbonifère.

Le terrain crétacé et les terrains tertiaires bordent le versant oriental de la chaîne appalachienne, et forment une large ceinture, ou plutôt un plateau, le long de la côte de l'Atlantique et du golfe du Mexique. Ces formations pénètrent dans le Texas, puis, se dirigeant vers le nord, forment une vaste zone entre la limite occidentale des formations paléozoïques et les Montagnes-Rocheuses.

On voit quelle simplicité présentent les traits généraux de cette vaste contrée, dont la description détaillée exigerait néanmoins des volumes.

Au milieu de la formation crétacée du Texas s'élève une montagne isolée, formée par des roches primitives ou métamorphiques, entourée par des roches de l'âge paléozoïque (voyez Rœmer, *Terrain crétacé du Texas*). On trouve dans l'Arkansas une bande étendue qui présente les mêmes caractères, près de la jonction des terrains paléozoïques et crétacés ; et le Missouri nous offre de même, au milieu des couches siluriennes et carbonifères, des montagnes de roches anciennes métamorphiques.

Les limites que nous venons de tracer et la nature des formations géologiques qu'elles embrassent peuvent servir à déterminer les ressources que présentent les divers États, soit à l'agriculture, soit à l'industrie minérale.

Quand on considère la barrière montagneuse qui borde vers l'est la grande plaine centrale et occidentale, il ne faut point s'imaginer qu'on ne puisse y parvenir qu'à l'aide de routes difficiles, ni que la contrée qui s'étend à l'ouest présente une surface dont la monotonie n'est jamais interrompue. Presque toute la frontière septentrionale est occupée par une suite de grands lacs, où les plus grands bateaux à vapeurs et vaisseaux peuvent naviguer, et qui présentent, seulement dans les États-Unis, une ligne de côtes de près de 3000 milles de longueur (sans parler de plus petites baies et de petites rivières qui communiquent avec eux). La jonction établie avec ces lacs et avec l'Atlantique rend les relations entre la côte et les contrées éloignées de l'intérieur aussi faciles qu'entre des villes maritimes qui seraient situées à la même distance. Les portions occidentales et centrales de cette grande surface sont traversées par des rivières navigables, dont les larges embouchures au sud, sur le golfe du Mexique, semblent appeler à elles le commerce et la navigation. Au N.-E. les bassins de ces rivières sont reliés par des canaux à ceux des grands lacs, et de nombreux chemins de fer traversent tout le territoire à l'est du Mississippi, et doivent s'étendre bientôt jusqu'aux Montagnes-Rocheuses.

On concevra aisément que, si l'on veut se rendre compte des ressources industrielles d'un pays si étendu et si nouveau, on acquerra une idée plus juste de sa richesse en étudiant ses principaux caractères physiques et géologiques, qu'en examinant dans une exposition des collections de roches et de minerais.

On peut énumérer dans l'ordre suivant les formations géologiques des États-Unis, en omettant, bien entendu, les subdivisions dont on trouvera le détail dans les rapports géologiques des divers États.

Alluvions et drift.
Terrains tertiaires.
Terrain crétacé.
Terrain jurassique.
New red sandstone (trias?).
Couches de houille.
Calcaire carbonifère et autres couches inférieures à la houille.
Old red sandstone ou terrain dévonien.
Silurien supérieur.

Silurien inférieur.
Roches cristallines ou métamorphiques de la chaîne appalachienne, appartenant à la période paléozoïque.
Roches anciennes, métamorphiques ou cristallines.
Roches trappéennes.

Alluvions et drift.

Les dépôts superficiels qu'on rapporte au drift sont distribués sur presque toute l'étendue des États-Unis. Ils sont ordinairement caractérisés par la présence de grands blocs erratiques ; mais, dans beaucoup de cas, ils manquent ou restent invisibles, sur de larges étendues où la contrée est recouverte par une puissante couche de matériaux remaniés par les eaux. Dans les régions montagneuses le dépôt superficiel, reposant près des roches anciennes, ne présente aucune stratification, et des fragments de toute dimension, plus ou moins usés, sont confusément assemblés, tandis que, dans une grande partie de la vaste plaine qui s'étend à l'ouest des Alleghanys, le dépôt est stratifié plus ou moins régulièrement, et il s'y est opéré une sorte de séparation entre les éléments.

Partout des matériaux de transport couvrent la surface des roches anciennes sous-jacentes. Dans toutes les localités les formations anciennes présentent les traces évidentes d'une puissante action érosive, et dans le nord des États-Unis où les blocs erratiques sont abondants, comme dans le sud où les matériaux sont moins volumineux, les résultats sont d'une nature semblable et semblent devoir être rapportés à l'action d'un agent commun.

En règle générale, dans les États du nord et du centre de l'Union, le caractère de ce drift dépend de la nature des couches immédiatement sous-jacentes, ou situées plus au N. à une petite distance. Bien que des blocs erratiques provenant des montagnes du nord puissent être observés à 100 ou 200 milles de leur origine, on ne peut néanmoins supposer qu'une proportion considérable de la masse superficielle ait subi un transport aussi considérable. Ce n'est que le long de quelques vallées ou thalwegs que de grandes quantités de matériaux usés par les eaux ont été transportés vers le sud. La surface qu'ils occupent est petite, quand on la compare à celle qui est recouverte par des matériaux qui sont en relation avec des roches très voisines. Ce dépôt superficiel, quand il n'a pas été dérangé et qu'il repose sur les roches anciennes qui en ont fourni les éléments, contient ordinairement une grande quantité de matériaux grossiers ; des changements ultérieurs, produits par des courants plus ou moins violents, ont modifié leur conditions primitives, séparé les éléments et opéré parmi eux une stratification régulière.

Durant une période encore plus récente, des eaux ont séjourné dans les petites dépressions de la formation ; une riche végétation a pris naissance, et les débris de sa décomposition ont été conservés dans des marécages avec des os d'animaux, des coquilles et autres débris fossiles. Les eaux marines ont aussi envahi la formation et déposé du sable, de l'argile et des débris organiques qui se sont mélangés avec les matériaux accumulés du drift.

Formations tertiaires.

Les formations tertiaires proprement dites consistent en couches marines, littorales et lacustres, argiles de toute espèce, lits calcaires et arénacés, et mélanges divers des matériaux précédents. Ces couches, qui contiennent des débris nombreux de coquilles, de polypiers, etc., de plantes et d'animaux terrestres, et de coquilles d'eau douce, sont très développées aux États-Unis. Elles commencent à l'extrémité N.-E. du Maine, en y comprenant le dépôt de la vallée du lac Champlain et du Saint-Laurent ; elles s'étendent ensuite au S.-O. et prennent leurs caractères les mieux définis dans le New-York, puis, se prolongeant au S. et au S.-O., occupent la zone qui s'étend entre l'Atlantique et le pied des montagnes. Les mêmes formations continuent à se développer au S. et à l'O., et, s'infléchissant au pied méridional des Highlands, s'étendent au loin sur la partie sud de la Géorgie, de l'Alabama, du Mississippi, et sur une grande étendue de la Louisiane et du Texas, tout le long du golfe du Mexique. Au nord du Texas, les formations tertiaires suivent la base des Montagnes-Rocheuses, et forment une large ceinture, bordée des deux côtés par le terrain crétacé. Les dépôts tertiaires atteignent même les sources du Missouri et se perdent dans les interminables prairies connues sous le nom des *Mauvaises Terres*.

A l'O. des Montagnes-Rocheuses, entre cette région et la Sierra Nevada, le capitaine Fremont et le capitaine Stransbury ont trouvé des fossiles tertiaires dans des localités très éloignées, ce qui permet de croire que cette formation est largement développée dans cette partie du continent.

Les dépôts tertiaires sont également très étendus sur les côtes occidentales, ainsi que Dana et d'autres nous l'ont appris (voyez *Exploration des États-Unis*, *Géologie*, par T.-D. Dana), mais on n'y a pas encore étudié la formation avec détail.

Terrain crétacé.

Le terrain crétacé, dans la partie occidentale des Etats-Unis, apparaît d'abord dans l'État de New-Jersey, où on l'a reconnu depuis longtemps et où les fossiles les plus caractéristiques de la craie ont été découverts par M. Morton. Il occupe une large étendue dans le New-Jersey et dans le Delaware ; en suivant les sinuosités du terrain tertiaire, il apparaît à divers intervalles dans la Virginie, la Caroline du Nord et du Sud, où les dénudations du terrain tertiaire l'ont mis au jour. Il traverse ensuite le nord de la Géorgie, le centre de l'Alabama, le Mississippi, et se dirige vers le nord, le long des roches paléozoïques, jusqu'au point où l'Ohio entre dans l'État d'Illinois ; de là il s'étend à l'ouest du Mississippi en suivant la direction des couches paléozoïques, traverse l'Arkansas, et prend une grande largeur dans le Texas. Dans leur étendue totale, les dépôts crétacés occupent, du N. au S., plus de 22 degrés, depuis le tropique du Cancer jusqu'au 48e degré de latitude N.

Terrain jurassique.

Les formations qui appartiennent à ce terrain et qui ont été reconnues jusqu'ici forment une ceinture de 10 à 12 milles de largeur et de 50 milles de longueur, à l'O. de Richmond (Virginie), et peuvent être comprises sous le nom de bassin de Chesterfield. Le bassin de la Rivière-Profonde, dans la Caroline du Nord, est d'une nature semblable et s'étend, sauf quelques interruptions, presque à travers tout cet État. Ces deux bassins doivent être considérés comme les parties séparées d'une seule et même formation (1). C'est là jusqu'aujourd'hui le seul groupe géologique que l'étude des fossiles ait autorisé à ranger avec certitude dans le terrain jurassique aux États-Unis.

Cette formation repose sur le gneiss ou le granite, et les poudingues et grès de la série ont été formés par les matériaux cimentés de ces roches anciennes. Plus haut, la roche prend les caractères d'un schiste micacé, qui, dans ses portions supérieures, devient très argileux.

---

(1) La question de l'âge de la ceinture non interrompue qui s'étend à l'O. et suit la formation de New-Jersey et celle des terrains de la vallée du Connecticut ne sont pas encore résolues.

Nouveau grès rouge, schistes et conglomérats.

Les conglomérats, grès et dépôts schisteux si bien connus, de la vallée du Connecticut, dans le Massachusetts et le Connecticut, et le long de la vallée de l'Hudson et qui traversent le New-Jersey, la Pennsylvanie, la Virginie et la Caroline du Nord, ont été ordinairement rapportés au nouveau grès rouge ou à la période du trias. Dans les deux États que nous venons de mentionner, les dépôts que nous considérons reposent dans des dépressions ou bassins isolés à l'O. de ceux que nous avons décrits, et en diffèrent par des caractères plutôt négatifs que positifs.

Cette formation a une grande largeur dans le New-Jersey et la Pennsylvanie, et diminue graduellement vers le S.-O.; elle ne forme plus, dans le Maryland et le nord de la Virginie, qu'une zone étroite, dont la continuité est interrompue dans ce dernier État. Ces bassins isolés sont rangés sur la même ligne que cette ceinture, et n'en sont, sans aucun doute, que des prolongements.

Dans la vallée du Connecticut, on trouve cette formation sur une longueur de 100 milles et une largeur de 19 à 20 milles. C'est dans cette vallée qu'elle a été étudiée avec le plus de soin et qu'on a trouvé la plus grande partie des fossiles les plus caractéristiques. Les restes organiques consistent en plantes fossiles du genre *Voltzia* et autres, en nombreux poissons fossiles; on y trouve aussi des empreintes de pas, dont la plupart paraissent devoir être rapportés à des oiseaux, et quelques-uns à des quadrupèdes.

Terrain houiller.

La série de dépôts sédimentaires qui constitue ce qu'on nomme généralement le terrain houiller, présente, aux États-Unis, les mêmes caractères qu'en Europe. A la base, on trouve ordinairement un grès grossier ou conglomérat, à cailloux de quartz. En quelques parties cette roche manque et ce sont des couches schisteuses avec charbon qui reposent sur le calcaire inférieur. Les couches suivantes consistent en grès à grain plus ou moins fin, parfois en conglomérats ou grès d'une nature analogue, calcaire, houille, argiles réfractaires. L'épaisseur énorme de cette formation en quelques points est réellement étonnante, et l'on demeure confondu en songeant aux changements qui ont été nécessaires pour produire la succession de couches d'une nature si distincte et d'une pareille puissance.

La grande formation carbonifère des États-Unis constitue l'un de ces traits géologiques les plus frappants, en même temps qu'elle se rattache aux conditions économiques de l'ordre le plus important. La région carbonifère de l'E. des Alleghanys se prolonge depuis la limite extrême N. de la Pennsylvanie jusqu'au centre de l'Alabama ; sa longueur est de plus de 750 milles, et sa largeur maximum dépasse 180 milles, sans compter les bassins anthraciteux qui forment des îlots le long du bord oriental de cette vaste surface, évaluée par M. H.-D. Rogers à près de 63000 milles carrés.

Le grand district occidental, ou, comme on l'appelle généralement, le district de l'Illinois, occupe la plus grande partie de l'État de l'Illinois, ainsi qu'une portion de l'Indiana et du Kentucky. Les derniers travaux du docteur D.-D. Owen en ont fait voir qu'il s'étend plus loin dans l'Iowa et le Missouri qu'on ne l'avait supposé auparavant.

Il est vrai que sa partie située à l'O. du Mississippi n'est pas seulement séparée de la partie orientale ou district de l'Illinois par le cours du fleuve ; celui-ci, en effet, coule le long d'une douce ligne anticlinale, dont l'élévation a contribué à séparer les deux portions de la région carbonifère de l'ouest.

Cette portion, située dans l'Iowa et le Missouri, occupe un espace dont la partie qui est à l'E. du Mississippi n'égale que les deux tiers. Suivant l'opinion de M. Hall et d'autres géologues marquants, ce district houiller s'étend beaucoup plus loin vers le S.-O. que les limites aujourd'hui encore restreintes des explorations géologiques faites avec soin, bien qu'il ne dépasse point sans doute la parallèle des régions siluriennes et métamorphiques qu'on rencontre au S. du Missouri.

Sa plus grande étendue du S.-E. au N.-O. ou des sources de la Rivière-Verte, en Kentucky, à sa limite N. sur la rivière des Moines, dans l'Iowa, est de plus de 550 milles ; sa largeur maximum à travers l'Illinois et le Missouri, dans la direction E.-O., est de plus de 400 milles ; et l'on peut compter 300 milles depuis sa limite septentrionale dans l'Iowa à ses limites actuelles méridionales sur la Rivière-Osage. Cette région de l'ouest, en y comprenant les deux côtés du Mississippi, a donc une plus grande superficie que celle qui est située à l'E. de la chaîne appalachienne. La portion seule qui est à l'E. du Mississippi est presque aussi grande que cette dernière, et celle qui est à l'O. la dépasse de plus de moitié.

Plus loin, vers le S., on sait qu'il existe un bassin houiller d'une étendue considérable dans l'Arkansas. Il est probable que dans l'origine il était dans le prolongement de celui de l'Iowa et du Missouri, et qu'il en a été séparé depuis par le soulèvement d'une ceinture de roches métamorphiques; peut-être aussi ces chaînes déjà soulevées formaient-elles une île allongée dans la mer, où se sont déposées les couches de la formation carbonifère. On connaît plus loin encore, au S. et à l'O., d'autres petits bassins, isolés ou en relation avec le bassin principal; on a retrouvé d'ailleurs la grande couche inférieure du calcaire carbonifère dans le Nouveau-Mexique et plus au N., en certains points, le long des Montagnes-Rocheuses. On peut donc en inférer qu'on retrouvera les couches de houille, sinon sur toute l'étendue, au moins par intervalles et dans des bassins isolés de cette vaste surface. En outre, il existe, dans le Michigan, un petit district jusqu'ici improductif, qui ne semble pas tirer aujourd'hui profit de sa position dans une plaine basse. Sa position géographique lui donne cependant de grands avantages : il est situé au-dessus du niveau des eaux navigables du lac Huron, et est plus accessible au mouvement commercial des grands lacs qu'aucun autre bassin houiller. On n'a encore aujourd'hui que trop peu de renseignements précis pour être à même de juger de l'importance que lui prépare l'avenir.

On trouve aussi dans le Rhode-Island et le Massachusetts, sur une petite étendue, des couches de houille qui n'ont subi presque aucune altération (1).

En étudiant la position relative de ces couches et de celles de la Nouvelle-Écosse en connexion avec l'âge et les caractères de la contrée où elles apparaissent, on est conduit à penser que ces portions détachées ont pu autrefois être en liaison avec le district des Alleghanys, que celui-ci même ne faisait qu'un avec celui de l'ouest, et qu'ainsi, dans leur condition primitive et avant qu'aucune action perturbatrice ne masquât leur unité, ces formations carbonifères pouvaient avoir recouvert toute la partie des États-Unis qui est située à l'E. des Montagnes-Rocheuses.

Outre l'immense et inépuisable région qui s'étend à l'E. de ces montagnes, des échantillons de houille ont été rencontrés

(1) Les couches de houille peu altérées auxquelles il est fait ici allusion sont celles que l'on exploite aujourd'hui; mais je suis convaincu qu'il existe sur une beaucoup plus grande étendue, dans la Nouvelle-Angleterre, des couches qui ont été plus ou moins métamorphisées.

par le capitaine Stransbury au delà du fort Laramie : on en a trouvé à Bellingham Bay, à Puget's Sound, dans le district de Washington ; mais on ne sait que peu de chose de son âge et des roches qui lui sont associées. Ces derniers échantillons ressemblent tellement aux charbons bitumineux des couches de houille ordinaire, qu'on ne saurait établir entre eux aucune différence, et que nous sommes ainsi fondés à espérer qu'on trouvera sur les côtes occidentales des États-Unis cette source importante de progrès et de richesse nationale.

Calcaire carbonifère.

La grande formation du calcaire carbonifère, inférieure aux couches de houille et qui, dans les districts houillers forme ordinairement le trait le plus apparent de la topographie de la contrée, perd toute son importance et son relief sur les bords de la plus grande partie de la région houillère des Alleghanys. Sur la limite orientale, en Pennsylvanie, c'est à peine si l'on peut dire qu'elle existe encore ; et l'on ne peut y rapporter que de minces lits calcaires associés au schiste rouge dont nous parlerons plus loin.

Dans la Virginie, cette formation prend un grand développement et devient très puissante dans le sud de cet État. Dans le Tennessée, et toujours dans la même direction, le calcaire forme une étroite ceinture le long du bord oriental du bassin houiller et se prolonge jusqu'à sa limite méridionale dans l'Alabama. Vers le N., cependant, il disparaît graduellement et se perd complétement vers la rivière de l'Ohio, ou n'occupe plus au milieu des grès qu'une place insignifiante.

Dans le district de l'Ouest ou du Mississippi, nous retrouvons la formation beaucoup plus puissante et plus étendue. Le calcaire occupe sur le bord oriental une zone étroite, qui traverse l'Indiana et le Kentucky ; il reparaît sur le côté occidental du bassin ou le long de la vallée du Mississippi ; le lit du fleuve y est creusé à une grande profondeur, sur la longueur de plusieurs milles, et ses bords sont souvent élevés comme des falaises. Le calcaire s'étend avec une largeur variable sur un des côtés du Mississippi, à une petite distance au-dessus de l'Ohio jusqu'au confluent du Rock River, sur une longueur de plus de 300 milles par conséquent, et vient affleurer à l'E. et à l'O. sous les couches de houille. Il paraît, à de nombreux intervalles, le long du Missouri, et partout où on l'a observé jusqu'ici, forme une large ceinture qui entoure toute la division occidentale du terrain houiller. Plus à l'O., il apparaît au fort Laramie, et forme le dépôt sédimentaire le plus important aux environs du grand lac Salé. On sait aussi qu'il se rencontre dans le Nouveau-Mexique et le Texas, mais ses limites n'y ont pas encore été tracées. Dans le Michigan, on sait peu de chose à l'égard de cette formation : on a seulement trouvé en plusieurs localités, sur le bord occidental du bassin houiller, un calcaire qu'on lui rapporte. La grande variété de caractères que présente ce calcaire dans l'O., jointe à d'autres considérations, a engagé le docteur D.-D. Owen à le séparer en deux étages, l'un inférieur, l'autre supérieur. La puissance totale est de 390 pieds, et l'on peut y observer douze couches, de caractères lithologiques différents.

Sur une vaste région dans l'Ouest, ce groupe de lits calcaires, avec leurs couches de marne ou de grès tendre argileux, et des intercalations marneuses avec lits de calcaire terreux, imprime à la contrée qui s'étend autour du district houiller un caractère de fertilité qu'on ne retrouve plus au même niveau géologique dans le reste des États-Unis. Sur les bords septentrionaux, occidentaux et orientaux du district de l'Est ou des Alleghanys, il existe une zone étroite où l'on ne trouve aucune formation calcaire importante, et il faut chercher les calcaires bien loin des affleurements houillers ou dans les lits qui sont intercalés entre les couches de charbon. L'effet de cette différence géologique apparaît, non-seulement dans les caractères différents des constructions, mais encore dans les conditions agricoles des deux contrées : dans l'une on ne trouve que des prairies aux environs de la houille, tandis que dans l'autre, à la même latitude et dans les mêmes conditions stratigraphiques, le sol, très calcaire, fournit des céréales de première qualité.

Ce ne sont pas seulement les ceintures extrêmes des deux bassins qui présentent un tel contraste : celui qui se manifeste entre leurs parties centrales n'est pas moins frappant. Cette formation calcaire, avec les lits nombreux de nature analogue qui alternent avec les couches de houille elle-même, fréquemment coupée sur toute sa profondeur par le lit des rivières et de profondes vallées, a servi à distribuer les matériaux calcaires avec abondance sur toute la surface de la contrée. C'est par ce motif, peut-être, et aussi par suite de l'accumulation des matériaux provenant de la destruction de couches calcaires des formations inférieures dans les districts du nord, que le bassin de l'ouest, avec ses grandes plaines et ses prairies, présente un aspect si différent de la partie de la contrée qui appartient au bassin de l'est, malgré l'identité de leur position géologique.

Il paraît nécessaire de signaler ici, en même temps que ce calcaire, une formation importante de schiste rouge qui est en relation avec lui, et qui, lorsque le calcaire manque, se trouve directement sous les conglomérats qui accompagnent les couches de houille. Cette formation (n° XI de la carte de la Pennsylvanie) a une épaisseur de 2949 pieds à Pottsville, et s'étend sur une grande largeur sur la limite orientale du bassin houiller dans la Pennsylvanie et la Virginie.

Terrain dévonien.

Le groupe de calcaires, schistes et grès, constituant la division du système paléozoïque qu'aux États-Unis on range généralement dans le terrain dévonien, forme un assemblage hétérogène. Dans l'ordre descendant, la série commence par la formation n° X de la carte de la Pennsylvanie ; on la compte actuellement dans le terrain dévonien, bien que sa position dans ce terrain ou le terrain carbonifère soit encore un sujet de débats (1). Nous y ajouterons les schistes et grès des monts Cattskill, et les groupes de Chemung et de Portage, ainsi que les grès et lits calcaires du groupe d'Hamilton dans l'État de New-York, ainsi que leur prolongement à l'O., connu sous le nom de schiste noir, et grès de Waverly, ou grès à grain fin de l'Ohio, l'Indiana, l'Iowa, etc. La série entière de ces dépôts repose sur un calcaire qui appartient à la même formation et qui occupe la même étendue qu'eux. Nous allons tracer d'abord les limites et les caractères des couches de la portion supérieure, puis nous décrirons la formation calcaire.

Premier groupe. — La série des couches que nous considérons est beaucoup plus puissante dans la partie orientale des États-Unis, dans l'État de New-York et la Pennsylvanie, que dans l'ouest. Dans la première région, elle forme un dépôt littoral, où les matériaux les plus grossiers se sont accumulés, tandis que vers l'ouest, où les plus ténus ont été transportés, l'accumulation est moins puissante. Pour ce motif, les ressources économiques que la formation peut fournir et son influence sur la topographie et la fertilité du sol varient d'une manière sensible dans ses diverses parties.

Aux points où le développement du groupe est le plus considérable, les lits inférieurs ont été formés par un limon fin impal-

(1) Cette formation de psammites micacés est rangée par M. de Verneuil dans le terrain carbonifère. A. L.

pable, et sont suivis par d'autres couches formées par un limon semblable, mais mélangé d'une grande quantité de matière calcaire; et parfois, quand celle-ci diminue, la roche prend un caractère arénacé. On trouve, en montant plus haut, de minces bandes calcaires, mais elles n'apparaissent pas constamment. Elles sont suivies de schistes tendres, ou de grès schistoïdes, puis de lits arénacés, qui, dans le haut, alternent avec des bancs de schiste plus ou moins épais. Viennent après des schistes rouges et verts et des grès dont la texture est parfois plus grossière que celle des dépôts inférieurs, mais qui sont souvent schistoïdes ou alternent avec le schiste. C'est cette formation qui constitue le massif des monts Cattskill. Il faut ajouter encore le n° X, qui est un grès grossier ou conglomérat, et parfois une masse schisteuse, avec mélange de conglomérat.

Cette formation peut être considérée comme occupant près de la moitié méridionale de l'État de New-York, depuis la rivière d'Hudson : elle s'étend le long de la limite septentrionale du bassin houiller de Pennsylvanie et forme le bord méridional du lac Érié, depuis les environs de Brussels (New-York), à Cleveland (Ohio). En ce point, elle se recourbe pour former une large ceinture qui borde le bassin sur toute sa limite occidentale, mais qui s'amincit graduellement vers le sud. A l'E., elle suit le bord du bassin houiller des Alleghanys, depuis la Pennsylvanie jusqu'à l'Alabama.

Le schiste noir et le grès fin entourent de la même manière le bassin houiller du Michigan et forment autour de lui une vaste zone de plusieurs centaines de pieds d'épaisseur, tout le long du bassin houiller de l'Illinois; ils affleurent partout sous le calcaire carbonifère nettement dessiné : parfois les deux roches semblent néanmoins passer de l'une à l'autre, ou leurs couches extrêmes présentent quelques alternances. Le long de la vallée du Mississippi, ces dépôts affleurent en beaucoup de points sous le calcaire carbonifère, et les grès fins, si répandus à l'O. de ce grand bassin houiller, sont probablement du même âge.

Second groupe. — A la base de toutes ces couches sédimentaires repose une immense formation calcaire, moins importante par sa puissance que par la surface qu'elle occupe et les traces évidentes qui permettent de reconnaître qu'elle formait autrefois un banc corallien continu et non interrompu, sur un espace qui n'a pas moins de 500,000 milles carrés.

Dans les limites des États-Unis, on l'a retrouvée dans sa position primitive dans les États de New-York, de New-Jersey et de Pennsylvanie, d'où elle s'étend ensuite vers l'E. Le trait le plus apparent qu'elle forme, cependant, s'étend depuis les montagnes d'Helderberg jusqu'à l'Hudson, près Albany, vers l'O , en traversant ainsi tout l'État de New-York. Elle franchit le Niagara aux Roches-Noires, s'étend, à travers le Canada supérieur, le long d'une ligne à peu près parallèle à la côte du lac Érié, et, traversant la partie septentrionale du Michigan, forme la limite occidentale du lac Huron vers le nord, au passage de Mackinac, et forme encore un immense dépôt isolé, qui recouvre l'île de Mackinac et Gross-Cep. Au S.-O. de ces points, elle occupe le lit du lac Michigan; elle a été dénudée avec les couches plus tendres qui lui sont inférieures, et excavée pour recevoir les eaux du lac.

L'axe anticlinal qui s'étend depuis l'ouest du lac Erié, à Cincinnati, à la partie méridionale du Tennessée, permet de voir ce calcaire qui remplit une large zone des deux côté des couches inférieures qui ont été soulevées. Dans la partie centrale de l'Ohio, il s'étend de Sandusky, au S., à la rivière d'Ohio, et de là au Kentucky. Il est également étendu dans l'est et le sud-est de l'Indiana, atteint l'Ohio aux chutes de Louisville, et, passant ensuite dans le Kentucky, où il occupe une surface considérable, il s'étend dans le Tennessée en circonscrivant le bord irrégulier de la formation silurienne de cet État.

Après avoir suivi une direction N.-E. dans le nord de l'Indiana, il court parallèlement à la limite septentrionale du bassin houiller de l'Illinois; et, s'étendant au haut du lac Michigan et au delà du Mississippi, il forme en partie le fond de la vallée de la rivière du Cèdre-Rouge dans l'Iowa.

Malgré la remarquable persistance de ce groupe, il n'atteint nulle part une grande épaisseur; sa puissance, dont le maximum est peut-être de 350 pieds, est généralement plus petite que 100 pieds. Dans son prolongement N.-O. dans l'Iowa, le docteur Owen estime qu'il n'a pas plus de 60 ou 70 pieds, et à sa limite S.-O. la plus éloignée, il possède à peine la moitié de cette épaisseur.

Troisième groupe. — Les géologues sont aujourd'hui très disposés, en Amérique, à étendre les limites de la formation dévonienne encore plus bas que le calcaire que nous venons de décrire; ils veulent y rattacher une série de grès et de roches arénacées comprise entre ce calcaire et un autre calcaire appartenant à la formation inférieure (1). Dans l'est de l'État de New-York, on y

(1) M. de Verneuil rapporte de même au terrain dévonien le Caudagalli-grit et le grès d'Oriskany.

distingue deux membres différents, le Caudagalli-grit et le grès d'Oriskany. Le premier forme une masse compacte, à grain fin et serré, d'une couleur sombre, et ne contient presque pas de fossiles ; le second est un grès grossier, poreux, friable, très riche en débris organiques.

Ces roches atteignent leur limite septentrionale à la montagne d'Helderberg, près de l'Hudson ; et dans leur plus grand développement vers l'ouest ne vont pas au delà de 200 milles et présentent de fréquentes interruptions jusqu'à cette distance. Leur puissance totale dans l'État de New-York n'atteint pas 500 pieds. Le membre inférieur de ce groupe est beaucoup plus persistant que le supérieur ; il s'étend au S. à travers la Pennsylvanie et dans la Virginie, et acquiert dans cette direction dans le premier de ces États, suivant le professeur H.-D. Rogers, une épaisseur de 700 pieds.

Terrain silurien supérieur.

Les formations qui, aux États-Unis, peuvent être comprises dans la partie supérieure du système silurien, consistent en calcaire, connu, dans l'État de New-York, sous le nom de groupe calcaire inférieur du Helderberg, qui renferme, en allant de haut en bas, les membres suivants : calcaire à Encrines, schiste calcaire à Delthyris, calcaire à Pentamères, calcaire d'eau douce. Vient ensuite le groupe salin d'Onondaga (New-York), et le groupe calcaire du Niagara, qui est compris dans le n° VI de la carte de la Pennsylvanie. On trouve encore plus bas une série de schistes, de grès, de conglomérats et de lits calcaires, constituant le groupe de Clinton (New-York) ; et le tout se termine par le grès de Médina. Ces deux dernières formations répondent au n° V de la carte de la Pennsylvanie.

Les différents membres ou groupes de la série ont une distribution géographique différente et méritent d'être traités part.

Groupe A. — Les calcaires inférieurs d'Helderberg ont leur plus grand développement dans les montagnes d'Helderberg, près de l'Hudson, et dans le New-Jersey, la Pennsylvanie et la Virginie. Leur puissance en Pennsylvanie est évaluée à 900 pieds, et est quatre fois plus grande que dans l'État de New-York. Dans la direction S.-O., et parallèlement à la chaîne appalachienne, ce groupe semble avoir la même étendue que les autres membres des formations paléozoïques. Cependant nous le voyons s'amincir dans la direction de l'O. ; et, à l'exception des membres inférieurs, on ne le connaît plus à 100 milles à l'O. de l'Hudson. Le membre inférieur ou calcaire d'eau douce s'étend jusqu'aux limites occidentales de l'État.

Groupe B. — Le second groupe ou groupe salifère d'Onondaga a une plus vaste étendue; il traverse entièrement de l'E. à l'O. l'État de New-York, et entre ensuite dans le Canada occidental, où il est souvent masqué par des dépôts superficiels. Il court au N.-O. vers le lac Huron, dont le lit est pratiqué dans une de ses dénudations, et apparaît de nouveau à la base de l'île Mackinac et de Gross Cep, dans le Michigan.

Il se dirige ensuite de nouveau dans la direction du lac Michigan, où il a été raviné ainsi que le calcaire qui lui est supérieur (1). Dans le nord de l'Ohio, à l'O. de la ville de Sandusky, et le long de la rivière du même nom, il prend une certaine importance; mais dans le sud de cet État et dans le sud-est de celui d'Indiana, où on le retrouve encore en beaucoup de points, son épaisseur devient insignifiante.

Le membre le plus persistant de cette partie du système silurien est le groupe du Niagara, qui consiste en schiste et calcaire ou en calcaire seulement, et est bien connu pour former la cataracte du Niagara. Bien que ce groupe soit extrêmement mince, et qu'on le reconnaisse avec difficulté, sur la limite de la série paléozoïque, il devient très important dans l'ouest de l'État de New-York. Il ne forme dans l'est de cet État qu'un lit mince de peu de conséquence; mais il augmente graduellement d'épaisseur vers l'O. et constitue un des traits les plus apparents de la topographie de la contrée sous forme d'une grande terrasse, ou d'une crête montagneuse, comme on l'appelle, qui s'étend de Rochester, près Lewiston (New-York), Queenstown et Saint-Weit (Canada oriental), à l'entrée du lac Ontario, se dirigeant ensuite vers l'O. Cette formation constitue le pays élevé, nommé Cabot's Head, entre le lac Huron et la baie de Géorgie, et, dans son prolongement, forme en grande partie les îles de Menitouline. De là, vers l'ouest, en passant par le détroit de Mackinac, elle constitue le bord septentrional du lac Michigan et la péninsule qui le sépare de la baie Verte, et, plus loin, la rive occidentale du lac Michigan près de son extrémité méridionale. Elle occupe, encore plus à l'ouest, les régions élevées du sud du Wisconsin et du nord de l'Illinois, et, s'étendant dans

(1) Voyez, sur l'origine et la position relative des vallées des lacs Michigan et Huron, un travail de James Hall dans le *Rapport de* Foster *et* Whitney *sur le district du lac Supérieur*.

l'Iowa, remplit une vaste surface entre le Mississippi et la rivière du Cèdre-Rouge.

L'axe anticlinal déjà mentionné, qui s'étend de Nashville au lac Erié, amène au jour ce calcaire dans les îles qui sont près du bord occidental de ce lac; il s'étend ensuite au S.-O., est interrompu par le terrain silurien inférieur, environ à 50 milles au nord de Cincinnati, et s'étend de là sur les deux côtés de l'axe dans les États de Kentucky et de Tennessée; les couches calcaires inférieures qu'on observe aux chutes de l'Ohio se rapportent à cette formation, ainsi que les calcaires fossilifères bien connus du comté de Perry (Tennessée).

Groupe C. — Nous pouvons pour le moment réunir les deux formations suivantes : le groupe de Clinton et le grès de Médina (de la carte de New-York) ; la série comprendra ainsi des schistes, des grès schisteux, des conglomérats, des couches calcaires, des minerais de fer, et une formation massive de grès rouge tendre à la base. Ce groupe, dans sa totalité, ou représenté par quelques-uns de ses membres, a presque la même étendue que les précédents. Il suit les mêmes contours, affleure sous lui, et sert souvent à supporter la saillie du calcaire supérieur qui forme ainsi une suite de proéminences hardies. Tous les membres du groupe sont bien développés dans la partie centrale de l'État de New-York, qui forme sa limite la plus septentrionale, si nous exceptons quelques parties du Maine, où, d'ailleurs, son étendue est mal connue, et les montagues Vertes où il est représenté par quelques roches métamorphiques. Depuis le point de son plus grand développement, aux environs d'Oneida, la formation diminue et devient plus calcaire en avançant vers l'O. ; et près du Niagara elle ne consiste plus qu'en un lit de schiste et un lit calcaire. Au N.-O., et en traversant le Canada, sa puissance décroît rapidement; elle suit la ligne du calcaire supérieur, et on la trouve dans les îles du lac Huron. On la reconnaît directement sur la côte orientale de la Baie Verte, et aussi dans le Wisconsin, qui est la limite la plus occidentale qu'on puisse lui assigner aujourd'hui. Quelques-uns des lits calcaires de ce groupe, si bien définis dans l'État de New-York, sont, à l'ouest, entièrement confondus avec la chaîne du Niagara. On peut dire la même chose des roches qui bordent le grand axe central, bien qu'ici on puisse à peine reconnaître les couches qui appartiennent au groupe.

Du S.-O. du mont Shawangunk, dans l'Etat de New-York, ce groupe s'étend à travers le New-Jersey, et dans la Pennsylvanie, où il atteint l'épaisseur énorme de 2000 pieds. Quelques-uns de ses membres se trouvent encore dans la Virginie et le Tennessée, et atteignent la limite méridionale de la région paléozoïque dans l'Alabama.

La base du terrain silurien supérieur est marquée par un immense dépôt de grès en couches épaisses, d'une structure tantôt fine, tantôt grossière, et de conglomérats puissants, avec lits minces de schistes ou de grès schistoïde. Ces conglomérats contiennent fréquemment des cailloux d'une grande dimension, et souvent les matériaux qui les composent ne sont que peu usés, et les fragments conservent à un degré frappant leur forme angulaire.

Malgré la nature grossière des matériaux dont la roche est ordinairement composée, ils sont cimentés au point de faire naître l'idée qu'une certaine partie de la matière siliceuse pouvait se trouver en dissolution au moment de l'agrégation de la masse. La forme moins arrondie des fragments, et la nature compacte de la roche, jointes aux particularités de ses stratifications et de ses autres traits, servent à distinguer ce conglomérat de ceux des terrains houillers.

La roche a été d'abord reconnue dans la vallée du Mohawk, au S. d'Utica, où elle forme une masse qui a moins de 20 pieds d'épaisseur, et qui se perd en s'amincissant à la fois vers l'E. et vers l'O., tandis que vers le S. elle plonge sous les formations supérieures. A 70 milles plus au S., cependant, elle apparaît brusquement avec une grande puissance dans le mont Shawangunk, qui commence à quelques milles de l'Hudson, près Kingston, et de là se dirige par les comtés Ulster, Sullivan et Arago, jusqu'au Delaware. La formation s'étend ensuite au S.-O., à travers New-Jersey, la Pennsylvanie, le Maryland et la Virginie, et imprime son caractère principal à la topographie de la contrée sur toute sa longueur. En y comprenant le mont Shawangunk, dans l'État de New-York, et son prolongement, le mont Kittatinny en Pennsylvanie, elle forme une chaîne de montagnes continue, depuis la rivière d'Hudson, dans l'État de New-York, à la partie septentrionale de la Virginie. Nous ne devons pas oublier d'observer que, bien que le mont Schawangunk se termine d'une manière abrupte, au N.-E., à quelques milles de l'Hudson, cependant les conglomérats et grès grossiers sont mélangés avec les schistes inférieurs de la période précédente, dans les roches contournées et métamorphiques de l'E. de l'État de New-York.

Cette formation forme ordinairement le trait le plus marqué dans les passages et la topographie de la contrée qu'elle traverse, et constitue la chaîne de montagnes que nous avons décrite. Sa surface est ordinairement stérile, et souvent complétement dépouillée de végétation sur une vaste étendue, et les grès blancs qu'on voit souvent sous forme de grande tache blanche donnent un aspect particulier à quelques parties de ces montagnes. Il est aussi digne de remarque que ces roches présentent tous les caractères d'une ancienne ligne de côtes, pendant la période des soulèvements, et marque de la manière la plus tranchée, dans le dépôt de la série entière des roches poléozoïques, la période à laquelle la faune de cette époque a subi les changements les plus considérables.

Terrain silurien inférieur.

Les formations géologiques qui aujourd'hui sont comprises dans l'étage inférieur du terrain silurien consistent, en allant de haut en bas, en une série de grès argileux et de schistes, connue sous le nom de formation III dans la carte de Pennsylvanie, ou groupe de la rivière d'Hudson, y compris le schiste d'Utica, dans l'État de New-York. Après eux viennent les calcaires de la formation II de la Pennsylvanie, ou calcaire de Trenton, Black-River, Bird's-eye et Chazy, de l'État de New-York. A la base du système vient la formation I, ou les grès calcifères, et grès de Potsdam. Ces trois divisions dans le terrain silurien inférieur nécessitent une courte description séparée, bien que, sauf peu de modifications, elles occupent à peu près la même étendue géographique.

Groupe A. — La division mentionnée en premier lieu, ou division supérieure, acquiert un grand développement vers le N.-E.; elle pénètre aux États-Unis, en venant du Canada. et suit continuellement le bord oriental du lac Champlain, ordinairement dans un état de métamorphisme partiel. Elle s'étend au-dessous de Withe-Hall, et forme une lisière sur les deux côtés de la vallée de l'Hudson, où elle prend peu à peu son caractère normal. Elle s'étend dans cette vallée jusqu'à Newburg, et suivant ensuite la direction du S.-O. continue à travers les Etats de New-Jersey, de Pennsylvanie, de Maryland et de Virginie, tantôt sur une seule ligne, tantôt sur deux ou trois qui reviennent plus loin se confondre. A son extrémité S.-O., elle suit la direction générale des Alleghanys, et ses contours correspondent aux sinuosités des précédentes formations. A l'O. et au N.-O. de la rivière d'Hudson, les couches passent sous les calcaires inférieurs de l'Helderberg, depuis Kingston, sur l'Hudson, jusqu'au Mohawk : elles continuent à suivre cette vallée avec de nombreuses ondulations, et on les voit s'élever sur les deux flancs, par suite du soulèvement des roches inférieures. Enfin elles finissent par s'étaler sur une vaste surface sur la côte orientale du lac Ontario. Dans le Canada, elles reparaissent sur la rive septentrionale du lac, et peuvent être suivies sans interruption jusqu'au fond de la baie de Géorgie, sur le lac Huron. Le bassin de cette baie et sa continuation dans l'espace compris entre les îles Manitouline et les terres ont été creusés en partie dans le groupe de la rivière d'Hudson, dont les lits supérieurs apparaissent au bord septentrional de quelques-unes de ces îles. A l'O., ce groupe atteint le bord septentrional du lac Michigan qui forme la Pointe aux Baies entre la grande et la petite baie de Noquet. Il est recouvert par les eaux de la Baie Verte, près du fond de laquelle, sur la côte orientale, on voit passer les portions supérieures du groupe sous les calcaires supérieurs. On la revoit sur la côte orientale du lac Winnebago et audelà il se perd peu à peu en s'amincissant. Depuis la baie de Géorgie, à l'O., la formation prend un caractère argilo-calcareux : les lits supérieurs consistent en calcaire impur, et les lits arénacés manquent presque absolument.

Cette formation est extrêmement développée dans la région qui entoure Cincinnati, dans l'Ohio, à une distance de 50 milles, et au S.-O. le long de l'axe qui s'étend à travers le Canada et le Tennessée. Dans cette partie des États-Unis, elle consiste en lits alternatifs de schiste tendre et de lits minces calcaires, avec traces accidentelles des grès schistoïdes qui forment un trait si constant et si caractéristique de cette formation dans son extension orientale.

Sur le Mississipi, et en beaucoup de points du Wisconsin et de l'Iowa, on trouve quelques lits minces de calcaire argileux avec couches de schiste, surmontant la roche riche en plomb, et occupant probablement la position du groupe de la rivière d'Hudson. En admettant cette dernière correspondance, nous pourrons suivre le groupe depuis la frontière N.-E. des États-Unis, au S.-O. dans l'Alabama, et nous avons de bonnes raisons de supposer, par suite de son apparition dans la région de Cincinnati et de Nashville, qu'il existe sous les formations supérieures sur toute l'étendue intermédiaire.

Groupe B. — La seconde division de la formation du terrain silurien inférieur, ou le groupe calcaire, consiste, dans sa partie orientale, en quatre membres distincts : le calcaire de Trenton, le calcaire de Black-River, le calcaire de Bird's-eye et le calcaire de Chazy. Celui de Trenton est le plus persistant et donne souvent son nom au groupe entier.

La série de ces calcaires pénètre, aux États-Unis, au N. par les vallées du Saint-Laurent et du lac Champlain, et on les reconnaît le long du Saint-Laurent, jusqu'à son embouchure. Nous les retrouvons, avec quelques légères interruptions à travers les îles de New-York, de New-Jersey, de Pennsylvanie, de Virginie et d'Alabama. Le groupe s'étend à l'O. de la partie centrale de l'État de New-York, le long de la vallée du Mohawk, par les chutes de Trenton et la rivière Noire à l'extrémité orientale du lac Ontario. De là, en franchissant le Saint-Laurent, il traverse le Canada, et apparaît sur les bords de la baie de Géorgie, et le long du bord septentrional du lac Huron ; il forme la plus grande partie de l'île de Saint-Joseph, et s'étendant à l'O., entre le lac Supérieur et le lac Michigan, forme la côte occidentale de la Baie Verte. De là il entre dans l'État de Wisconsin, et continue à peu près parallèlement au bord occidental du lac Michigan, jusqu'à 50 milles de la frontière méridionale de l'État : il s'infléchit à l'O., et occupe une large surface dans la partie méridionale du Wisconsin et le N. de l'Illinois. Il s'étend dans l'Iowa, et, suivant le cours du Mississipi, se montre par intervalles au niveau des chutes de Saint-Antoine.

Le long de l'axe de soulèvement qui s'étend de Nashville au lac Erié, on retrouve ces calcaires sur les bords de la rivière des Pierres dans le Tennessée, et à Francfort dans le Kentucky ; mais ils n'apparaissent plus au N. de l'Ohio.

Les calcaires de la période du terrain silurien inférieur occupent la même étendue que les schistes qui leur sont supérieurs dans l'E. des États-Unis, tandis que dans l'O., ou dans la vallée du Mississipi, ils s'étendent au delà de leurs limites positivement reconnues.

Ce groupe calcaire, soit qu'on y retrouve les diverses sous-divisions, soit qu'on n'en forme qu'une seule masse, est partout très important. Dans toutes les localités de l'E. et du S.-E., il consiste en calcaires, parfois avec intercalation de silex et de lits minces schisteux. A l'O., les schistes augmentent d'épaisseur, et en certains endroits, comme aux chutes de Saint-Antoine, ils prennent un tel développement, qu'une grande partie de la roche n'est plus propre à aucun emploi.

On observe, en avançant vers l'O., une diminution générale dans la puissance et un changement dans les caractères des calcaires de ce groupe ; et dans cette direction commence à s'intercaler un autre membre, qui n'est encore que peu important avant d'arriver dans le Wisconsin, mais qui se développe alors beaucoup dans cet État, et devient très intéressant, à cause du gisement qu'il renferme (calcaire à galène) dans le Wisconsin, l'Illinois et l'Iowa, et qui a été autrefois représenté dans la carte de cette contrée sous le nom de *Cliff limestone*.

Groupe C. — La dernière division du terrain silurien inférieur est formée par le grès calcifère et le grès de Potsdam. Le premier étage est représenté par le calcaire magnésien inférieur dans les États qui bordent le haut Mississippi. Ces deux roches alternent fréquemment à leur point de jonction, et en beaucoup de localités on trouve un dépôt friable arénacé succédant au grès calcifère, qui est, sans aucun doute, dû à la répétition des mêmes actions auxquelles il faut rapporter les lits arénacés inférieurs.

La division supérieure est, pour la plus grande partie, dépourvue de matière calcaire ; cependant, dans sa partie occidentale, elle contient quelques bancs calcaires. Le grès supérieur ou calcifère a reçu originairement ce nom, parce que partout on y observe à différents degrés la combinaison des éléments calcaires et arénacés. Cette formation pénètre dans les États-Unis par la vallée du Saint-Laurent, au N.-E., et traverse le N. de l'État de New-York, jusqu'au bord du lac Ontario. On peut la suivre en descendant la vallée du lac Champlain, et au delà, sans interruption, à travers les États de New-Jersey, de Pennsylvanie, de Maryland, de Virginie et de Tennessée. A l'O., elle traverse le Canada, circonscrit le bord septentrional, et se développe davantage sur le lac Supérieur ; du bord occidental du lac, elle prend une grande extension dans le N. du Wisconsin, et s'étend le long des bords du Mississippi, en formant les roches élevées et pittoresques qu'on voit entre la prairie du Chien et les chutes de Saint-Antoine.

Roches métamorphiques supérieures.

Il est extrêmement difficile de trouver une ligne de séparation entre les dernières roches fossilifères et les roches métamorphiques supérieures, parce que les formations qui séparent l'Hudson de l'Atlantique sont formées par les diverses roches paléozoïques déjà décrites, tantôt en présentant les caractères des terrains de transition, tantôt parfaitement cristallines et dépourvues de restes organiques. Toutes ces roches, cependant, quel que soit le degré de l'action métamorphique qu'elles ont subie, peuvent être classées sous le nom de roches métamorphiques supérieures, et distinguées d'une série plus ancienne de roches cristallines dont la stratification est discordante, comme on peut l'observer aux Petites Chutes (New-York), le long de la vallée du lac Champlain et du Saint-Laurent et dans beaucoup d'autres localités. Les roches métamorphiques supérieures comprennent des roches quartzeuses ou grès modifiés, des calcaires à grain fin souvent micacés, blancs ou colorés, des roches schisteuses, variant des schistes tendres et fissiles, verts ou rougeâtres, aux schistes micacés et talqueux auxquels les premiers passent successivement, ainsi que diverses roches gneissiques, syénitiques et hornblendiques, qui toutes peuvent être clairement reconnues comme dérivant de roches stratifiées fossilifères de la période paléozoïque (1).

On a reconnu ces formations dans une grande partie de la Nouvelle-Angleterre, et ce n'est que sur de petites surfaces que les roches métamorphiques anciennes, si encore elles ont bien ce caractère, se rencontrent entre leurs limites. Le prolongement de la même formation au S.-O., dans une direction parallèle aux anciennes roches fossilifères, fait voir une large surface, couverte par des roches de cet âge, occupant la même étendue que les formations paléozoïques du N. au S.

La direction et l'inclinaison des couches qui composent cette formation sont ordinairement bien constantes, bien qu'en certains points et sur de petites étendues leur étude soit rendue difficile et obscure par des dislocations, des plissements et des contournements nombreux.

La plus grande partie de cette formation est formée de roches plus anciennes que celles de la période carbonifère, et nous leur trouvons superposées, dans l'E. de la Nouvelle-Angleterre, ces roches carbonifères métamorphiques. Dans beaucoup de localités, on peut suivre, d'une manière nette, le passage des roches métamorphiques aux couches fossilifères, et les restes organiques disparaissent graduellement à mesure que les lits prennent une structure plus cristalline, par suite d'une action métamorphique de plus en plus puissante. Toute espèce de doute sur la cause de l'oblitération des fossiles se trouve ainsi levée.

Les roches de cet âge forment le trait physique le plus marqué dans la topographie de la contrée qu'elles occupent. Quand on entre dans les États-Unis au N. par la vallée du Saint-Laurent, on trouve les roches paléozoïques, d'abord dans leur condition normale et sans aucune altération, ne présentant que de légers changements lithologiques; ces changements deviennent ensuite plus marqués, et les couches deviennent plus ou moins cristallines,

---

(1) Le granite, ordinairement rapporté à une origine ignée, occupe relativement un petit espace.

les fossiles s'oblitèrent graduellement, les lits s'inclinent parfois jusqu'à la verticalité, et souvent présentent une série de plissements ; enfin, les roches s'élèvent en nombreuses collines et en chaînons, qui finissent par prendre le caractère de chaînes de montagnes, telles que les Montagnes Vertes du Vermont et les Montagnes Blanches du New-Hampshire.

Roches dévoniennes métamorphiques. — Les roches altérées de la période des groupes de Chemung et de Portage qui forment une partie du terrain dévonien ont été reconnues en grandes masses à Gaspé, par MM. Logan et Hunt, et l'on sait qu'elles sont très développées dans l'Etat du Maine. Ces roches circonscrivent la partie orientale de la Nouvelle-Angleterre, en formant une partie des Montagnes Blanches, et, en plongeant sous la formation carbonifère du Massachusetts et de Rhode-Island, constituent, avec cette dernière, la plus grande partie des formations métamorphiques et cristallines à l'E. de la rivière du Connecticut.

Roches siluriennes métamorphiques. — Dans la portion de la contrée située entre la rivière d'Hudson à l'O. et la rivière du Connecticut à l'E., les roches plus ou moins métamorphiques appartiennent aux formations que nous avons décrites déjà dans le terrain silurien. Dans la partie occidentale de cette ceinture, les quartzites ou grès durs, les calcaires cristallins et diverses roches schisteuses reposent sur des lits de gneiss grossier et de hornblende, reconnus pour être de simples modifications et altérations du grès de Potsdam, du groupe du calcaire de Trenton et du groupe de la rivière d'Hudson, avec des lits accidentels et interrompus de conglomérat du Shawangunk, mélangé avec les schistes et formant les sommets des éminences les plus élevées.

Les roches siluriennes supérieures ont été suivies par MM. Logan et Hunt, depuis les parties non altérées, dans la vallée du Saint-Laurent, jusqu'aux limites de l'Etat de Vermont, où elles prennent le caractère de calcaires micacés, alternant avec des schistes micacés.

La partie inférieure de cette formation métamorphique, ou celle qui se compose principalement de schistes siluriens inférieurs associés à des lits arénacés, constitue la zone aurifère proprement dite, comme il a été abondamment prouvé par l'étude de leur prolongement septentrional ; et les nombreuses observations qui ont été faites dans les Etats du Sud autorisent à attribuer le même âge à la zone aurifère de Virginie, des deux Carolines et de la Géorgie. Il est impossible aujourd'hui de décider si le gisement aurifère de Californie est du même âge que les roches métamorphiques supérieures des Etats-Unis de l'Est.

Roches metamorphiques anciennes ou inférieures.

Dans beaucoup de parties des États-Unis il existe de larges espaces recouverts de roches cristallines, que leur âge et leurs caractères lithologiques distinguent entièrement des précédentes. On a proposé d'adopter, pour les désigner, le nom de roches métamorphiques anciennes ou inférieures, pour les distinguer des roches altérées qui appartiennent évidemment à la période paléozoïque. On sait qu'elles sont d'une date antérieure, et, dans beaucoup de cas, sinon toujours, elles ont acquis leur structure cristalline et pris leur caractère métamorphique avant le dépôt des formations sédimentaires environnantes qu'il faut ranger parmi les roches anciennes paléozoïques.

Dans beaucoup de localités, le grès de Potsdam, qui est la roche fossilifère la plus ancienne que l'on connaisse aux États-Unis, s'appuie dans une position horizontale, ou peu s'en faut, contre les masses inclinées, ou repose sur les hautes cimes formées par les roches de cet âge, dont la date antérieure est ainsi établie d'une manière évidente.

La formation consiste en une longue série de syénites, gneiss, schistes divers hornblendiques, micacés et talqueux, etc., avec calcaire très cristallin. Toute la série conserve encore des traces évidentes de stratification, qui permettent de croire que les roches primitivement sédimentaires ont été métamorphosées par des agents ignés. Comme témoins et preuves de ce changement, nous trouvons d'énormes masses éruptives de granite et d'autres roches généralement regardées comme d'origine ignée, qui ont pénétré les roches cristallines stratifiées dont se compose en grande partie le système. Il est de même traversé par de nombreux dykes trappéens, qui sont quelquefois très puissants et très étendus. La roche principale et caractéristique de cette formation est un gneiss syénitique, d'une texture généralement ferme et compacte. Les plans de stratification sont souvent obscurs et voilés par la structure granitique ou syénitique. L'une des substances qui accompagne de la manière la plus constante, et qui caractérise cette formation, est le minerai de fer magnétique.

Elle compose principalement la chaîne de montagnes qui s'étend au nord du Saint-Laurent, parallèlement au fleuve jusqu'aux mille îles (*Thousand Island*). On la retrouve à travers le

Canada jusqu'aux bords du lac Supérieur ; on l'observe sur les deux bords du lac, et de là elle se dirige à l'O. jusqu'aux sources du Mississippi.

Quelques portions de la chaîne de montagnes du Maine et du New-Hampshire appartiennent peut-être à cette formation ; mais leur étendue et leurs limites n'ont pas été bien définies, et les roches qui les composent n'ont pas été nettement distinguées des roches avoisinantes d'origine plus moderne (1).

Dans le nord de l'État de New-York, cette formation forme des montagnes qui atteignent 5000 pieds d'élévation, mais dont la hauteur est généralement moindre. Dans cette région, elle couvre un espace d'une forme à peu près triangulaire, qui a environ 150 milles de longueur du N. au S., et 125 milles de l'E. à l'O., ou du lac Champlain au fleuve du Saint-Laurent. Les points les plus élevés sont au centre de cet espace. La formation expire graduellement au N. et au S., non sans produire néanmoins quelque dérangement parmi les roches siluriennes qui la recouvrent. Dans la partie méridionale de l'État de New-York, elle reparaît et forme des crêtes et des montagnes élevées qui, cependant, atteignent à peine la moitié de la hauteur de celles du nord. De la limite sud de cet État, elle pénètre dans celui de New-Jersey, en formant une rangée presque continue de montagnes, depuis la frontière jusqu'au Delaware. De là à la frontière du Maryland, les roches métamorphiques anciennes s'étendent à travers le S.-E. de la Pennsylvanie, en formant deux ceintures étroites, l'une continue, l'autre formée seulement de chaînons séparés les uns des autres par des formations de date plus récente.

Les roches de cette période continuent à conserver la même direction dans le Maryland, la Virginie et la Caroline du nord, et sont connues sous le nom de *Montagnes Bleues* ou *Montagnes du Sud :* les points les plus élevés en Virginie atteignent la hauteur de 4000 pieds. La formation qui continue de s'étendre au S.-O. se prolonge probablement avec des interruptions plus ou moins

(1) Au Canada, M. Logan a reconnu une autre série de roches métamorphiques qui reposent sur celles que nous venons de décrire, et d'une date antérieure à nos roches métamorphiques supérieures, ce qui constitue trois divisions. Cette formation moyenne se compose de grès, de conglomérats avec fragments de jaspe rouge, lits de schiste, et calcaire. « Les schistes sont verdâtres, très chlorités, et contiennent souvent de l'épidote ; souvent ils prennent le caractère de conglomérats, par suite de la présence abondante des fragments de syénite. »

prononcées jusqu'aux limites méridionales des formations métamorphiques de l'Alabama.

Nous l'avons déjà montrée sur le bord du lac Supérieur : elle y forme une large zone qui s'étend, sauf de légères interruptions, sur toute la côte, depuis le voisinage de la rivière Camp, jusqu'à Point-Abbaye ; elle s'étend ensuite à l'O., et sur une largeur de plus en plus croissante, dans le nord du Wisconsin, et dans la partie centrale de cet État descend vers le S. jusqu'à la hauteur du bord du lac Winnebago. La formation se dirige ensuite vers le N.-O., et atteint le Mississippi au N., aux chutes de Saint-Antoine. On en trouve un rameau isolé dans le Missouri, un autre dans l'Arkansas, et un troisième dans le Texas.

Formation trappéenne.

Trapps rouges les plus modernes. — La formation de cette nature, la plus apparente dans les États-Unis de l'E., est celle qui est associée au nouveau grès rouge, dont nous avons déjà décrit l'étendue. Les roches basaltiques ou trappéennes accompagnent le grès dans la vallée du Connecticut, et sont assez puissantes pour marquer les traits les plus apparents de la topographie de la contrée. Le mont Holyoke et le mont Tom, dans le Massachusetts, en sont des exemples frappants. Les roches de l'E. et de l'O., près New-Haven, forment aussi des points élevés dans le pays, et appartiennent à la formation trappéenne.

Les trapps forment de nombreux dykes et des crêtes moins apparentes, souvent parallèles, et de longues ceintures continues qui suivent toute la longueur de la vallée, jusqu'à 100 milles au-dessus de son embouchure. Les palissades de la rivière d'Hudson, et de nombreuses éminences trappéennes dans le New-Jersey, appartiennent toutes à la même époque et présentent les mêmes relations géologiques. Enfin, cette formation peut être encore suivie dans la Pennsylvanie, le Maryland, la Virginie et la Caroline du sud, et partout accompagne les grès.

Une des matières qui se trouve toujours associée à ces trapps est le cuivre natif ; mais, sur toute leur étendue, on ne l'a jamais trouvé en quantité exploitable. On trouve fréquemment des minerais de cuivre dans les grès qui sont au contact des trapps ; et il semble y avoir quelque rapport entre la présence de ces amas minéraux et l'éruption des dykes. Les roches trappéennes qui appartiennent à cette période du nouveau grès rouge présentent souvent la structure en colonnes et sont divisées naturellement par des pans verticaux. Ce caractère est moins apparent dans les roches que nous allons décrire.

Trapps du lac Supérieur. — La deuxième grande formation trappéenne est celle du lac Supérieur, qui est en relation avec le grès du terrain silurien inférieur. Elle est développée de la manière la plus frappante à Keweenaw-Point, et s'étend de là, en occupant une grande largeur, à l'E. et à l'O., en suivant généralement les contours du lac. La formation est flanquée au N. par des conglomérats, et au S. par des grès, et une ceinture étroite du trapp traverse le conglomérat sur toute sa longueur, parallèlement à la formation centrale principale. Vers l'extrémité occidentale du lac Supérieur, la formation trappéenne est interrompue et ne se présente plus qu'en dykes isolés ou en crêtes qui percent le grès. L'île Royale présente à peu près les mêmes traits que Keweenaw-Point; le trapp y forme une rangée semblable et parallèle, mais le conglomérat y est peu développé. Sur le bord septentrional du lac, dans les limites des Etats-Unis, le trapp s'étend en nombreux rameaux étroits et en dykes qui tous sont dirigés vers l'O. Plus loin, en différents points situés entre le lac et le Mississipi, au-dessus des chutes de Saint-Antoine, on trouve encore fréquemment des lits de trapp.

Au delà de ce fleuve, on sait peu de chose de leur étendue et de leurs caractères : mais, d'après les informations que l'on possède, il paraît probable que la même formation se retrouve par intervalles entre le Mississippi et les Montagnes-Rocheuses.

Dans toute cette étendue, les associations sont si semblables qu'il est permis de rapporter avec certitude tous ces trapps à la même formation. Sur les bords de la rivière Sainte-Croix, ils forment de grandes protubérances et des nœuds à travers les couches à Lingules du grès de Potsdam ; et, dans le nord du Wisconsin, on les trouve à peu près dans une position semblable, plus ou moins en relation avec le grès inférieur, mais fréquemment associés à de la syénite ou à du gneiss syénitique.

La formation trappéenne du lac Supérieur, là où elle atteint son principal développement, contient des quantités immenses de cuivre natif qui, dans cette région, caractérise toujours une certaine variété de trapp.

Dans une carte de Keweenaw-Point, publiée récemment par M. Whitney, géologue des États-Unis, nous trouvons la description suivante : trapp amygdaloïde, contenant du cuivre natif; trapp porphyrique, contenant du cuivre sulfuré ; et trapp cristallin, ou les filons ne sont point productifs.

La principale partie de la formation, en ce point, est formée par la variété précédente : une ligne mince qui passe par le centre représente la portion cristalline ; et au S. une étroite bordure, qui s'unit au grès, appartient à la première variété.

Trapps de l'ouest. — On trouve des trapps et des basaltes à des intervalles fréquents, depuis la frontière du Texas jusqu'au pays des Indiens Blackfoot, et plus loin encore vers le N. A l'O., dans la grande plaine qui s'étend entre les Montagnes-Rocheuses et la Sierra-Nevada, des roches du même caractère se rencontrent fréquemment, mais leur position géologique n'est point connue. On trouve de même souvent et sur de vastes étendues des roches basaltiques et volcaniques au N.-O., dans les Montagnes-Rocheuses, et le long de la Colombie jusqu'à la côte de l'océan Pacifique.

Trapps de la série métamorphique. — Un système de roches trappéennes d'un âge différent, consistant en trapps, diorites, porphyres, etc., traverse les roches métamorphiques de la période que nous avons précédemment indiquée. Elles n'affectent pas généralement d'une manière bien sensible la topographie de la contrée, et ne paraissent pas avoir de relation avec le soulèvement de la chaîne montagneuse où on les rencontre. Elles la traversent en différentes directions, et remplissent des fissures ou joints, qui sont souvent parallèles entre eux. On rencontre partout ces dykes dans les États-Unis, mais ils sont plus nombreux sur le bord septentrional du lac Huron et du lac Supérieur.

Dans la partie orientale de la Nouvelle-Angleterre, et surtout près de la côte, ces dykes traversent les couches, et des ceintures plus étendues de roches semblables occupent des portions de la région métamorphique. Les bordures porphyriques semblent parfois, le résultat d'un métamorphisme extrême, qui aurait réduit en fusion des roches préexistantes.

ON CERTAIN TRAINS OF ERRATIC BLOCKS ON THE WESTERN BORDERS OF MASSACHUSETTS UNITED STATES

WEEKLY EVENING MEETING,

Friday, April 27.

THE DUKE OF NORTHUMBERLAND, K.G. F.R.S. President, in the Chair.

SIR CHARLES LYELL, F.R.S.

On certain Trains of Erratic Blocks on the Western borders of Massachusetts, United States.

ON the western borders of the State of Massachusetts, in Berkshire, and on the eastern confines of the adjoining State of New York, a great number of erratic blocks are seen, remarkable for their large size and their distribution in long parallel trains, each of them continuous in nearly straight lines over hill and dale, for the distance of five, ten, or twenty miles or more. These trains are of geological interest, not only from their length, and the size of the blocks, but also from the precision with which they can be traced to their starting points, and the low latitudes in which these starting points are situated. The area alluded to occurs in lat. 42° 25′ south, corresponding to that of the north of Portugal; and the western borders of Berkshire, where they join the State of New York, are about 130 miles from the Atlantic coast, in a direction due west of the city of Boston, in Massachusetts.

In the accompanying plan, Fig. 1, it will be observed that the mountain ranges A B C run N.N.E. and S.S.W., whereas the trains of erratic blocks (from No. 1 to No. 7 inclusive,) have a direction nearly transverse to these ranges, and consequently to the intervening valleys, their direction being about N.W. and S.E. In one sense we may affirm that the course of the stones has no relation whatever to the present configuration of the country, because the present drainage or flow of the rivers is quite in a different direction; but in another point of view we shall find that a close relation can be made out between the actual inequalities of hill and dale, and the course and mode of dispersion of the erratics; so that there is good reason to infer that the superficial inequalities were very nearly what they are now, before any of the trains originated.

Distance in a straight line between the mountain ranges **A** *and* **C**, *about eight miles.*

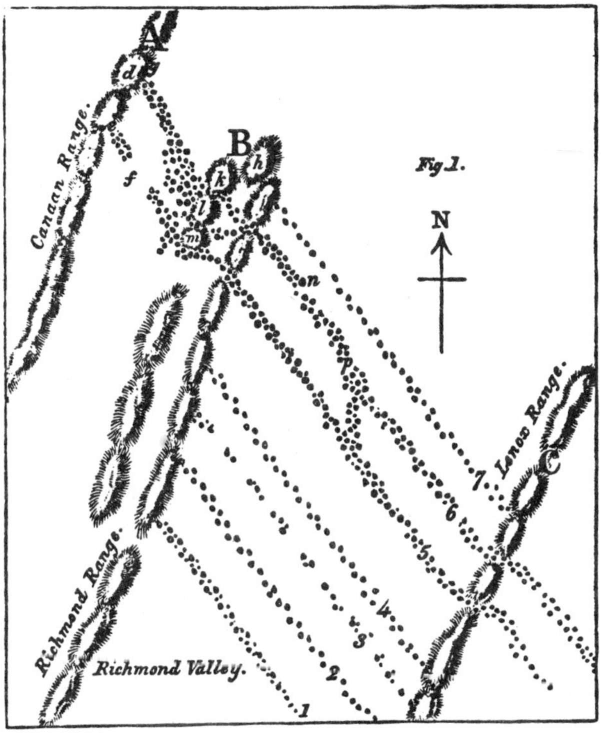

Map, showing the relative position and direction of seven trains of erratic blocks in Berkshire, Massachusetts, and in part of the State of New York.*

A. Canaan range, in the State of New York. The crest consists of green chloritic rock.

B. Richmond range (part of the Taconic range of Hitchcock?) the western division of which consists in Merriman's Mount of the same green rock as A, but in a more schistose form, while the eastern division is composed of slaty limestone.

C. The Lenox range, consisting in part of mica-schist, and in some districts of crystalline limestone.

---

* As the description of the trains could not be understood without a ground plan, the above copy of a diagram, exhibited during the discourse, is thought necessary; but it makes no pretensions to geographical accuracy, the speaker and his companion Prof. Hall having been unable to procure a good map of the district, on which to lay down in detail the results of their observations.

d. Knob in the range A, from which most of the train No. 6, is supposed to have been derived.

e. Supposed starting point of the train No. 5, in the range A.

f. Hiatus of 175 yards, or space without blocks.

g. Sherman's House.

h. Perry's Peak.

k. Flat Rock.

l. Merriman's Mount.

m. Dupey's Mount.

n. Largest block of train, No. 6.

p. Point of divergence of part of the train No. 6, where a branch is sent off to No. 5.

No. 1. The most southerly train examined by Messrs. Hall and Lyell, between Stockbridge and Richmond, composed of blocks of black slate, blue limestone, and some of the green Canaan rock, with here and there a boulder of white quartz.

No. 2. Train composed chiefly of large limestone masses, some of them divided into two or more fragments, by natural joints.

No. 3. Train composed of blocks of limestone and the green Canaan rock; passes south of the Richmond Station on the Albany and Boston railway; is less defined than Nos. 1 and 2.

No 4. Train chiefly of limestone blocks, some of them 30 feet in diameter, running to the north-west of the Richmond Station, and passing south of the Methodist Meeting-house, where it is intersected by a railway cutting.

No. 5. South train of Dr. Reid, composed entirely of large blocks of the green chloritic Canaan rock; passes north of the Old Richmond Meeting-house, and is three-quarters of a mile north of the preceding train (No. 4).

No. 6. The great or principal train (north train of Dr. Reid), composed of very large blocks of the Canaan rock, diverges at *p*, and unites by a branch with train No. 5.

No. 7. A well-defined train of limestone blocks, with a few of the Canaan rock, traced from the Richmond to the slope of the Lenox range.

Dr. Reid, the agriculturist of Berkshire, first called attention in 1842 and 1845 to these phenomena. Professor Hitchcock contributed many valuable observations in 1844, and Professors Henry D. and William B. Rogers treated of the same subject in 1846.* The district was re-examined in October 1852, by Professor J. Hall and Sir Charles Lyell, by whom some of the data referred to in this discourse were ascertained. Within the area particularly referred to, the trains Nos. 5 and 6 are the most conspicuous, by their length and by the magnitude and frequency of the blocks composing them. These fragments consist of a green chloritic rock, remarkably tough, sometimes compact, but occasionally schistose. It is met with in place at *d*, in the highest crest of the Canaan ridge, and reappears in its more slaty form in the western division of the Richmond ridge B. at *l*, or Merriman's Mount. It passes on the one hand into a quartzose conglomerate,

* "Boston Journal of Natural History," for June 1846.

and on the other, when most metamorphic, into a crystalline rock, in which sometimes chlorite, sometimes hornblende and felspar are developed. A large proportion of the green fragments, in trains 5 and 6, have evidently come from the ridge A, and a large proportion of the whole from its highest summit *d*, upon which fragments often 30 feet in length may now be seen, some of them having probably constituted for years the exposed crest of the ridge, and having in that position acquired a smoothed and rounded outline so characteristic of the protuberances of hard rock in regions where erratics and glacial striæ abound. Such dome-shaped masses are called "roches moutonnées," on the borders of Swiss glaciers. Several of the fragments having this shape, and lying on the crest of A, have been slightly removed from their original position, as if just ready to set out on their travels. They are angular in their lower parts, where they exhibit such an outline as the jointed rock would possess if a great fragment fell from an undermined cliff.

To the westward of the ridge A no similar green blocks are to be found, not even a small number, such as we might have expected to roll down to the base of a hill having so steep a western declivity. It is evident, therefore, that the propelling power, whatever it was, acted exclusively in a south-easterly direction.

Dr. Reid has traced the train No. 5 for more than ten, and No. 6 for more than twenty miles to the south-east, crossing the Richmond and Lenox mountains B and C, and probably extending beyond the points to which they have already been followed. Messrs. Hall and Lyell found both trains extremely well-defined after they emerge from the Richmond range, but by no means so distinct in their passage over the first valley between A and B. A great number of blocks have collected at the base of *d*, Fig. 1, or the highest knob before alluded to of A, particularly around *g*, or Sherman's House. From this point to the Richmond range, a nearly continuous stream may be traced, and the blocks are seen to pass through a gap or depression, in the eastern division of the ridge B, between Flat Rock and Merriman's Mount, *k* and *l*. But when we attempt to follow the other train, No. 5, from its supposed point of origin *c*, (a spot about half a mile distant from *d* before alluded to,) we find at *f* an hiatus, not less than 175 yards long, where there are no erratics. This break is not caused by the stones having been used up for building, no such materials being observable in the walls enclosing fields, or in the farm-houses in the neighbourhood. A vast number of blocks seem to have crossed the valley in a direct line between A and B, and to have accumulated on the north-western slope of Merriman's Mount *l*, as well as to the south of it, around Dupey's Mount *m*; and they seem to have crossed the Richmond ridge by depressions both to the north and south of Dupey's Mount, those to the north proceeding westward to join the train No. 6. The number of large blocks lying on the west slope of Dupey's Mount, and many of them to the south of the line which would connect the southern train, No. 5, with its supposed starting point *e*, is very great. One of these, 24 feet long, is poised upon another which is about 19 feet in length. The largest of all, composed like the rest of the green Canaan rock, lies on the west flank of Dupey's Mount, and is called "the Alderman." Dr. Reid measured it, and ascertained that it is above 90 feet in diameter, and not much under 300 feet in circumference. At some points about 40 or 50 blocks, the smallest of them larger than a camel, may be seen at once. Among the larger masses the best known, in consequence of its proximity to the Richmond Meeting-house, belongs to train No. 6, and is that marked *n* on the plan, Fig. 1. According to the measurement of Messrs. Hall and Lyell it is 52 feet long, by 40 in width, its height above the drift in which it is partially buried being 15 feet. At the distance of several yards occurs a smaller block, three or four feet in height, 20 feet long, and 14 broad, composed of the same compact chloritic rock, and evidently a detached fragment from the bigger mass, to the lower and angular part of which it would fit on exactly. This erratic (*n*) has a regularly rounded top, worn and smoothed like the roches moutonnées before mentioned, but no part of the attrition can have occurred since it left its parent rock, the angles of the lower portion being quite sharp and unblunted.

After the two great trains, Nos. 5 and 6, have crossed the ridge B, and entered the Richmond valley, which is about four miles broad, and about 800 feet deep below the crests of A and B, each train is exceedingly well defined. They are about half a mile apart, the train No. 6 varying in width from 100 to 300 feet, the space intervening between them usually very free from erratics, but here and there a solitary large straggler being visible. At one point *p*, Fig. 1, part of the train No. 6 diverges and forms a branch uniting with No. 5.

The average size of the blocks of all the seven trains laid down on the plan lessens sensibly in proportion as we recede from their point of departure, yet not with any regularity, a huge block recurring here and there in the midst of a train of smaller ones. Many which have wandered farthest from their parent rock retain their angles extremely fresh and sharp. Almost everywhere beneath the trains is a deposit of sand, mud, gravel, and stones, for the most part unstratified, and resembling the "northern drift" of Great Britain and parts of the north of Europe. It varies in thickness from 10 to 50 feet, being of greatest depth in the valleys. The uppermost portion is occasionally, though rarely, stratified; and where stratification occurs, it seems as if the mass first thrown down had been acted upon by currents, and partially rearranged. This drift has been well exposed in some recent railway cuttings, where it is occasionally seen to be 20 or 30 feet thick, immediately under several of the trains before alluded to. The stones in general are more rounded than the erraticsal ready described; occasionally some are seen with one or more flattened, smooth, striated, or furrowed sides. They consist invariably, like the seven trains before mentioned, of kinds of rock only met with in the region lying to the north-west. In one cutting, the drift below the main train No. 6 is 30 feet thick, and contains one or two angular blocks of the green Canaan rock, of considerable, but not of the largest size. There are no appearances here or elsewhere warranting the conclusion that the trains owed their origin to the removal of an upper portion of the covering of drift, the lighter materials having been washed away, and the heavier made to stand out in relief. On the contrary, the erratics of each train, whether large or small, look as if they had dropped down over the linear spaces where they are now strewed, on the surface of hill and valley, equally where the drift is thickest, as where it is very thin or wanting. As a rule, the drift contains no blocks of the first magnitude, although a few occur, and some of the biggest are partially buried in drift, showing that the transport of the heavier, and of a certain portion of the lighter materials, was contemporaneous.

Almost in every place where the removal of the superficial detritus has exposed the underlying rock, a polished, striated, and furrowed surface is seen, like that underneath the modern glaciers of the Alps. The direction of the rectilinear furrows or grooves has been proved by a multitude of observations, made by Professor Hitchcock in this and other adjoining parts of Massachusetts and New York, to be from N.W. to S.E. or similar to the course of the large erratics.* The same geologist has pointed out that such ridges as A B and C are smooth and furrowed, not only on their tops, but sometimes 100 or 200 feet below, on their north-western sides; whereas, on their south-eastern declivities, if steep, there has been no such action, although these also are grooved and polished where their slope is gentle. The furrows, which are from an inch to a foot in depth, usually cross the strike of the highly inclined slates and slaty limestones at a considerable angle, and in such a manner as to demonstrate that the strata of the ridges A B and C, and of the intervening valleys, bent as they are into a series of anticlinal and synclinal folds, were already in their present position, and had even suffered aqueous denudation before the drift was thrown down upon them, and therefore long before the distribution of the erratics above described had taken place.

Although the trains, Nos. 5 and 6, are the most conspicuous, several of the others are well defined, and contain limestone blocks 30 feet in diameter. Thus No. 7 was seen on the western flank of the Lenox range, and followed across the Richmond valley to the eastern side of the ridge B, where the limestone, which has supplied the travelled masses, is in situ. In like manner, Messrs. Hall and Lyell observed, half a mile south of Pittsfield, enormous

* "Final Report on Geology of Massachusetts," p. 383, *et seq.* 1841.

blocks of mica-schist, from 30 to 50 feet in their longest diameter, on the south-east side of the Lenox range C; whereas no similar fragments of mica-schist, whether large or small, are found in any part of the Richmond valley, or on the ridge B, or indeed anywhere between A and the Hudson river.

Some boulders of white quartz rock, two or three feet in diameter, make a part of almost every train, as well as of the subjacent drift, and these may be traced to hills, between the Canaan ridge and the Hudson river, where the Potsdam sandstone has been altered into quartzite.

Sir Charles then proceeded to explain his theory. He believes that all the large erratics have been transported to the places they now occupy by floating ice,—not by icebergs, nor by terrestrial glaciers, but by coast-ice. The hypothesis of glaciers is out of the question; because, even if we could imagine that in lat. 42° 30′ N., the ridges A B C, now only from 1000 to 2000 feet above the sea, once rose, and that at a period, geologically speaking, very modern, to such an elevation as to enable them to generate glaciers, still such glaciers could not have descended from the higher regions in one direction only. They would have radiated from a centre, carrying as many blocks westward as eastward. Their course, moreover, would have been principally S.S.W., or down the valleys now separating the ridges, instead of being south-east, or almost at right angles to the valleys. If, on the other hand, we assume that the country was lower instead of higher, so as to have been submerged beneath the waters of a sea, in which icebergs floated annually from arctic regions, these bergs might bring with them gravel and stones of northern origin, but could not without the aid of coast-ice become freighted with blocks derived from the very region referred to in this discourse, (lat. 42° N.) The northern ice might aid, by chilling the waters of the ocean, and increasing the quantity of coast-ice in a low latitude, but it could do little more.

Fig. 2.

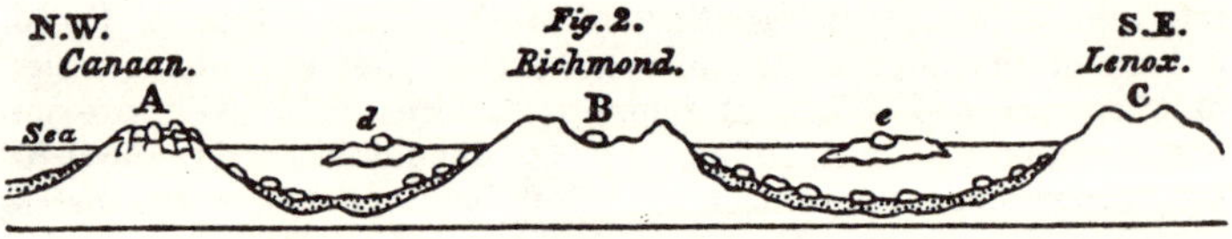

d, e Masses of floating ice carrying fragments of rock.

Suppose the highest peaks of the ridges A B C, in the annexed diagram, to be alone above water, forming islands, and *d e* to be masses of floating ice which drifted across the Canaan and Richmond valleys, at a time when they were marine channels, separating islands, or rather chains of islands, having a N.N.E. and S.S.W. direction. A fragment of ice, such as *d*, freighted with a block from A, might run aground, and add to the heap of erratics at the N.W. base of the island B, or passing through a sound between B and the next island of the same group, might float on till it reached the channel between B and C. Year after year two such exposed cliffs in the Canaan range as *d* and *e* (Fig. 1), undermined by the waves, might serve as the points of departure of blocks, composing the trains Nos. 5 and 6. It may be objected that oceanic currents could not always have had the same direction; this may be true, but during a short season of the year when the ice was breaking up the prevailing current may have always run S.E.

If it be asked why the blocks of each train are not more scattered, especially when far from their source, it may be observed, that after passing through sounds separating islands, they issued again from a new and narrow starting point; moreover, we must not exaggerate the regularity of the trains, as their width is sometimes twice as great in one place as in another; and No. 6 sends off a branch at *p*, which joins No. 5. There are also stragglers, or large blocks, here and there in the spaces between the two trains. As to the distance to which any given block would be carried, that must have depended on a variety of circumstances; such as the strength of the current, the direction of the wind, the weight of the block, or the quantity and draught of the ice attached to it. The smaller fragments would, on the whole, have the best chance of going farthest; because, in the first place, they were more numerous, and then being lighter, they required less ice to float them, and would not ground so readily on shoals, or if stranded, would be more easily started again on their travels. Many of the blocks, which at first sight seem to consist of single masses, are found, when examined, to be made up of two, three, or more pieces, divided by natural joints. In case of a second removal by ice, one or more portions would become detached and be drifted to different points further on. Whenever this happened the original size would be lessened, and the angularity of the block previously worn by the breakers would be restored, and this tendency to split may explain why some of the far-transported fragments remain so angular.

In the ravine between Merriman's Mount and Flat Rock (*k* and *l*, Fig. 1), the erratics, instead of following a N.W. and S.E. course, run within 10 or 15 degrees west and east; and Messrs. Hall and Lyell observed that the glacial furrows there on the exposed rocks below deviated in like manner from the normal direction, and were directed like the train of erratics nearly west and east. They were told that the like deflection, both of trains and furrows, was observable where Nos. 5 and 6 cross the Lenox range; and this deflection is so represented on the plan (Fig. 1); although the speaker had not an opportunity of verifying the fact. The direction of floating ice, when threading the sounds separating islands, would be governed by the shape of the land and the marine channels, and might be expected to differ from the direction of currents flowing in the open sea between chains of islands.

Sir C. Lyell endeavoured in 1842, when explaining the origin of drift, boulders, and glacial furrows in the neighbourhood of the Falls of Niagara, both in Canada and in the State of New York,* to show that the whole region, with its elevated platforms and its valleys, had first gone down gradually, and had then been re-elevated during the glacial period. All geologists who are acquainted with Berkshire, Massachusetts, are agreed that the position of the erratics cannot be explained by the subsequent unequal elevation of the mountain ranges, as if B, for example, had been uplifted to a greater height than A, after the great boulders had been stranded on B. It is clear that the ridge B has intercepted many erratics on its north-western side, as it would do now, if submerged, and the blocks have chiefly crossed through gaps or depressions in B. The glacial furrows also are such as would be made on the fixed rocks, after they had already assumed their actual position, and when the present hills and valleys existed.

Although the principal mass of drift had accumulated before the trains, yet we see some of the biggest blocks partially buried in drift. This we might have expected, according to the hypothesis above suggested, for coast-ice, such as forms annually in the Gulf of St. Lawrence and along the coast of Labrador, does not merely bear away great stones but also mud, sand, and gravel.

The speaker exhibited a drawing of a large angular block of sandstone, about eight feet in diameter, which he and Mr. J. W. Dawson saw in 1852, stranded on the mud-flats near the mouth of the Petitcodiac estuary, where it joins the Bay of Fundy, and where the water is salt. The ferrymen stated that this block was brought down by ice three years before, from a cliff several miles up the river. About the year 1850 much larger blocks of sandstone were removed by coast-ice from the base of a perpendicular cliff at the South Joggins, in the Bay of Fundy, in salt water, and floated for about half a mile, where they dropped or were grounded on one side of the pier built for loading vessels with coal. The vessels being thereby prevented from nearing the pier it was found necessary to blast these ice-borne rocks, at low tide, with gunpowder. All this occurred in latitude 46° N., corresponding to that of Bordeaux; and when we bear in mind that the Bay of Fundy opens towards the south, and is therefore never invaded by icebergs, such as are stranded occasionally near St. John's, Newfoundland, or such as are annually drifted far to the south of the Bay of Fundy in the open ocean, we may well imagine that, with a different configuration of the land, coast-ice may once have exerted great power as far south as the latitude of Berkshire. The buoyant power of sheets of ice, even of moderate thickness, is so great that the magnitude of erratics depends more on the dimensions of the fragments into which rocks undermined by the sea happen to split, than on the peculiar intensity of the frost in that region.

* "Travels in North America," Vol. I., p. 99; and Vol. II., p. 48.

It has been objected to the theory of submergence that the great train, No. 6, has climbed a part of the ridge B, higher than its supposed starting point in A. But there are no exact barometrical or trigonometrical data for this assertion. Messrs. Hall and Lyell could only estimate roughly the relative heights of the knob *d* in A, and that of the pass between *k* and *l* in B, by means of a spirit-level, as they stood on the Canaan knob. It appeared to them that the gap in B, or the ravine between Merriman's Mount and Flat Rock, is 50 or 100 feet below the highest crest of A, in which case the objection falls to the ground. Some of the erratics of No. 6, in the Richmond valley, have probably come from an elevated point in Merriman's Mount, and therefore present no difficulty, since that mount consists of the green chloritic rock, but others have come from the Canaan ridge, and have crossed the ridge B, as before stated. If it could be shown that some of these stones repose on B, at points actually higher than the crest of A, the fact would be important, but by no means inexplicable by the glacial theory. Mr. C. Darwin has shown, that during the gradual subsidence and submergence of a coast situated in high latitudes, packed ice with stones frozen into it is continually driven up on the sea-beach above high-water mark, and if the land be going down at the rate of a few inches in a year, the boulders, by being simply kept up to the sea level, will slowly climb up the hills higher and higher, so that a ridge after sinking may, when it re-emerges, have stones lodged upon it at levels above those of the lands from which the same stones were derived.*

The drift of Berkshire and of New England in general has a great resemblance to the terminal moraines of glaciers, being unstratified and containing fragments of rock, some angular, others rounded. But the proportion of rounded boulders is far more considerable in the drift than in an ordinary glacier moraine, in which last, as Mr. D. Sharp has lately shown in reference to some Swiss glaciers, the rounded blocks are quite the exception to the rule. Want of stratification is the natural result of the melting of matter out of stationary ice, the light particles and the heavy stones dropping down together, and no current of water sorting the materials, and carrying those of less specific gravity to greater distances. Stones frozen into coast-ice may have been rounded, some by rivers, others by the waves of the sea. The dearth of marine shells is sometimes urged as an argument against the hypothesis of submergence beneath the sea, and it is certainly strange that marine shells should be so rare in drift. They are, however, extremely scarce in parts of New England, such as Vermont and Maine, where a few, nevertheless, do occasionally occur; and this holds good in Canada, as also in Ireland and other parts of the North of Europe. As we cannot doubt that the formation accu-

---

* C. Darwin, Geol. Quart. Journ. Vol. VI., p. 315. 1848.

mulated in certain cases under water, we are at liberty to assume the same in all others, where such an explanation will best accord with the facts.

The Saxicava rugosa, Mya truncata, Natica clausa, and other recent species of mollusca, are common to the drift of North America and Europe, and constitute part of a fauna, characteristic of a climate somewhat colder than that of the latitudes where the fossils are now met with. The Caplan also, or Mallotus villosus, a fish swarming in the seas of Greenland, Labrador, and wherever ice abounds in the North Atlantic, is found fossil in clayey concretions or claystones, in the drift of Maine and Canada, as well as in Greenland. It appears that in the glacial period, as now, the isothermal lines, or rather the lines of equal winter temperature bent many degrees farther south on the west than on the east side of the Atlantic, so that the monuments of glacial action extend some eight or ten degrees farther south in North America than they do in Western Europe. Large erratics and glacial furrows are rare south of latitude 48° or 50° in Europe, and are seldom seen south of 40° in North America ; but where mountain chains like the Alps or Himalaya have formed independent sources of cold, we find exceptions. In Madeira and the Canary Islands, between latitudes 28° and 33° N., Sir C. Lyell searched in vain for glacial striæ, and other concomitant phenomena, although some of the mountains there are of great height. In the southern hemisphere all the manifestations of the agency of ice, which are wanting in the equatorial zone, reappear in full force, when we reach Chiloe, Patagonia, and Tierra del Fuego.

If ice-islands, running aground on the bed of a sea or on a a coast, can smooth and furrow the subjacent floor of the ocean by pushing before them or pressing down under them loose sand, pebbles, and stones, the size of such islands is certainly sufficient to afford as much friction and mechanical force as we require. Some of them, measured in the southern hemisphere, exceed 10 miles in diameter, and their height out of the water is from 100 to 200 feet, the mass of ice below being about eight times in volume that rising above the sea-level ; if such masses when they run aground are moving only at the rate of one or two miles an hour, their momentum must be enormous.

The area now subject to the action just alluded to in both hemispheres, is many times greater than that over which terrestrial glaciers descend. In like manner, in the period of the "Northern drift," the submarine were far more extensive than the supramarine glacial operations ; and since we have evidence of much sea having been converted into land since the glacial period, we must expect to find more space bearing the imprint of subaqueous ice than of space exhibiting evidence of the movement of ice over dry land.

In conclusion, Sir C. Lyell observed, that as the great majority of mankind live now as they have always lived, near the equator, or in countries not more than 25 degrees distant from it, they can never behold ice or snow. We may imagine, therefore, a nation to have made considerable progress in science without knowing anything of the causes appealed to in this discourse. If such a people were told by travellers of the geological appearances above described, how great would be their perplexity! They might at first ascribe the transport of erratics to floods of extraordinary violence, but they would scarcely be able to hazard a reasonable conjecture in regard to the coincidence between the direction of glacial furrows and that of the trains of erratics. A stone, not held fast by ice, but merely pushed along in mud, could not scoop out a long rectilinear furrow, one inch, or sometimes a foot deep, in a hard rock. Still more mysterious would be the discovery of a connection between the former migration southwards of an arctic fauna, and the conveyance of large erratics to the same regions. If the glacial hypothesis afforded no more than a plausible explanation of the association of so many distinct and independent classes of phenomena, it would deserve greater favour than has been shown to it by some modern geologists. The inclination evinced by many to introduce catastrophic action, as peculiarly applicable to the case of drift, arises in a great degree from the absence of stratification in drift. The usual geological proof of successive accumulation, and of the lapse of time, is here wanting; hence the sudden uplifting and sinking of land, the displacement of the sea, and the raising of gigantic waves of translation, rolling over continents at the rate of fifty miles an hour, accompanied by rapid gyrations of the marine fluid, have been imagined. The rate of movement, suggested by the glacial hypothesis, is singularly opposed to these views. Blocks carried by glaciers travel for centuries at an average rate of less than an inch per hour, and those borne along by floating ice float a mile or a mile and a half an hour. The observer of icebergs can seldom tell whether their course be north or south, east or west. In like manner the submergence and re-emergence of land will best account for the appearances above described, if the movement were slow, as now in Greenland and Scandinavia; in other words, if it be such as would be insensible to any human inhabitant. Yet the power of the machinery appealed to in both cases is equally vast; for it must be capable of uplifting and depressing continents, and removing from place to place the great volume of superficial materials found in the drift. The real difference of opinion consists in the amount of time during which the force is supposed to have been developed.

[C. L.]

[LETTER ON AGE OF RICHMOND COAL FIELD, VIRGINIA]

In order to verify still further the specific determinations of fossil plants made by Emmons, I placed in the hands of my friend, Professor Oswald Heer, a part of the collection of plants made by me in April 1849 at the coal-mines of Chesterfield county, Virginia, and also the descriptions and drawings published by Emmons, W. B. Rogers, Bunbury and Lyell; aud I give below the result which Mr. Heer was kind enough to send me.

(TRANSLATION.)

Zurich, July 25th 1857.

* * * * * * *. The *Tæniopteris magnifolia* W. B. Rogers, is a species of the genus *Strangerites*, that differs by its large size from all the European species.

Acrostichites oblongus Emmons. This species, at first considered by Bunbury and W. B. Rogers as the *Pecopteris Whitbyensis*, is entirely different. In the first place the form of the *sari* is not the same; then the American plant has a *nervatio reticulata*, or at least its secondary nerves are united to each other, while in the English plant the secondary nerves are digitated and dichotomous, and are never united laterally.

Pecopteris falcatus Emmons. Resembles so much the *Laccopteris germinans* of Gœppert (See: *Fossil Flora*, livr. 1. and 2., pl. VI, fig. 8.) that it is very probable it belongs to this species. The *Laccopteris germinans* is found according to Braun, in the Upper Keuper of Baireuth, Franconia.

Pecopteris Carolinensis Emmons. This species belongs to the genus *Gutbieria.*

Pecopteris (Aspidites) *bullatus* Bunbury. Resembles so much the *Pecopteris Stuttgartiensis* of Brongniart, that I do not hesitate to regard it as belonging to the same species. It is surprising that Bunbury has not recognized its great resemblance, at least, to the European species, so abundant in the Keuper of the environs of Stuttgart and Basle, and in Bavaria.

Neuropteris linnæfolia Bunbury. Does not belong to the genus *Neuropteris*, but is a *Cyclopteris* similar to the *Cyclopteris pachyrachis* of the German Lias.

* *

Finally some species, such as: *Pterozamites longifolius*, *Equisetum columnare*, *Calamites arenaceus* and *Pecopteris Stuttgartiensis*, are fossils characterizing the Keuper of Germany and of the Swiss Jura. Some other species are similar to species found in Europe in the Keuper and Lias, although specifically entirely different; so that I am led to regard this coal flora of Chesterfield county, Virginia, and of Deep river, North Carolina, as contemporaneous with that of the Keuper of Wurtemberg, of Switzerland and of Bavaria; and I find nothing in it to indicate the Jurassic epoch.

Very truly yours

OSWALD HEER.

The authority of Professor Heer in fossil botany is like that of Agassiz for fossil fishes; and Lyell who until now (See: *Supplement to the fifth edition of a manual of Elementary Geology*, page 30; second edition, revised; London, 1857.) had regarded the strata of the Richmond coal-field in Virginia as belonging to the lower part of the Jurassic series, has not hesitated to change his opinion in consequence of Heer's determination; and during a visit to Zurich, in August 1857, he recognized these coal-fields of Virginia and North Carolina as of the age of the Keuper, and not liassic or oolitic as he had previously thought. Further, in a letter dated Florence, Tuscany, 5th October 1857, Sir Charles Lyell says:

«Another subject on which I am desirous of writing to you, is the Virginia Keuper. My brother-in-law, C. Bunbury, observes in a letter on this point, referring to his paper in the *Quarterly Journal of the Geol. Soc.* on the Richmond plants, that he said: «The formation might belong to the Jurassic (in which he included the Lias) or to the Triassic periods, and that it might with almost equal probability be referred to either».

«He, Bunbury, adds «at the time I wrote this the Basle and Baireuth beds were supposed to be Lias. The only specimen I saw of the supposed *Pecopteris Whitbyensis* was very imperfect indeed»».

As the Basle and Baireuth beds are recognized now by every geologist as belonging to the Keuper, it will appear that Bunbury has never intended to put the Virginia coal field in the true Jurassic of England. So that we all agree to regard the Red Sandstone of Virginia and North Carolina as *Keuper*; and the brothers Rogers and James Hall stand alone in their opinion of *Great Oolite*, *Inferior Oolite* and *Lias.*

ON THE CONSOLIDATION OF LAVA
ON STEEP SLOPES
AND ON THE ORIGIN
OF THE CONICAL FORM
OF VOLCANOES

ART. XXV.—*On the Consolidation of Lava on Steep Slopes, and on the Origin of the Conical Form of Volcanoes*; by Sir CHARLES LYELL, M.A., D.C.L., F.R.S.*

DURING two recent excursions made in the autumns of 1857 and 1858 to Mount Etna, Sir C. Lyell had an opportunity of examining sections of lava-currents of known date, which had descended steep slopes, and had consolidated thereon in tabular and stony masses, the inclination of which sometimes exceeded 30°. This fact has an important bearing on the theory of "craters of elevation," it having been affirmed by geologists of high authority, that when lavas congeal on a declivity exceeding 5° or 6°, they never form continuous beds of compact stone, but consist entirely of scoriaceous and fragmentary materials.

The origin of such mountains as Etna and Vesuvius had of old been referred to the cumulative effect of a long series of ordinary eruptions, it being seen that reiterated showers of ashes and streams of lava were often poured out from a permanent central vent. This opinion was advocated by Mr. Scrope in his work on volcanoes in 1825, and by Sir C. Lyell in his *Principles of Geology*, after his exploration of Vesuvius and Etna in 1828; both authors considering the injection from below of

* Proc. Roy. Inst. of Great Britain, April, 1859.

melted matter, in the shape of dykes, as part of the cone-making process.

But in place of this simple explanation of the phenomena, Von Buch substituted the following hypothesis: that a vast thickness of horizontal or nearly horizontal sheets of lava and scoriæ, having been first deposited, an expansive force operating from below, exerts a pressure both upwards and outwards, from a central axis towards all points of the compass, so as suddenly to uplift the whole stratified mass, making it assume a conical form; giving rise at the same time, in many cases, to a wide and deep circular opening at the top of the cone, an opening called a "crater of elevation."

In all great volcanoes of which sections can be obtained, there are some layers of compact stone, inclined at angles of 10°, 20°, and sometimes much higher angles, and these beds are often among the uppermost, or last formed of the whole series. Hence it was logically inferred, when once the law above laid down respecting the consolidation of melted matter had been accepted, that every mountain containing such inclined and compact layers, must owe its conical form almost exclusively to the development of mechanical force exerted at the close of the volcanic operations, or after all the alternating lavas and scoriæ were heaped up. The hypothesis of a sudden and violent movement was perhaps the more readily embraced, because it relieved its advocates from the necessity of making unlimited drafts on past time, thousands of centuries being required if lofty cones, like Mount Etna, are to be built up by successive eruptions of ordinary intensity. The magnitude also of certain craters or "calderas" (implying, probably, one or more great explosions, followed by aqueous erosion), and the occasional steepness of the dips of certain lavas, beyond that which is found on the flanks of ordinary cones, (many of which might have been assigned to local dislocation), afforded additional arguments in favor of the new hypothesis. The lecturer then gave a rapid review of the controversy respecting "craters of elevation," stating the objections made to it by English and continental writers, including the late M. Constant Prevost; and he went on to observe that the principal object of this discourse was to show that the law laid down by M. E. de Beaumont, and by the late M. Dufrenoy, as governing the cooling and solidification of lava currents, on steep slopes, has no foundation in fact. Signor Scacchi had already, in 1855, seen and described a compact stony lava which in that year had flowed down the flanks of Vesuvius from near the margin of the great crater to the base of the cone in the Atrio del Cavallo, having a thickness of from 1½ in the upper to 4½ in its lower part, and dipping at angles varying from 32° to 38°. The interior of this current was laid open to view by a rare accident, namely, the sinking down in the same year (1855) of a certain portion of the north flank of the cone, whereby one side of the new lava stream was engulphed, and a section of the remainder rendered visible. Although this current had cooled on an average declivity of 35°, it was as compact and as free from vesicles as many lavas which have congealed on level ground at the foot of Vesuvius.*

The first exemplification of a similarly inclined stony lava of known date on Mount Etna, described by the lecturer, and of which a pictorial representation was given, occurs in a ravine called the Cava Grande, near Milo, about 17 miles north of Catania, and 7 from the sea, above the level of which it is elevated about 2000 feet. A branch of the lava current of 1689 descending from the Val del Bove, cascaded over the right bank of that ravine 220 feet high, and on cooling, formed a tabular mass more than 16 feet in thickness, inclined at an average angle of about 35°, and concealing the face of the precipice for a width of about 400 feet. The internal structure of this new lava has been exposed to view by the falling down and partial removal of its scoriaceous crust on the left side; a removal caused by the annual waste of the steep bank of the ravine produced by the action of rain, and the torrent which flows at the bottom. The ravine intersects alternating beds of tuff, scoria, and lava, slightly inclined to the east, or seawards, being a series of the older products of Etna. This new and steeply inclined lava consists of three parallel layers, an upper fragmentary and scoriaceous mass about 8 feet thick; a central stony layer, 5 feet thick; and a lower bed consisting of thin strata of fragmentary scoriæ, in all three feet thick, but the bottom part of which is not visible. The compact central portion is a dolerite or trachi-dolerite, containing crystals of feldspar with some olivine, and is of the ordinary specific gravity of trap. It is divided by joints, 9 or 10 feet apart, so that among the fragments detached by denudation, and strewed over the sloping bank and bottom of the ravine, are angular masses of huge size, with a fracture like that of many ancient igneous rocks. The normal thickness of this bed of compact dolerite is 5 feet, where it dips at 32° and 35°, but near the top where it first enters the ravine, and where the inclination increases to 45° and 47°, the thickness is reduced to one-half or 2½ feet; yet when dipping at 47°, it is still not only stony and compact, but there is no breach whatever of continuity in the mass, and not more joints than in the less inclined portion. This branch of the lava of 1689, which has given a new facing to part of the right bank of the Cava Grande, exhibits but slight in-

* This section, seen by Signor Scacchi in 1855, was looked for by Sir C. Lyell, in company with Signor Scacchi in 1857, and found to be totally buried and concealed by the lavas poured out in the early part of that year.

equalities on its surface, appearing almost even when contrasted with the main current of the same date, from the surface of which many parallel and longitudinal ridges project prominently, sometimes 40 feet above their base, and with very steep sides sloping at angles of from 35° to 70°. The dip of the main current is between 10° and 16° east. From this and other examples, it is inferred that wherever the slopes are excessive (between 25° and 45°) the surfaces of the cooling lavas are less rugged than where the melted matter has congealed on more level ground.

Allusion was next made to some lavas which have cascaded over sea-cliffs 500 feet high, between Aci Reale and Santa Tecla. One of these at a place called the Scalazza of Aci Reale, exhibits a longitudinal section of a tabular mass of stony rock 20 feet thick, inclined at angles of 23° and 29°, which is connected uninterruptedly with the main body of the same lava resting on the gently sloping platform above, of which the sea-cliff is the abrupt termination. The above-mentioned highly inclined stony lava is covered as usual by a parallel layer of scoriæ (in this case 12 feet thick) and its base consists of another bed of scoriæ of slight thickness.

Several other sections of modern lavas of Etna, which have not been disturbed in their position since the day of their formation, and which are inclined at angles exceeding 30° were then enumerated. For a detailed account of those, reference was given to a paper by the lecturer, recently published in the *Philosophical Transactions* (Part 2 for 1858, p. 703). Among them is a current, inclined at 35°, occurring in the Cava Secca, a deep valley near Zafarana; and another reposing on the face of the great precipice at the head of the Val del Bove, under the sunk space called "The Cisterna." This remarkable current has a mean inclination of 35°, and the central stony layer is seven feet thick. Above and below are parallel overlying and underlying masses of scoriæ five and seven feet thick respectively. The flanks of the stream have been undermined and denuded by that constant waste which makes the innumerable dikes to stand out in relief on all the precipices surrounding the Val del Bove. Perhaps, also, in this instance, the lateral excavation of the lava may have been assisted by a rush of water like that of 1755, commonly called Recupero's flood, which descended the same precipice, the "Balzo di Trifoglietto." Suggestions were then offered on the probable cause of that singular inundation, which swept in a few hours from near the summit of Etna through the Val del Bove to the sea. The Canon Recupero traced its course, a few months after the event, by following the line of sand and boulders which it had left in its track; and calculated that the volume of water was so great, that, had all the snows of the top of

Etna been melted instantaneously, they could not have furnished enough water for the deluge. He, therefore, concluded that the water was vomited forth from the summit crater itself. Sir C. Lyell conjectures that there may have been masses of ice in the cone during the eruption which is recorded to have accompanied the flood of 1775, and the ice may have been suddenly melted by hot vapors and injected lava. In support of this hypothesis, he mentioned his having ascertained the continued existence, in 1858, of the same glacier which was alluded to by him in the first edition of his *Principles of Geology*, as occurring at the base of the cone, which had been quarried before 1828. This mass of ice the Catanians again quarried, four years ago, to a depth of four feet, without reaching the bottom. It is covered by ten feet of volcanic sand, and this again by lava. The tale of the mountaineers, who assured Recupero that the water of the flood of 1775 was hot, may have been correct, if the origin here assigned it be true.

Some account was next given of the lavas of 1852–53, which were still hot, and emitting columns of vapor at the time of Sir C. Lyell's last visit. They were more voluminous, perhaps, than any ever poured forth from Etna in historical times, except those of 1669, which overflowed a great part of the city of Catania. The narrative of the people of Zafarana, of the manner in which the frontal wall of lava, 30 feet high, and inclined at an angle of 37°, had crept slowly over green pastures and vineyards, and overwhelmed habitations in the suburbs of that town, reminded Sir Charles of similar tales which he had listened to seven weeks before in the Alpine valley of Zermatt, where the great glacier had, in the preceding spring, been pushing onwards with irresistible force, an equally steep mound of stony fragments, forming the frontal moraine by which green meadows, gardens, and chalets had been overwhelmed. A description was then given of the changes brought about by the lavas of 1852–53 in the scenery of the Val del Bove, and in that of the lower Valley of Calanna, in the interval since 1828, when the lecturer first visited Etna. These changes are very striking; the fresh currents having run from the head of the Val del Bove both in a northeast and in a southeast direction for a distance of six miles, with a breadth in each case of two miles, and having been piled up one over the other in some places (as at the Portella of Calanna) to a depth of more than 100 feet. The longitudinal and nearly parallel ridges on the surface of this new lava field are from 20 to 70 feet high; and there is now a black and monotonous wilderness in many places, where, in 1728, there were verdant forests.

One branch of this lava of 1852 cascaded over a precipitous declivity 500 feet high, at the head of the Valley of Calanna, and consolidated at angles of 35°, 45°, and even 49°. The scoriaceous crust having been partially washed off, the surface of a continuous crystalline and stony mass is exposed to view, only moderately vesicular, and having the steep inclinations above alluded to. This same current rests on an older one, that of 1819, which passed down the same steep cliff, and which has at some points a dip of more than 40°.

[The author continues with facts and reasonings similar to what is published in his paper of last year (this Journal, vol. xxvi, p. 214)]:

In conclusion, the lecturer gave a brief sketch of the series of geological events which he supposed to have occurred on the site of Etna since the time of the earliest eruptions, events which may have required thousands of centuries for their development. The first eruptions are believed to have been submarine, occurring probably in a bay of the sea, which was gradually converted into land by the outpouring of lava and scoria, as well as by a slow and simultaneous upheaval of the whole territory The basalts, and other igneous products of the Cyclopean Islands were formed contemporaneously in the same sea, the molluscous fauna of which approached very near to that now inhabiting the Mediterranean; so much so, that about nineteen-twentieths of the fossil species of the sub-Etnean tertiary strata still live in the adjoining seas. Hence, as that part of Etna which is of subaerial origin is newer than such fossils, the age of the mountain is proved to be, geologically speaking, extremely modern. During the period when the volcano was slowly built up, a movement of upheaval was gradually converting tracts of the neighboring bed of the sea into land, and causing the oldest volcanic and associated sedimentary strata to rise, until they reached eventually a height 1200 feet (and perhaps more) above the sea level. At the same time the old coast line, together with the alluvial deposits of rivers, was upraised, and inland cliffs and terraces formed at successive heights. The remains of elephants, and other quadrupeds, some of extinct species, are found in these old and upraised alluviums. Fossil leaves of terrestrial plants also, such as the laurel, myrtle, and pistachio, of species indigenous to Sicily, have been detected in the oldest subaerial tuffs. At first the cone of Trifoglietto, and probably the lower part of the cone of Mongibello, was built up; still later the cone last-mentioned, becoming the sole centre of activity, overwhelmed the eastern cone, and finally underwent in itself various transformations, including the truncation of its summit, and the formation of the Val del Bove on its eastern flank. Lastly, the phase of lateral eruptions began, which still continues in full vigor.

[RATE OF FORMATION OF DELTA OF THE MISSISSIPPI]

Sir Charles Lyell had already suggested to Mr. Croll that, assuming the accumulation of ice at the Pole depressing the centre of gravity of the earth, the submergence that would have resulted had the quantity of water in the sea remained the same would, to some extent, be counteracted by the reduction in volume consequent on the formation of the ice. With regard to the delta of the Mississippi, the data on which he argued had considerably altered since first he wrote on the subject, inasmuch as recent calculations had doubled the estimated volume of water flowing into the sea, and thus it was capable of producing the same effect in half the previously calculated time. The progress of the delta at any spot was of necessity variable, as the position of the mouth changed. The American engineers had allowed only 40 feet as the depth of the fluviatile deposits, whereas from boring Sir Charles had concluded it to be at least 500 or 600 feet. There was now reason to suppose that it was much more, possibly as much as 1500 feet. This being the case, notwithstanding the amount of work done by the river being doubled, his calculation as to the time required for the formation of the delta might not after all be so excessive.

History of Geology

An Arno Press Collection

Association of American Geologists and Naturalists. **Reports of the First, Second, and Third Meetings of the Association of American Geologists and Naturalists, at Philadelphia, in 1840 and 1841, and at Boston in 1842.** 1843

Bakewell, Robert. **An Introduction to Geology.** 1833

Buckland, William. **Reliquiae Diluvianae:** Or, Observations on the Organic Remains Contained in Caves, Fissures, and Diluvial Gravel. 1823

Clarke, John M[ason]. **James Hall of Albany:** Geologist and Palaeontologist, 1811-1898. 1923

Cleaveland, Parker. **An Elementary Treatise on Mineralogy and Geology.** 1816

Clinton, DeWitt. **An Introductory Discourse:** Delivered Before the Literary and Philosophical Society of New-York on the Fourth of May, 1814. 1815

Conybeare, W. D. and William Phillips. **Outlines of the Geology of England and Wales.** 1822

Cuvier, [Georges]. **Essay on the Theory of the Earth.** Translated by Robert Kerr. 1817

Davison, Charles. **The Founders of Seismology.** 1927

Gilbert, G[rove] K[arl]. **Report on the Geology of the Henry Mountains.** 1877

Greenough, G[eorge] B[ellas]. **A Critical Examination of the First Principles of Geology.** 1819

Hooke, Robert. **Lectures and Discourses of Earthquakes and Subterraneous Eruptions.** 1705

Kirwan, Richard. **Geological Essays.** 1799

Lambrecht, K. and W. and A. Quenstedt. **Palaeontologi:** Catalogus Bio-Bibliographicus. 1938

Lyell, Charles. **Charles Lyell on North American Geology.** Edited by Hubert C. Skinner. 1977

Lyell, Charles. **Travels in North America in the Years 1841-2.** Two vols. in one. 1845

Marcou, Jules. **Jules Marcou on the Taconic System in North America.** Edited by Hubert C. Skinner. 1977

Mariotte, [Edmé]. **The Motion of Water and Other Fluids.** Translated by J. T. Desaguliers. 1718

Merrill, George P., editor. **Contributions to a History of American State Geological and Natural History Surveys.** 1920

Miller, Hugh. **The Old Red Sandstone.** 1857

Moore, N[athaniel] F. **Ancient Mineralogy.** 1834

[Murray, John]. **A Comparative View of the Huttonian and Neptunian Systems of Geology.** 1802

Parkinson, James. **Organic Remains of a Former World.** Three vols. 1833

Phillips, John. **Memoirs of William Smith, LL.D.** 1844

Phillips, William. **An Outline of Mineralogy and Geology.** 1816

Ray, John. **Three Physico-Theological Discourses.** 1713

Scrope, G[eorge] Poulett. **The Geology and Extinct Volcanos of Central France.** 1858

Sherley, Thomas. **A Philosophical Essay.** 1672

Thomassy, [Marie Joseph] R[aymond]. **Géologie pratique de la Louisiane.** 1860

Warren, Erasmus. **Geologia**: Or a Discourse Concerning the Earth Before the Deluge. 1690

Webster, John. **Metallographia**: Or, an History of Metals. 1671

Whiston, William. **A New Theory of the Earth.** 1696

White, George W. **Essays on History of Geology.** 1977

Whitehurst, John. **An Inquiry into the Original State and Formation of the Earth.** 1786

Woodward, Horace B. **History of Geology.** 1911

Woodward, Horace B. **The History of the Geological Society of London.** 1907

Woodward, John. **An Essay Toward a Natural History of the Earth.** 1695